Bernadette Filotas

PAGAN SURVIVALS, SUPERSTITIONS AND POPULAR CULTURES IN EARLY MEDIEVAL PASTORAL LITERATURE

Is medieval pastoral literature an accurate reflection of actual beliefs and practices in the early medieval West or simply of literary conventions inherited by clerical writers? How and to what extent did Christianity and traditional pre-Christian beliefs and practices come into conflict, influence each other, and merge in popular culture?

This comprehensive study examines early medieval popular culture as it appears in ecclesiastical and secular law, sermons, penitentials and other pastoral works – a selective, skewed, but still illuminating record of the beliefs and practices of ordinary Christians. Concentrating on the five centuries from c. 500 to c. 1000, *Pagan Survivals, Superstitions and Popular Cultures in Early Medieval Pastoral Literature* presents the evidence for folk religious beliefs and piety, attitudes to nature and death, festivals, magic, drinking and alimentary customs. As such it provides a precious glimpse of the mutual adaptation of Christianity and traditional cultures at an important period of cultural and religious transition.

Studies and Texts 151

Pagan Survivals, Superstitions and Popular Cultures in Early Medieval Pastoral Literature

by

Bernadette Filotas

Pontifical Institute of Mediaeval Studies

This book has been published with the help of a grant from the
Canadian Federation for the Humanities and Social Sciences,
through the Aid to Scholarly Publications Programme,
using funds provided by the
Social Sciences and Humanities Research Council of Canada.

Library and Archives Canada Cataloguing in Publication

Filotas, Bernadette, 1941-
Pagan survivals, superstitions and popular cultures in early medieval pastoral literature / by Bernadette Filotas.

(Studies and texts ; 151)
Includes bibliographical references and index.
ISBN 0-88844-151-7

1. Pastoral literature, Latin–History and criticism. 2. Christian literature, Latin (Medieval and modern)–History and criticism. 3. Popular culture–Europe–Religious aspects–Christianity. 4. Popular culture–Europe–History–To 1500. 5. Popular culture in literature. 6. Paganism in literature. 7. Paganism–Europe–History–To 1500. 8. Superstition in literature. 9. Superstition–Europe–Religious aspects–Christianity. I. Pontifical Institute of Mediaeval Studies. II. Title. III. Series: Studies and texts (Pontifical Institute of Mediaeval Studies) ; 151.

BL690.F54 2005 274′.03 C2005-900801-6

Pontifical Institute of Mediaeval Studies
59 Queen's Park Crescent East
Toronto, Ontario, Canada M5S 2C4

www.pims.ca

Printed in Canada

For Thomas, Sophie, Joseph and Pascale

Contents

Acknowledgements

I am happy to have this opportunity to acknowledge my great debt to M. Pierre Boglioni, for his generous help and advice, and for his untiring encouragement and kindness. My grateful thanks also to Mr. William D. McCready and M. Joseph-Claude Poulin for their helpful comments and proposed corrections.

I should also like to express my appreciation to the Département d'histoire of the Université de Montréal for its financial support, and to the staff of the Bibliothèque des lettres et sciences humaines for the friendliness and good-humoured efficiency which made the many hours I spent in the library so rewarding and pleasant an experience. I also thank Mr. Larry Eshelman for making the library of St. Paul's University, Ottawa, available to me.

My special gratitude for their help and encouragement goes to Mary Ann Kim, Margaret Moore, Barbara and Rae Brown, Andrea Blanar, Ava Couch and, most particularly, to Charles Giguère, to whom I owe more than I can ever repay.

Abbreviations

AASS	J. Bolland, *Acta Sanctorum quotquot toto orbe coluntur, vel a catholicis scriptoribus celebrantur.* Ed. G. Henschenius *et al.* Paris, 1863-.
Bächtold-Stäubli	Hanns Bächtold-Stäubli, *Handwörterbuch des deutschen Aberglaubens.* Ed. Hoffman Krayer. 10 vols. Berlin, 1987.
Barlow	C. W. Barlow, ed., *Martini Episcopi Bracarensis opera omnia.* New Haven, 1950.
Bieler	*The Irish Penitentials.* Trans. and ed. Ludwig Bieler. Dublin, 1963.
Blaise	Albert Blaise, *Dictionnaire latin-français des auteurs chrétiens.* Rev. Henri Chirat. Reprint, 1993.
Caspari	C. P. Caspari, ed., *Eine Augustin fälslich beilegte Homilia de sacrilegiis.* Christiania, 1886.
CCCM	*Corpus Christianorum Continuatio Medievalis*
CCSL	Corpus Christianorum Series Latina
Concilia Germaniae	C. F. Schannat and J. Hartzheim, *Concilia Germaniae.* 1759. Reprint, Aalen, 1970 -.
CSEL	*Corpus Christianorum Ecclesiasticorum Latinorum*
DA	*Deutsches Archiv*
DACL	*Dictionnaire d'archéologie chrétienne et de liturgie*
Daremberg and Saglio	C. Daremberg and E. Saglio, eds. *Dictionnaire des antiquités grecques et romaines d'après les textes et les documents.* 5 vols.Paris, 1877-1919.
DHGE	*Dictionniare d'histoire et de géographie ecclésiastiques*
DTC	*Dictionnaire de théologie catholique*
Dict. étymologique	A. Ernout and A. Meillet, *Dictionnaire étymologique de la langue latine. Histoire des mots.* 4th ed., Paris, 1967.
ERE	J. Hastings, ed., *Encyclopedia of Religion and Ethics.* 13 vols. 4th imp., Edinburgh, 1959-1961.
Etymologiae	Isidore of Seville, *Etymologiae librorum XX.* In *Etimologias. Edición Bilingüe*, ed. José Oroz Reta and Manuel-A. Marcos Casquero. Madrid, 1985.
Haddan and Stubbs	A. W. Haddan and W. Stubbs, *Councils and Ecclesiastical Documents Relating to Great Britain and Ireland*, 3 vols. 1869-1873. Rep., Oxford, 1964.

HE	Bede, *Historia ecclesiastica gentis Anglorum.* Ed. Bernard Colgrave and R. A. B. Mynors. Oxford, 1969.
HF	*Historia Francorum libri X.* Ed. B. Krusch and W. Levison. *MGH SRM* 1. Editio Altera.
JEH	*Journal of Ecclesiastical History*
Mansi	J. D. Mansi, *Sacrorum conciliorum nova et amplissima collectio*
MGH	*Monumenta Germaniae Historica*
CapEp	Capitula Episcoporum
CapRegFr	*Capitularia Regum Francorum*
Ep	Epistolae
EpSel	*Epistolae Selectae*
SRM	*Scriptores Rerum Merovingicarum*
SS	*Scriptores*
Mordek	H. Mordek, ed., *Kirchenrecht und Reform im Frankenreich: Die Collectio Vetus Gallica, die älteste systematische Kanonensammlung des fränkischen Gallien.* Berlin, 1975.
PG	J. P. Migne, *Patrologia Graeca*
PL	J. P. Migne, *Patrologia Latina*
Quicherat and Daveluy	L. Quicherat and A. Daveluy, *Dictionnaire latin-français.* Rev. Émile Chatelain. 54th ed., Paris.
RB	*Revue Bénédictine*
REAug	*Revue des études augustiniennes*
RHE	Revue d'histoire ecclésiastique
RHÉF	*Revue d'histoire de l'Église de France*
SC	*Sources chrétiennes*
Schmitz	H. J. Schmitz, *Die Bußbüche.* 2 vols. 1898. Reprint, Graz, 1958.
SSAM	Settimana di studio sull'alto medioevo, Spoleto.
VSH	Charles Plummer, *Vitae Sanctorum Hiberniae.* 1910. 2 vols. Reprint, 1968.
Vives	*Concilios Visigóticos e Hispano-Romanos.* Ed José Vives with Tomás Marín Martínez and Gonzalo Martínez Díez. Barcelona and Madrid, 1963.
Wasserschleben	F. W. H. Wasserschleben, *Die Bußordnung der abendländischen Kirche.* 1851. Reprint, Graz, 1958.

Introduction

The purpose of this work is to analyse early medieval Latin pastoral literature for evidence of the aspects of popular religion and culture that the ecclesiastical hierarchy perceived as survivals of paganism and superstitions and to provide a systematic inventory of the data found.[1]

It is limited geographically to Latin Christendom, specifically to the parts of western Europe which had formed the northwestern half of the Roman Empire, together with Ireland and the region beyond the Rhine. This area was inhabited by peoples of Latin, Celtic and Germanic stock. With the exception of the Irish, all had shared directly or indirectly in the experience of Roman rule and had benefited or suffered from the effects of the disintegration of the Empire; all had been converted to Christianity in the Latin rite. As a result, they shared certain common experiences and traditions, and give at least the appearance of forming a cultural entity. The impression is strengthened by the nature and bias of our most important sources for their history and culture, namely, writings produced by men whose preoccupations, perceptions and language were determined by their position in the Latin Church.

Proposing chronological limits for a process such as Christianisation or any phase of it is always problematic and historians choose dates according to their special interests.[2] Although it draws on both earlier and later material, the present study concentrates on the five centuries stretching from c. 500 to c. 1000 which achieved the formal conversion of the Celts and the continental Germans. This period is bracketed by the episcopacy of Caesarius of Arles (c. 502-542) and the compilation of Burchard of Worms' *Decretum* (1008-1012). Caesarius set the tone for the Christian polemic against pagan survivals; generations of medieval missionaries and pastors repeated his themes and echoed his very words. It would be difficult to overstate his importance for the history of the relations between the early medieval Church and popular culture in Western Europe. The penitential

[1] For the detailed discussion of the basic concepts, see chapter 1.

[2] *E.g.* Ramsay MacMullen, *Christianity and Paganism in the Fourth to the Eighth Centuries* (New Haven and London, 1997; Jean Chélini, *L'Aube du moyen âge. Naissance de la Chrétienté occidentale. La vie religieuse des laïcs dans l'Europe carolingienne* (750-900) (Paris, 1991); Peter Brown, *The Rise of Western Christendom. Triumph and Diversity, AD 200-1000* (Cambridge, Mass., 1996); André Vauchez, "The birth of Christian Europe. 950-1100," in Robert Fossier, ed., *The Cambridge Illustrated History of the Middle Ages, 2, 960-1250* (trans. Stuart Airlie and Robyn Marsach; Cambridge, 1997), 80-119.

contained in the *Decretum*, the *Corrector sive medicus* is not only the last great reiteration of that polemic but also the most complete collection of contemporary popular customs.

These dates, moreover, mark a separate phase in western European history. The fairly amorphous groupings of tribes that had moved into Western Europe and settled there were forming themselves into more or less stable states. Politically, the period spans the Merovingian and Carolingian eras, roughly from Clovis' conversion to Catholicism and baptism in 496 to the death in 987 of the last Carolingian king, Louis V. It includes the rise and fall of Visigothic Spain and the heyday of the Anglo-Saxon kingdoms. From the perspective of culture, it was the period of the emergence of vernacular cultures. From the perspective of the history of Christianity, this was the period which saw the appearance of the distinctive characteristics of the medieval Church, acquired as a result of adjustments to new political realities and of the interaction between different pre-Christian traditions and religious ideas and structures imported from the Mediterranean world.[3]

Until the last few decades, historians have paid little attention to early medieval popular culture. Traditional history has tended to treat popular culture as marginal to the true business of history, the grand lines of social, economic, political and intellectual movements. This is true of works on religion also, even of those dealing with the more or less long periods of transition during which the work of conversion was completed.[4] The study of the abolition of paganism and establishment of Christianity has concentrated on institutional developments, political struggles and controversies over doctrine.[5] The Church's rise first to influence,

[3] Brown's *The Rise of Western Christendom* shows that the distinctive characteristics of Christianity in the West developed from the varied responses of Christians in Gaul, Ireland, Anglo-Saxon England and Francia to their separate social and political environments, and that similar developments were occurring simultaneously among Eastern Christians. The revived and increasing interest in the development of ethnic identities in the Middle Ages is signalled by *After Empire. Towards an Ethnology of Europe's Barbarians*, ed. Giorgio Ausenda (San Marino, 1995) and *Strategies of Distinction. The Construction of Ethnic Communities, 300-800*, ed. Walter Pohl and Helmut Reimitz (Leiden, 1998). See also Jean Dhondt, *Le haut moyen âge (VIIIe-XIe siècles)* (1968; revised by Michel Rouche, Paris, 1976) and Michael Richter, *The Formation of the Medieval West. Studies in the Oral Culture of the Barbarians* (Dublin and New York, 1994).

[4] See, for example, J. Gaudemet, *L'Église dans l'empire romain* (Paris, 1958); Ferdinand Lot, *The End of the Ancient World and the Beginnings of the Middle Ages*, trans. P. and M. Leon (reprint with new material, New York, 1961); J. Delumeau and H.-I. Marrou, *Nouvelle histoire de l'Église*. V. 1. *Des origines à Saint Grégoire le Grand* (Paris, 1963); K. Baus *et al.*, *The Imperial Church from Constantine to the Early Middle Ages* and Friedrich Kemp *et al., The Church in the Age of Feudalism,* vols. 2 and 3 of *History of the Church*, ed. Hubert Jedin and John Dolan, London, 1980-).

[5] Let the first volume, covering the period 350-814, of Gustav Schnürer's valuable handbook, *Church and Culture in the Middle* Ages, trans. George J. Undreiner (Paterson, N. J., 1956) serve as example: out of 540 pages, ten are dedicated to what may be termed popular culture,

then to power, the intro-duction of monasticism, the conversion of different groups, the struggles against heresies have been traced often. The transition from paganism to Christianity in the beliefs and practices of ordinary persons has not received the same attention. A few general remarks about sacred trees and springs, amulets and charms, soothsayers and enchanters, weather magic and love magic, were often thought to be sufficient to describe what was a long, difficult and not always successful process.[6]

The lack of attention to popular religion and culture is not surprising. The material for the history of public institutions, leading figures, and dominant ideas, although incomplete, is presented directly in the documents, while the material for the history of the ideas and customs of anonymous men and women is usually missing. Where it exists, it does so almost by accident, scattered through and hidden in documents with quite a different focus. To some extent, folklorists filled the vacuum left by historians. In the 19th century, the romantic movement and the new discipline of anthropology fuelled an interest in the beliefs and customs found in traditional, pre-industrial culture. The great collections of folklore are rich in data, rarely set into historical context, about popular medieval beliefs.[7] Among historians, German scholars were the first to work intensively in the field of popular religion. R. Boese identified a number of superstitions prevalent in 6th-century Gaul from the writings of Caesarius of Arles. Hans Vordemfelde analysed the barbarian codes to form a picture of Germanic folk religion. Wilhelm Boudriot

including popular Christianity. This book first appeared in 1923, predating the modern preoccupation with the subordinate classes but from this point of view it compares favourably with more recent works. In *Histoire du christianisme des origines à nos jours*, ed. Jean-Marie Mayeur *et al.* (Paris, 1980?-), a major enterprise of historiography still in the process of completion, this subject occupies barely two pages of the approximately 250 given to "Le christianisme en Occident" in vol. 4, which covers most of our period and is fittingly titled *Évêques, moines et empereurs (610-1054)* (ed. G. Dagron, Pierre Riché *et al.,* Paris, 1993). In Jedin and Dolan's *History of the Church* (see n. above), the two volumes dealing with the early Middle Ages and the feudal period mention popular practices only incidentally and in passing. The masses of Christians have found virtually no place in these works. Even Chélini's study of the religious life of the laity is concerned almost exclusively with ecclesiastically sanctioned forms of piety, and allocates one short chapter to resistance, within Christian territory, to the Church's monopoly of religious expression (*L'aube du moyen âge,* 101-110).

[6] It is instructive to look at a modern missionary's account of the problems of conversion. See, for example, J.-É. Monast, *On les croyait chrétiens: Les Aymaras* (Paris, 1969).

[7] *E.g.*, Jacob Grimm, *Teutonic Mythology* (1835), translated from the 4th ed. (1887) by J. S. Stallybrass (4 vols., New York, 1966); W. G. Wood-Martin, *Traces of the Elder Faiths of Ireland* (2 vols., London, 1902); Paul Sébillot, *Le folk-lore de France* (4 vols., Paris, 1904-07); Hanns Bächtold-Stäubli, *Handwörterbuch des deutschen Aberglaubens* (ed. E. Hoffman Krayer, 1927-1942; reprint, 10 vols., Berlin, 1987); *Encyclopedia of Religion and Ethics*, 13 vols., ed. J. Hastings (4th imp., Edinburgh, 1959-1961); Wilhelm Mannhardt, *Wald- und Feldkulte* (Berlin 1875-77); *idem, Baumkultus der Germanen*; A. Dieterich, *Mutter Erde. Ein Versuch über Volksreligion* (Leipzig and Berlin, 1925; reprint, Stuttgart, 1967).

and Dieter Harmening examined the official literature of the Church to test its reliability as evidence of actual superstitions.[8]

Since the 1960's, historians of mentalities have increasingly turned their attention to popular religion and culture, and have recognised the role of the laity in the formation of medieval Christianity. Raoul Manselli's *La religion populaire au moyen âge. Problèmes de méthode et d'histoire* was one of the first to consider popular religion in detail.[9] Oronzo Giordano's *Religiosità popolare nell'alto medioevo,* Jean-Claude Schmitt's monumental chapter on superstition in the *Histoire de la France religieuse* and Aron Gurevich's work on popular culture are works of synthesis.[10] Lines of interpretation of the documents, based on the split between the clergy and laity or on distinctions between town/country and men/women, have been proposed by Jacques Le Goff and Pierre Boglioni respectively.[11]

Regional and thematic studies also testify to current interest in these subjects. Joyce Salisbury's *Iberian Popular Religion, 600 B.C. to 700 A.D: Celts, Romans and Visigoths*, Yitzhak Hen's *Culture and Religion in Merovingian Gaul A.D. 481-751*, and Felice Lifshitz's *The Norman Conquest of Pious Neustria* are regional studies of Christianisation and popular culture and religion (in the case of Neustria, of the historiographic presentation of these phenomena).[12] Michel Meslin wrote on the pre-Christian roots of medieval New Year's customs, and Karen Louise Jolly on Anglo-Saxon popular culture as a middle ground between elite and folk culture.[13] Early medieval magic has attracted particular attention. In addition to monographs by Valerie Flint and Eugene D. Dukes,[14] there are briefer analyses of its different

[8] R. Boese, *Superstitiones Arelatenses e Caesario Collectae* (Marburg, 1909); Hans Vordemfelde, *Die germanische Religion in den deutschen Folksrechten* (Giessen, 1923); Wilhelm Boudriot, *Die altgermanische Religion in der amtlichen kirchlichen Literatur des Abendlandes vom 5. bis 11. Jahrhundert* (1928; reprint, Darmstadt, 1964); Dieter Harmening, *Superstitio: Überlieferungs- und theoriegeschichtliche Untersuchungen zur kirchlich-theologischen Aberglaubensliteratur des Mittelalters.* (Berlin, 1979). See also Holger Homann, *Der Indiculus superstitionum et paganiarum und verwandte Denkmäler* (Göttingen, 1965).

[9] Paris and Montreal, 1975. But see review by R. C. Trexler in *Speculum* 52 (1977) 1019-1022.

[10] Giordano, *Religiosità popolare nell'alto medioevo.* Bari, 1979; Schmitt, "Les 'superstitions'," in J. Le Goff and R. Rémond, eds., *Histoire de la France religieuse*, v. 1, (Paris, 1988), 416-551; Gurevich, *Medieval Popular Culture: Problems of Belief and Perception*, trans. J. M. Bak and P. A. Hollingsworth (Cambridge, 1990).

[11] Le Goff, "Culture cléricale et traditions folkloriques dans la civilisation mérovingienne," in *Pour un autre moyen âge* (Paris, 1977), 223-235; Boglioni, "La religion populaire dans les collections canoniques occidentales de Burchard de Worms à Gratien," in N. Oikonomides, ed., *Byzantium in the 12th Century. Canon Law, State and Society* (Athens, 1991) , 335-356.

[12] New York and Toronto, 1985; Leiden, New York, Cologne, 1995; Toronto, 1985.

[13] Meslin, *La fête des kalendes de janvier dans l'empire romain,* Latomus 115 (Brussels, 1970); Jolly, *Popular Religion in Late Anglo-Saxon England. Elf Charms in Context* (Chapel Hill, N.C. and London, 1996).

[14] Flint, *The Rise of Magic in Early Medieval Europe* (Princeton, 1990); Dukes, *Magic and Witchcraft in the Dark Ages* (Lanham, Md., and London, 1992). Flint's work is discussed below.

aspects, such as women's participation, healing magic, the multiple uses of magical herbs.[15] Penitentials provide the raw material for studies of the development of norms of sexual conduct and of alimentary regulations.[16] Richard E. Sullivan's account of the correspondence between Pope Nicholas I and Boris I of Bulgaria gives a rare and highly instructive insight into the conceptual difficulties faced by new converts.[17]

Recent studies, in addition to two separate *Settimane di studio sull'alto medioevo*, emphasise the cultural dynamics of the development of Christianity.[18] Michael Richter noted that "the necessity to formulate the Christian message in language and concepts familiar to the recipients results almost inevitably in an enculturation."[19] But more than that, the evangelising bishops and missionaries of the early middle ages themselves did not stand outside the societies in which they worked. They shared in general beliefs and attitudes. Ramsay MacMullen made this point forcefully in his description of the cultural shift during Late Antiquity and the early middle ages that resulted in Christianity's adopting forms acceptable to the majority of the population, both urban and rural.[20] William E. Klingshirn's study of late antique Gaul suggests that the masses of newly baptised Christians also played an active role in shaping the religious life that became the norm in a given society. The critical element in the successful implantation of Christianity was the response of "local culture," the laity especially in the countryside, to the teachings of the ecclesiastical hierarchy. Acceptance or rejection of these teachings depended

[15] H. R. Dienst, "Zur Rolle von Frauen in magischen Vorstellungen und Praktiken–nach ausgewählen Quellen," in *Frauen in Spätantike und Frühmittelalter: Lebensbedingungen–Lebensnormen–Lebensformen. Beiträge zu einer internationalen Tagung am Fachbereich Geschichtswissenschaften der Freien Universität Berlin 18. bis 21. Februar 1987*, ed. W. Affeldt (Sigmaringen, 1990), 173-194; Daniela Gatti, "Curatrici e streghe nell'Europa dell'alto Medioevo," in *Donne e lavoro nell'Italia medievale*, ed. Maria Giuseppina Muzzarelli *et al* (Turin, 1991), 127-140; Karen Louise Jolly, "Magic, miracle and popular practice in the early medieval West. Anglo-Saxon England," in *Religion, Science and Magic in Concert and in Conflict*, ed. J. Neusner, E. S. Frerichs and P. V. McC. Flesher (New York and Oxford, 1989), 166-182; Franco Cardini, "Le piante magiche," in *L'ambiente vegetale nell'alto medioevo* (SSAM 37, Spoleto, 1990) 623-658.

[16] Pierre Payer, *Sex and the Penitentials. The Development of a Sexual Code, 550-1150* (Toronto, 1984); Maria Giuseppina Muzzarelli, "Norme di comportamento alimentare nei libri penitenziali," *Quaderni medievali* 13 (1982) 45-80.

[17] "Khan Boris and the conversion of Bulgaria: A case study of the impact of Christianity on a barbarian society" (1966; reprinted in *Christian Missionary Activity in the Early Middle Ages*, Variorum Collected Studies Series 431, Aldershot, 1994).

[18] *La conversione al cristianesimo nell'Europe dell'alto medioevo* (SSAM 14, Spoleto, 1967) and *Cristianizzazione ed organizzazione ecclesiastica delle campagne nell'alto medioevo: espansione e resistenze* (SSAM 28, Spoleto, 1982).

[19] "Models of conversion in the early Middle Ages," in *Cultural Identity and Cultural Integration. Ireland and Europe in the Early Middle Ages*, ed. Doris Edel (Dublin, 1995), 116-128; here, 123.

[20] *Christianity and Paganism*, 74-102.

on the extent to which they filled the needs of the community.[21]

The problem of the mutual adaptation and acculturation of traditional societies and an alien religion has been the subject of important works by Valerie Flint and James C. Russell. Flint presented clerics as offering the newly converted peoples of Western Europe acceptable Christian forms of their now-forbidden traditional institutions and practices (shrines, "witch-doctors," divination, incantations, amulets, *etc.*) in order to ease the transition from paganism to Christianity.[22] Russell posited that in their attempts to modify the "beliefs, attitudes, values and behavior" of the Germans, missionaries followed a policy of what they had expected to be a temporary accommodation with Germanic culture (in his words, a "deliberate misrepresentation of Christianity in Germanic terms").[23] In the end, he claims, it was the Germanic world-view that prevailed, and the official religion of the middle ages in Western Europe became in essence a Christianised version of Germanism.

In fact, a pristine, culturally uncontaminated form of Christianity had hardly existed previously. Christianity had never been impervious to the influence of the cultural experiences and assumptions of those who preached its message and of those who received it. Despite considerable theological rigidity, it early showed itself able to adapt to its surroundings in matters of discipline, organisation and ideas. Thomas D. Hill pointed out that "Christian Latin culture itself [involved] the assimilation of an ancient and originally pagan culture by Christianity which was originally the cult of Aramaic and Greek Jews in the eastern Mediterranean."[24] The apostolic Church abandoned circumcision and Jewish dietary rules in concession to the gentiles, Ireland developed a monastic system better suited to local conditions than the diocesan system prevailing in the Mediterranean world, and the distrust of sexuality even within the context of marriage can be traced, James

[21] *Caesarius of Arles; The Making of a Christian Community in Late Antique Gaul* (Cambridge, 1994).

[22] Flint's book has evoked both admiration and criticism; see the extensive discussions in Alexander Murray, "Missionaries and magic in dark-age Europe," *Past and Present* 136 (1992), 186-205; Richard Kieckhefer, "The specific rationality of medieval magic," *The American Historical Review* 99 (1994), 813-834. See also Giselle de Nie, "Caesarius of Arles and Gregory of Tours. Two sixth-century Gallic bishops and 'Christian magic'," in *Cultural Identity and Cultural Integration,* 170-196; and Rob Meens, "Magic and the early medieval world view," in *The Community, the Family and the Saint. Patterns of Power in Early Medieval Europe*, Joyce Hill and Mary Swan (Turnhout, 1998), 285-295. Dukes, unlike Flint, maintains that the Church steadfastly refused to give its sanction to magic (*Magic and Witchcraft*).

[23] *The Germanization of Early Medieval Christianity. A Sociohistoric Approach to Religious Transformation* (New York and Oxford, 1994); here, 211. For conflicting evaluations of this work, see Thomas F. X. Noble's critique in the *American Historical Review* 11 (1995), 888-889, and Robin Chapman Stacey's review in *The Medieval Review* (1994) on-line.

[24] "The *æcerbot* charm and its Christian users," *Anglo-Saxon England* 6 (1977), 213-221; here, 221.

A. Brundage has shown, to the influence of pagan philosophy on the early Fathers, especially St. Jerome and St. Augustine.[25] The integration within medieval Christianity of elements of ethnic cultures and of magic continued a trend that had been present since the beginning.

The sources for the study of the popular religion of the early middle ages are not extensive. Onomastics and toponymy provide hints as to the popularity of cults and the location of cultic centres.[26] Despite the difficulty of deducing ritual and beliefs from objects unaccompanied by explanatory texts, archaeology is an indispensable tool for verifying, correcting and adding to information found in written documents.[27] The spread of monastic foundations and the building of churches corroborate the evidence of saints' lives on missionary activity. Roadside crosses testify to the Christianisation of the countryside. The orientation of graves, cremation or the presence of grave goods in Christian cemeteries are no longer considered necessarily to reflect pagan survivals, but burial grounds are still a generous source for the history of the process of Christianisation and of resistance to

[25] "The married man's dilemma: Sexual mores, canon law and marital restraint," in *Life, Law and Letters. Historical Studies in Honour of Antonio García y García*, ed. Peter Lihehan (Rome, 1998), 149-169.

[26] For example, see Paul Lefrancq, "Les données hagiographiques de la toponymie cadastrale de l'ancien diocèse d'Angoulême et du département de la Charente," and Sylvianne Lazard, "Tradition ancienne et influence chrétienne dans l'anthroponymie ravennate du Xe siècle," in *La piété populaire au moyen âge. Actes du 99e Congrès National des Sociétés savantes, Besançon, 1974* (Paris, 1977), 401-409 and 445-454. A brief discussion of current research in prosopography is to be found in Jörg Jarnut, "Nomen et gens. Political and linguistic aspects of personal names between the third and the eighth century. Presenting an interdisciplinary project from a historical perspective," in *Strategies of Distinction,* 113-116.

[27] See Bryony Orme, *Anthropology for Archaeologists. An Introduction* (Ithaca, N. Y., 1981), 218-254, Ralph Merrifield, *The Archaeology of Ritual and Magic* (New York, 1987) 1-21, and Patrick Geary, "The uses of archaeological sources for religious and cultural history," in *Living with the Dead in the Middle Ages* (Ithaca, N. Y., and London, 1994), 30-45; Alain Dierkens, "Quelques aspects de la christianisation du pays mosan à l'époque mérovingienne," in *La civilisation mérovingienne dans le bassin mosan*, ed. M. Otte and J. Willems (Liège, 1986), 29-83, and "The evidence of archaeology," in *The Pagan Middle Ages,* ed. Ludo J. R. Milis, trans. Tanis Guest (1991; Woodbridge, Suffolk, 1998) 39-64. See also H. Roosens, "Reflets de christianisation dans les cimetières mérovingiens," *Études Classiques* 53 (1985), 111-135; Günter P. Fehring, "Missions- und Kirchenwesen in archäoligischer Sicht," in *Geschichtswissenschaft und Archäologie: Untersuchungen zur Siedlungs-, Wirtschafts- und Kirchengeschichte*, ed. Herbert Jankuhn and Reinhard Wenskus (Sigmaringen, 1979), 547-591; Helen Geake, "Burial practice in seventh- and eighth-century England," in *The Age of Sutton Hoo. The Seventh Century in North-Western England*, ed. M. O. H. Carver (Woodbridge, 1992), 83-94; Sabine Racinet, "Recherches archéologiques et textuelles sur les traces de la christianisation de la Picardie," *Mélanges de Science Religieuse* 53 (1996), 43-60; Édouard Salin, *La civilisation mérovingienne d'après les sépultures, les textes et le laboratoire,* esp. vol. 4 (Paris, 1959).

it.[28] The burial of pre-Christian dead in Christian tombs, the location of cemeteries in churchyards, erection of chapels in cemeteries and internment of bodies within the church itself signal the integration of ethnic and Christian concepts about the relationship between the dead and the living. Although the archaeological record is undoubtedly richer for the social elite than for the subordinate classes, one may extrapolate their values and concepts to some extent from those of more privileged groups. The ornamentation of churches, the marginalia of liturgical works, and even textiles reflect popular culture.[29] Folklore, which incorporates the oral culture of the past, may serve as a clue to all but forgotten myths and rituals.[30]

But the essential source remains the written documents produced by clerics for ecclesiastical and secular purposes: pastoral literature, hagiography, the liturgy, theological works, histories and chronicles, scientific treatises, poetry and legends.[31] The data found here is dispersed, incomplete, repetitive, sometimes difficult to identify and often difficult to evaluate, always hard to manage. However, except for Boudriot's and Harmening's analyses of the value of normative pastoral literature, there has been no large-scale attempt to assess the reliability of the written documents.

The need for such a study is pressing especially from the point of view of practices opposed by the Church. Grimm's great work on German mythology, beliefs and rituals, which relied on written documents for its medieval component, is still an essential tool for historians and ethnologists, but the scholarship of the last century and half has made his collection, valuable as it is, out-of-date. Moreover, Grimm paid considerably less attention to chronology than to the geographic diffusion of the beliefs and practices he described, and tended to accept at face value the evidence presented in his documents.[32] In effect, Boudriot and Harmen-

[28] See Tania M. Dickenson and George Speake's suggestion that the cremation of a pagan Anglian prince may perhaps be understood best "as an act of 'transculturation', the incorporation of elite Anglian burial fashions in order to resist external domination" by neighbouring Christian princes ("The seventh-century cremation buried in Asthall Barrow, Oxfordshire: A reassessment," in *The Age of Sutton Hoo*, 94-140; here, 123).

[29] See C. Gaignebet and J. D. Lajoux, *Art profane et religion populaire au moyen âge* (Paris, 1985); Michael Camille, *Image on the Edge. The Margins of Medieval Art* (London, 1992); C. R. Dodwell, "The Bayeux Tapestry and the French secular epic," in *The Study of the Bayeux Tapestry*, ed. Richard Gameson (Woodbridge, Suffolk, 1997) 47-62, esp. 60-61

[30] See Jacques Le Goff, "Culture ecclésiastique et culture folklorique au moyen âge: saint Marcel de Paris et le dragon," in *Pour un autre moyen âge,* 236-279.

[31] See Boglioni, "Pour l'étude de la religion populaire au moyen âge. Le problème des sources," in *Foi populaire, foi savante. Actes du Ve Colloque du Centre d'études d'histoire des religions populaires tenu au Collège dominicain de théologie (Ottawa)* (Paris, 1976), 93-148.

[32] These weaknesses are not always avoided by contemporary historians. Flint, for example, generally gives her documents equal value and ignores the influence of literary tradition on her authors. Both she and Hen treat the period they are studying as a whole, without giving enough consideration to the effect of changes over time.

ing have cast doubt on the reliability of the documentation for Germanic religion and superstition. They maintained that the pagan survivals and superstitions mentioned in pastoral literature composed on Germanic territory were largely based not on Germanic customs and beliefs but on the customs and beliefs prevalent centuries before in the Mediterranean region, which had been described by Caesarius of Arles and, before him, by St. Augustine. This criticism, a serious attack on one of the principal bases of the historiography of early medieval popular culture, has not been met on its own terms except by Rudi Künzel in a brief essay outlining a critical method for the assessment of medieval testimony as to paganism and superstition.[33]

It can be argued plausibly that the continual reuse in pastoral literature of the earlier texts is proof of the persistence of the cultural phenomena described, as well as of the Church's attitude toward them.[34] This, however, applies only to generalities. It does not respond to the problem of identifying regional particularities or changes in behaviour over time for it is evident that the hierarchy used the same vocabulary and formulations to condemn a wide variety of practices. Pastoral literature may convey significant information about such changes and particularities, but the information becomes accessible only if the documents are examined systematically for continuities, variations, abandonment of certain themes and introduction of others.[35] It is, therefore, essential to organise the available material on pagan survivals and superstitions. The present study is intended to do so for pastoral literature.

I have privileged this type of document (that is, the Latin texts), specifically those written for normative purposes, because it seems to me to be the most important source for the popular culture of the early middle ages. It is the only form of literature concerned directly, if seldom, with the beliefs and rituals of ordinary men and women. As such, pastoral literature presents the policies, and reveals the attitudes, of the hierarchy with respect to the traditions of the local communities during and immediately following the period of conversion. Pastors saw a continuation of the old cults or a perversion of the new in various popular customs, and incorporated their strictures against them in legislation, penitentials, sermons, letters and tracts written to combat specific practices.

[33] "Paganisme, syncrétisme et culture religieuse populaire au haut moyen âge," *Annales ESC* 4-5 (1992) 1055-1069.

[34] *Cf.* Gurevich, *Medieval Popular Culture*, 37. In his review of Harmening's book, Schmitt pointed out that the repetition of the texts is in itself a document of the evolution of the Church's attitude toward superstition, and, moreover, that the information contained in these texts is borne out by other types of contemporary written documents, for example, saints' lives, *exempla*, vernacular literature (*Archives des sciences sociales des religions* 53[1982], 297-299).

[35] See Michel Lauwers, "'Religion populaire', culture folklorique, mentalités. Notes pour une anthropologie culturelle du moyen âge," *Revue d'Histoire Ecclésiastique* 82 (1987), 221-258; here 223.

Each of the different types of documents has a particular focus and value. Church councils, one may assume, dealt with customs that were both fairly widespread and public, but their references to them were usually cursory and vague. Penitentials, on the other hand, went into considerable detail but since they were concerned with private behaviour, it is difficult to judge how common a given belief or practice was. Most of the relevant sermons (they are few in number) give the impression of being written in response to actual, local pastoral problems, as do tracts and letters. The information from these sources can at times be supplemented from other roughly contemporary documents (barbarian codes, histories, hagiography and archaeology).

But pastoral literature has major drawbacks for the study of popular culture. It is inherently hostile toward the practices that it describes. Laws forbid or impose behaviour; penitentials deal with sins not good deeds; preachers usually try to correct, not congratulate, their flocks. This means that the authors of the documents describe objectionable behaviour and beliefs almost exclusively. Almost as important, the texts represent mainly the perspectives of the clerical elite, whose literary training (in the Bible, the writings of the Church Fathers and, to a certain extent, the classics) and professional preoccupations tended to make them value literary tradition as highly as practical observation and inclined them to focus on certain groups in society (notably clerics, peasants and, in some cases, women) and on certain types of behaviour at the expense of others. The experience and outlook of the lower ranks, especially of the parish clergy in remote rural areas, must have been different but was reflected only rarely in the documents. How accurate the descriptions of the practices and beliefs or how prevalent they were is therefore doubtful, and it is certain that many actual practices and beliefs never found a place in the documents.

An exhaustive survey of the data on pagan survivals and superstition in this source, with identification of the date and geographic origin of the evidence, seems all the more important in light of the interest in the middle ages shown by anthropologists and ethnohistorians, since the material is difficult of access to all but specialists in the field of early medieval culture. It is not completely familiar even to medievalists whose field is a later period. The material is scattered through masses of works of all sorts, in specialised collections and in isolated works, some of which are out of print and difficult to obtain, and the Latin in which it is couched is often obscure, sometimes to the point of incomprehensibility.

Moreover, it has become evident to me that it is the assemblage in its entirety of the relevant pronouncements of churchmen which reveals the value of pastoral literature as a source for the history of popular culture, its strengths and limits. It is only through the systematisation of such an assemblage according to chronology and geography that one may perceive continuities, shifts in emphasis over time, new problems or problems newly confronted, and local variations.

The approach taken is thematic and descriptive. The basic concepts, historical context and the sources are presented in chapter 1. The beliefs and practices identified as survivals of paganism or superstition are classified under the following headings: idolatry (chapter 2), nature cults (chapter 3), the cult of time (chapter 4), sacred space (chapter 5), beneficent magic (chapter 6), ambivalent and destructive magic (chapter 7), death and dying (chapter 8) and, finally, alimentary restrictions (chapter 9). Since it is often difficult to determine under which heading a belief or practice belongs, the same passage may appear in several different contexts.

This focus is on terminology, chronology, and the origin and diffusion of descriptions of practices and beliefs. Texts have been presented generally in translation, but pains have been taken to inventory the terms used, together with variations in spelling that might betray that a new nuance or wholly new meaning had been added to a familiar word. The material is arranged chronologically, to establish the date of the first appearance of a belief or practice and to trace continuities, significant variations and innovations, shifts of emphasis, and abandonment of topics. To the extent possible, the geographical origin of each text is identified, as well as the regions where the text was repeated. Again when possible, the general context in which a document was produced is taken into consideration. Unfortunately, in many cases only the most general indications can be given (*e.g.,* early 7th century or late 8th, continental or insular).

Interpretation has been a secondary consideration in view of the dangers of embarking on it with often compromised data drawn from over five centuries and half a continent. Nevertheless, I have suggested explanations or partial explanations when the material allowed, for example, in the matter of lunar beliefs, weaving magic, drinking customs and alimentary restrictions. In some cases, an abundant bibliography is available. This is true particularly of pagan cults that have been the object of numerous studies. Certain practices have also been covered thoroughly in monographs and articles. But some of the themes found in pastoral literature have aspects which are ignored wholly or in part by histories of medieval culture and religion and I have had to look elsewhere for guidance, especially to anthropology. In some cases, however, I have not been able to find any useful reference.

This study was prompted by certain questions. The most fundamental is the extent to which pastoral literature may be taken to reflect actual beliefs and practices rather than a literary tradition. But there are others also. What kind of image of popular religion emerges from these texts? Do they differentiate between clerical and lay, townspeople and countryfolk, men and women? What beliefs and practices can be identified as belonging to specific groups? What groups and types of practices are missing from the texts? Finally, to what extent did Christianity influence and merge with traditional, pre-Christian beliefs and practices in popular religion?

1
Concepts, Contexts, Sources

1.1 CONCEPTS

Paganism, superstitions, pagan survivals and popular culture are controversial, nebulous concepts suspect in the eyes of many modern historians The ethnocentrism and value judgments implicit in such pejorative expressions as paganism and superstition make it preferable to think instead in terms of alternative belief systems. The notion of cultural survivals is under attack from those who point out that all cultural phenomena in a given society perform a current function in that society. The definition of popular culture turns on the complex question of who constitute "the people" whose culture is being discussed. Nevertheless, understood within the context of the historical situation and of the sources, these concepts provide legitimate, even inescapable, categories for the examination of the religious and cultural life of the early Middle Ages.[1]

1.1.1 *Paganism and superstition*

In modern usage, paganism and superstition are generally two distinct concepts. The *Shorter OED* defines paganism in part as "pagan belief and practices; the condition of being a pagan" and as "pagan character or quality; the moral condition of pagans," with a pagan being "one of a nation or community which does not worship the true God" or "a person of heathenish character or habits." It defines superstition, again in part, as the "unreasoning awe or fear of something unknown, mysterious or imaginary, especially in connection with religion; religious belief or practice founded upon fear or ignorance." Here, paganism is the opposite of the true, revealed religion, while superstition can form part of any religion. But the dictionary also gives another meaning to superstition which blurs this distinction: it is an "irrational religious system; a false, pagan or idolatrous religion." This meaning of the word applies best to "superstition" as it appears in pastoral literature, with added secondary meanings of "obsolete" and "superfluous."[2]

[1] See Ian Wood, "Pagan religions and superstitions east of the Rhine from the fifth to the ninth century" and the following discussion, in *After Empire*, 253-302.

[2] *S.v.v.,* "Paganism," "Pagan," "Superstition." "Obsolete" and "superfluous" practices were also superstitious: medieval authors treated Jewish practices and feasts as *superstitio*, *e.g.,* Whoever has recourse to auguries our enchangments is to be separated from the assembly of the church, likewise those adhering to Jewish superstitions and feasts (*Statuta Ecclesiae Antiqua* [c. 475] 83 [LXXXIX], *CCSL* 148, 179). References to Jewish practices have not been included in this

Denise Grodzynski suggested that the two terms represent a shift in point of view: paganism was religion in the 1st century A.D., superstition in the 5th, and magic later in the Middle Ages.[3] Our literature, however, gives only slight signs of this altered viewpoint. "Paganism" and "superstition" (together with "idolatry" and "sacrilege") were applied indiscriminately to the same types of behaviour. There was more stress on paganism and pagan survivals in the early phases of Christianisation or re-Christianisation–in southern Gaul and Galicia in the 6th century, in the Rhineland from the mid 8th century on–than later,[4] but pastors continued to connect the behaviour of their charges to pagan customs and beliefs to the very end of our period.

I make no attempt to establish a distinction not made by the authors of the texts, nor to formulate a uniformly acceptable definition of these terms. For the sake of convenience, I have accepted as pagan and superstitious any beliefs or practices condemned in pastoral literature which explicitly or implicitly entailed a reliance on powers not coming from God and not mediated by the Church–according to the Council of Tours of 567, "that which did not belong to the Church's way." Included with the cults of deities, nature and the dead, and with magic, therefore, are the cults of dubious saints and angels, the unauthorised use of sacramentals and other syncretic practices.

This pragmatic approach offers important advantages. It acknowledges that the evidence is taken from an institutional perspective and is biased by institutional interests, that it deals with what clerics thought existed, not necessarily with what did exist and, therefore, that it has to be treated with scepticism. It enables us to discriminate between behaviour perceived to have theological content (such as divination or idolatry) and behaviour that was not so considered (such as contumely, fornication or murder). It avoids the problem of assessing devotional practices acceptable to contemporary churchmen but questionable to the modern mind (for instance, aspects of the cult of saints and relics, and the use of the Creed and Lord's Prayer as *incantationes*) or institutions such as the judicial ordeal. It saves us, therefore, from the danger of imposing, in the guise of detachment and historical distance, modern categories of thought and modern judgments on situations far removed from our own.

study. St. Boniface used the word *superstitio* in the sense of "superfluous" when he urged the Archbishop of Canterbury to forbid "the superstition in unnecessary vestments hateful to God." He feared that the introduction of overrich garments into the cloister had a bad effect on morals, not that wearing them entailed wrong belief (Boniface to Cuthbert, [747], Haddan and Stubbs 3, 382).

[3] "Superstitio," *Revue des études anciennes* 76 (1974), 36-60; here, 54. But Michele R. Salzman showed that *superstitio* had a triple meaning (paganism, divination, magic) even in the 4th century. The resulting ambiguity was exploited by Constantine and his still pagan administrators to tolerate the continuation of pagan cults while attempting to suppress divination and magic ("'Superstitio' in the Codex Theodosianus and the persecution of the pagans," *Vigiliae Christianae* 41 [1987], 172-188).

[4] Robert Markus, *The End of Ancient Christianity* (Cambridge, 1990), 209-210.

Above all, this approach enables us to chart the changes in clerics' perceptions and preoccupations from time to time and region to region, through the abandonment of certain themes, reprise of older ones abandoned for a while, and introduction of new ones. Christianity, not as an abstract set of dogmas and moral principles without a human, historical and geopolitical context, but as a living religion in the form that was handed down to Western Europe, came into being during the period stretching between roughly 500 and 1000 A.D. It defined itself through a dual process of inclusion and exclusion of a host of traditional customs, rites and beliefs. Certain ones among them were interpreted and adapted in such a way that they could be integrated into acceptable practice; others were seen as being "not Christian" or incompatible with Christianity as it was understood at the moment, and were rejected. Focus on these rejected elements of traditional culture allows us to perceive an important aspect of the formation of medieval Christianity and, at the same time, to catch glimpses of rites and myths that have otherwise left few traces on the historical record.

Paganism is a notoriously amorphous notion. It has no content in itself, and does not describe a coherent set of beliefs and practices. The history of the Christian usage of the word illustrates its lack of precision. It was applied to the polytheism of classical Greek and Roman religion, then to that of the barbarian tribes with whom Christian missionaries came into contact. "Pagan" came to include all those, except Jews and, later, Muslims (but sometimes even them), who did not subscribe to Christianity. It has been used to describe primitive religions such as shamanism and animism as well as highly sophisticated systems, including a religion as far removed from polytheism as Buddhism and philosophies such as Confucianism and Marxism. During the Reformation, Protestants condemned Catholic practices as pagan. In common modern usage, paganism is often synonymous with materialism, hedonism, irreligion and even atheism. It has been, above all, a Christian concept, with entirely negative significance until very recently, when various New Age groups have seen in paganism a respect for nature and (oddly enough) for women missing from Christianity, and have been proud to call themselves pagan.

From the beginning, the notion of "paganism" had connotations in which the religious and the cultural were mixed. Various theories have been put forward to explain the origins of the Christian sense of the word pagan. According to one, it came into the Christian vocabulary by way of its meaning as countryman, the inhabitant of the *pagus* (country district). Paganism was the religion of the *pagani,* the rural population being identified with the peasantry who were slower in accepting Christianity than urban populations. Another theory holds that the Christian usage derived from the secondary meaning of *paganus* as civilian in contrast to soldier, because the pagan was not a "soldier of Christ." Christine Mohrmann advanced still another explanation, that the Christian meaning was based on the argot of gladiators, for whom the *paganus* was an outsider, a non-gladiator.[5] In all of these, the *paganus* was the "other," divided from the in-group by ways of life, culture, perceptions and values, not merely by adhesion to a different religion.[6] The element of peculiarly religious exclusiveness was drawn from the distinction made in the Bible between Israel, God's Chosen People, whose heirs Christians claimed to be, and other nations, worshippers of strange gods.[7]

The paganism presented in pastoral literature is a broadly cultural rather than strictly religious manifestation–in effect, the ethnic traditions and folklore of newly Christianised peoples.[8] Although the formal aspects of pre-Christian religion (idolatry, the worship of more or less clearly conceptualised divinities, sacrifices to them and rituals in their honour) did not disappear completely, they dwindled over generations of Christianity. What remained important were practices and techniques based on an inarticulate sense of the interconnection of the supernatural and the natural. This interconnection could be manipulated by individuals and the community to serve their own needs as they had been by previous generations, without reference to the institutional Church, although individual priests and clerics were often implicated in the process.

The word pagan and related terms were applied to beliefs and practices which did not in fact entail a rejection of Christianity, let alone an organised cult. It is

[5] See J. Zeiller, *Paganus* (Fribourg, Paris, 1917), and Christine Mohrmann, "Encore une fois: Paganus" (in *Études sur le latin des chrétiens* [Rome, 1965] 3, 279-289). See also M. Roblin, "Paganisme et rusticité," *Annales ESC* 8 (1953), 173-183; Henri Maurier, *Le paganisme* (Paris, Ottawa, 1988) 11-24; J. Toutain, "Pagani, pagus" in Daremberg and Saglio 4, 273-276; Henri Leclercq, "Paganus, pagana, paganicum," *DACL* 13, 275-280); P. Siniscalco, "Païen, paganisme," *Dictionnaire encyclopédique du christianisme ancien* (Fr. adapt. by François Vial) 2, 1852-1853); *A Catholic Dictionary of Theology* (London, 1971), *s.v.* "Paganism."

[6] The English equivalent, "heathen," is derived from a word which first appeared in Ulphilas' translation into Gothic of the line "There was a gentile woman, of Syro-Phoenician race" (Mark 7, 26) as *haithno*. It seems likely that Ulphilas (c. 311-383), who evangelised the Goths and translated the bible in a somewhat bowdlerised version for their benefit, did not use a native Gothic word, but instead transliterated an Armenian word *het'anos*, itself probably from the Greek *ethnos*. If so, "heath" is a reverse derivation of "heathen." See Zeiller, *Paganus*, 59-64, and the unabridged *OED*, *s.v.v.* "Heath" and "Heathen."

[7] For the biblical concepts of the difference between Israel and others, see *Dictionary of the Bible*, ed. J. Hastings (New York, 1914), *s.v.* "Nations"; Robert Young, *Analytical Concordance of the Holy Bible* (8th ed., rev., London, 1939); *Encyclopedic Dictionary of the Bible,* ed. A. van der Born, adapted and trans. Louis F. Hartmann (New York, Toronto, London, 1963) *s. v. v.* "Gentiles" and "Hellenes;" *Encyclopaedia Judaica* (New York, 1971), *s. v.* "Gentile;" Duane L. Christensen, "Nations" in *The Anchor Bible Dictionary*, ed. D. N. Freeman (New York, 1992) 1037-49. For a discussion of idolatry in the Old Testament, see Moshe Halbertal and Avishai Margalit, *Idolatry*, (Cambridge and London, 1992), 9-30.

[8] Folklore in the sense of Theodor Gaster's definition as "that part of a people's culture which is preserved consciously or unconsciously, in beliefs and practices, customs or observations of general currency, in myths, legends and tales of common acceptance, and in arts and crafts which express the temper and genius of a group rather than of an individual" (*Standard Dictionary of Folklore, Mythology and Legend* [New York, 1949], *s.v.,* "Folklore").

true that an anti-Christian reaction often followed the initial phase of conversion, but pastoral literature was not concerned with it.[9] Rather, it dealt with the behaviour of people who thought of themselves as Christian. We are given no reason to doubt the sincerity of their belief in the teachings of the Church or of their willingness to participate in her rites. On the contrary: the literature suggests that, in general, they embraced both dogmas and rites wholeheartedly–as they understood them. They interpreted and made use of the sacramentals and rituals of their new religion in the light of tradition: they armed themselves with amulets containing relics, celebrated Christian feasts as they had seasonal feasts of the past, got drunk in honour of the sainted dead, used holy chrism as a healing charm, read the future in communion vessels. At the same time, they clung to many of their old customs with equal sincerity.[10] In effect, a genuine Christian piety flourished side by side and intertwined with the traditional beliefs and practices labelled pagan by the clerical authors.[11]

While paganism was essentially a Christian concept, the notion of superstition was inherited from judgments made in pre-Christian religions about practices seen as irrational, stupid, fear-driven or excessive.[12] In the classical world, although a distinction might be made between *religio* (reverence of the gods) and *superstitio*

[9] The death of the prince under whom Christianity was first accepted sometimes gave the signal for a return to traditional religions, for example in Kent after Ethelbert's death, Northumbria after Edwin's and Hungary after St. Stephen's. Often the reaction was led by the ruler's yet unbaptised heirs; see Arnold Angenendt, "The conversion of the Anglo-Saxons considered against the background of the early medieval mission," in *Angli e Sassoni al di qua ed al di là del mare* (SSAM 28, Spoleto, 1982) 747-781, here 747-754.

[10] This is the stage which A. D. Nock called "adhesion" rather than conversion which he defined as "the reorientation of the soul of an individual, his deliberate turning from indifference or from an earlier form of piety to another, a turning which implies a consciousness that a great change is involved, that the old was wrong and the new is right." At this stage, there had not yet been "any definite crossing of religious frontiers, in which an old spiritual home was left for a new once and for all;" the individual still had "one foot on each side of fence which was cultural and not creedal." The new religion was seen as a useful supplement but not a substitute, and "did not involve the taking of a new way life in place of the old" (*Conversion. The Old and the New in Religion from Alexander the Great to Augustine of Hippo* [Oxford, 1933], 6-7).

[11] Felice Lifshitz argues that in the early Middle Ages, accusations of paganism were in fact frequently over different interpretations over what it meant to be Christian, over "disagreement over definitions of Christianity rather than 'survivals' of paganism" (*The Norman Conquest of Pious Neustria* [Toronto, 1995], 29, n. 35).

[12] For an analysis of this word and its use in the Middle Ages, see Grodzynski, "Superstitio;" Dieter Harmening, *Superstitio*, 14-48; Schmitt, "Les 'superstitions'," 421-453; Marina Montesano, "*Supra acqua et supra ad vento." "Superstizioni," Maleficia e incantamenta nei predicatori francescani osservanti (Italia, sec. XV)* (Rome, 1999) 5-44; Micheline Laliberté, "Religion populaire et superstition au moyen âge," *Théologique* 8 (2000), 19-36. Even Germanic pagans, asserted Ian Wood, "recognised a distinction between a proper and an improper relationship with the supernatural" ("Pagan religions and superstitions," 261).

(fear of them), and foreign religions be dismissed as superstitious,[13] superstition was not so much a negation but a distortion of right religion. As Grodzynski pointed out, being religious was the necessary condition for being superstitious.[14] Christian polemicists used the word "superstition" indiscriminately for a variety of condemned practices. Some may be viewed as a negation of Christianity (for example, the worship of pagan deities), others lacked a clear-cut theological content (mourning and burial customs), and still others had a distinctly Christian cast (the use of relics and Scripture as talismans). The behaviour described as superstitious in our texts ranges from idolatry or "false religion" to foolish and unnecessary religious or semi-religious customs.[15] The concept of superstition, therefore, was one of the essential elements in medieval churchmen's negative judgment of the behaviour of their charges.

Despite the tendency of some modern scholars to avoid it in favour of more neutral expressions,[16] superstition is a useful concept allowing distinctions not made in pastoral literature. In discussing the practices of Christianised Germans east of the Rhine, Ian Wood visualises non-Christian rites and actions "along a spectrum ranging from a community to an individual religion, effectively from public to private ... Essentially the distinction between public and private is that between formal pagan practice and individual superstition," with some "grey areas" especially in the realm of family religion.[17] The grey areas may be rather extensive, if they take in wakes and funeral processions, the noisy celebration of vigils, and communal meals in the woods and by streams. Nevertheless, the distinction between public and private is implicit in numerous studies of medieval religion, from E. Vacandard's classic essay on idolatry in Merovingian Gaul, in which paganism consists of public manifestations of forbidden rituals,[18] to hist-

[13] L. F. Janssen, however, suggests that *superstitio* had a sinister meaning for Romans, as being not only alien but actively hostile to Rome. *Religio*, moreover, was communal religion, meant to ensure the welfare of society as a whole, while *superstitio* was intended only for private security, at the expense of the rest ("'Superstitio' and the persecution of the Christians," *Vigiliae Christianae* 33 [1979], 131-159).

[14] "'Superstitio,'" 41.

[15] For Harmening, "Götzendienst" (*Superstitio*, 41) and "Aberglaube," for Montesano, "cattiva religione" and *vana observatio* ("*Supra Acqua et supra ad vento*," 14). Harmening listed the varied meanings of *superstitio* in Christian usage as heathenry, idolatry, demon-worship, false and inadequate knowledge, false religion, outdated and unnecessary systems and customs, excess, and excessive, quasi-religious esteem for the goods of this world (*op. cit.*, 40).

[16] For example, Lifshitz rejects it and words like it as being "ideological and exclusionary concepts" (*Pious Neustria,* 18 n. 1); John D. Niles prefers to consider the pagan Anglo-Saxons as "possessed of animistic beliefs" rather than as superstitious ("Pagan survivals and popular beliefs," in *The Cambridge Companion to Old English Literature*, ed. Malcolm Godden and Michael Lapidge [Cambridge, 1991], 126-141; here, 132).

[17] "Pagan religions and superstitions east of the Rhine," 261-262.

[18] "L'idolâtrie en Gaule au VIe et au VIIe siècle," *Revue des Questions Historiques* 65 (1899), 424-454.

ories which set aside the opinion of contemporary churchmen and refer to local Christianities, such as "Neustrian Christianity," "community Christianities" or "parochial Catholicism,"[19] that came into being after the formal conversion of the population in a given area. It is clear that many religious customs condemned by medieval churchmen should not be considered to be paganism or pagan survivals; these may, for the sake of convenience, be termed "superstitions."

1.1.2 *Pagan survivals*

The concept of cultural survivals was introduced by E. B. Tylor in *Primitive Cultures.*[20] Survivals are "processes, customs, opinions, and so forth, which have been carried on by force of habit into a new state of society different from that in which they had their original home, and they thus remain as proofs and examples of an older condition of culture out of which a newer has been evolved."[21] Some of these survivals have no relevance to the present, while the social function of others has shifted from the "serious business" of society to the folkloric: superstitions, legends, games, riddles, proverbs, rituals of courtesy.[22] People cling to outmoded beliefs and practices because they fail to understand clearly the new context in which they live.

But do survivals in Tylor's view as the debris of the past really exist? Anthropologists of the following generation denied that folk practices and beliefs remained unchanged by historical experience. The important questions were why some survived and others did not, why in one place but not elsewhere.[23] Modern scholars argue that all actual practices are relevant to the community in which they are found. The American anthropologist Clyde Kluckhohn insisted that cultural forms survive only if "they constitute responses that are adjustive or adaptive in some sense, for the members of the society or for the society." For Jean-Claude Schmitt, the very notion is invalid when applied to popular culture. A belief or rite is coherent and relevant, not a combination of heterogeneous relicts and innovations–"rien n'est 'survécu' dans une culture, tout est vécu ou n'est pas."[24] A practice that appears meaningless and obsolete to an observer standing outside a given culture has meaning and relevance within the context of that culture for its members.

[19] Respectively, Lifshitz, *Pious Neustria,* 18 n. 1; William E. Klingshirn, *Caesarius of Arles,* 2; Gurevich, *Medieval Popular Culture,* 100.

[20] 2 vols., London, 1871, reissued with an introduction by Paul Radin as *The Origins of Culture* and *Religion in Primitive Culture* (New York, 1957). References are to the latter edition. For the response to Tylor's theory of survivals and the subsequent development of the idea, see Margaret Hodgen, *The Doctrine of Survivals* (1936; Folcroft, Pa., 1977).

[21] Tylor, *Origins of Culture,* 16.

[22] Tylor, *Origins of Culture,* 70-159.

[23] See Hodgen, *The Doctrine of Survivals,* 140-174.

[24] Kluckhohn, *Navaho Witchcraft* (Boston, 1944), 79; Schmitt, "'Religion populaire' et culture folklorique," *Annales ECS* 31 (1976), 946.

Moreover, the term "survival" implies misleadingly that practices were carried over unaltered from a previous cultural context. In fact, customs were adapted to fit new ideas so thoroughly that their very nature was changed, as Burchard of Worms illustrated in descriptions of infant burial customs in 11th-century Hesse. Hermann Bausinger and Rudy Künzel maintained that it is more accurate to think in terms of "continuities" and adaptation than of the survival of pagan beliefs and practices.[25]

In a practical sense also, the word "survivals" is misleading when applied to reprobated practices in the early Middle Ages. Except for Roman religion, there is little information available about those pagan rituals and beliefs that left no traces in archaeology and for which the only source is the writings of Christian clerics. It is often impossible to prove that a given practice was not of a later origin.[26] Caesarius of Arles (d. 542) and Atto of Vercelli (d. 961) give rare glimpse of the development of new practices in their dioceses, but there must have been many others. It is unknown to what extent pastoral literature itself and the efforts of missionaries resulted inadvertently in spreading into new areas the very practices that they wished to eradicate.

Historians have found the term "survivals" a useful shorthand to describe these traditional elements which, while having religious aspects, were not a part of the system approved by the Church. This is true not only of older works,[27] but of more recent studies as well.[28] A modern author speaks of the "twilight world of pagan survivals and syncretism" prevailing in Saxony a generation after it had supposedly become Christian.[29] It is implicit in other terms referring to aspects of some of the same phenomena, such as the "Germanisation" of Christianity, or

[25] Bausinger, "Zur Algebra der Kontinuität," in *Kontinuität? Geschichtlichkeit und Dauer als volkskundliches Problem*, ed. H. Bausinger and W. Brückner (Berlin, 1969),, 9-30; Künzel, "Paganisme, syncrétisme et culture religieuse populaire au haut moyen âge," 1056-1057.

[26] *Cf.* Ronald Hutton, *The Pagan Religions of the Ancient British Isles* (Oxford and Cambridge, 1991), 284-341.

[27] *E.g.,* Stephen McKenna, *Paganism and Pagan Survivals in Spain up to the Fall of the Visigothic Kingdom* (Washington, 1938); Francis P. Magoun, "On some survivals of pagan belief in Anglo-Saxon England," *Harvard Theological Review* 40 (1947), 33-46. Wilfrid Bonser dedicated a chapter to "Survivals of paganism" in *The Medical Background of Anglo-Saxon England. A Study in History, Psychology and Folklore* (London, 1963), 117-157.

[28] *E.g.*, Michel Meslin, "Persistances païennes en Galice, vers la fin du VIe siècle," in *Hommages à Marcel Renard* (Brussels, 1969) 2: 512-524; Ruth Karras, "Pagan survivals and syncretism in the conversion of Saxony," *The Catholic Historical Review* 72 (1986): 553-572. See also Hans Kuhn, "Das Fortleben des Germanischen Heidentums nach der Christianisierung," in *La conversione al Cristianesimo*, 743-757; Timothy Gregory, "The survival of paganism in Christian Greece," *Journal of American Philology* 107 (1986): 229-242; Odette Pontal, "Survivances païennes, superstitions et sorcellerie au moyen âge d'après les décrets des conciles et synodes," *Annuarium Historiae Conciliorum* 27/28 (1995/96): 129-136; Niles, "Pagan survivals and popular beliefs."

[29] David F. Appleby, "Spiritual progress in Carolingian Saxony. A case from ninth-century Corvey," *The Catholic Historical Review* 82 (1996): 599-613; here 603-604.

"traditional culture" and "pagan resistance."[30] There can be little doubt that after the initial, often traumatic, phases of conversion, the overwhelming majority of the populations of western Europe identified themselves as Christian. They brought their children to be baptised, received the sacraments, attended the vigils of the saints and trusted in their relics, buried their dead as close to the church as possible. At the same time, however, some (there is no knowing what percentage) continued to practice religious or quasi-religious rituals, both communal and private, that had their roots in the pre-Christian past. In that sense, they can be seen as relics of paganism, even if the participants themselves saw no conflict between them and their new religion.

Finally, the notion of pagan survivals imposes itself because it is a point of view reflected and often explicitly stated in pastoral literature. Again and again the authors condemned practices as being left over from paganism or coming from pagan customs. Whether the customs in question were pagan in any religious sense may be debated. Robert Markus emphasised the large secular component of daily life in pre-Christian society. He protested against the use of the term "pagan survivals" to mean whatever resisted "the efforts of the Christian clergy to abolish, to transform or control" and insisted on the "sheer vitality of non-religious, secular institutions and traditions and their power to resist change."[31] Earlier Hans Kuhn had dismissed as "inessentials" all the customs associated with pagan religion except for belief in the gods, their worship and acts of cult.[32] There is, however, no question about the attitude of the early medieval Church. Given her understanding of her own role, as the one, divinely-instituted intermediary on earth between God and man, the expression of the New Covenant, with the authority and duty to set standards, sanction techniques, determine symbols and articulate myth, it is not surprising that churchmen saw practices or beliefs that challenged this as survivals from the religions of darkness and ignorance.

1.1.3. *Terminology*

From Caesarius of Arles to Burchard of Worms, the most common terms for

[30] Russell, *The Germanization of Early Medieval Christianity*. See also Stefano Gasparri, *La cultura tradizionale dei Longobardi; Struttura tribale e resistenze pagane* (Spoleto, 1963); Raoul Manselli, "Resistenze dei culti antichi nella pratica religiosa dei laici delle campagne," in *Cristianizzazione ed organizzazione,* 57-108; *Pagans and Christians. The Interplay between Christian Latin and Traditional Germanic Cultures in Early Medieval* Europe, ed. T. Hofstra, L. A. J. R. Houwen and A. A. MacDonald (Groningen, 1995).

[31] *The End of Ancient Christianity,* 9.

[32] "Das Fortleben des Germanischen Heidentums," 744. *Cf.* Hutton's equation of "pagan survivals" with the "memory of, and reverence for, the old deities" (*The Pagan Religions,* 289). But Gregory also included the "traditional practices related to healing, death and the family" in his definition of paganism ("The survival of paganism," 230 and 241). For McKenna also, paganism entailed "not only the worship of the pagan gods, but also the practices associated with pagan worship, such as astrology and magic" (*Paganism and Pagan Survivals in Spain,* vii).

the condemnation of certain types of behaviour used involved some variation of the words meaning pagan–*paganus, gentilis* or, rarely, *ethnicus.*[33] The faithful were told in so many words that they followed pagan traditions, that they behaved like pagans, that a custom was a survival of pagan observances, a part of pagan traditions or "the filth of paganism" or outright pagan observances; that they wasted their time in pagan idleness or performed rites invented by pagans or acts similar to the crime of pagans. Feasts were held at the "loathsome sites of pagans" while some people participated in pagans' diabolical games, songs and dances.

The term *superstitio* was applied to the same type of activity, though more seldom, and often together with a word signifying paganism. The pagan custom of displaying and swearing on the heads of beasts was *superstitio*, according to a mid 6th-century Council of Orleans. Somewhat later, Martin of Braga assured Galician peasants that divination was *vana superstitio.* Boniface V warned the King of Northumbria against the "most pernicious superstition of idolatry." Christians around Reims ate "superstitious food in the company of pagans" in the first decades of the 7th century. The mid 8th-century *Indiculus superstitionum et paganiarum* contains thirty articles taking in the cults of deities, the dead, nature and saints, inappropriate rituals involving the church, divination, protective and healing magic, races, effigies, and beliefs concerning the relationship of women to the moon. In the 9th century, Carolingian rulers took measures against *superstitiones* practiced in different parts of the empire during funerals and Christian feasts, while a contemporary penitential declared that the observances of the Calends of January were a superstition to be avoided by Christians.

From the authors' point of view, superstition was not necessarily belief in non-existent entities or irrational and ineffective rituals.[34] It was sinful to offer worship to false gods (idols) because they were the forms adopted by demons, not because they did not exist. Similarly with magic: it was often efficacious, but it was efficacious because of the intervention of demons.[35]

Sometimes terms for idolatry (*idololatria, daemonia, cultus* or *cultura idolorum* or

[33] *Ethnicus* appears three times in relevant documents: in a 7th-century Spanish penitential, Christians fast and *ethnici* feast at the New Year (*Homiliare Toletanum, Hom.* 9, *PL* Suppl. 4, 1942); according to Burchard of Worms' *Decretum* (1008-1012), a person who refuses to mend his ways after mutilating others or burning down houses should be treated *tanquam ethnicus et publicanus* (11.30, *PL* 140: 88); the Latin translation of a 10th-century English penitential forbids Christians to perform divination as do the *ethnici* (*Confessionale Ps.-Egberti* 2.23, *PL* 89, 419).

[34] Harmening, *Superstitio*, 43.

[35] This does not mean that all authors invariably shared the beliefs that they described. Some turned a skeptical eye. Maximus of Turin and Caesarius of Arles, for example, tried to convince their flocks that the attempt to save the eclipsed moon by magic was silly since the eclipse was a natural phenomenon. Agobard of Lyons ridiculed those who believed in weather-magicians. Several of the questions in Burchard of Worms' penitential begin with "did you believe?" In these, the sin lay not in being a weather-magician or werewolf, addling people's minds, stealing magically, flying through the air, or making zombies, but in believing in the existence of such things or in trying to perform such feats.

daemonorum, idolorum servitus) and idolaters (*cultores idolorum/daemonum, veneratores, immolantes*) were used to describe the same kind of behaviour. These words appeared in three contexts. One was literally the cult of deities or demons, effigies or natural objects, expressed in acts of worship, that is, rituals and offerings (*sacrificia, oblationes, immolationes*). Here there can be no doubt that the authors thought that they were describing clearly defined actions that were being performed by their often Christian contemporaries.

Terms for idolatry also appear in rosters of capital sins and works of the devil. In these the relevance of the terms to actual practice is difficult to determine.[36] Idolatry is generally not given a special emphasis; it appears as one sin among many, as it had in St. Paul's letter to the Galatians (5, 19), copied faithfully into Charlemagne's *Admonitio generalis* of 789. It lists the cult of idols and sorcery among a variety of sins such as fornication, brawling "and the like."[37] In an 8th-century tract, the works of the devil are "cults and idols, lots and omens, processions and theatrical shows, theft and fraud, murder and fornication, pride and boasting, banqueting and drunkenness, dances and lies."[38] Idolatry here is neither more nor less serious a sin than any other, and the authors do not seem to make a causal connection between it and the rest. The word may even have been used in a figurative sense, as it was by Pope Gregory I when, echoing *Eph* 5,5, he equated avarice with idolatry.[39]

On the other hand, idolatry was explicitly linked at times with acts of cult or magic. Having catalogued it among other capital sins, an 8th-century English penitential added a separate item concerning "the use of things pertaining to idols, that is, omens and the rest of it."[40] Rabanus Maurus (d. 856) compiled a long roll of vices and sins ending with "the entire cult of idols and demons, namely, omens and those who sacrifice to (or in the vicinity of–*ad*) stones, trees and springs, and who perform enchantments or divinations and so on, are all sacrilege."[41] Two anonymous continental sermons of roughly the same period treated idolatry as the first of the capital sins, a sacrilege of which all manifestations were also sacrilegious:

[36] Such lists go back to the first Christian centuries, see Cyrille Vogel, *Le pécheur et la pénitence dans l'Église ancienne* (Paris, 1966), especially 13-14 and 22-24. See also Aimé Solignac, "Péchés capitaux," in *Dictionnaire de spiritualité* 12: 853-862.

[37] 82, *MGH CapRegFr* 1: 61.

[38] *Ordo de catechizandis rudibus* (796) 60, Bouhot, 223. Similar lists are to be found in Ps.-Eligius, *De rectitudine catholicae conversationis*, 2, *PL* 40: 1170; *Conc. Cabilonense* (c. 650?) Mansi 10: 1197; *Dicta Pirmini* (724-753) 29: 188; Ps. Boniface (8th/9th century?) S. 15.1, *PL* 89: 870; Burchard of Worms, *Decretum* (1008-1012) 2.165, *PL* 140: 637.

[39] Gregory I (601), *MGH Ep* 2: 320. This is the only figurative use of idolatry in the texts; the related sins of usury and the falsified weights and measures were put under the heading *De sacrif.* in the *Poen. Sangallense Simplex* (8th century, 1st half) 13, *CCCM* 156: 120.

[40] *Poen. Egberti* (before 766) 1.1, Schmitz 1: 575.

[41] *Hom.* 67, *PL* 110: 127.

> These are capital sins. The sacrilege that is called the worship of idols. However, all the sacrifices and soothsayings of the pagan are sacrileges, as are the sacrifices of the dead around [?] corpses or over tombs, or omens, or amulets, or the offerings made on stones, or to springs, trees, Jupiter, Mercury or the other pagan gods, because they are all diabolic; and many other things which would take too long to list are all, according the judgment of the holy fathers, sacrileges to be avoided and detested by Christians, and they are recognised to be capital sins.[42]

Penitential texts dealing with forbidden foods are the third context in which idolatry is mentioned. A ban on eating certain kinds of flesh, first appearing in the *Penitential of Theodore* (668-756) and then copied in essence some dozen times down to the 11th century, was justified by a reference to the New Testament: "It is certain that if birds and other animals were strangled in a snare or if they were killed by a hawk and were found dead, they are not to be eaten by men, because four laws of the *Acts of the Apostles* give these commands: abstain from fornication, blood, what is strangled, and idolatry."[43] But *idolatria*, the word used in this passage, normally carries a wider significance than the *idolis immolatum* ("what was sacrificed to idols") of the Vulgate. Since Theodore did not refer either to idolatry or idols elsewhere in the penitential, his meaning is impossible to judge. It is not clear, therefore, if he meant to equate eating forbidden flesh with idolatry, or simply to present it along with idolatry and fornication as distinct, equally reprehensible practices. Burchard of Worms, on the other hand, included a passage in his *Decretum* that made a distinction between the easily forgiven sin of eating immolated food in the company of pagans and outright idolatry, which he put on a par with murder and fornication as a sin requiring formal, public penance.[44]

Sacrilegium (in its technical sense, the profanation or misuse of sacred objects, persons or places[45]) also sometimes meant practices elsewhere labelled pagan or superstitious, notably in the late 8th-century *Homilia de sacrilegiis*. The cults of idols, of the dead and of nature, praying and making offerings anywhere except in church, astrology, singing and dancing through the countryside on Sunday, eating and drinking near shrines, animal sacrifices and other offerings, divination and the consultation of soothsayers and enchanters, attempts to protect the moon during an eclipse, lighting torches and "needfires," the use of amulets against sickness and other dangers–these and related practices and beliefs were all sacrilege because they violated the vows made at baptism. Let he who does such things, warns the

[42] Ps. Boniface (9th century?) S. 6.1, *PL* 89: 855. There is no preposition linking "sacrifices of the dead" and "corpses" in this text. The Anonymous sermon (late 8th/9th century), Scherer, 439, probably gives the correct version: "sacrifices of the dead around [*circa*] dead bodies." In both sermons, this passage is followed by an enumeration of other sins.

[43] *Poen. Theodori* (668-756) II, 11.2, Schmitz 1: 544.

[44] Burchard of Worms, *Decretum* (1008-1012) 10.37, *PL* 140: 839.

[45] *E.g.*, [S]acrilege, that is, the theft of sacred objects (*Poen. Egberti* [before 766] 1.1, Schmitz 1: 575).

Homilia be aware that he has forfeited his faith, and is a pagan, not a Christian: "sciat, se fidem perdere, non esse christianus, sed paganus."

Sometimes these terms were used not because the author suspected a lapse into paganism or superstition, but to drive home the heinousness, or merely unseemliness, of other offences. This is evidently the case in St. Boniface's remarkable claim that drunkenness was a vice peculiar to the English and pagans, and his unfavourable comparison of English to pagan sexual mores.[46] That point is made even more strongly in a 9th-century penitential: "If anyone has committed adultery with his mother, he should acknowledge that he is a pagan, because such a crime is sacrilege."[47] These terms were used even in criticisms of hair and clothing style. The 7th-century Welsh canons demanded the exclusion of any Catholic who let his hair grow or hang down in the barbarian fashion. In 787, two papal legates upbraided the English for wearing clothes in the pagan style, of the kind that their forefathers had cast away, thus imitating "the life of those whom you have ever hated."[48] A few years later, Alcuin scolded Ethelred King of Northumbria and his subjects for copying the pagan Northmen in haircut, beard and clothing.[49] Such statements were not motivated by fears of paganism, but by the resolve to have Christians differentiated from pagans in both internal and external disposition.[50]

There were, however, other types of behaviour which the authors did not identify explicitly as being pagan or superstitious or sacrilegious but which contained magical or ritual elements. This is particularly true with respect to attempts to gain illicit access to the graces controlled by the Church–chrism, the sacred species, altar vessels, sacraments, the precincts of the church for the burial of one's dead. Another case is certain forms of compulsory drunkenness, toasting and drinking contests that had their origin in sacrificial feasts for the gods and the dead. The point where the uncertain boundary between religion and culture is obliterated altogether is reached with medieval alimentary restrictions. These

[46] Boniface to Cuthbert, Archbishop of Canterbury (747), Haddan and Stubbs 3: 382; Boniface to Aethelbald of Mercia (746-747) *MGH EpSel* 1: 150; Boniface to the priest Herefrid (746-747) Ep. 74, *ibid.,* 156. In his letter to Aethelbald, Boniface lauded pagan Saxon women's ferocious chastity and the marital love and fidelity, extending even to suttee, customary among "the most loathsome and degenerate race of men," the pagan Wends, to emphasise the disgrace of the king's promiscuity–one of the very few cases where pagan customs were held up for praise.

[47] *Poen Merseburgense* c (9th century or earlier) 10, Wasserschleben, 436.

[48] *Canones Wallici* (7th century) 61, Bieler, 148; *Legatine Synods–Report of the Legates George and Theophylact of their proceedings in England* (787) 19, Haddan and Stubbs 3: 458. But George and Theophylact also criticised the English for the "pagan superstition" of facial scarring, mutilating their horses and eating horseflesh (as well as settling conflicts by lots "in the pagan fashion") –practices that may indeed have been based on ideas not easily reconciled with Christianity.

[49] *MGH Ep* 4: 43.

[50] On the other hand, the leader of the pagan reaction in Hungary after St. Stephen's death was said to have promptly proclaimed his return to the ancestral religion by shaving his skull so as to leave three pigtails (*Chronicon Pictum,* ed. L. Mezey, [Budapest, 1964], 110).

occupy a puzzlingly important place in the penitentials.[51] The influence of the Old Testament, desire to draw a distinction between human and animal behaviour and fear of inadvertent cannibalism have been plausibly advanced in explanation, but these do not account fully either for the large number of texts (between three and four hundred separate although highly repetitive clauses) nor for their precision. It may be postulated that other considerations, such as fear of non-Christian cultic practices and magic, had played a role as well.

Pastoral literature, then, presents a notable variety of practices and beliefs of baptised Christians as idolatrous, pagan, superstitious or sacrilegious: the cult of deities and nature, celebrations of the natural cycle of the year and unauthorised rituals to celebrate the liturgical cycle, reverence shown to certain places, recourse to cunning men and women, divination and other forms of magic, mortuary rituals and even alimentary customs.

1.1.4. *Popular culture*[52]

But this literature reveals only limited aspects of the cultures of various communities – limited, because their permitted manifestations are generally ignored. References to sanctioned forms of religious devotion are incidental. The use of prayers in approved circumstances, the legitimate cults of saints, relics and miracles are almost entirely absent. The popularity, seemingly immense, of saints' festivals appears only in condemnations of improper behaviour during the vigils, the pious custom of making pilgrimages to Rome only in St. Boniface's plea to the archbishop of Canterbury to keep Englishwomen home. Marriage customs and the obligations of kinship and friendship, however deeply they were rooted in tradition, are not treated as part of *observatio gentilium*. Purely secular matters are mentioned only if they are an occasion for a moral problem such accidental death or injury occurring during sports or work, or dishonesty in commercial transactions. Only those aspects of culture in which the hierarchy sensed an infringement on the Church's monopoly over religious matters were so categorised.[53]

[51] *E.g.,* in the late 7th-century *Canons of Adomnan* (Bieler, 176-181). See chapter 9.

[52] See John Story, *An Introductory Guide to Cultural Theory and Popular Culture* (Athens, Ga., 1993), especially 1-19; David Hall's "Introduction" to Stephen L. Kaplan, *Understanding Popular Culture. Europe from the Middle Ages to the Nineteenth Century* (Berlin, New York, Amsterdam, 1984), 5-18; Michel Lauwers, "Religion populaire," in *Catholicisme* 12: 835-849; *idem*, "'Religion populaire', culture folklorique, mentalités;" Karen Louise Jolly, *Popular Religion in Late Saxon England*, 6-34. For discussion of the pertinent historiography see Peter Biller, "Popular religion in the central and later Middle Ages," in *Companion to Historiography*, ed. Michael Bentley (London and New York, 1997), 221-246, and, with emphasis on Italian and French works, Montesano's introduction to her "'*Supra acqua et supra ad vento*'," vii-xviii. See also Michel Meslin, "Du paganisme aux traditions populaires," in *Mythes et croyances du monde entier, 1, Le monde indo-européen*, ed. A. Akoun (Paris, 1985), 153-200.

[53] On the Church's monopolisation of the realm of the sacred, with the consequent denigration of popular culture and its effect on churchmen's attitudes, see Caterina Lavarra, "Il

Can these aspects of culture be considered "popular" in the sociological sense of belonging to certain disadvantaged groups: peasants, the poor, the unfree, the illiterate who formed "the people" as opposed to the privileged classes, the clerics and the lay nobility? At first sight, pastoral literature justifies this view. Sermons exhorted masters to correct their servants, and ecclesiastical and lay authorities called upon landowners to compel their dependents to abandon their customary observances. We are told that it was the common horde who worshipped trees and raised a clamour at wakes. Rustic, ignoble, ignorant, uncouth, stupid–these are words freely applied to those who indulged in such customs.

But closer examination shows that many forbidden practices were not restricted to the lower classes but were shared by their superiors. Jean-Claude Schmitt insisted that during the Middle Ages, folklore was limited neither to a rural environment nor to any particular social class. This is seldom articulated in our sources, in part perhaps because of close social and familial connections between clerical and secular elites, in part also because of the leading role often taken by princes and the aristocracy in furthering the work of missionaries.[54] Agobard of Lyons was exceptional in stating outright that all men, nobles and commoners, townsmen and countryfolk, old and young, believed utterly in the existence of weather magicians–an exaggeration no doubt, but nonetheless evidence of a belief that transcended class. There are other, unmistakable indications as well. The reluctance of landowners to enforce decrees to destroy shrines and effigies suggests that they might still have adhered to the old cults. Penalties imposed for the same offence varied to fit the social standing of the transgressor–whippings for the unfree, fines for their masters. The distinction drawn by Jacques Le Goff between "culture cléricale" and "culture folklorique"[55] disappears altogether when it comes to magic, practised by clerics and laity alike. Churchmen in the highest positions, such as Gregory of Tours and Hincmar of Reims, believed wholeheartedly in sorcery. Penitentials in particular repeatedly identified priests and other clerics as soothsayers and enchanters, so much so that Dieter Harmening maintains that the magic described is not popular and peasant in origin at all, but learned clerical magic adopted by the folk.[56]

sacro cristiano nella Gallia merovingia tra folklore e medicina professionale," *Annali della Facoltà di Lettere e Filosofia* (Bari) 31 (1988): 149-204.

[54] Schmitt, "Les traditions folkloriques dans la culture médiévale; quelques réflexions de méthode," *Archives de sciences sociales et des religions* 52 (1981): 5-20; here, 7; Karl Ferdinand Werner, "Le rôle de l'aristocratie dans la christianisation du nord-est de la Gaule," *RHÉF* 62 (1976): 45-73, and Richter, "Practical aspects of the conversion of the Anglo-Saxons," in *Irland und die Christenheit*, ed. Proinséas Ni Chathain and Michael Richter (Stuttgart, 1987), 362-376; here 363-368.

[55] "Culture cléricale et traditions folkloriques dans la civilisation mérovingienne," 223-235. Carlo Ginzburg's analysis of the interaction between women accused of witchcraft and their judges or inquisitors demonstrates the ground common to both ("Witchcraft and popular piety: Notes on a Modenese Trial of 1519," and "The inquistor as anthropologist," in *Clues, Myths and the Historical Method*, trans. John and Anne C. Tedeschi; [Baltimore, 1992], 1-16 and 156-164).

[56] "'Contra Paganos'–'Gegen die vom Dorfe'? Zum theologischen Hintergrund ethnologischer

Many of the practices and beliefs condemned as superstitious and pagan, then, were parts of the common heritage which bound all classes together, as much as did the oral traditions of secular culture[57] and the forms of popular piety encouraged by the Church: the cult of the saints (which Peter Brown argued had been developed among the highest intellectual and social levels of Christian society[58]), the cult of relics of the sort so warmly recommended by Gregory of Tours, and the belief in and expectation of miracles (such as those recounted by Gregory the Great in his *Dialogues*). Where the elements of this shared culture originated is less important than the fact that they were accepted, consciously or otherwise, by the majority of all groups in society. They formed the popular culture which, Karen Jolly observed, should be seen "as a meeting ground between elite and folk cultures and not as the antithesis of 'high' culture."[59] Here, the notion of popular culture loses much of its sociological significance and becomes the culture shared by most of the members of society, chiefly by the laity but, to varying degrees, by many of the clergy as well, particularly by those serving in rural parishes. For this reason, Michel Lauwers prefers to think in terms of folkloric rather than popular culture, that is, culture conceived as an ethnological rather than sociological phenomenon.[60]

If popular culture was the possession of the community as a whole, why talk of "popular cultures" in the plural? The answer lies partly in the reality of the situation of the clergy and in their perception of their role in society. Through their

Begriffe," *Jahrbuch für Volkskunde* 19 (1996): 487-508. This interpretation seems extreme even when applied to the early penitentials, and is altogether untenable for Burchard of Worms' penitential. By contrast, Eugene Dukes asserts the primacy of peasant magic, stating that the penitentials in essence merely "re-echo what was learned firsthand by an ordinary cleric in his role of confessor" (*Magic and Witchcraft*, 211-212). The truth probably lies somewhere between these positions, with a considerable amount of cross-fertilisation between learned and folkloric sources. For insights on learned, clerical magic, although of a considerably later period, see Richard Kieckhefer, *Forbidden Rites. A Necromancer's Manual of the Fifteenth Century* (University Park, Pa., 1997).

[57] "[T]here was, largely, one culture, shared by clerics and lay people, but this was the traditional cultural which apparently held infinite enticement ... [there are] indications that this type of culture was not bound to a particular social class; its attraction went beyond the aristocracy and would thus have functioned as a formidable social cohesive" (Richter, *The Formation of the Medieval West,* 144. See also 255-256 and *passim*). The use of popular motifs and folkloric themes in the decorative arts commissioned and paid for by the rich, both secular and clerical, shows that popular culture was a common possession. See Gaignebet and Lajoux, *Art profane et religion populaire au moyen âge*, Camille, *Image on the Edge* and Dodwell, "The Bayeux Tapestry and the French secular epic," especially 60-61. Gurevich concluded rather ambiguously that medieval "learned and popular culture represented different traditions within the context of one culture" (*Historical Anthropology of the Middle Ages,* ed. Jana Howlett [Chicago and Cambridge, 1992], 39-40).

[58] *The Cult of the Saints: Its Rise and Function in Latin Christianity* (Chicago, 1981).

[59] *Popular Religion in Late Saxon England,* 12. See also her "Magic, miracle and popular practice."

[60] "'Religion populaire', culture folklorique, mentalités," 228 and *passim.*

literacy, professional preoccupations, privileged status and the discipline to which they were subjected, they (especially the members of the higher clergy) participated in another, elite, clerical culture as well as the common one. They identified themselves primarily with this culture which they saw as standing apart from and above the rest of society. This might have been, as Michael Richter claimed in a slightly different context, "wishful thinking ... rather than a reflection of how things were in life."[61] Nevertheless, such an attitude enabled them to "folklorise" the elements of popular culture that they condemned by ascribing them to subordinate groups labelled contemptuously "rustics", "the rabble", "the ignorant or stupid", or "worthless women" (*mulierculae*).

But this popular culture was not monolithic. The texts allow us to distinguish beliefs and customs associated principally with certain social groups: spinning- and weaving-women's cults, herdsmen's and farmers' magic, military cults, even monks' and clerics' drinking customs and commemorative rituals. A host of details about cults, rites and magic centring on fertility and reproduction, protection, love and black magic indicate the existence, at least in the minds of churchmen, of a subculture belonging specifically to women. Traces of distinct ethnic cultures also emerge in the use of a vernacular term or name, more often in nonstereotyped injunctions directed at individual ethnic groups. Descriptions of Galicians' unusual divinatory rituals, the alimentary habits of Irish outcasts, Frankish and Lombard Yuletide rituals, Anglo-Saxons' mutilations of themselves and their horses, Saxon funerary practices and healing practices unique to Spaniards reinforce evidence found in legal codes, oral tradition and, to a certain extent, archaeology of the diversity of ethnic cultures.[62] Even purely local cultures appear, although very rarely, in our texts, as in Agobard of Lyons' account of the cult of an otherwise unknown saint.[63] It is more accurate, therefore, to think in terms of a plurality of popular cultures rather than of a single, all-embracing culture.[64]

1.2. Context

1.2.1 *Western European paganisms*

The use of the single, general word "pagan" to describe non-Christian reli-

[61] *The Formation of the Medieval West,* 254. Richter was talking of the distinction between literate and oral cultures.

[62] It should be noted that at this period ethnic identities were highly flexible, allowing for the inclusion of different groups in the *gens*. The studies collected in *Strategies of Distinction* emphasise the difficulty of using the usual signifiers (language, weapons, hair, names, burial customs, *etc.*) to determine ethnicity. For the emergence of ethnic communities in the Roman world, see Patrick Geary, "Barbarians and ethnicity," in *Late Antiquity. A Guide to the Postclassical World*, ed. G. W. Bowersock, Peter Brown and Oleg Grabar (Cambridge, Mass., and London, 1999), 107-129.

[63] *De quorum inlusione signorum* (828-834) 1, *CCCM* 52: 237.

[64] This is recognised in studies such as Gasparri's *La cultura tradizionale dei Longobardi,* Lifshitz's *Pious Neustria,* and Jolly's *Popular Religion in Late Saxon England.*

gions and customs should not obscure the differences in belief and practice among various social and ethnic groups–differences almost universally ignored by the authors of medieval pastoral literature. Classical, especially Roman, paganism provided their terms of reference, although the knowledge of Roman paganism came to most at second hand, through the writings of St. Augustine and Isidore of Seville. The same formulas were used almost everywhere in Western Europe throughout our period: cults of Roman deities, descriptions of idols, recourse to soothsayers, love magic, spells, alimentary taboos. However, it was a heterogeneous reality that underlay these terms. A deity might have had a Latin name but was not identical with the Roman god; Germanic idols did not resemble the statues of Roman deities; soothsayers used different techniques; love magic was determined by the institutional form that marriage took in different cultures; spells differed in form and purpose; the meaning of alimentary practices varied from place to place. Ian Wood insists on the existence of a plurality of Germanic paganisms "determined by geographical factors and by contrasting social and political structures."[65] The same may be said of other traditional religions as well. Moreover, layers of paganism were superimposed one upon another wherever the movement of populations, settlers, armies, merchants and refugees brought different peoples into contact. The form that paganism took was highly dependent on locality and time.

Medieval missionaries and pastors, therefore, found themselves faced with indigenous religious concepts and attitudes that had many common features, but which also showed marked particularities. Understanding their critique of their flocks requires that we look to the pre-Christian religions for the origins of beliefs and practices that remained pertinent in the early centuries of Christianisation.[66]

Prehistoric religions:[67] Relatively little is known about prehistoric European religions, and much of what is surmised is based on preconceived notions and imaginative recreations of states of mind, cultic practices and social organisation.[68]

[65] "Pagan religions and superstitions east of the Rhine," 264.

[66] See Ken Dowden, *European Paganism; The Realities of Cult from Antiquity to the Middle Ages* (London and New York, 1990). For pre-Christian religions in the Roman Empire, with special emphasis on archaeology, see Jules Toutain, *Les cultes païens dans l'empire romain*, 3 vols. (1905-1907; Rome, 1967). This work, despite its drawbacks (it is almost a century old, it has a minimum of interpretation, it touches Great Britain only marginally and Ireland, of course, not at all, and the lack of an index makes access to the vast amounts of information it contains difficult) is still an invaluable survey of the subject. For a chronological analysis of pre-Christian religions in Great Britain, see Hutton, *The Pagan Religions.*

[67] See Karl J. Narr, "Approaches to the religion of early paleolithic man," *History of Religions* 4 (1964): 1-22; E. O. James, "Prehistoric religion," in *Historia religionum: Handbook for the History of Religions,* 1, *Religions of the Past,* ed. Claas Juoco Bleeker and Geo Widengren (Leiden, 1969), 442-94; Marija Gimbutas, "Prehistoric religions," in *Encyclopedia of Religion* (New York, 1987) 11: 505-515; Dragoslav Srejovic, "Neolithic religion," trans. V. Kostic, *ibid.,* 10: 352-360.

[68] See, for example, the different views of the existence of and evidence for a generalised

What evidence exists has survived mostly in caves, where wall-paintings, deposits of bones and footmarks point to magical and/or religious rituals, and in burial-sites implying a belief in an afterlife and a cult of the dead. Excavations of dwellings in the Balkans suggest that domestic religion centred around the hearth, with storage areas also being considered sacred; rituals focused on activities such as weaving and grinding grain. Women undoubtedly played an important role in these, and perhaps in communal rituals also. Their nature must be extrapolated from written descriptions of religious practices from a later period, such as offerings of wine and grain on the family hearth in Roman domestic religion, the magical protection of the bounds of houses and settlements, the spells used well into the Middle Ages to ensure the success of weaving, or the sacrificial rituals of Romans, Celts and Germans.

Roman religion: Of these, Roman religion is the best documented.[69] This also was composed of different layers–primitive Roman religion (which Herbert Rose characterised as "polydaimonism" rather than polytheism[70]) and the religions of the family, the city and the state, culminating in the imperial cult. Added to these were components drawn from Etruscan divinatory techniques and Oriental mystery religions. Roman administrators, soldiers, colonists and merchants, together with their womenfolk and servants, carried these with them to all parts of the

neolithic cult of "mother-goddesses" in the region stretching from the Indus Valley to north-western Europe, *e.g.,* James, "Prehistoric religion;" *idem, The Cult of the Mother Goddess* (New York, 1994); Gimbutas, "Prehistoric religions;" A. Fleming , "The myth of the Mother Goddess," *World Archaeology* 1 (1969): 247-261; P. Ucko, "The interpretation of prehistoric anthropomorphic figurines," *Journal of the Royal Anthropological Institute* 92 (1962): 38-54.

[69] The sources for Roman religion are archaeological and literary, the writings of poets (*e.g.,* Ovid, Horace), historians, (Varro, Polybius) and statesmen (Cato, Cicero). Standard works are: W. Warde Fowler, *The Religious Experience of the Roman People from the Earliest Times to the Age of Augustus* (1911; New York, 1971); Georges Dumézil, *Archaic Roman Religion* (trans. Philip Krapp; Chicago and London, 1966); Herbert Jennings Rose, *Ancient Roman Religion* (1948) reissued in *Religion in Greece and Rome* (New York, 1959), 157-312; Jean Bayet, *Histoire politique et psychologique de la religion romaine* (Paris, 1969); R. M. Ogilvy, *The Romans and their Gods* (New York, 1969); R. Schilling, "The Roman religion," in *Historia religionum* 1: 442-494. Despite re-evaluation of his approach and some of his findings, Franz Cumont's work, e.g., *Oriental Religions in Roman Paganism* (1911; trans. from the French; New York, 1956) is still a major influence on the study of religious syncretism in the late Empire, see Walter Burket in *Ancient Mystery Cults* (Cambridge, Mass. and London, 1987) and the papers presented in *Les Syncrétismes religieux dans le monde méditerranéen. Actes du Colloque International en l'honneur de Franz Cumont à l'occasion du cinquantième anniversaire de sa mort. Rome Academia Belgica, 25-27 septembre 1997*, ed. Corinne Bonnet and André Motte (Brussels, 1999). For mystery cults, see also M. J. Vermaseren, "Hellenistic religions," in *Historia religionum* 1: 495-532. Ramsay MacMullen, *Paganism in the Roman Empire* (New Haven and London, 1981). For the imperial cult, see D. Fishwick, *The Imperial Cult in the Latin West*, 2 vols. (Leiden, 1991) and Robert Turcan, "Le culte impérial au IIIe siècle," in *Aufstieg und Niedergang der römischen Welt*, 36 vols., ed. Hildegard Temporini and Wolfgang Haase, (Berlin and New York, 1972-), 16.2: 996-1084.

[70] *Ancient Roman Religion*, 172. See also Fowler, *Roman Religious Experience,* 116-123.

Empire, to merge with local cults.

The main function of Roman religion was the protection of the state and the family through the establishment of the right relationship with the gods or *numina.* Although state and private religion were quite separate, they exercised a degree of mutual influence. Some aspects of state religion were simply extensions of domestic religion: the public rites of Vesta, Janus and the state *penates* and *lares* may have originated as cults of the king's household, while the official calendar set aside dates for such private rites as the *Parentalia* and *Compitalia.* The essence of religion was the accurate performance of ritual; its principal ethical component was *pietas*, the scrupulous fulfillment of duty toward the gods, the state and the family. Precision was extremely important: the deity had to be addressed with exactly the right titles, prayers formulated in exactly the right words, and the terms of the agreements between man and god spelled out in the smallest detail.

In addition to the high gods and goddesses, with their own priests or priestesses and a public cult, and the *Di Manes*, the deified dead, there were dozens of gods and goddesses of varying importance, some known only by name and function, without priests of their own. Other deities could be added as convenient.

The Roman liturgical calendar, reformed during the early 7th century B.C. by Numa and then by Julius Caesar in 46 B.C., was based on a solar year of twelve months. Days were identified according to their sanctity, and most forms of secular activity were banned for at least part of days sacred to the gods (*dies nefasti*). The greatest feast of the year was celebrated on the Calends of January, which combined official, public and private celebrations. Feasts of the dead occurred in February and May. Numerous festivals celebrated the cycles of the warfaring and growing seasons. Principal features of most were a public procession to the shrine of the appropriate deity to offer sacrifices, the slaughter of ritual beasts of which the entrails were burned on the altar and the flesh either eaten in a communion banquet or sold on the market,[71] and in some cases games, which simultaneously provided honour to the god, self-advertisement for the rich and entertainment for the poor.[72]

Originally, the Romans appear to have worshipped not in temples, but in or by a holy place such as a wood, clearing or fountain, where an altar made of turf or stone was surrounded by a roofless enclosure, the *fanum.* The altar and enclosure remained characteristic of Roman shrines even when the Romans began to erect roofed temples, probably in imitation of the Greeks.

Private religion, concentrated on the family and the home, was intended to protect the *familia*, the household composed of blood relatives living together, with

[71] This was the main source of meat available in the markets, and created a problem of conscience for Christians who did not wish to eat idol-food.

[72] On games, see R. Auguet, *Cruauté et civilisation. Les jeux romains* (Paris, 1970); Paul Veyne, *Bread and Circuses: Historical Sociology and Political Pluralism*, trans. B. Pearce (abridged; London, 1990); Georges Ville, "Les jeux des gladiateurs dans l'empire chrétien," *Mélanges d'Archéologie et d'Histoire* 72 (1960): 273-335.

their dependents, servants and belongings, all under the authority of the *paterfamilias*. Like the heart of the city itself, homesteads were circumscribed by a sacred boundary. This, the cemetery, the spring and the house were the points of "special anxiety" around which domestic religion revolved.[73]

The Roman attitude toward the dead was ambivalent, compounded of loving reverence and superstitious dread. On the one hand, dead ancestors joined the *penates*, the deities honoured daily in the household. On the other, the very sight of a dead man and of anything to do with death was polluting. The concept of life after death was not comforting: unless disposed of properly the dead would, as *larvae* and *lemures*, walk and make mischief for the living. Originally, the body was buried within the house, but in historical times the corpse was interred in a cemetery or burned. The body, having been borne out of the house (feet first, to keep it from finding its way back), was carried in procession to its final destination. In the case of the rich, this could be a macabre affair, with trumpets and horns, dancers, mimes, clowns, keeners, masks of dead ancestors and effigies of the dead man. A funeral banquet was eaten at the graveside then and nine days later, after which the dead man was considered assimilated to the *Di Manes*.

Roman citizens may have continued to practice their family cults privately but by the late Empire, Oriental mystery religious had largely superseded the public cults of the Republic and early Empire. These had first entered Rome by way of Greece, where they had been introduced by Alexander the Great. Despite early resistance in Rome (at first only Cybele was allowed a place within the heart of the city) and continued caution and suspicion on the part of the religious elite, they established themselves increasingly in the city and the empire, until their popularity eventually overwhelmed the old cults. The most important were the cults of Isis and Serapis from Egypt, Cybele from Phrygia, Mithra from Persia and *Sol Invictus* from Syria. Their main appeal seems to have lain in the intense personal rapport that believers could achieve with the deity through conversion and participation in spectacular and emotionally satisfying ritual, in the discipline that they imposed, and in the hope that they offered of salvation and a personal afterlife. The feeling of belonging to a select group of initiates into sacred mysteries and the comradeship that comes from belonging to such a group must have exercised great appeal especially for those who were traditionally excluded from positions of power and prestige: women, foreigners and slaves.

Of the Oriental religions, the only one to achieve empire-wide official status was the cult of *Sol Invictus*, brought into Rome from Syria. The initial attempt of the emperor Heliogabalus (218-222) to replace Jupiter by this foreign deity failed, but in 274, the emperor Aurelian proclaimed *Sol Invictus* to be the official god of the empire. Other gods, although still tolerated, were made subordinate to him.

[73] Fowler, *Roman Religious Experience*, 68-73

Celtic paganism:[74] On the European continent, the largest group with whom the Romans came into contact were the tribes known as the Celts. "Celt" was a general term used by classical authors for the barbarians of northwestern and central Europe, irrespective of their ethnic and cultural traditions. Insofar as it is indicated by burial customs, sacrifices and ornamental motifs, Celtic religion was not static but evolved with changes in material culture and with contact with other peoples: Greeks, Romans, Germans, indigenous populations, even the Scythians. The exact nature of Celtic paganism is vague since archaeological materials are difficult to interpret without the aid of written documents, and contemporary Celts left no written accounts of their own.[75] Surviving contemporaneous accounts were written by outsiders to Celtic culture (e.g., Julius Caesar, Pliny, Strabo, Dio Cassio), while Welsh and Irish literature was compiled well after the events described, by monks whose knowledge was at best partial and whose conscious or unconscious bias and misunderstandings coloured the narratives. Nevertheless, the testimony of classical writers concerning the role of the druids, wood- and water-cults and head-hunting is borne out by archaeology and Celtic literature, and it has been claimed that some of the Irish epics copied by monkish scribes show "ways of thinking and behaviour more archaic than anything Homer sung of."[76] Hagiography is one of the richest, if most difficult to evaluate, sources for Celtic paganism.[77]

Given the wide dispersion of the Celts (from Galatia in Asia Minor and the sources of the Danube to the Atlantic Ocean and the British Isles), one may expect significant variations from region to region, but certain features appear to have been generally present: nature cults, the cult of mother-goddesses, triplism, a belief in some kind of after-life, use of solar and sky motifs in art. Topography provided the Celts with their holiest spaces. Favoured locations were mountains,

[74] For Celtic religion, see J. A. MacCulloch, *The Religion of the Ancient Celts (*1911; London, 1992); Toutain, *Les cultes païens* 3: 121-470; Françoise Le Roux, *Les druides* (Paris, 1961); A. Maartje Draak, "The religion of the Celts," in *Historia religionum* 1: 629-646; J. de Vries, *La religion des Celtes*, trans. L. Jospin (Paris, 1963); Nora K. Chadwick, *The Druids* (Cardiff, 1966); Hilda R. E. Davidson, *The Lost Beliefs of Northern Europe* (London and New York, 1993). See also Miranda Green, *The Gods of the Celts* (1986; reprint, Dover, N. H., 1993), *Symbol and Image in Celtic Religious Art* (London and New York, 1989, 1992) and *Dictionary of Celtic Myth and Legend* (London, 1992); Enrico Campanile, "Aspects du sacré dans la vie personnelle et sociale des Celtes," in *L'homme indo-européen et le sacré*, ed. R. Boyer *et al.,* (Aix-en-Provence, 1995), 155-182.

[75] For the sources for Celtic and Germanic religions, see Davidson, *Lost Beliefs,* 11-86. The Gundestrup Cauldron presents a good example of the kind of problems encountered in archaeology. Neither its date nor its origins are known (estimates vary from the 4th century B.C. to the 3rd century A.D., from northern Gaul to the Danube valley) and, while it is generally accepted that the scenes depicted have religious and mythological significance, it is impossible to interpret them for lack of explanatory texts.

[76] Draak, "The religion of the Celts," 630.

[77] Charles Plummer found numerous pagan elements in the Lives of Irish saints; see the introduction to *VHS,* lxxxiv-clxxxviii.

hilltops and other rock formations (particularly among the Iberocelts), trees and groves and, above all, water in all its forms.

Relatively little is known about Celtic religious rites, except for what is suggested by remnants of offerings and ex-votos. Altar furnishings indicate that they offered libations, though rarely. More common were other kinds of offerings such as first fruits, figurines, lamps, vessels of all description, axes, keys, torques, wheels, coins, weapons, antlers, all of which have been found in man-made and natural sanctuaries. Statuettes of swaddled infants, children and adults or carvings of different parts of the body were left at healing shrines. Representations of animals, domesticated or wild, some with mythological attributes (three-horned bulls, crowned serpents) are also common. Animals were offered in sacrifice either to be burned in a holocaust or, less frequently than among the Greeks, Romans and Germans, eaten as part of a communal feast. There is written and archaeological evidence for the practice of human as well as animal sacrifices, although the former were probably an exceptional part of public religion. The officiants may have been Druids, reputedly a highly trained professional class, whose members combined the functions of priest, seer, royal adviser and sorcerer. In addition, there is reason to believe that there were Celtic priestesses who performed sacrifices and acted as seers and healers.[78]

According to Pliny, the Celts followed the lunar calendar. Even in the supposedly Druidic Coligny calendar (first century A.D.?), which attempted to fit the lunar months into the solar year, the first half of the month, when the moon waxed, was considered lucky, and the second, when it waned, unlucky.[79] Seasonal celebrations were held in November, May and August (known in Ireland as *Samhain*, *Beltene* and *Lughnasadh* respectively), and also at midsummer, a few days after the solstice.

Divination was widely practiced. Druids or *vates* were said to find portents in entrails, the flow of blood and the death throes of victims. Omens could be read in the flight and song of birds as well. Dreams were prophetic, especially if dreamt while sleeping on a grave, or wrapped in cowhide, or after eating the flesh of sacred animals. Other methods involved shaman-like trances, pieces of wood, chewing acorns or hazelnuts, or the adoption of special postures–the antlered figure engraved on the Gundestrup Cauldron is shown sitting in a typically shamanlike pose, and it has been suggested that it represents a shaman, not the god Cernunnos. Curses were a Druidic specialty, but were also much practiced pri-

[78] Strabo reported that the priestesses of a small island in the mouth of the Loire destroyed the roof of their temple every year and rebuilt it that same day to the accompaniment of the sacrifice of one of the their number–a rite, de Vries suggested, meant to a open a new period of time (*La religion des Celtes*, 225-227). There is no mention of Druidesses in the classical period, and the *dryades* gifted with prophetic powers who are mentioned in the 3rd century A. D. appear to have been merely fortune-tellers or sorceresses (Chadwick, *The Druids*, 78-80 and 99; Toutain, *Les cultes païens* 3: 407).

[79] See J. A. MacCulloch, "Calendar (Celtic)," *ERE* 3: 79-82.

vately, as in the case of theft. Some were inscribed on lead tablets (*defixiones*), usually asking for the restoration of stolen goods and for punishment of enemies.[80]

The Celts who came under the influence of the Roman Empire adopted many of the forms of Roman religion, though with characteristic features, *e.g.*, temples composed of two concentric structures, monuments such as Jupiter columns and the merging of the names and attributes of Roman and Celtic divinities. The upper classes enthusiastically embraced the imperial cult, which gave them access to rank and power in the administration of the empire. This and other cults of Roman divinities largely obscure the older Celtic religions, but the extent to which they affected the actual beliefs and practices of ordinary persons cannot be gauged.

Germanic paganism:[81] Archaeology, toponymy, etymology, the writings of more or less contemporary authors, both classical (especially Julius Caesar and Tacitus[82]) and medieval (sermons, letters, the *Vitae* of missionary saints, legal texts, and Histories such as those written by Bede, Saxo Grammaticus and Adam of Bremen), mythology (much of it by way of sagas originating in Christianised Iceland) and folklore provide the sources for Germanic religion.[83] The possible ignorance, bias and misunderstandings on the part of the authors make the reliability of these sources questionable. However, unlike Celtic religion, which was already falling under Roman influence when classical authors took note of it and which showed little resistance to Christianity, Germanic paganism continued to flourish and evolve well into the Christian era. It knew a period of exceptional vigour during

[80] *E.g.* on a cursing tablet found in London: "I curse Tretia Maria and her life and memory and liver and lungs mixed up together, and her words, thoughts and memory, thus may she be unable to speak what things are concealed" (Green, *Dictionary of Celtic Myth and Legend*, *s.v.* "Curse").

[81] For Germanic paganism, see Julien Ries, *Pensée religieuse indo-européenne et religion des Germains et des Scandinaves (Louvain-la-Neuve*, 1980); Georges Dumézil, *Gods of the Ancient Northmen*, ed. and trans. Einar Haugen *et al.*; Berkeley, 1973); R. L. M. Derolez, *Les dieux et la religion des Germains*, trans. F. A. L. Cumen (Paris, 1962); *idem,* "Les Germains" in *Religions du monde* (Paris, 1965) 11: 61-137; Régis Boyer and Eveline Lot-Falck, *Les religions de l'Europe du Nord* (Paris, c. 1974); Herbert Schutz, *The Prehistory of Germanic Europe* (New Haven and London, 1983); David Wilson, *Anglo-Saxon Paganism* (London and New York, 1992); Régis Boyer, "Le sacré chez les Germains et les Scandinaves," in *L'homme indo-européen*, 185-213. See also Davidson, *The Lost Beliefs of Northern Europe* and *idem,* "Germanic religion," in *Historia religionum* 1: 611-27.

[82] What little Caesar had to say about German religion seems very much off the mark but, although he did not claim personal knowledge of all that he described, much of what Tacitus wrote about religion is supported by other sources (*Germania*, 27). For a discussion of Tacitus' sources and credibility, see Clarence W. Mendell, *Tacitus. The Man and his Work* (1957; reprint, New Haven, 1970), 199-222.

[83] For a discussion of the sources, see Ries, *Pensée religieuse,* 52-60, and Davidson, *The Lost Beliefs*, 11-86. Dumézil based himself almost entirely on mythology and language for his analysis of German religion; by contrast, Ries relies heavily on written documents, and Schutz on archaeology.

the Viking period centuries after the continental Germans were converted. But what the pagan Germans themselves thought is unknown for they, like the Celts, left no written account of their own to throw light on their beliefs. The written evidence comes from two different areas, periods of time, and points of view. On the one hand, material dealing with the continental Germans dates as far back as the beginning of the Roman Empire; it is presented from the Roman and then the Christian point of view. It gives a good deal of information about practices and ritual, but little about mythology. The Scandinavian material, dating from a much later period, is rich in mythology, less so in descriptions of cult. Nevertheless, archaeology suggests strongly that the religion of the continental Germans and the Scandinavians had many features in common.[84]

Disentangling the religion of the Celts from that of the Germans, especially those of the western part of the European continent (for example, the Frisians, Franks and Saxons), is difficult.[85] The mythologies are distinctly different, but religious practice appears very similar. Both worshipped their gods in nature, paid cult to trees and placed sacrificial offerings in bogs and lakes, had mother-goddesses (it is not clear, for example, whether Nehelannia was a Celtic or Germanic goddess of abundance), practiced a cult of the dead. Often the difference seems to be one of degree–for example, the cult of the head is generally accepted as being part of Celtic religion, but there is also evidence for it, though on a much lesser scale, among the Germans. Celtic, continental Germanic and Scandinavian art also have much in common in content, motif and style.

To Caesar and Tacitus, the Germans had astonishingly few gods. According to the former, they knew of only three: the Sun, Vulcan and the Moon; according to the latter, Mercury, Hercules and Mars. In addition, claimed Tacitus, Isis was worshipped by some of the Suebi, Nerthus by several minor tribes in Jutland and the Alci ("Castor and Pollux") by the Nahanarvali.[86] In fact, the Germans had a whole host of more or less important deities divided into two categories: the "high gods" of mythology, very few of whom enjoyed any discernible cult, and a large

[84] Ries, following J. De Vries (*Altgermanische Religionsgeschichte*, 3rd ed., 1970), insists that the study of the religions of the south and north Germans cannot be separated as they are the expressions of a single Germanic religious thought-system (*Pensée religieuse*, 62). For the sagas, see Régis Boyer, *Sagas islandaises* (Paris, 1992).

[85] Ries pointed out that the term Germanic refers to a linguistic not a racial concept (*Pensée religieuse*, 35). The identification of German ethnicity was due to the Romans who applied the name *Germani* to certain groups living to their north. Caesar defined the Rhine as the boundary between the Celts and the Germans on the basis of strategic and political, not ethnic, considerations. According to Schutz, this was a self-fulfilling prophecy: "Had Caesar not identified certain tribes as Germanic, no one would have concluded that they were anything but Celts." After Caesar, those living west of the Rhine became increasingly Celticised, those east, Germanised: "The undeniable effect of Caesar's conquests was the splitting-up of the old world of the La Tène, contributing to the compacting of western Europe into two distinct cultural areas" (*The Prehistory of Germanic Europe*, 337-347).

[86] Caesar, *De bello gallico* 6, 21; Tacitus, *Germania* 40, 43.

number of other divinities of whom neither myth nor sometimes even name is known, but to whom cult was evidently paid. The principal deities whose cult survived well into the Christian era were Thor and Woden, roughly identified with Jupiter and Mercury. Numerous other beings (giants, dwarves, kobolds, nixes) also figure in myth and folklore. The gods were not supreme; like men, they were subject to Destiny (*wyrd*), sometimes represented as a goddess with weaving implements, sometimes as three sisters, the Norns.

Tacitus claimed that the Germans refused to imprison their gods in man-made temples, or even to represent them in human guise; instead they worshipped them, as abstractions, in sacred woods and groves.[87] Trees were the link between men and gods. Jean-Louis Brunaux noted that, unlike the Celts for whom a post or column could satisfactorily symbolise a sacred tree, the Germans venerated the tree in its natural form only.[88] Mountains and bodies of water also were sacred. Tombs, burial mounds and other places haunted by the dead, such as crossroads, were also holy as being the threshold between this world and the next.

The most widely celebrated festival was held at Yuletide. It was a feast of the dead, the dangerous time when the Furious Host, the army of the dead, rode out in search of new recruits. It was the custom in all Teutonic countries to plough around the field at this time, perhaps as a form of protective magic. The meaning of other seasonal festivals varied according to region: the spring festival opened the agricultural year for the continental Germans and the seafaring and warfaring season for the Scandinavians; in Saxony, the autumn festival celebrated the dead, but in Scandinavia it was a harvest feast, a time particularly propitious for marriages.

The principal components of German religious practice were sacrifice, feasting and divination. The favoured sacrificial beasts were stallions, bulls and boars, "fighting animals, appropriate offerings for warrior peoples."[89] Their flesh was shared between gods and men in communal feasts. Human sacrifices are mentioned in the literature and substantiated by archaeology but, as with the Celts, they must have been quite uncommon and restricted to times of crisis.[90] Many cases of apparent sacrifice admit of other explanations: punishment for crime,

[87] *Germania*, 9.

[88] "Les bois sacrés des Celtes et des Germains," in *Les bois sacrés. Actes du Colloque International organisé par le Centre Jean Bérard de l'École Pratique des Hautes Études (Ve section), Naples, 23-25 Novembre 1989* (Naples, 1993), 57-65; here, 64. The symbolic importance of trees is one of the best-attested elements of Germanic religion both in mythology and in Christian literature. In the *Eddas*, man and woman were created from an elm and an ash; the cosmic tree Yggdrasil was the very centre of Asgard, the meeting-place of the gods; the oak was sacred to Thor; Odin had to hang on a tree for nine days. Christian missionaries and churchmen fought relentlessly against the cult of trees: Caesarius of Arles preached against it, St. Boniface felled the sacred oak of the Hessians, Charlemagne had the world-pillar Irmensul chopped down.

[89] Davidson, *Lost Beliefs,* 90.

[90] See Hilda R. E. Davidson in "Human sacrifice in the late pagan period in north western Europe," in *The Age of Sutton Hoo,* 331-340.

murder, unknown burial-rituals, mutilation after death. Blood was not always shed, for the sacrifice could be hanged, strangled or drowned. Festivals, tribal assemblies and funerals were celebrated with banqueting, sports and games, racing and, sometimes, animal combats. The most important element of the feast was drinking. Beer was specially brewed, toasts were offered to the gods and to the dead, and drunkenness was a sacred obligation, particularly at funeral feasts.

The Germans had both priests and priestesses, although apparently these did not constitute a priestly caste similar to the Druids. German women enjoyed a higher status than did their contemporaries elsewhere, so much so that a shocked Tacitus recounted that the Sitones had "degenerated" to the level of submitting to feminine rule.[91] Although there are indications that priestesses were involved in human sacrifices, Derolez believed women's prestige came from their personal charisma rather than from official status, and that their principal role was in family rituals and in the cult of fertility deities.[92]

Divination was important in both public and domestic religion. Here again women had a major role, which gave some of them considerable influence in political matters. Specialised knowledge was passed from mother to daughter. Divinatory techniques included casting lots, observing currents of water and the behaviour of animals, and interpreting dreams.

1.2.2. *Christianisation and conversion*[93]

The great achievement of the Church during the early Middle Ages was the conversion or re-conversion of most of Western Europe.[94] Christianity had been

[91] *Germania, 45.*

[92] *Dieux et religion*, 186, 195.

[93] The studies of conversion collected in *Conversion to Christianity. Historical and Anthropological Perspectives on a Great Transformation*, ed. Robert W. Hefner (Berkeley, Los Angeles, London, 1993) provide useful insights as to the problems faced by medieval Christians. Of particular interest are Charles F. Keyes' description of the missionary activity of a Buddhist abbot in Thailand ("Why the Thais are not Christians: Buddhist and Christian conversion in Thailand," 259-283) and David K. Jordan on the mechanisms of conversion and "faith maintenance" in China ("The glyphomancy factor: Observations on Chinese conversions," 285-303).

[94] For surveys of the history of the Christianisation of Western Europe, see Richard Fletcher, *The Conversion of Europe; From Paganism to Christianity (371-1386 AD)* (London, 1997); Eugen Ewig, "The missionary work of the Latin Church," in *History of the Church* 2: 517-601; *Évêques, moines et empereurs*, 607-670 and 718-725. For the personnel and methods of the missions, see J. T. Addison, *The Medieval Missionary: A study of the Conversion of Northern Europe, A. D. 500-1300* (1936; Philadelphia, 1976). For different approaches to the problems of conversion, see Agnès Boulouis, "Références pour la conversion du monde païen aux VIIe et VIIIe siècles; Augustine d'Hippone, Césaire d'Arles, Grégoire le Grand," *REAug* 33 (1987): 90-112; Alain Dierkens, "Pour une typologie des missions carolingiennes," in *Propagande et contre-propagande religieuse*, ed. J. Marx (Brussels, 1987), 77-93; J. M. Hillgarth, "Modes of evangelisation of Western Europe in the 7th century," in *Irland und die Christenheit*, 311-331; Ian Wood, "Pagans and holy men, 600-800," *ibid.*, 347-361; Michael Richter, "Practical aspects of the conversion of the Anglo-Saxons;"

well-established in Roman Spain, Gaul and Britain, but the Church had lost ground during the Germanic invasions of the 5th and 6th centuries. Only Ireland, which had never formed a part of the Roman Empire and was just in the process of being converted, was spared the experience of invasion. Elsewhere a kind of folk-paganism re-emerged, especially among the rural populations who had little access to the services of priests during the time of troubles. Even among some of the educated, there were signs of a pagan revival: in Rome itself during the pontificate of Gelasius I (492-296), a senator attempted to celebrate the *Lupercalia* (15 February) against the plague.[95]

In the relative peace under the rule of the new barbarian kingdoms, the Church had to win over the Arian Germans to Catholicism, convert the other Germanic tribes to Christianity, and strengthen the faith and discipline the morals of the faithful. These tasks were achieved at varying rates.[96] With the rulers' support, the conversion of the Visigothic and Suevan Arian Christians went rapidly after the first setbacks. Where the Church had to convert a people as yet relatively untouched by Christianity, success was sometimes slower in coming.[97]

Robert A. Markus, "Gregory the Great and a papal missionary strategy," in *The Mission of the Church and the Propagation of the Faith* ed. G. J. Cuming, (Cambridge, 1970), 29-38. Richard E. Sullivan's studies of various aspects of missionary work, reprinted in *Christian Missionary Activity*, are indispensable for the history of this period. See also Élisabeth Magnou-Nortier, "La christianisation de la Gaule (VIe-VIIe siècles). Esquisse d'un bilan et orientation bibliographique," *Mélanges de science religieuse* 53 (1996): 5-12; Martine De Reu, "The missionaries: The first contacts between paganism and Christianity," in *The Pagan Middle Ages*, 13-38.

[95] *Lettre contre les lupercales*, ed. G. Pomarès, *SC* 65 (Paris, 1959). See also A. W. J. Holleman, *Pope Gelasius I and the Lupercalia* (Amsterdam, 1974).

[96] *La conversione al cristianesimo* presents regional studies of different aspects of conversion in Western Europe. Only a few of many other studies are mentioned here: For Ireland, Kathleen Hughes, *The Church in Early Irish Society* (London, 1966); Dáibhí Ó Cróinín, *Early Medieval Ireland 400-1200* (London and New York, 1995), 14-40. For England, Henry Mayr-Harting, *The Coming of Christianity to Anglo-Saxon England* (3rd. ed., London, 1991); James Campbell, "Observations on the conversion of England," in *Essays in Anglo-Saxon History* (London, 1986), 69-84; Ian Wood, "The mission of Augustine of Canterbury to the English," *Traditio* 69 (1994): 1-17. For Iberia, E. A. Thompson, "The conversion of the Visigoths to Catholicism," *Nottingham Mediaeval Studies* 4 (1960): 4-35; *idem*, "The conversion of the Spanish Suevi to Catholicism," in *Visigothic Spain: New Approaches*, ed. Edward James (Oxford, 1980), 61-76. For Gaul, Lucien Musset, "De St. Victrice à St. Ouen: La christianisation de la province de Rouen d'après l'hagiographie," *RHÉF* 62 (1976): 141-15; C. E. Stancliffe, "From town to country: the christianisation of the Touraine, 370-600," in *The Church in Town and Countryside. Papers Read at the Seventeenth Summer Meeting and the Eighteenth Winter Meeting of the Ecclesiastical History Society* ed. D. Baker (Oxford, 1979), 43-59; Werner, "Le rôle de l'aristocratie"; Nancy Gauthier, *L'évangélisation des pays de la Moselle. La province romaine de Première Belgique entre Antiquité et Moyen-Âge, IIIe-VIIIe siècles* (Paris, 1980). See also Russell, *The Germanization of Early Medieval Christianity*; Manselli, "Resistenze dei culti antichi."

[97] Personal (notably marital), political and commercial contacts were the usual preliminaries to conversion. Moreover, old-established Christian enclaves probably existed in the midst of largely pagan territory, as they did in England before St. Augustine's arrival in Kent, see Camp-

The conversion of the Franks was inaugurated by the baptism of Clovis (c. 496) but as late as the mid 7th century public signs of pagan cults were being openly displayed in the countryside.[98] The Angles and Saxons of England were converted during the 7th century, thanks to separate initiatives on the part of Gregory I and Irish monks from Lindisfarne.[99] During the same period, under the influence of Luxeuil and the monks who had followed St. Columban to the continent, the first, mostly unsuccessful efforts were made to convert the Frisians and the tribes beyond the Rhine, a task taken up systematically at the end of that century by English missionaries encouraged and protected by Frankish rulers. By St. Boniface's death (754), the Alamannians, Hessians and Thuringians had been converted to Christianity, but it was only near the end of the century that the Saxons reluctantly accepted Christianity. In the territory beyond, among the Scandinavians, Slavs and Hungarians, paganism continued to thrive despite missionary inroads.

This first phase of conversion affected chiefly the exterior forms of religious life. The populace was baptised, the shrines, temples and public cults of the old deities were suppressed and the structures and rituals of the Christian Church were set in their place. But among many, the attitudes and beliefs that had informed private cults and domestic religion persisted. This was at least in part a result of the method of conversion generally adopted, which emphasised group loyalty rather than personal conviction. Undoubtedly there were numerous individual conversions, the work of hermits and itinerant preachers.[100] But, given the numbers of people to be converted, and the conceptual and linguistic gulf separating them from the Christian missionaries, the more highly organised missionary campaigns were directed at tribal kings and leaders, not at the masses. The means varied. They ranged from preaching, as in Kent and Northumbria, to preaching backed by political pressure, as in Frisia and Hesse, to outright force in which missionaries played a role secondary to the military, as among the Saxons and Wends. Once the rulers had been prevailed upon to accept the new faith, the conversion, or, at least, the Christianisation, of their subjects usually followed upon a very rudimentary catechesis.[101]

bell, "Observations on the conversion of England," 14-16. The missionary activity of St. Columban was among lapsed Christians, not pagans.

[98] See Vacandard, "L'idolâtrie en Gaule au VIe et au VIIe siècle."

[99] But see Rob Meens on British missionary activity among the Anglo-Saxons before St. Augustine arrived in Kent, in "A background to Augustine's mission to Anglo-Saxon England," *Anglo-Saxon England* 23 (1994): 5-17.

[100] Many of such conversions were the work of the Irish *peregrini*, for whom "activity in conversion was really a by-product of their lives as monastic exiles and pilgrims" (Addison, *The Christian Missionary,* 13). Their achievements depended on their personal charisma, their piety, ascetism, miracles and preaching ability, but often they failed to follow up their initial successes with the establishment of the permanent structures necessary to safeguard the faith.

[101] Richter distinguishes between conversion and Christianisation and points out "that the

The case of the Franks may serve as an example. In Gregory of Tours' account, Clothilde and St. Remigius had been attempting to persuade Clovis of the truth of Christianity with scant success until he was finally convinced by his victory over the Alamannians at Tolbiac in 496.[102] Thereupon "all the people" announced that they, too, rejected their old gods. Over three thousand of Clovis' warriors followed him to the baptismal font. Not the smallest suggestion is made that St. Remigius had tried to proselytise among them previously, or had provided them with any form of religious instruction before administering the sacrament.

That people converted thus have a profound understanding of the implications and requirements of their new religion was not to be expected.[103] Public paganism was eliminated relatively easily because, suggests Ian Wood, it was tied to social and political structures that changed with the acceptance of the new religion.[104] Private customs and beliefs were another matter. Oronzo Giordano maintained that the old religion was not replaced; rather, a new layer had been added to a pre-existing religiosity.[105] One historian went so far as to doubt whether most supposed Christians in the early Middle Ages could be considered as Christian in anything but name.[106] Real education in the meaning of Christian life

latter can take place without the former in the narrow religious sense," that is, conversion as defined by Nock ("Models of conversion in the early Middle Ages," 122).

[102] On the state of Clovis' actual beliefs, see William M. Daly, "Clovis: How barbaric, how pagan?" *Speculum* 69 (1994): 619-664.

[103] For an analysis of the problem of the change from paganism to Christianity, see Lifshitz, *The Norman Conquest*, especially 1-36; Robert A. Markus, "From Caesarius to Boniface: Christianity and paganism in Gaul," in *Le septième siècle. Changements et continuités / The Seventh Century: Changes and Continuity. Actes du Colloque bilatéral franco-britannique tenu au Warburg Institut les 8-9 juillet 1988*, ed. Jacques Fontaine and Jocelyn N. Hillgarth (London, 1992), 154-172; Karras, "Pagan survivals and syncretism in the conversion of Saxony;" Appleby, "Spiritual progress in Carolingian Saxony." For the recognition by contemporary churchmen of conversion as a gradual process, see Olivier Guillot, "La conversion des Normands peu après 911. Des reflets contemporains à l'historiographie ultérieure (Xe-XIe s.)," *Cahiers de civilisation médiévale* 24 (1981): 101-116 and 181-219. Paolo Golinelli maintained that the methods of conversion adopted, from above or with force, resulted in the bipolarisation of Christianity between the educated class and the masses ("La religiosità popolare tra antropologia e storia," in *La terra e il sacro. Segni e tempi di religiosità nelle campagne bolognesi*, ed. Lorenzo Paolini [Bologna, 1995], 1-9).

[104] "Pagan religions and superstitions east of the Rhine," 264.

[105] *Religiosità popolare,* 13. This is a point also made by Richter: conversion was not the "substitution of one set of beliefs by another" but "the acceptance and importation of new ideas and ways of life into previously existing modes" (*The Formation of the Medieval West,* 40).

[106] L. G. D. Baker, "The shadow of the Christian symbol," in *The Mission of the Church*, 17-28; here, 25. Baker dismissed medieval forms of rural piety as "Christian paganism" (*ibid.,* 27). *Cf.* a modern missionary's opinion of the spiritual condition of the Aymaras of South America, who had been formally converted to Christianity in the 15th century, and who considered themselves to be *muy catholicos*, that the main contribution of Christianity had been to furnish the indigenous population with a new vocabulary that allowed them to express their traditional beliefs in Catholic terms (Monast, *On les croyait chrétiens*, 17). Many of the magical or religious practices deplored by medieval missionaries have exact parallels in the Aymara rituals described by Monast.

often came after baptism as the missionary phase of conversion gave way to the pastoral, with the establishment of a network of monasteries and of parishes subject to periodic visits of inspection by the bishop. Force was still sometimes advocated; particularly when a territory was officially Christian, it was an attractive option.[107] In time, other methods became more important: the elaboration of the liturgy, preaching, and the use of confession to educate the faithful. The focus, however, remained on external conformity to ritual.

1.3 SOURCES–PASTORAL LITERATURE

Pastoral care may be defined broadly as all the measures taken under the aegis of the Church for the spiritual, moral and even physical welfare and doctrinal orthodoxy of the faithful.[108] For the early Middle Ages, the essence of pastoral care, as explained by Gregory the Great in his enormously influential *Regulae pastoralis liber*, was teaching the faithful in the terms best adapted to the particular character and circumstances of each individual.[109] In practice, this meant that those charged with the cure of souls were obliged to observe the behaviour of the members of their flock closely, and to develop means to combat the failings that they found. Pastoral literature is the written form in which this care and these observations are recorded.

1.3.1. *Value and limitations*

On its face, this type of literature is of questionable value as a source for popular religion and culture since the clerical authors were, by virtue of their position, hostile witnesses to any form of belief and practice that did not conform to the norms set by the Church. It is, however, the principal source of information about popular mentality and behaviour during the early Middle Ages. Archaeology and art provide data valuable for material culture but are difficult to interpret with respect to beliefs and rituals. Most forms of written documents have little or nothing to say about the lives of ordinary persons, who left no written records of their own and to whom others paid scant attention. Histories and chronicles concentrated on the unfolding of God's design through the doings of great men,

[107] Caesarius of Arles exhorted landowners to destroy shrines and sacred trees on their estates, and heads of families to beat their children and slaves if they persisted in pagan practices. Gregory I instructed the bishop of Sardinia in the methods to be used against idolaters and the clients of soothsayers: beatings for slaves and close confinement for the free. King Erwig threatened Visigothic landlords with the loss of property rights if they failed to punish their slaves for participating in pagan cults. Martin of Braga, on the other hand, appears to have relied solely on the effects of instruction and argument, while St. Aidan and St. Cuthbert, according to Bede, preached assiduously to the inhabitants in remote areas.

[108] See J.-P. Bagot, "Pastorale," in *Catholicisme* 10: 765-774 and William A. Clebsch and Charles R. Jaeckel, *Pastoral Care in Historical Perspective. An Essay with Exhibits* (1964; New York, 1967).

[109] *PL* 77: 13-128.

while the best minds of the age devoted themselves to theological controversy, exposition of the Bible, the means of achieving perfection and the development of the liturgy. The humble appear in hagiography and the Germanic legal codes, but merely as objects of the miraculous powers of the saint and of the demands or punitive force of the law. Folk songs and folk tales, such as those that Charlemagne is said to have collected, have virtually disappeared, at least in their vernacular form.[110]

An immense body of contemporary works contains scraps of relevant material, often of a kind and precision missing from normative texts. St. Patrick's Confession bears a trace of an Irish fosterage ritual in the account of the saint's refusal "to suck the breasts" of the Irish sailors who carried him to freedom.[111] The *Medicina antiqua* describes the virtues of specific kinds of herbs together with the proper rituals for their use.[112] The witches' cauldron is found in the Salic Law.[113] A soothsayer known as an *umbrarius* appears in another Germanic Law, and a *librarius* in the *Life of St. Samson of Dol.*[114] Even theological works and exegeses may contain folkloric elements. In an exposition of Scripture, Caesarius of Arles warned his hearers against believing that the prophet Elisha used his staff for magic (*augurium*). It was meant as an aid to walking, he assured them, not for other, suspect purposes–a caveat which makes sense only if 6th-century Provençal magicians used their staffs as magic wands.[115] But in all of these, popular culture occupies a small, almost incidental, place. It was only in pastoral literature that, from time to time, the focus was turned on ordinary men and women in themselves.[116]

None even of these sources give direct access to the religious culture of the illiterate masses–all are filtered through the minds and words of the clerical authors. But pastoral literature, normative literature in particular, offers the best avenue of approach. By their nature, laws (including secular laws affecting reli-

[110] Einhard recorded Charlemagne's attempt to collect and have written down "the barbaric and very ancient songs which celebrated the deeds and wars of the kings of old" (*Vita Caroli Magni* 29, *MGH Scriptores Rerum Germanicarum* 25: 33). For an example of heroic Carolingian poetry, see *MGH Poetae Latini Aevi Karolini* 1: 109-110, and a translation by Paul Edward Dutton in *Carolingian Civilization: A Reader* (Reprint; Peterborough, 1999), 48. For oral vernacular literature, see also Richter, *The Formation of the Medieval West,* 231-254.

[111] *Confessio*, 18 in *Confession et Lettre à Coroticus,* ed. and trans. Richard P. C. Hanson *SC* 249 (Paris, 1978), 88. See John Ryan, "A difficult phrase in the 'Confession' of St. Patrick: *Reppuli suggere mammellas eorum*," *The Irish Ecclesiastical Record* (1938): 293-299.

[112] *Medicina antiqua: Libri quattor medicinae. Fac-similé du Codex Vindobonensis 93, conservé à Vienne à la Bibliothèque nationale d'Autriche.* French trans., Marthe Dulong; studies by Charles H. Talbot and Franz Unterkirchen 2 vols. (Paris, 1978).

[113] *Pactus Legis Salicae* 64.1, *MGH Leg* 1.4.1: 230.

[114] *Edict of Theodoric* 108, *MGH Leg* 5: 164; *Vita Samsonis* 2-5, *AASS* Iul 6: 574-575. I am obliged to J.-C. Poulin for this reference.

[115] *S.* 128.7, *CCSL* 103: 529-530.

[116] See Boglioni, "Pour l'étude de la religion populaire au moyen âge."

gion), penitentials, to some extent sermons, and other documents concerning pastoral problems were intended to deal with practices and beliefs which their authors were convinced to be prevalent among their contemporaries.

But to argue for the authors' good faith is not necessarily to argue for their accuracy. One may well ask how right they were in this conviction. Their interpretations of the behaviour they castigated as pagan or superstitious is, as has been seen, open to question. In an age when political and religious considerations were closely intertwined, it is certain that political motivations were mixed with the purely pastoral, and the exact weight to give one or the other is debatable.[117] The authors may have erred in attributing pagan meanings to customs and used tendentious definitions of Christianity. What they wrote, however, presents less of a problem than what they did not write. The limited number of themes which our texts touch upon and the absence of reference to numerous areas of life raise the question as to how well even the most conscientious member of the hierarchy knew his charges. The parish clergy might have known them very well, but the amount of feedback between them and their superiors during parish visitations and regular diocesan assemblies is doubtful: only one of the ninety-six questions that Regino of Prüm recommended bishops to ask parish clergy concerns the laity's conduct–a striking contrast to the numerous detailed questions found in penitentials.[118] Nevertheless, among the authors of pastoral literature were some of the ablest men of their generation, surely most of whom took their spiritual responsibilities seriously. Their conceptual world was conditioned not only by training, professional biases and political pressures, but also by the society within which they lived and carried out their duties. A close reading of their writings opens a window not only on their mentality but on that society as well.

The question of bias aside, this literature has other serious drawbacks as a source specifically for paganism, pagan survival and superstition, and for popular culture as a whole. In the first place, even it has relatively little to say on such

[117] For example, it is generally agreed that legislation aimed at uprooting traditional Saxon practices was motivated by the determination to integrate Saxony forcibly into the Carolingian empire and destroy local particularities; see Bonnie Effros, "*De partibus Saxoniae* and the regulation of mortuary custom. A Carolingian Campaign of Christianization or the suppression of Saxon identity," *Revue Belge de philologie et d'histoire* 75 (1997): 267-286. On the other hand, while both Lifshitz and Markus agree that the *paganiae* attacked by St. Boniface and the early Carolingians were practices tolerated as Christian by the Merovingian Church, they differ in their interpretation of the dynamics of the campaign against these practices. The former considers Boniface, "prelate sidekick" of ambitious princes, largely as a tool manipulated by the Carolingians to justify their having seized power (*Pious Neustria*, 58-62); the latter sees Boniface as struggling to eliminate genuine abuses in the Merovingian Church and, at the same time, to introduce his own ascetic ideal "of what being a Christian involved" (*The End of Ancient Christianity*, 211).

[118] *De synodalibus causis* (c. 906) 73: 24. For parish visitations as a method to raise the standards of Carolingian parish clergy, see E. Vykounal, "Les examens du clergé paroissial à l'époque carolingienne," *RHE* 14 (1913): 81-96.

subjects. Its focus was primarily on theological questions and matters of institutional concern, such as clerical discipline, the responsibilities of bishops, priestly celibacy, and church property. The laity was a lesser consideration, and references to popular beliefs are sporadic at best. Of the over five hundred and fifty canons enacted by Gallican Councils between 511 and 695, barely over thirty, about six percent, concern what might be called paganism and superstition. Less than ten percent of the three hundred and thirteen clauses of the influential *Penitential* of Theodore deal with this topic. The overwhelming majority of medieval sermons never touched it at all. This is true even of the sermons of Caesarius of Arles, one of the principal sources for the early medieval attitude to and interpretation of popular paganism and superstition. Only forty-four (about eighteen percent) of the almost two hundred and fifty sermons or fragments of sermons that have been identified as his contain even a passing reference to them. At best, this literature provides only isolated glimpses of popular culture and religion.

Moreover, although many modern historians accept without hesitation that this data, scanty as it is, is authentic evidence for actual beliefs and practices,[119] their validity is debatable. They are found in texts written over five centuries, under varied political and social circumstances, in areas far removed from each other, with varied climatic and topographical characteristics; they were intended ostensibly for peoples of different backgrounds, customs and languages. It cannot be doubted that throughout the period and area being considered, Christians indulged in practices that struck their pastors as pagan survivals and/or superstitions, but one might expect them to have varied from place to place and time to time. Yet, when it comes to these subjects, the same themes appear again and again. The very words in which they are expressed are often the same. Often we must ask whether these stereotyped words and phrases were a convenient form of shorthand used by clerics (whose precarious grasp of Latin might not have allowed them to depart from the usual patterns) to draw the attention of diocesan and parish clergy and civil authorities to well-known local practices, or merely formulas emptied of their original meaning, with no relevance to actual conditions.

Incongruous juxtapositions of material and the complex relationships between the sources add to doubts about the reliability of the information they contain. Canons from 4th-century Asia Minor, Africa and Spain were reproduced in Carolingian decrees. An 8th-century English council included among its decisions those of a Frankish council reported to it by St. Boniface. Practices first described in 6th-century sermons intended for the inhabitants of southern Gaul and northwestern Iberia were assembled, scissors-and-paste fashion, in sermons presumably delivered to the Alamannians and northern Franks of the 8th century, and elements drawn from Caesarius of Arles turned up in 10th-century Anglo-Saxon

[119] *E.g.*, Dukes, *Magic and Witchcraft* and Giordano, *Religiosità popolare*. See also Russell, *The Germanization of Early Medieval Christianity*, 109-110 (especially n. 9) and 201-202.

texts.[120] Penitentials in particular are notoriously derivative. The *Penitential* of Theodore of Canterbury took some of its dietary prohibitions from the Irish *Penitential of Cummean*, and the *Old Irish Penitential* returned the compliment by citing Theodore's authority. The compilers of continental penitentials copied with abandon from their insular predecessors and from each other, sometimes combining several different works into one. Copying crossed genres as well, so that practices that first appeared in conciliar decrees found their way into sermons, or vice versa, and material from both into penitentials. Such a welter of borrowings and cross-borrowings compromises the credibility of the texts. Under these circumstances can they be used at all as sources for popular religion?

In effect, Wilhelm Boudriot and Dieter Harmening have said no. They argued that the authors copied one from another (especially from Caesarius of Arles) to such an extent that the texts often prove nothing more than the persistence of literary tradition and give little credible information about the customs of the people living at the time and in the places where they were written.[121] For Boudriot, the documentation was unsatisfactory as evidence specifically for pre-Christian Germanic religion; he accepted its validity for the superstitions ("Aberglauben") of southern Gaul in the 6th century.[122] For Harmening, it was of doubtful value for medieval European superstitions altogether, since it was based on concepts drawn from the Mediterranean world of late Antiquity and largely borrowed by Caesarius himself, from St. Augustine.[123]

These assessments are surely too pessimistic. Jean Gaudemet considered the frequent repetition of the same prescriptions in general as proof of noncompliance with the canons.[124] Aron Gurevich, Jean-Claude Schmitt and William E. Klingshirn saw them as proof (in Gurevich's words) of "the stability of the vital pheno-

[120] See Joseph B. Trahern, "Caesarius of Arles and Old English literature: Some contributions and recapitulations," *Anglo-Saxon England* 5 (1976): 105-119.

[121] Boudriot, *Die altgermanische Religion*; Harmening, *Superstitio*. Others (*e.g.*, Boese, *Superstitiones Arelatenses*,20-56) had noted Caesarius' influence on subsequent pastoral literature without perceiving the implications.

[122] Boudriot accepted only a handful of texts without reservation: some of Gregory the Great's letters, St. Boniface's correspondence with the papacy and his English friends, Charlemagne's legislation for the Saxons, two Old German baptismal formulas, resolutions issuing from the Council of Neuching, and many of the questions in Burchard of Worms' penitential. The legislation associated with the Carolingian reforms and empire which contained material going back to an earlier period of the Frankish Church he found of more questionable value. The sermons of Pirmin of Reichenau, Burchard of Würzburg and Rabanus Maurus and the *Homilia de sacrilegiis* he rejected as utterly worthless as evidence of German paganism (*Die altgermanische Religion*; 7-8). This, with some modifications, is also Dowden's position, *European Paganism*, 149-166.

[123] For the relationships between medieval texts and their literary sources, both Late Antique and medieval, see *Superstitio*, 320-337. For a summary of the sources used by Caesarius in his sermons, see G. Morin, *Sancti Caesarii Arelatensis Sermones, CCSL* 103, xix.

[124] *Les sources du droit de l'Église en Occident du IIe au VIIe siècle* (Paris, 1985), 110.

mena" described in the literature.[125] Even the highly repetitive penitentials were, according to Gurevich, "practical guides and not exercises of abstract learning devoid of any connection with the time when they were composed."[126] Cyrille Vogel went farther. Since these booklets were intended to help confessors in their pastoral tasks, they were "the reflection, incomplete perhaps, but faithful, of the moral and spiritual atmosphere" which surrounded Christians at the place where and time when each penitential was written. Their chaotic organisation and crudeness, he claimed, vouched for their value as historical documents, more than would have tidy lists compiled by scribes without personal knowledge of pastoral problems.[127] As for councils and synods, it is unlikely that the participants spent much time debating matters of no immediate concern to them. The practical nature of most of their other decisions testifies to the relevance of our texts. The testimony of busy administrators and conscientious pastors must be taken seriously, if with reservations, especially since they did not copy previous material wholesale, but selectively.

Occasionally, we are given direct evidence that a stereotyped description or text was in fact relevant to the actual situation. In a letter written c. 847 to settle a debate about the permissibility of a certain type of divination (*sortes*), Leo IV referred the bishops of Brittany to a frequently cited canon issued by the Council of Ancyra (314), concerning divination and the lustration of houses by soothsayers. He pointed out that the *sortes* were similar to the practices described by Ancyra, and the same judgment applied to them.[128] Here a formulation adopted in Asia Minor in the 4th century was being applied to a ritual (similar, therefore not identical) practiced in 9th-century Brittany. In the mid 8th century, St. Boniface wrote to the pope complaining of eye-witness reports of scandalous celebrations in Rome using much the same terms in which Caesarius of Arles had castigated the revellers in the Narbonnaise some two centuries earlier; the pope was not able to refute the accusation in his reply.[129] Archbishop Hervé of Reims' early 10th-century account of the consumption of sacrificial foods by the recently converted Normans, echoed in John IX's letter to him, is an almost word-for-

[125] Gurevich, *Medieval Popular Culture,* 37; Schmitt, "Les 'superstitions'," 450-451; Klinghshirn, *Caesarius of Arles,* 283-284.

[126] Gurevich, *Medieval Popular Culture*, 37. Yitzhak Hen, however, thought that these documents expressed a "mental reality rather than a practical one," that is, the mind-set of the authors rather than the lay culture surrounding them (*Culture and Religion in Merovingian Gaul*, 171) See also *idem*, "Paganism and superstition in the time of Gregory of Tours: 'une question mal posée'," in *The World of Gregory of Tours*, ed. Ian Wood et Kathleen Mitchell (Leiden, 2002), 229-240. For Harmening's response to Gurevich, see "Anthropologie historique ou herméneutique littéraire: Une critique ethnographique des sources médiévales," *Ethnologie française* 27 (1997): 445-456.

[127] *Le pécheur et la pénitence* (Paris, 1969), 40.

[128] Leo IV to the bishops of Brittany (847-848) *Ep.* 7, *MGH Ep* 5: 594.

[129] Boniface to Zacharias (742) *Ep.* 50, *MGH EpSel* 1: 84; Zacharias to Boniface (743) *Ep.* 51, *ibid.,* 90.

word reiteration of phraseology standard in the literature from the earliest period on, that they ate of the food which they had sacrificed to idols.[130]

Boudriot's and Harmening's criticisms have hardly received the attention that they merit,[131] yet the arguments and counter-examples above only prove that customary formulations may have corresponded in some general way to actual behaviour. It remains true that the testimony for individual practices can be accepted only if they can be authenticated independently. Rudy Künzel proposed nine criteria for identifying descriptions based on actual observation: (1) the existence of a description of a similar practice in texts of a different genre (for example, hagiography and the Germanic legal codes); (2) the presence of a new, nonstereotypic term in the midst of a series of standard terms; (3) the description of the same practice, in the same area, in two independent texts written at two periods long removed from each other; (4) different interpretations in the texts of the same practice; (5) a new element added, plausibly given the historical context, to a description of an old rite; (6) the use of a vernacular term in the Latin text; (7) the purpose of the text (for example, baptismal vows designed for a specific group); (8) the degree of conformity to a stereotype–the less stereotypic a document is, the more likely it is to present authentic information; (9) the existence of an element given an incongruously Christian interpretation (for example, the characterisation of a poltergeist as a demon).[132] To these may be added as tenth criterion the omission of an important element from an otherwise stereotyped list of terms (such as the disappearance of the *caragius* from lists of cunning men after the mid 9th century); this implies, although it does not prove, that the text was written under the influence of actual conditions. By applying these criteria through a systematic study of the documents we arrive at a limited number of texts which, Carlo Ginzburg has shown, are "more rewarding than the massive accumulation of repetitive evidence" which makes up the bulk of the documentation.[133]

But even when the information given is not stereotypical, it cannot be accepted automatically as representing real beliefs or customs in the place or time to which it supposedly applies. We seldom know the source on which the authors themselves relied. The descriptions of Anglo-Saxon and Saxon idols drawn by Gregory I and Gregory II respectively may well have been based more on memories of Mediterranean cults than on precise knowledge of Germanic practices. The most detailed source for 6th-century Iberian popular culture and paganism is Martin of Braga's model sermon *De correctione rusticorum*. But Martin

[130] John IX, *Ep.* 1, *PL* 131: 29.

[131] Flint, for instance, in her important study of the Church's response to popular culture, acknowledges the influence of Caesarius of Arles' sermons on subsequent literature, but minimises its importance: "Caesarius's long shadow reached into many corners, and perhaps on occasion convenience, rather than true contemporary feeling, prompted the repetition of his vehement words" (*The Rise of Magic,* 43).

[132] "Paganisme, syncrétisme et culture religieuse au haut moyen âge."

[133] "The inquisitor as anthropologist," in *Clues, Myths and the Historical Method,* 164.

of Braga (520-580) was a Pannonian by birth; he had travelled extensively in the East before arriving c. 550 in Galicia where he became abbot, then, in 556, bishop, of Dumio and later Archbishop of Braga. He was involved in the conversion of the Suevi to Catholicism, in theological controversies and in the redaction of an important set of canons (the *Canones ex orientalium patrum synodis*, also known as the *Canones Martini*) and of various other works. How was he able to familiarise himself with the private practices of Galician "rustics"–offerings of wine and grain on the hearth, invocation of Minerva by spinning- and weaving-women, secreting of rags and crumbs in boxes? If he relied on his own observations, did he make them in Galicia or elsewhere in his youth or while on his travels, and then put them into his sermon on the assumption that such were the universal practices of countryfolk?

This brings us to the two-fold problem of language. On the one hand, we may question how well the clergy understood the speech and customs of their flock, an issue of which Carolingian reformers were quite aware.[134] In general, practices are usually only mentioned in the texts, not described.[135] No doubt, this was due in part to the authors' disdain for anything that they considered pagan, and their unwillingness to examine it closely or describe it accurately. In part, also, they may have expected their readers to be familiar with the matter at hand and to be able to apply the lesson without having the details spelled out. But this must have been due at least sometimes to an inadequate knowledge of the language spoken by their charges. Could Martin of Braga really converse with the Iberic and Suevan laity of his diocese? Was Caesarius of Arles, who may have been Burgundian rather than Gallo-Roman by birth, able to understand the speech of all the different groups that made up the multi-ethnic population of his diocese–Gallo-Romans, Syrians, Greeks, Egyptians, Jews, Burgundians, Goths, Franks? The Irish and Anglo-Saxon missionaries may have learned the native languages well enough to preach intelligibly to Frisians, Franks and other Germans, but did they comprehend fully the daily, homely vocabulary in which their flock spoke to them? The early 9th-century Poen. Valicellanum I exceptionally glosses six terms dealing with reprobated practices: *mathematicus, fana, veneficium, sacrilegium, cervulus aut vetula* and *sortes sanctorum* (astrologer, shrines, sorcery, sacrilege, masquerades, the lottery

[134] On the difficulty of translating unfamiliar concepts into the language of a different cultural milieu, and resulting misunderstandings, see Edward Evan Evans-Pritchard, *Theories of Primitive Religion* (reprinted with corrections, Oxford, 1989), 11-14. For the early medieval period, see Rosamond McKitterick, *The Frankish Church and the Carolingian Reforms 789-895* (London, 1977), 80-114, and Francesco Delbono, "La letteratura catechetica di lingua tedesca. Il problema della lingua nell'evangelizzazione," in *La conversione al cristianesimo,* 697-741. For missionaries' problems of communicating with the Anglo-Saxons, see Richter, "Practical aspects of the conversion of the Anglo-Saxons," 370-371, 375. See also Pierre Boglioni, "Les problèmes de langue dans les missions du haut moyen âge d'après les sources hagiographiques," (work in progress).

[135] The poverty of early medieval documents in this respect is strikingly evident when contrasted with the wealth of detail found in later texts, such as the Franciscan sermons examined by Montesano in "'Supra Acqua et supra ad vento'."

of the saints). Every one of these is standard, found regularly in penitentials, yet a conscientious scribe could fear that they were unfamiliar to the very men for whose guidance they were intended. At the beginning of the 11th century, Burchard of Worms found it necessary to explain vernacular terms for the benefit of confessors: XL *dies continui quod vulgus carinam vocat*, *homo in lupum transformari quod teutonica Werewulff vocatur*, *agrestes feminae quas sylvaticas vocant*, *herba jusquiamus quae Teutonice belisa vocatur* (forty days of fasting, werewolf, woodwives, henbane)–this in an area converted, or reconverted, by St. Boniface two hundred and fifty years earlier.[136]

On the other hand, the largely uniform Latin vocabulary of the texts does not do justice to the diversity of the cultures and periods described.[137] A word that had one meaning in the Romanised world of the Mediterranean had quite a different meaning when applied to the culture of the northern Celts and Germans. The *magus* of a 7th-century Irish canon had little in common with the Persian magicians of late Antiquity, nor the *pitonissa* of an 8th-century Frankish sermon with the Witch of Endor or the priestess of Delphi. The "wicked songs" (*cantica turpia*) of 9th-century Mainz were not likely to be the same even in spirit as the "wicked songs" of 6th-century Arles. Documents from every century and every region of the period and area covered in this work forbade the faithful in almost identical terms to offer vows to trees (*vota ad arbores*), without indicating the differences that must have existed from place to place and time to time as to the kinds of tree, vows and ritual involved. The monotonous reiteration of words and phrases creates a deceptively homogeneous appearance of an undifferentiated folk paganism stretching from the Mediterranean to the Irish Sea and the Rhine, from late Antiquity to the end of the Carolingian Empire.

Other kinds of vocabulary problem are rarer. Some words are used once only, without explanation. What was the nature of the sorcery known as *canterma* in the Sicilian dialect at the end of the 6th century? or of *dadsisas*, a ritual practiced over graves at the middle of the 8th around Hainaut? or of the *maida* under which candles were burned in northern Italy during the winter festival at the end of that century? The difficulty may come from the use of a familiar term in an unfamiliar context. How to interpret an Iberian canon forbidding clerics and pious laymen from participating in feasts with *conferti*? Does this really mean sausages, or is it a mistake for confraternities, *confratriae*? Or, when another Spanish text consistently gives the spelling *monstruosa* in passages where it must mean "menstruous," can we be wholly sure that it does not mean "menstruous" again in a passage where "monstrous" is normally expected? In other cases, the entire text is phrased so

[136] See Burchard of Worms, *Decretum* (1008-1012) 19, 5.1, 5.151, 5.152 and 5.194, Schmitz 2: 409, 442 and 452.

[137] *Cf.* Richter's observation that "the Latin language is a most inadequate tool for grasping aspects of early medieval cultures outside the sphere of Latin" (*The Formation of the Medieval West*, x).

confusingly as virtually to defy understanding.[138]

Slight differences in standard passages also pose a difficulty for interpretation. Do they represent a spelling variation only, or did the author or scribe intend a difference in meaning? Is an *emissor tempestatum*, for example, identical to an *immissor tempestatum*, or is one a malign sorcerer who summons a devastating hailstorm, the other, a benign wizard who chases it away? Do all the passages condemning the notorious New Year's practice of going as or with a *vetula, vecula, vecola, vegula, vehicula, uicula, vetulus, feclus,* refer to the same thing? How much was due to the authors' observation of local practices, and how much to scribal errors repeated by the parish priest during confession or while preaching at the altar, to be taken up and put into practice or adapted by the faithful?

Under these circumstances, it is often difficult to make a categorical statement about the exact meaning of a text or the prevalence, even the existence, of a practice or belief at any given period or in any given place. If I have relied heavily on equivocations ("they might have," "it may be assumed," "conceivably," "possibly," "presumably," *etc.*), it is in recognition of the ambiguousness of the material.

1.3.2 *Typology*

Legislation: The term legislation is taken to include both ecclesiastical and secular laws. The laws of the Church are expressed in decisions of church councils and synods, diocesan regulations, and canonic collections (penitentials are considered separately).[139] The secular laws considered here are the regulations

[138] An extreme example is this passage from a Spanish penitential: Si quis mulier qui uiros ad benedicentes barbas succenderint, sibe qui capillos in sola fronte benedictos tonserint, et postea, quod absit, ad deformitate(m) peruenerint, agenda sit eis penitentia annos VII (*Poen. Cordubense* [10th century] 129, *CCSL* 156A: 64). *Deformitas* here means fornication, according to the editor, Francis Bezler (private communication), but even with this help, the exact meaning of the text is impossible to decipher. On the subject of the psychoanalytical and anthropological significance of hair, see Edmund Leach, "Magical Hair," *Journal of the Royal Anthropological Institute* 88 (1958): 147-164.

[139] The bibliography for ecclesiastical legislation is immense. Only a small selection is given here. For an inventory of sources, see Augustus Potthast, *Repertorium fontium historiae Medii Aevii primum ab Augusto Potthast digestum, nunc cura Collegii historicorum e pluribus nationibus emendatum et auctum*, 7 vols. (Rome, 1962; 1997). Jean Gaudemet presents a systematic organisation of the different forms of law in *Les sources du droit de l'Église en Occident* and *Les sources du droit canonique* (Paris, 1993), 17-41. For general histories of councils, see Hefele-Leclercq, vols. 3 and 4, and Piero Palazzini, *Dizionario dei concili*, 6 vols. (Rome, 1963-1968). A. G. Cicognani, *Canon Law*, trans. J. M. O'Hara and F. Brennan (2nd revised ed., Philadelphia, 1935) is still useful as a model of clarity despite its outdated vocabulary. For summaries, see James. A Brundage, *Medieval Canon Law* (London and New York, 1995), 18-43; Constant van de Wiel, *History of Canon Law* (Louvain, 1991) 11-75; David Knowles with Dimitri Obolensky, *The Christian Churches, 2, The Middle Ages* (London 1969), 138-145. For the value of such sources for social history, see Walter Ullmann, "Public welfare and social legislation in the early medieval councils," in *The Church and the Law in the Earlier Middle Ages* V, Variorum reprints (London, 1975); Franz J. Felten, "Konzilakten als Quellen für die Gesellschaftsgeschichte des 9. Jahrhunderts," in *Herrschaft, Kirche, Kultur. Beiträge*

affecting religion, chiefly embodied in Carolingian capitularies. Capitularies, strictly speaking, are the collections of the edicts of Carolingian rulers, from 779 to the beginning of the 10th century, but the term is also applied to the edicts of Merovingian rulers. Unlike the Germanic laws which were personal and ethnic, that is, the particular customary laws, prerogatives and obligations of each national group (for example, the Alamannians), Carolingian edicts emanated from the will of the sovereign, were territorial in scope and limited to the lifetime of the sovereign. However, after Charlemagne was crowned emperor in 800, his legislation began to infringe on the rights enshrined in customary law.

The decisions of councils and synods, bishops' *capitula* and capitularies form the most authoritative part of pastoral literature from the point of view of popular culture. The date and the location of councils and synods, often even the names of the participants, are known. The date and the purpose for which capitularies were issued are also well defined. Bishops' regulations or *capitula* (usually quite short) are likewise clearly identified. It is fair to assume that most of these were in response to specific problems experienced at the particular time and in the particular place where it was issued.[140]

The value of collections as records of actual practices and beliefs is more doubtful. Nevertheless, since the majority of them (such as the *False Decretals*) allowed such subjects only a very small place while others (such as the *Collectio Anselmo dedicata*) ignored them altogether, the very fact that the compiler chose to include them may be significant. In some–maybe most–cases, the inclusion of such canons may be the result of the desire to be as complete as possible. How else is one to understand the repetition in the *Epítome hispánico* and the *Hispana* of clauses concerning Christians who lapsed during the persecutions before the Peace of the Church? But when a clearly spurious canon is attributed to the distant past, it is evident that the compiler had in mind some current problem. The most striking example of this is the well-known *Canon Episcopi* concerning Diana's cavalcade, which was attributed by Regino of Prüm (c. 906) and Burchard of Worms (1008-1012) to the fourth-century Council of Ancyra. The intention was evidently to add weight to a contemporary canon, by invoking the authority of a prestigious council of the remote past.

Penitentials and penitential texts:[141] Penitentials were manuals meant to provide

zur Geschichte des Mittelalters, ed. Georg Jenal and Stephanie Haarländer (Stuttgart, 1993), 177-201.

[140] For synods and bishops capitularies, see Guy Devailly, "La pastorale en Gaule au IXe siècle," *RHÉF* 59 (1973): 23-54. For evidence for a considerable amount of familiarity with such legislation among the parish clergy, see Yitzhak Hen, "Knowledge of canon law among rural priests. The evidence of two Carolingian manuscripts from around 800." *Journal of Theological Studies* 50 (1999): 117-134.

[141] Penitentials assign a specific penance for each sin; penitential texts lay down rules for discipline, but do not prescribe a specific penance. For the history and development of penitentials, Gabriel Le Bras, "Pénitentiels," *DTC* 12.1: 1160-1179, is still valuable; see also Allen J. Frantzen, "Bussbücher," *Lexikon des Mittelalters* 2: 1118-1123. Frantzen's *The Literature of Pen-*

guidelines to priests in the administration of private penance. In essence, they were more or less detailed catalogues of sins, with an appropriate penance suggested for each, depending on the sin and the status of the sinner.[142] This system of penance according to a price list ("la pénitence tarifée") had its origins in 6th-century British and Irish Celtic monastic communities that practiced private confession rather than the public penance customary in the early Church and on the continent.[143] Since private penance implied that some of the sins to be confessed were not generally known to the community, questions touched on the most intimate thoughts and deeds.

On the continent, where they were first introduced by Irish, then Anglo-Saxon, missionaries in the 7th and 8th centuries, the penitentials enjoyed a widespread popularity.[144] Since many parish priests did not have the skill and knowledge necessary to administer the sacrament effectively and to mete out appropriate penance, the authors or compilers of penitentials provided them with a minimal tool to help them in their pastoral duties when once a year, before the beginning of Lent, they were required to summon their flocks to the sacrament of penance.[145] In addition, traditional Germanic law, based on the principle of compensation or *weregeld*, provided a favourable environment, despite the opposition of bishops who feared that these often ill-organised and anonymous booklets of foreign provenance allowed too much autonomy to priests at the expense of episcopal authority.[146] The use of penitentials reached its peak between approx-

ance in Anglo-Saxon England (New Brunswick, N. J., 1983) outlines the history of penitentials and provides an excellent analysis of Irish, English and Frankish penitentials. A discussion of penitentials and translations of some of the most important ones are found in John T. McNeill and Helena Gamer, *Medieval Handbooks of Penance: A Translation of the Principal 'Libri Poenitentiales' and Selections from Related Documents* (1938; reprint, New York, 1990), 3-71. Martine Azoulai, *Les péchés du Nouveau Monde. Les manuels pour la confession des Indiens (XVIe-XVIIe siècle)* (Paris, 1993) throws a light on missionaries' perception of native culture, and is useful for purposes of comparison.

[142] All penance, according to one of the earliest of the penitentials, that of Cummean (before 662), was to be adapted to take into account how long the sinner continued in the sin, his education, the temptation to which he was subject, his moral fibre, his degree of repentance, and the compulsion under which he sinned, for "Almighty God who knows the hearts of all and has bestowed diverse natures will not weigh the weights of sins in an equal scale of penance" (Bieler, 132-3; Bieler's translation).

[143] For the history of penance, see Cyrille Vogel, *Le pécheur et la pénitence dans l'Église ancienne* and *Le pécheur et la pénitence au moyen âge*. See also O. D. Watkins, *The History of Penance*, 2 vols. (reprint, New York, 1961), especially vol. 2. The origins of private penance are discussed in Frantzen, *The Literature of Penance*, 22-26.

[144] See Raymund Kottje, "Überlieferung und Rezeption der irischen Bußbücher auf dem Kontinent," in *Die Iren und Europa im früheren Mittelalter*, ed. H. Löwe (Stuttgart, 1982), 511-524.

[145] Rosamond Pierce (McKitterick), "The 'Frankish' Penitentials," *Studies in Church History* 11 (1975): 31-39; here, 38.

[146] See, for example, the *Lex Saxonum* (*MGH Legum* 5: 47-84). For the opposition of the Frankish hierarchy to the penitentials, see Pierce, "The 'Frankish' Penitentials," and the critique of this article by Allen J. Frantzen, "The significance of the Frankish Penitentials," *Journal of*

imately 700 and 950, then gradually declined. From the 12th century onward, the new theology of penance put the emphasis increasingly on interior contrition rather than external disciplinary practices; accordingly, the penitentials lost their function and came to be replaced by *Handbooks for Confessors* better adapted to the new approach.

Focusing as they do on each individual's actions and thoughts, penitentials cover the widest range of reprobated practices and beliefs: public actions such as masquerades and processions, semi-public ones such as mourning rites and the consultation of cunning folk, and the most private ones, such as dream adventures and the concoction of love potions. Almost every penitential contains material that may be considered to deal with pagan survivals and superstitions in some form. Nevertheless, such documents were accepted only slowly as a source for early medieval and cultural history since, despite the wealth of information that they contain, the claim of most penitentials to represent reality accurately stands on shakier ground than does the claim of either legislation or sermons.[147]

Some scholars found the subject matter of the penitentials distasteful or implausible. Charles Plummer, displaying a concern which does not greatly afflict modern historians, could not see how "anyone could busy himself with such literature and not be the worse for it."[148] Nora Chadwick questioned their credibility when it came to the more extreme articles, and ascribed them to the excessive conscientiousness and over-active imagination of the monastic authors.[149] However, Pierre Payer argued forcefully that if there is no reason to doubt the accuracy of the penitentials when they dealt with such a workaday sin as murder, it is unreasonable to doubt it when they came to other, more exotic sins.[150] This applies to idolatrous and magical practices as well as the type of practices treated by Payer. There is, of course, a difference between more or less public acts, or those of which the effects are publicly visible, and secret acts and beliefs. The testimony of the penitentials on the former must be accepted–it is as sure that people participated in drunken wakes as that they stole each other's livestock. It may not be equally sure that they concocted and administered love philtres. A measure of uncertainty about such secret practices is unavoidable. But there is no need to assume that the clergy frequently resorted to invention, since there cannot be many beliefs and practices listed in the penitentials which anthropologists have not encountered in other cultures.[151]

Ecclesiastical History 30 (1979): 409-421.

[147] Thomas P. Oakley was one of the first to recognise the value of this once despised form of document for historical research; see "The penitentials as sources for medieval history," *Speculum* 15 (1949): 210-223. For penitentials as a "mirror" of popular culture, see Gurevich, *Medieval Popular Culture*, 78-103.

[148] Introduction to *Venerabilis Baedae opera historica* (Oxford, 1896), clviii.

[149] *The Age of the Saints in the Early Celtic Church* (London, 1961), 149.

[150] See also Pierre Payer, *Sex and the Penitentials* (Toronto, 1984), 13.

[151] See, for example, the range of typical magical beliefs listed in Raymond Firth, "Reason and

Whether these beliefs and practices were current at a given time and place is another question altogether. Yitzhak Hen dismissed the penitentials of the Merovingian period as having little basis in contemporary practice. They were influenced not by conditions around them but by "literary conventions together with [the] real fears and anxieties" of the clergy, "encouraged by external influences on the Frankish Church." These were the example of the Anglo-Saxons and the Irish, and the paganism of the border areas of Merovingian territory which were at the moment undergoing St. Boniface's missionary efforts: "[T]he forbidden practices must not be taken as an accurate reflector of reality, but as a reflector of the mental preoccupation of the Christian authorities."[152]

Hen possibly overestimates the extent to which the process of Christianisation was complete in Merovingian Gaul at least in the 6th and early 7th centuries.[153] But it is indubitable that in no other source is the literary tradition so overwhelmingly evident as it is in the penitentials. Large sections were copied word for word from other, sometimes from several other, penitentials, so that the same sin may have received different penances in different parts of the booklet. The types of sin described were virtually unchanged throughout the period–the same cultic acts and magic are described in penitential after penitential, with significant additions being rare until the end of our period. For example, a clause concerning the introduction of cunning men into houses to discover and get rid of hexes, originally from the Council of Ancyra (314) and quoted by Martin of Braga (572), persisted in penitentials up to the time of Burchard of Worms.[154] On the other hand, the illicit use of chrism, a source of great concern to Frankish councils and synods throughout the 9th century, is found in only three penitentials, none of which were Frankish.

Nevertheless, there were independent elements even in Merovingian penitentials. The frequently-repeated clause of the *Penitential of St. Columban* concerning the practice of holding feasts in the vicinity of shrines because of ignorance, greed or idolatry has no parallel in any other Irish nor in any English penitential. There is, however, a parallel to be found in Caesarius of Arles' sermons, which blamed such sacrilegious feasts on naiveté, ignorance or (more plausibly, according to Cae-

unreason in human beliefs," in *Witchcraft and Sorcery*, ed. Max Marwick (Middlesex, 1970), 38-40. Examples of almost every kind of magical cure mentioned in the penitentials can be found in the account of 19th and 20th-century folkloric medicine by Wayland D. Hand, *Magical Medicine; The Folkloric Component of Medicine in the Folk Belief, Custom and Ritual of the Peoples of Europe and America* (Berkeley, Los Angeles and London, 1980).

[152] *Culture and Religion in Merovingian Gaul*, 188-189.

[153] E. Vacandard's "L'idolâtrie en Gaule au VIe et au VIIe siècle" has not yet been refuted concerning the extent of open idolatry existing in Merovingian Gaul well into the 7th century. See also J. M. Wallace-Hadrill, *The Frankish Church* (Oxford, 1983) 17-36.

[154] Martin of Braga, *Canones ex Orientalium Patrum Synodis* 71, Barlow, 140; Burchard of Worms, *Decretum* [1008-1012] 19, 5.60, Schmitz 2: 422.

sarius) greed.[155] In this case at least, it is evident that either Columban's penitential was based on a Merovingian text or it took into account rituals observed in Merovingian territory. Other customs cited in penitentials of the period also have a continental origin. The critique of the cult of trees and springs, singing and dancing around churches, and the rituals of the Calends was not drawn from Irish or English sources but was based on practices for which independent continental evidence is available, not only Caesarius' sermons, but also in civil law, letters and hagiography.

Penitentials are most credible when they introduce new material: the *Poen. Vinniani* on clerical and love magic, the *Poen. Columbani* on ritual meals at shrines, the *Poen Cummeani* on dietary practices, the *Canones Hibernenses* on mourning rituals, and the *Poen. Theodori* on magic, especially of the domestic variety. The Spanish penitentials, while they reiterate many of the old articles, contain striking proof that Mozarabic mentality and practice differed widely from what prevailed east of the Pyrenees. Several other penitentials introduce more ore less significant variations on the originals, which suggests that some independent thought and observation went into their compilation.

As a source, Burchard of Worms' *Corrector sive medicus* stands alone. The one hundred and ninety-four questions that make up the *interrogationes* were not copied wholesale from earlier penitentials or from the canon law recorded so voluminously in the other nineteen volumes of the *Decretum*. They almost certainly reflect practices prevalent in the Rhineland of the 11th century.[156] This penitential leaves out some of the practices found in its sources, presumably those that did not apply, elaborates on standard articles and adds new material. It modifies the penances traditionally assigned to certain sins, sometimes drastically, and thus bears witness both to the persistence of practices and to a shift in the interpretation of those practices. In some cases where his predecessors had seen real danger

[155] *Poenitentiale Columbani* (c. 573) 20, Bieler, 104; Caesarius of Arles (502-542) *S.* 54.6, *CCSL* 103: 239-240.

[156] There is nothing known about the nonliterary sources of Burchard's penitential. Burchard belonged to an aristocratic Hessian family, he was educated in his youth at the monastery of Lobbes, and his early career was in ecclesiastical and civil administration. He was ordained priest only after his nomination to the bishopric. Under the circumstances, it is doubtful that he or his two assistants, one a bishop, the other to become an abbot, had much personal experience on the parish level. For Burchard's biography, see Albert M. Koeniger, *Burchard I. von Worms und die deutsche Kirche seiner Zeit* (Munich, 1905); Wolfgang Metz, "Zur Herkunft und Verwandtschaft Bischof Burchards I. von Worms," *Hessisches Jahrbuch für Landesgeschichte* 26 (1967): 27-42; G. Allemang, "Burchard I de Worms," *DHGE* 10: 1245-1247. For popular paganism in the *Corrector*, see François Alary, "La religion populaire au XIe siècle: Le prescrit et le vécu d'après le *Corrector sive medicus* de Burchard de Worms," *Cahiers d'Histoire* (Université de Montréal): 13 (1993) 48-64, and Cyrille Vogel, "Pratiques superstitieuses au début du XIe siècle d'après le *Corrector sive medicus* de Burchard, évêque de Worms (965-1025)," in *Mélanges E.-R. Labande. Études de Civilisation Médiévale (IXe-XIIe siècles). Mélanges offerts à Edmond-René Labande* (Poitiers, 1974), 751-761.

of idolatry or truck with demons, Burchard saw only trivial misbehaviour, ignorance and folly. Some practices are described in minute detail, down to the very direction in which magic corn is ground or the very toe to which a magical herb is tied. Burchard's unusual emphasis on belief as well as practice, expressed in the questions to be asked (*credidisti?* not merely *fecisti?*), reveals a sensitivity to the mentality of his charges as well as to their external behaviour. Finally, the inclusion of vernacular terms clearly indicates familiarity with local beliefs and practices and vocabulary.

Sermons:[157] As a source for popular religion, sermons fall between conciliar legislation and penitentials. While councils and synods presumably dealt principally with behaviour general to the ecclesiastical region, and penitentials concentrated on the most private thoughts and actions of individuals, sermons ideally focused on the needs of the parish. This is where one might expect to find the beliefs and practices prevailing in the community to which the preacher addressed himself. Sermons, therefore, should be an invaluable source for the customs peculiar to each parish as a social unit.[158] However, this is not the case for the early Middle Ages.

In general, this period was a low point in the history of preaching, especially from the mid 6th century to the Carolingian reforms at the end of the 8th. Even after the reforms, sermons, as opposed to readings from the Gospels, appear to have been rare. The right to preach was reserved to bishops except in the Narbonnaise where the Council of Vaison (529), under Caesarius of Arles' leadership, authorised parish priests to preach and, in their absence, gave permission to deacons to "recite" the homilies of the Fathers of the Church.[159] Many were reluctant to exercise this right, at least partly from feelings of inadequacy and the pres-

[157] For early medieval sermons, see Jean Longère, *La prédication médiévale* (Paris, 1983), 35-54; for the content of sermons in Germanic territory, F. R. Albert, *Die Geschichte der Predigt in Deutschland bis Luther*, 2. vols. (Gütersloh, 1892-1893); for missionary preaching, Wilhelm Konen, *Die Heidenpredigt in der Germanenbekehrung* (Düsseldorf, 1909). For preachers' techniques, see Thomas L. Amos, "Early medieval sermons and their audience," in *De l'homélie au sermon: Histoire de la prédication médiévale*, ed. Jacqueline Hamesse and Xavier Hernand (Louvain-la-Neuve, 1993), 1-14. R. Emmet McLaughlin examined the evolution of the medieval sermon and the reasons for the decline of preaching after Caesarius of Arles in "The word eclipsed? Preaching in the early Middle Ages," *Traditio* 46 (1991): 77-122. For the Carolingian period, see McKitterick, *The Frankish Church*, 80-114, and Milton McGatch, *Preaching and Theology in Anglo-Saxon England: Aelfric and Wulfstan* (Toronto and Buffalo, 1977), 29-39; for popular sermons at this period, see T. L. Amos, "Preaching and the sermon in the Carolingian World, " in *De Ore Domini: Preacher and Word in the Middle Ages*, ed. Amos *et al.* (Kalamazoo, 1989), 41-60.

[158] *Cf.* Gregory I's *Regula pastoralis* (*Pastoral Care*, trans. Henry Davis [Westminster, Md., 1950]), of which Bk. 3, comprising well over half of the work (134 out of 217 pages), is dedicated to preaching. In fact, most of the admonitions prescribed by the pope are such that they would appear to be given more appropriately privately in confession, rather than publicly in a sermon.

[159] 2, *CCSL* 148A: 78-79. Significantly, of the reforming councils of 813, only that of Arles upheld the right of priests to preach (10, *MGH Concilia* 2.1, 251).

sure of other responsibilities. There is no reason to doubt the sincerity of the pleas advanced by Caesarius' fellow-bishops when he exhorted them to preach. Some thought they lacked the eloquence to preach their own sermons, some were unable to commit their great predecessors' sermons to memory, some found that their other pastoral duties absorbed all their energies.[160] If this was the case at a time when and in a region where the traditions of literacy and rhetoric were still strong, it must have been all the more so later during the Merovingian period, when educational standards had declined further and the pressure of ecclesiastical and secular responsibilities on bishops remained high.[161] It is not to be expected that parish clergy would have been more competent or more willing to preach than their superiors even had they had the right to do so.

Most of the sermons preserved in homiliaries appear to have been preached to a clerical audience, or used for private devotions, rather than preached to a lay congregation. Even those of Caesarius of Arles, which had been prepared with a mixed audience in mind, were "soon reabsorbed by the monastic tradition."[162] In the form that has survived, such sermons were rarely applicable to the laity. They are in Latin, which may have been incomprehensible to rustic audiences even in Caesarius' Provence, and which was certainly incomprehensible to the laity and no doubt even to some of the clergy later on.[163] Moreover, the sermons did not usually concentrate on the behaviour of the faithful but on that of the clergy and on expositions of dogma and biblical exegesis, subjects of meditation for monks and other clerics, but probably of scant interest to the ordinary church-goer.[164] The

[160] *S.*1, 12, *CCSL* 103: 3; *S.*1.20, *ibid.*, 16; *S.*1.9, *ibid.*, 9.

[161] Pierre Riché, *Éducation et culture dans l'Occident barbare (VIe–VIIIe siècles)* (Paris, 1962), 311-320.

[162] McLaughlin, "Preaching in the early Middle Ages," 105. As for the intended audiences for Caesarius' sermons, sermons 1 and 2 were directed to the diocesan clergy (*S.* 2 being meant to be read, not preached), *sermons* 233-238 to monks, *sermons* 16, 17, 19, 22 and 151 to the outlying parishes. The audience intended for the others cannot be so precisely identified but since many deal explicitly with the manners and morals of the laity (*e.g.*, drinking, concubinage, alms-giving, social justice and business morality, as well as pagan survivals), most must have been delivered before a congregation containing a substantial number of lay people.

[163] Gauthier found it significant that a cultured and aristocratic youth such as Gregory of Utrecht was able to read the Latin Scriptures, but did not understand them well enough to translate them; this reflects poorly on the general level of comprehension of Latin texts (*L'évangélisation des pays de la Moselle*, 437). For the low level of clerical literacy in the Carolingian period, see Michel Banniard, *Viva Voce. Communication écrite et communication orale du IVe au IXe siècle en Occident latin* (Paris, 1992), 395-397.

[164] This appears to have been the case even when the sermon was preached in the vernacular, a practice that A. Lecoy de la Marche believed was quite widespread in the early Middle Ages. He described a sermon, composed in Celtic by an Irish monk and predating the Carolingian reforms, of which the contents were theological in tone, and which must have gone over the heads of a rural congregation (*La chaire française* [2nd ed., Paris, 1886], 235-238).

majority of sermons, therefore, contain little material, or none, about popular culture.[165]

The lack of sermons suitable for the laity does not necessarily imply an indifference to the moral and spiritual education of the laity. Other means at hand were better adapted to their needs and perhaps to the aptitudes of many of their pastors. J. Goering observed the principal characteristic of early medieval pastoral care was "the safeguarding of God's presence among the faithful ... through rites and rituals." It took the forms of rituals of healing (prayers, relics, exorcisms) and sustaining (prayers of monastics, seasonal liturgies), education and guidance "through recurrent ritual actions, through art and especially through the oral traditions of poetry and story-telling), and reconciliation."[166] R. Emmet McLaughlin tied the decline of preaching at this period to liturgical changes.[167] The expanded and elaborated liturgy of the mass reduced the time available for preaching, but at the same time took over much of its educational function through scriptural readings, prayers and the singing of hymns in which the congregation was expected to join. The vigils of saints, celebrations of the dedication of churches, and liturgical processions during Rogation days served to inculcate ideas of the majesty of God and the virtues of the saints. Participating in such rituals was more satisfying and more stirring emotionally than listening to sermons, and the lessons learned thus were more easily retained.[168] The essentials of dogma were taught in the baptismal responses, and in the Lord's Prayer and Creed, which each person was required to memorise. Prayers were composed in alliterative form to make them easy to remember.[169] Private confession provided the parish priest, armed with his penitential, with the opportunity to give moral instruction tailored to the requirements

[165] For example, see the sermons in *XIV Homélies du IXe siècle d'un auteur de l'Italie du Nord*, ed. Paul Mercier, *SC* 161 (Paris, 1970), and *Abbo von Saint-Germain-des-Prés. 22 Predigten, Kritische Ausgabe und Kommentar*, ed. Ute Önnerfors (Frankfurt-am-Main, 1985). See also a missionary sermon mistakenly attributed to St. Gall, which deals exclusively with the history of salvation and disregards completely the lives of the putative audience (*Der konstanzer Predigt des Heiligen Gallus. Ein Werk des Notker Balbulus*, ed. Wilhelm Emil Willwoll [Freiburg, 1942], 5-17).

[166] "Medieval Church, pastoral care in," in *Dictionary of Pastoral care and Counseling*, ed. Rodney J. Hunter *et al.* (Nashville, 1990), 698-700; here, 699. For practical illustrations of this, see H. G. J. Beck, *The Pastoral Care of Souls in South-East France during the Sixth Century* (Rome, 1950) and Guy Devailly, "La pastorale en Gaule au IXe siècle," *RHÉF* 59 (1973): 23-54.

[167] "Preaching in the early Middle Ages," 102. See also McKitterick, *The Frankish Church,* 115-154, for the changes in the liturgy and their effect during the Carolingian era, and Pierre Riché, *Éducation et* culture, 542-547, for alternate methods of instructing the faithful.

[168] People found sermons dull. Even Caesarius of Arles, who had taken pains to adapt his sermons in length, vocabulary and subject matter to the capacities of his audience, had been obliged to run after his flock to keep them from leaving before the sermon, and eventually to lock the church doors to prevent them from escaping while he preached (*Vita Caesarii* 1, 27, *MGH SRM* 3: 466-466). One may also relate the tendency of Carolingian landowners to establish private masses to their desire to escape the sermons imposed by the reform movement.

[169] Cyril Edwards, "German vernacular literature. A survey." In *Carolingian culture. Emulation and Innovation*, ed. Rosamond McKitterick (Cambridge, 1994), 141-170; here, 146.

of each individual. The paintings that decorated churches also played a didactic role, as Gregory I noted.[170] Benedict Biscop decorated the church of St. Peter with pictures of the Virgin Mary, the apostles, and the visions of the *Apocalypse* for the benefit of the illiterate.[171] The example of Caedmon shows that Bible stories were turned into vernacular song for the delight and edification of all.[172]

In fact, there can be no doubt that the faithful were exposed at least sometimes to sermons intended for their use. St. Caesarius of Arles preached tirelessly in town and country.[173] St. Augustine of Canterbury and St. Paulinus of York preached to the Kentish and Northumbrian courts, St. Cuthbert to countrymen in remote hamlets.[174] The *Lives* of missionary saints on the continent, such as St. Gall, St. Vulframm, St. Boniface and St. Anskar, describe them as preaching,[175] as well as using other, perhaps more effective, methods: miracles, violence and political pressure from powerful lay patrons. A few sermons give signs of having been written in immediate response to behaviour just observed, for example, on the eve of the feast of St. John Baptist or during a lunar eclipse. Sometimes the very awkwardness of the phraseology and highly practical admonitions combined with a fairly primitive theology testify that the author was a parish priest of limited learning, struggling to find words in which to instruct his flock.

Accounts of popular outbursts present added though indirect evidence for sermons. The Council of Rome of 745 had to deal with Aldebert and Clement, two *pseudoprophetae* active in the northern parts of the Frankish state, clerics who, by means of their teachings and (in the case of Aldebert) miracles, were able to win over a large number of supporters, not only "feeble women" and the uncouth (*rustici*), but even bishops. The outbreaks of hysterical mass movements recorded by Gregory of Tours, Agobard of Lyons and Atto of Vercelli, which seem to have had some kind of a Christian core, also must have been fuelled by inspirational addresses, probably from clerics practiced in popular preaching.[176] This suggests that large numbers of people, many of them quite simple, were accustomed and receptive to sermons.

[170] *MGH Ep* 2: 270.

[171] Bede, *Vita sanctorum abbatum monasterii,* 6, in *Opera Historica,* ed. Loeb, 2: 404. See Paul Meyvaert, "Bede and the church paintings at Wearmouth-Jarrow," *Anglo-Saxon England* 8 (1979): 63-77.

[172] Bede, *HE* 4, 24: 414-420.

[173] He preached every Sunday, every feast day and daily throughout Lent and the octave of Easter, as well as several times a week at Lauds and Vespers, especially during Advent and Lent (A. Malnory, *Saint Césaire Évêque d'Arles* [Paris, 1894], 32).

[174] Bede, *HE,* 1.25: 72; 2. 9: 164; 4.27: 432.

[175] McGatch points out, however, that *praedicatio* and *praedico* are ambiguous terms, "which seem as often to mean enunciation of doctrine, or teaching, or even reading from the Fathers ... as preaching" (*Preaching and Theology in Anglo-Saxon England,* 35).

[176] *Conc. Romanum* (745) 6, *MGH Concilia* 2.1: 37-44; Gregory of Tours, *HF* 10. 25, *MGH SRM* 1: 517-519; Agobard of Lyons, *De quorundam inlusione signorum* (828-834) 1, *CCCM* 52: 237; Atto of Vercelli (d. 961) *Ep.* 3 (*ad Plebem Vercellensem*), *PL* 134: 104-105.

The number of surviving sermons relevant to our subject is small: excluding those of Caesarius of Arles, not much more than forty, from a period of almost six hundred years. I have gone beyond the chronological period of this study to include 5th-century sermons from northern Italy. They were preached to mixed populations similar in culture and tradition to that of 6th-century Gaul, and their wealth of descriptive detail casts light on medieval descriptions of popular practices. I have also included works which, like Martin of Braga's *De correctione rusticorum*, were probably never preached in the form that they were written. As confessors were meant to use penitentials as a guide only, and to put only those questions that bore a relation to the individual penitent's life, so preachers were meant to use only the sections applicable to their audience.

To what extent may these sermons, few as they are, be taken at face value as evidence? There is little doubt that the sermons of Maximus of Turin, Peter Chrysologus and even Caesarius of Arles can be accepted as truthful descriptions of the customs prevailing in their dioceses during their lifetime. With Martin of Braga's sermon *De correctione rusticorum*, which also appears to be based generally on direct observation, there may be some questions as to its applicability to 6th-century Galicia. Can as much be said for other medieval sermons?

Although the influence of the sermons of Caesarius of Arles and, to a much lesser degree, of Martin of Braga is clearly evident in the references to pagan survivals and superstitions, almost every sermon which refers to such customs contains some material which is independent not only of these models, but of legislation and penitentials as well. The homilies of Burchard of Würzburg, cobbled together entirely from passages taken from Caesarius' sermons, are the outstanding exception; it is their merit that they highlight the originality of the rest. The other sermons then must be accepted as providing authentic evidence for popular culture; their independent elements support their claims to a measure of trust in material copied from earlier sources.

1.3.4 *Incidental literature*

A number of documents are particularly valuable because they were written explicitly in response to immediate, pastoral concerns. The author, date, place and circumstances, therefore, are clearly defined, and it may usually be assumed that the texts were based on at least second-hand eyewitness accounts of actual practices. Most important among these are letters written by popes to missionaries and others engaged in work among pagans or Christians of questionable orthodoxy.

1.3.5 *Complementary works*

Documents with a non-pastoral focus provide occasional glimpses of daily life and popular beliefs and practices, and throw important light on passages in pastoral literature. No attempt was made to sift such works systematically but, where possible, additional material was drawn from different types of texts.[177] A few of these predate the Middle Ages (notably, Tacitus' *Germania*) but most are early medieval. Among these are Germanic legal codes, histories (such as Gregory of Tours' *History of the Franks* and Bede's *Ecclesiastical History of the English Nation*), scientific works (*e.g.,* Isidore of Seville's *Etymologies* and the anonymous *Medicina antiqua*), treatises (Ratherius of Verona's *Praeloquium* and Gilbert of Nogent's *De sanctis et eorum pignoribus*), autobiography (Gilbert of Nogent's *Monodiae*) and travellers' tales (an Arab diplomat's description of Rus idolatry and burial rites).

Hagiography[178] is particularly valuable as a source for popular culture, since the saints' virtues and prowess are often demonstrated in their dealings with pagans and ordinary Christians. It is a source difficult to use by virtue of the number of hagiographic accounts extant and of the riches of individual *Lives* in terms of evidence varying from the ostensibly historical to the miraculous to the out-and-out fabulous.[179]

The hagiographer is not a historian, declared Jean-Pierre Laporte, but a partisan,[180] and the uses of hagiography for propaganda has been amply discussed by Felice Lifshitz in her *Pious Neustria.* However, the value of hagiography for a student of popular culture does not lie necessarily in factual depictions of customs and beliefs. The author of a *Life* may describe accurately, misrepresent, misinterpret or even make up an episode, but he does so in terms of his own cultural experiences and expectations. His work may or may not be a true account of the past, but in some fashion it mirrors the time and the circumstances in which it was written–as a painting of the Crucifixion may depict the armour, clothing and hairstyle of its own age, not that of Christ.[181] St. Eligius's sermon, for example, might not tell us much about the practices prevailing around Merovingian Noyon

[177] But, for a criticism of the value of such works as a source for paganism, see R. I. Page, "Anglo-Saxon paganism: The evidence of Bede," in *Pagans and Christians: The Interplay between Christian Latin and Traditional Germanic Cultures in Early Medieval Europe*, ed. Tette Hofstra, L. A. J. R. Houwen and A. A. MacDonald, (Groningen, 1995), 99-129. I am indebted to Karen Jolly for bringing this article to my attention.

[178] For the historical value of hagiography, see Baudouin de Gaiffier, "Hagiographie et historiographie," in *La storiografia altomedievale,* SSAM 17 (Spoleto, 1970), 139-166 and discussion, 179-196. See also Réginald Grégoire, *Manuale di agiologia* (Fabriano, 1987); R. Aigrain, *L'hagiographie, ses sources, ses méthodes, son histoire* (Paris, c. 1953); R. Aubert, "Hagiographie," *DHGE* 23: 40-56.

[179] By miraculous, I mean standard miracles of curing the sick, immobilising enemies, destroying pagan shrines, *etc.*; by fabulous, stories and miracles with especially strong mythic and folkloric elements, of a kind found in many Lives but in particular abundance in those of Irish saints–an extreme example is the *Life of St. Brendan of Clonfert* (*Bethada Náem nÉrenn. Lives of Irish Saints*, ed. and trans. Charles Plummer [Oxford, 1922], 44-92).

[180] "La reine Bathilde, ou l'ascension sociale d'une esclave," in *La femme au moyen âge*, ed. Michel Rouche, (Maubeuge, c. 1990), 147-167; here, 167.

[181] Therefore the importance given by Thomas Head to "the institutional and intellectual contexts" of the composition of hagiography and the cults of saints and their relics (*Hagiography and the Cult of Saints. The Diocese of Orleans* [Cambridge, 1990] here, 19).

in the first half of the 7th century, but it tells us a good deal about what concerned Carolingian clerics in the 8th. In the same way, the insertion into the 13th-century version of the *Life of St. Germanus of Auxerre* (d. 448) of an episode which did not appear in the first *Life*, written c. 480 by Constantius of Lyons, is a notable indication of issues thought important in the 13th century, not in the 5th.[182]

By contrast, a *Life* written close to the period of the saint's own life has a greater chance of reflecting his own concerns and experiences as well as that of his biographer's. The *Vita Martini* and the *Vita Vulframmi* are good examples of these. They were composed reasonably soon after the death of the saints, and may be accepted to some degree as realistic depictions of the era. They also have the advantage of containing data that can be verified from pastoral texts and, on occasion, from other sources as well. Sulpicius Severus' *Life of St. Martin of Tours* describes Martin as suspecting a ritual circumambulation with a deity when he met a procession of peasants. In fact, it was a funeral procession; the fluttering veil with which the body was covered had deceived him. This tends to authenticate and to be authenticated by the evidence of pastoral literature on both funeral customs and the circumambulation of fields at later periods.[183] Descriptions of human sacrifices in the *Life of St. Vulframm* are borne out to some extent by the discovery of strangled corpses buried in bogs not far from his area of activity. This lends credibility to the account in the same *Vita* of the rejection of the Christian heaven by a Frisian *dux* on the grounds that it would separate him from his pagan ancestors. That in turn may explain the appeal of a heretic who declared that pagans would go to heaven as well as Christians.[184]

Most of the *Lives* cited here were chosen not only for their relevance to pagan survivals and superstitions during the early Middle Ages (there are many such), but also because their subjects were contemporary missionaries, bishops and abbots who encountered the kinds of situations suggested in pastoral literature, and because they were written during the period under consideration. Others were used, but rarely, to illustrate or cast further light on particular points, such as the persistence of certain themes, the anachronistic introduction of a new theme in the *Life* of an early saint, or different interpretations of practices.[185] The difficulty

[182] Jacob of Voragine, *Legenda aurea*, ed. G. P. Maggioni (Florence, 1998), 689-694; *Vita Germani Autissiodorensis* (*Vie de saint Germain d'Auxerre* ed. René Borius, *SC* 112 [Paris, 1965]).

[183] Vita Martini 12.1, *Vie de saint Martin,* ed. Jacques Fontaine, *SC* 133 (Paris, 1967), 279; *Conc. Claremontanum seu Arvernense* (535), 7, *CCSL* 148A: 107; *Indiculus superstitionum et paganiarum* (743?) 28, *MGH CapRegFr* 1: 223.

[184] *Vita Vulframn*i 6-9, *MGH SRM* 5: 665-668; *Conc. Romanum*, *MGH Concilia* 2.1: 40. For this *Life* as historical document, see Stéphane Lebecq, "Vulfran, Willibrord et la mission de Frisie: Pour une relecture de la 'Vita Vulframni'," in *L'évangélisation des regions entre Meuse et Moselle et la fondation de l'abbaye d'Echternach (Ve-IXe siècles)* , ed. Michel Polfer (Luxembourg, 2000), 429-451.

[185] I have gone outside the chronological limits of this work for the *Life of St. Malachy* by St. Bernard of Clairvaux and the *Lives* of saints Germanus of Auxerre, Bernard and Dominic (in the *Golden Legend*). Saints Enda of Aranmore and Ciaran of Saigir were early medieval saints, but their *Vitae* were written down in the high Middle Ages, that of the former from varied sources,

was not to find *Lives* that could have provided useful information of this sort, but to limit the number used so that the hagiographic evidence would not assume disproportionate importance. Their purpose here is to indicate the background against which the authors of our documents worked, and I have made no attempt either to study all the possibly relevant *Lives*, or to sift even the ones used for all pertinent material.

that of the latter from sources of considerable antiquity (Kenney, *The Sources of the Early History of Ireland*, 374, 316).

2
Idolatry, Gods and Supernatural Beings

Idolatry, strictly speaking, is the worship paid to idols, usually defined as images or objects worshipped as representing deities and, by extension, the deities or false gods themselves, whom Christian writers identified with demons or whose worship they credited to the influence of demons.[1] The texts suggest that the ecclesiastical authorities were seriously concerned about this kind of worship of false gods. Terms for idolatry (*idolatria, idololatria, cultura idolorum/ diaboli/ daemonum, cultus idolorum/ diaboli/ daemonum, superstitio idolatriae, idolorum servitus, honor daemonum, honor simulacrorum, sacrilegium,*[2] *veneratio creaturae*), idolatrous objects (*idolothytes*–usually food offerings), and idolaters (*idolatri idolorum cultores, idola colentes, idolis/ idolothicis servientes, veneratores lapidum*) abound throughout our period. Less often, we are told about representations of supernatural beings (*simulacra, picturae, idola, lapides*[3]), their shrines (*fana, casulae*) and their sacrificial altars (*arae*).

A variety of supernatural beings appear in our texts. Deities (*dii, numina*) usually have Latin names (*Diana*, the *Genius*, *Janus*, *Juno*, *Jupiter*, *Liberus*, *Mars*, *Mercurius*, *Minerva*, *Neptunus*, *Orcus*, the *Parcae*, *Saturnus* and *Venus*); only a few have names drawn from another tradition (*Thunaer*, *Woden*, *Saxnote*, *Herodias*, *Holda/ Frigaholda*). Several lesser spirits (*aquaticae, dianae, dusioli, lamiae, mavones, nymphae, pilosi/ satyri, silvaticae*) appear. Expressions such as "too many others to describe" and "other nonsense of this sort" indicate the existence of numerous others. The devil

[1] St. Paul made the equation between idols and demons in denouncing pagan offerings in I *Corinthians* 10, 20. In St. Augustine's explanation of *Ps.* 95, 5 and 96, 11-12, all the pagan gods and their statues are by definition demons and representations of demons (*Enarrationes in psalmos*, *CCSL* 39: 1347-1348 and 1362-1365). In *The City of God*, 7.18, Augustine approved the explanation that the gods "were men, and that to each of them sacred rites and solemnities were instituted, according to his particular genius, manners, actions, circumstances; which rites and solemnities, by gradually creeping through the souls of men, which are like demons, and eager for things which yield them sport, were spread far and wide; the poets adorning them with lies, and false spirits seducing men to receive them" (7.18, trans. Marcus Dods and J. J. Smith [New York, 1950], 224). For Isidore of Seville, the gods were originally powerful men and the founders of cities; their statues (*simulacra*) were erected by their friends for consolation, but gradually the demons persuaded them that the statues themselves were gods (*Etymologiae* VIII, 11.4: 718).

[2] In continental texts, sacrilege sometimes signifies idolatry, as it does in two 9th-century sermons which call on the faithful to avoid "the sacrilege which is called the cult of idols," or a variety of traditional practices, as it does in the late 8th century *Homilia de sacrilegiis*.

[3] Other terms were *statua, effigies* and *figmenta,* see Agobard of Lyons, *De picturis et imaginibus* 31, 33, 34, *CCCM* 52: 179-181).

(*diabolus, zabulus, Satanas*) and anonymous demons also figure as objects of false worship.[4] Even angels (St.Michael and others) and saints (St. Martin of Tours, St. Stephen and doubtful claimants, living or dead, to holiness) could receive inappropriate cults. One might therefore expect to find pastoral literature rich in information about forbidden cults of deities. This is not so. A number of reasons may explain this lack.

First, either our authors did not know much about what the people actually believed about the gods nor how they honoured them, or they thought it inappropriate to write about them. The numerous indigenous divinities who left records of their names in classical literature and epigraphy are virtually absent from our texts although some memory of them must have survived in folk culture.[5] This is indicated most tellingly in an 8th-century Saxon formula for baptism which required the candidate to forsake *thunaer ende woden ende saxnote ende allum them unholdum* (accursed beings).[6] Burchard of Worms somewhat grudgingly provided a popular designation, that of the deity "whom the stupidity of the vulgar here names Holda." About others, the official texts are silent; evidence for their cults must be sought elsewhere.

Acts of cult (prayers, offerings) are described for some half dozen (Jupiter, Mercury, Minerva, the *Parcae* and the *pilosi/satyri*), and mentioned vaguely for a few more; specific beliefs are given concerning only Diana and the *mavones*, *Parcae* and *silvaticae*. The names, however, are not a reliable indication of the deity's nature: the Jupiter honoured by the weaving women of 6th-century Arles was, for example, unlikely to be the same Jupiter whose rites and feasts were celebrated in Austrasia two centuries later. The precise characteristics of other gods are unknown.

Secondly, the texts are often vague when referring to idols in general. In many passages, an *idolum* may be a deity, a man-made effigy or a natural object such as a tree. Generic terms for rituals and devotions ("sacrifice to idols,"make offerings to demons") give little clue as to actual practices. Even when an author gives concrete details, the validity of his information may be doubted. A case in point is Gregory I's well-known letter of 601 to Mellitus. Mellitus is to tell Augustine,

[4] Other texts prove that demons had names. In the 9th century *Translatio et miracula SS. Marcellini et Petri*, a *spiritus erraticus* gives Wiggo as his name, see Pierre Boglioni, "Un Franc parlant le syriaque et le grec, ou les astuces de Jérôme hagiographe," *Memini* 3 (1999): 127-153. Toward the end of the 10th century, Raoul Glaber related an anecdote concerning three demons in the guise of Vergil, Horace and Juvenal; see *Heresies of the High Middle Ages*, selected and trans. Walter L. Wakefield and Austin P. Evans; New York, 1961), 13.

[5] Draak cites the research of Marie-Louise Sjoestadt (*Gods and Heroes of the Celts*, trans. and additions by Myles Dillon, [London, 1949]) who found a total of 374 names of continental Celtic gods in such sources ("The religion of the Celts," 629-647; here, 632). The evidence of folklore is less reliable, as in the case of Charles Godfrey Leland's *Etruscan Roman Remains in Popular Tradition* (London, 1892)=*Etruscan Magic and Occult Remedies* (New Hyde Park, N. Y., 1963). Leland based his lists of divinities on tainted evidence and faulty etymology; see Raffaele Corso, "Presunti miti etruschi nel folklore della Romagna-Toscana," *Il folklore italiano* 4 (1929): 1-11.

[6] *Interrogationes et responsiones baptismales*, *MGH CapRegFr* 1: 222.

the head of the mission to the Anglo-Saxons, that after careful thought Gregory had come to the conclusion

> that the idol temples of that race should by no means be destroyed, but only the idols in them. Take holy water and sprinkle it in these shrines, build altars and place relics in them. For if the shrines are well built, it is essential that they should be changed from the worship of devils to the service of the true God ... And because they are in the habit of slaughtering much cattle as sacrifices to devils, some solemnity ought to be given them in exchange for this. So on the day of the dedication of the festivals of the holy martyrs, whose relics are deposited there, let them make themselves huts from the branches of trees around the churches which have been converted out of shrines, and let them celebrate the solemnity with religious feasts. Do not let them sacrifice animals to the devil, but let them slaughter animals for their own food to the praise of God, and let them give thanks to the Giver of all things for His bountiful provision ...[7]

This important letter outlining the policy of assimilation and substitution to be followed by missionaries (which Jacques Le Goff termed "obliteration")[8] is full of factual information: there are temples (some well-built, maybe even of stone); they contain effigies; the people frequent them and hold feasts during which they sacrifice numerous oxen to the gods.[9] But whether all this is based on private knowledge of Anglo-Saxon religion or on ideas drawn from the temples and sacrifices of classical religion is debatable. The pope writes nothing that does not apply to Mediterranean paganism at least as well, nothing to show special familiarity with the customs that he ostensibly describes. He mentions the names of no gods and seems unaware that the Anglo-Saxons, like their continental cousins, preferred to worship in the open.[10] No doubt he had in mind the systematic appropriation by

[7] Bede, *HE* 1.30: 106-109. Colgrave and Mynor's translation.

[8] This letter contradicts the pope's to the Kentish king Ethelbert, in which he urged him to destroy the buildings used as shrines and to lead his people to Christianity by a mixture of force and persuasion (*HE* 1.32: 112). Although the latter comes after the one to Mellitus in Bede's *Ecclesiastical History,* Robert A. Markus has argued convincingly that it was written first, and that the instructions to Mellitus were the result of second thoughts on the pope's part ("Gregory the Great and a papal missionary strategy," in *Studies in Church History* 6 [1970]: 29-38). No evidence has yet been found that Gregory's counsel about converting pagan temples in England into churches was followed; see Hutton, *The Pagan Religions,* 271. The English Church may have accepted his advice as adapting traditional modes of celebration, since English pastoral literature, in contrast to its continental counterparts, did not condemn singing and dancing on religious feasts or in front of churches. The reference to Le Goff is from "Culture cléricale et traditions folkloriques dans la civilisation mérovingienne," in *Pour un autre moyen âge* (Paris, 1977), 320.

[9] Gregory appears to suggest that the Anglo-Saxons should model their Christianised festivals on the Jewish Feast of the Tabernacles: his vocabulary echoes *Lev* 29, 40-42.

[10] This and the German distaste for effigies of the gods were recorded already by Tacitus in *Germania* 9 and 43. On Anglo-Saxon religion, see Wilson, *Anglo-Saxon Paganism* and Hutton, *The Pagan Religions*, 264-284. On the evidence of place names, the most popular gods appear to have

Christians of Italian pagan sites (temples, baths, amphitheatres, military barracks, *etc.*) to use for their own cults.[11] If such evidence cannot be discounted, neither can it be accepted at face value.

Thirdly, the texts give far more importance to the cult of nature and of numinous places such as trees, groves, springs and stones, crossroads, the bounds of the settlement and the family hearth, even man-made shrines, than to deities. At one time these may have been the home of supernatural beings, but the texts give little indication of it.[12] People were said to make vows to trees and pray to springs directly or in their vicinity (*vota vovere ad arbores, orare ad fontes*), not to spirits in them. Shrines and altars, which still retained the aura of holiness and still drew erring Christians, were already despoiled and ruined. Objects and places were sacred in and of themselves; whatever power existed in them was faceless and nameless. On one occasion, Caesarius of Arles spoke revealingly of *arbores fanatici,* trees that were either within the shrine or the shrine themselves.[13] If Fraser was right that animism preceded deism then there had been a regression, and animism had replaced the gods; in Salin's words, "c'est la préhistoire qui reparaît."[14]

been Woden, Thunor and Tiw. Wilson notes that gods' names are usually combined with a term for some natural or man-made feature (grove being particularly common), but never with the words meaning temple or shrine. The shrines or temples may not have been dedicated to any one deity, but simply provided general "facilities" for worship (*cf.* Bede's account of the temple of Redwald, the king of Essex, which contained two altars, on to be used for the Christian mass, the other for pagan sacrifices [*HE* 2.15: 190]). On the other hand, there is some support for the pope's description of shrines: ruins of a large wooden building have been found at Yeavering in Northumbria, which may have been a temple. It shows no signs of domestic use but there are a large number of animal bones on the site. So far, this is the only possible Anglo-Saxon temple to have been excavated. In addition, a deposit of the skulls of more than a thousand oxen has been found in Kent at Harrow (the name is derived from *hearh* or *heargh*, meaning hilltop sanctuary). This, notes Hutton, "argues for something more than just a slaughterhouse" (*op. cit.*, 270). No statues have been found that can be identified as Anglo-Saxon, but that may be because, being made of wood, they had decayed or been burnt.

[11] See Jan Vaes, "Nova construere sed amplius vetusta servare. La réutilisation chrétienne d'édifices antiques (en Italie)," *Actes du XIe Congrès international d'archéologie chrétienne* (Vatican, 1989): 399-321.

[12] In *Germania* 9, Tacitus leaves open the question whether the Germans worshipped gods in groves or only gave the name of gods to the hidden essence of the groves actually worshipped. The ancient Celts also believed in presences immanent in natural phenomena rather than in personal deities. Draak ("The religion of the Celts," 631-635) doubts that originally the Celts had "real *Gods.*" The identification of the presences with gods seems largely to have occurred late, probably under Roman influence. Often they had names, but some three-quarters of them were so closely tied to the locality that their name appears only once.

[13] *S.* 53.1, *CCSL* 103: 233. St. Martin of Tours appears to have chopped down another such tree; the pagans, who had watched the destruction of a temple unmoved, tried to save the tree (*Vita S. Martini* 13.1, *SC* 133: 280).

[14] James Fraser, *The New Golden Bough*, ed. Theodor H. Gaster (New York,1959), 423; Salin, *La civilisation mérovingienne* 4, 470. The resurgence in the European countryside of more primitive, pre-Roman techniques, social structures and ways of thinking as a result of the barbarian

Lastly, the concept of idolatry was not clearly defined in our texts. Often the words were used figuratively to denote a body of customs: obstinate clinging to a calendar based on the natural cycle, traditional methods of celebrating and mourning, time-honoured techniques such as divination and magic–in short, an entire way of life which resisted the Church's all-encompassing control and which she denounced as coming from pagan custom but which could not be connected with specific divinities.

2.1 SUPERNATURAL BEINGS BELONGING TO THE PRE-CHRISTIAN TRADITION

Continental texts identified some deities which enjoyed cults, but left others unnamed–"misnamed powers" *(falsidica numina)* called gods by the pagans of old but known to be inhabited by demons, "other pagan gods" who, along with Jupiter and Mercury, received offerings, anonymous gods to whom bulls and goats were sacrificed or, most contemptuously, "other such nonsense."[15] Other texts spoke of demons not gods. Charlemagne threatened heavy fines for any Saxon, noble, freeman or slave "who paid his vows to springs, trees or groves, or carried anything there in the pagan fashion and ate it in honour of demons."[16] The death penalty recommended by Regino of Prüm for those "who celebrated demons' nocturnal sacrifices or invoked demons with spells" implies a bloody, indeed a human, sacrifice. Burchard of Worms' version suggests a lesser offering: three years' penance for summoning demons "by some art" to fulfill one's wishes.[17] In these cases, "demons" does duty for the names of anthropomorphic deities, the objects of genuine sacrificial offerings.

On the other hand, two 9th-century continental penitentials seem to use "demons" figuratively for nature cults. One requires a penance of only a year for sacrificing to them in trivial matters (*in minimis*), "that is, to/near springs or trees." Sacrificing in important ways (*in machinis,* surely a mistake for *in maximis*) means belief in their "extremely filthy illusions," divination and the use of amulets; these call for up to ten years of penance. It is therefore doubtful that the author thought that the offerings placed by springs and trees were meant for deities. The other penitential imposes three years of penance on "whoever makes diabolical incantations, sacrifices to demons, makes vows to springs or wells or in any place other than church, or who practices any form of lot-casting."[18]

invasions has been pointed out by Jacques Le Goff in *Pour un autre moyen âge*, 139 and 227.

[15] Gregory II to the Old Saxons (738-739) *Ep.* 21, *MGH EpSel* 1: 35; Ps. Boniface (8th century?) *S.* 6.1, *PL* 89: 855; Zacharias to Boniface (748) *Ep.* 80, *MGH EpSel* 1: 174 (see also *Benedicti capitularium collectio* [mid 9th century] 2.27, *PL* 97: 756); *Vita Eligii* (c.700-725), *MGH SRM* 4: 706-707. See also an anonymous 8th-century sermon, Nürnberger, 44.

[16] *Capitulatio de partibus Saxoniae* (775-790) 21, *MGH CapRegFr* 1: 69.

[17] Regino of Prüm, *De synodalibus causis* (c. 906) 2.362: 351; Burchard of Worms, *Decretum* (1008-1012) 10.31, *PL* 140: 837.

[18] *Poen. Ps.-Gregorii* (2nd quarter, 9th century) 27, Kerff, 183-184; Poen. Parisiense I (late 9th century or later) 14, Schmitz 1: 683.

2.1.1 *Supernatural beings honoured in cults*

The usual source for the cult of gods is sermons: those of Caesarius of Arles, Martin of Braga, Eligius of Noyon, Atto of Vercelli and two anonymous authors of the 8th and 9th centuries. Deities also appear in papal letters to St. Boniface and in the contemporary *Indiculus superstitionum*, as well as in the collections of Pirmin of Reichenau, Regino of Prüm and Burchard of Worms. They are virtually ignored by penitentials and only two councils, both from the 6th century, even hint at the cult of a pagan god.

Jupiter. The documents attest to cults of two different deities identified as Jupiter. Caesarius of Arles took note of a practice apparently new to his diocese: "We have heard that the devil has so deluded some people that the men will not work nor the women make cloth on the fifth day;" "They say that there are still certain wretched women, not only elsewhere but in this city as well, who refuse to weave or spin" on the fifth day (that is, Thursday, Jupiter's day), in honour of Jupiter.[19] A distant connection exists between Jupiter and the *nundinae*, days of rest for peasants,[20] but the connection between him and such an exclusively feminine occupation is unexpected and (as far as I know) unique. Caesarius may in fact have been talking of the cult of a foreign god brought in by slaves. New technology in weaving made its appearance in Gaul at approximately this time with the introduction of the horizontal loom with pedals, a machine already well known in Egypt, particularly in Alexandria. Given the central position of Arles on trade routes between the Mediterranean world and the European hinterland and its traditional importance in the textile trade, the custom of observing Thursday as a holiday may have been introduced, together with the new loom, by Alexandrian artisans.[21] Or the day may have been honoured because of prevailing planetary influences rather than because of the god.

The dozen other documents decrying the observance of Thursday as a day of rest did not ascribe the practice to women in particular. The *Homilia de sacrilegiis* scolded those "who honour the day called Jupiter's because of Jupiter and do no work that day."[22] Six English and continental penitentials imposed fasts of varying length on laymen or clerics who "observe the fifth day in Jupiter's honour in the pagan fashion." The usual penance was from three to six years, but Burchard of Worms had evidently re-evaluated the custom; he suggested only forty days.[23] A

[19] *S.* 13.5, *CCSL* 103: 68; *S.* 52.2, *ibid.*, 230-1). See also *S.* 19.4, *ibid.*, 90.

[20] C. Jullian, "Feriae," in Daremberg and Saglio 2.2: 1047-1048.

[21] For the textile industry in Gaul, see Geneviève Roche-Bernard, *Costumes et textiles en Gaule romaine* (Paris, 1993) 77, 131-133. For trade and trade routes within Gaul and between Gaul and the rest of the world, see Salin, *Civilisation mérovingienne* 1: 120-204.

[22] 12, Caspari, 8.

[23] *Liber de remediis peccatorum* [early 8th century?] 3, Albers, 411 (see also *Poen. Egberti* (before 766) 8.4, Schmitz 1: 581; *Double Penitential of Bede-Egbert* (9th century?) 30.3, Schmitz 2: 694-695); *Poen. Ps.-Bedae* (9th century?) 30.3, Wasserschleben, 272; *Poen. Ps.-Theodori* (mid 9th century) 27.24, Wasserschleben, 598; Burchard of Worms, *Decretum* (1008-1012) 19, 5.92, Schmitz 2: 429.

9th-century penitential attributes this and similar customs to sorcerers and sorceresses.[24]

The stereotyped wording of these relies wholly on Caesarius of Arles. Other texts, however, introduce original elements. They do not connect the custom of idling on Thursday explicitly with Jupiter's cult; some situate the custom socially; some hint that other days also might have been so honoured. Martin of Braga's words suggest that in Galicia, only pagans continued this practice: "It is bad and vile enough that those who are pagans and are unacquainted with the Christian religion honour Jupiter's day or some other demon's and abstain from work."[25] But a little later, the Visigothic bishops meeting at Narbonne found this "execrable custom" generalised among Catholics, both men and women, slave and free; those who dared to be idle on ordinary Thursdays were to be subjected to harsh penalties according to their civic status.[26] St. Eligius' sermon also touches on the topic: "Unless it is a holy day, no one should spend Jupiter's day nor [any day? Jupiter's day?] in May nor at any time in idleness."[27] The custom persisted: a canonic letter of the Carolingian period chastises priests for giving communion to people who had disregarded multiple admonitions against observing Thursday and Friday (Venus' day).[28]

None of these is evidence for an active cult of Jupiter. Nevertheless there is convincing evidence that the peoples coming under Frankish influence during the 8th and 9th centuries continued to celebrate the rituals of his local German or even Celtic counterpart.[29] Gregory II's correspondence with St. Boniface (c. 732) shows that, within the last generation, bloody sacrifices had been offered to this god and sacrificial feasts held in his honour in Hesse and Thuringia. Boniface had sought the pope's advice on the validity of baptisms administered by a priest who had sacrificed to Jupiter and consumed sacrificial flesh. The use of the present

Burchard also included this in another volume of his *Decretum* (10.33, *PL* 140: 837-838).

[24] *Poen. Ps.-Gregorii* (2nd quarter, 9th century) 23, Kerff, 180. This edition's *feria in honore Iovis uel kalendas Ianuarii* ("feast in Jupiter's honour or the Calends of January") corrects the incomprehensible version in Wasserschleben, 543: *ut frater in honore Jovis vel Beli aut Jani* ("brother in honour of Jove or Bel or Janus"). Mansi 12: 293, also gives *ut frater.*

[25] *De correctione rusticorum* (c. 572) 18, Barlow, 202-203.

[26] *Conc. Narbonense* (589) 15, *CCSL* 148A: 257.

[27] *Vita Eligii* (c.700-725), *MGH SRM* 4: 706.

[28] *Epistola Canonica sub Carolo Magno* (c.814) 5, Mansi 13: 1095. Friday, being under Venus' protection, was a lucky day for marriages, but this observance appears to be connected solely to the planets and not to the goddess.

[29] The German equivalent of Jupiter is Thor (Donar, Thunor, Thunaer). The Celtic sky-, thunder- and weather-god Taranis also is sometimes identified with Jupiter. The remnants of about a hundred and fifty "Jupiter-Giant" columns have been identified, many of them along the Moselle and in the Rhineland; see Green, *Dictionary of Celtic Myth and Legend* under all the headings for Jupiter, and *idem, Symbol and Image in Celtic Religious Art* (London and New York, 1992), 123-130.

participle ("immolating," "consuming") suggests strongly that the man functioned during the same period as a priest in both Christian and pagan rites.[30] The question of baptisms performed by such priests came up again in 748, because of doubts that they had used the proper formula. Pope Zacharias authorised Boniface to rebaptise those who had been christened by sacrilegious priests "who used to immolate oxen and goats to the gods of the pagans, ate the flesh, and held sacrifices for the dead."[31] This agrees with Tacitus' remark seven centuries earlier that the Germans conciliated Hercules and Mars (*interpretatio romana* of Thor) by offering selected animals and, to some extent, with Adam of Bremen's hearsay description of the sacrifices offered at Uppsala in the 11th century.[32]

The *Indiculus superstitionum* (c. 743) paired Jupiter with Mercury in two clauses: concerning their rites and "the days (or feasts–*feriae*) which they made" for them.[33] The first must be considered to deal with the kind of bloody sacrifices and sacrificial meals mentioned by Boniface's correspondents–if Tacitus' observations were still valid, Jupiter was probably satisfied with livestock.[34] The *feriae* is more puzzling. That of Jupiter may be explained as the usual Thursday holiday from work, but there is no other evidence for any special significance to Wednesday. Possibly, *feriae* should be understood in this context to mean festivals rather than weekdays. Two anonymous sermons of the 8th or 9th century echo the *Indiculus* by condemning, among other sacrileges "too long to enumerate," the sacrifices offered "over stones, at springs and trees, to Jupiter or Mercury or to other gods of the pagans." One also includes the observance of the days sacred to them (*feriati dies*).[35]

Mercury: Thor's companion in these texts is surely the German not the Celtic Mercury, that is, Woden (Wotan, Odin), who rivalled Jupiter/Thor in importance and popularity.[36] According to Tacitus, he received the highest honour among the

[30] *Ep*. 28, *MGH EpSel* 1: 51.

[31] *Ep*. 80, *MGH EpSel* 1: 174-175.

[32] "They offer nine head of every kind of male animal with whose blood it is the custom to placate the gods. As for their bodies, they are hanged in a grove hard by the temple. For this grove is so sacred to the pagans that every tree of it is believed to be divine on account of the death or rotting away of the animals immolated. Dogs and horses hang there even now, together with men–a certain Christian told me that he had seen the bodies hanging [hanged?] indiscriminately together. Of the many other shameful incantations [*neniae*] which are usually performed during this kind of libatory rite, it is better to be silent" (Adam of Bremen, *Gesta Hammaburgensis ecclesiae pontificum usque ad a. 1072*, 4.27, *MGH SS* 7: 380).

[33] 8 and 20, *MGH CapRegFr* 1: 223. See Homann, *Der Indiculus Superstitionum,* 58-66 and 111-113.

[34] Germania, 9.

[35] *Anonymous sermon* (9th century, 1st half), Scherer, 439. See also Ps. Boniface (8th century?) S. 6.1, *PL* 89: 855.

[36] Grimm ranks him "the highest, the supreme divinity, universally honoured" of all the gods among the Teutonic races (*Teutonic Mythology*, 131). As god of war and the dead, he filled his warriors with battle-frenzy and their enemies with panic. Adam of Bremen identified Woden

gods, and his rites "on certain days" consisted of or included human sacrifices. These may have continued well into the Christian era–a letter written by Gregory II (732) responds to St. Boniface's report that some of the faithful had sold their slaves to pagans to be used for sacrifice.[37]

The Celtic or Roman Mercury appears only in Martin of Braga's sermon. It describes a practice which may have been linked to Mercury's role as psychopomp although Martin associated it with his character as "inventor of every form of theft and fraud": when "greedy men" traverse crossroads, they throw stones to make cairns [*acervi*] by way of sacrifice to him "as it were to the god of money."[38] The use of the present in a passage in which all the other verbs are in the past tense indicates a current practice. Given that written, epigraphical and archaeological evidence all proclaim the importance of Mercury in the religion of the continental Celts, this sparseness of information is surprising.

Minerva: In the texts, the cult of Minerva, the Roman goddess of war and wisdom, was practiced primarily by women. References to her cult in the 8th and 12th century depended on Martin of Braga's claim that women were accustomed to name her while weaving.[39] St. Eligius' sermon gives a general injunction against believing in or invoking the name of demons or Neptune, Orcus, Diana, Minerva, the Genius or "similar nonsense," but further on, it returns to women: "Let no woman dare to name Minerva or other ill-omened personages while weaving or dyeing or doing any other work."[40] The choice of the term "name" rather than

as *furor* and described his statue as showing him armed "like our Mars" (*Gesta Hammaburgensis ecclesiae pontificum* 4.26, *MGH SS* 7: 379). Though god of learning and magic, he could be fooled: in Paul the Deacon's *History of the Lombards*, Woden (Godan) was tricked by his wife Frea into giving victory in battle to the Lombards over the Vandals (*Historia Longobardorum* 1.8, *MGH Scriptorum Rerum Longobardicarum et Italicarum,* 52). He was more popular among the aristocratic and military classes than among peasants. Georges Dumézil noted that in Scandinavia and Iceland few men or places were named after Odin (*Gods of the Ancient Northmen*, 33), but Wilson identified at least thirty place names in England containing the element "Woden" or of "Grim," one of his by-names; these are usually associated with some topographical feature (*Anglo-Saxon Paganism*, 11, 20-1).

[37] Ep. 28, *MGH EpSel* 1, 51. *Germania*, 9. There is a considerable amount of documentary evidence for human sacrifice among the Germans. In chapter 40, Tacitus recounted the murder by drowning of the slaves serving the *Terra Mater* Nerthus. In the 8th century, Radbod's Frisians also apparently favoured drowning but used gladiatorial combat, hanging and strangling as well (*Vita Vulframmi* 6-8, *MGH SRM* 5: 665-667). See also Adam of Bremen's account (n. 32, above). Sacrificial victims were also buried alive in bogs; see P.V. Glob, *The Bog People*, trans. R. Bruce-Mitford (reprint, London, 1971).

[38] *De correctione rusticoru*m 7, Barlow, 187. In Burchard of Worms' penitential, the practice of building cairns (*aggeres*) is not connected to a cult of Mercury (*Decretum* 19, 5.94, Schmitz 2: 430).

[39] Martin of Braga, *De correctione rusticorum* (572) 16, Barlow, 198. See also Dicta Pirmini (724-753) 22: 172-173; *Anonymous sermon to the baptised* (12th century), Caspari, *Kirchenhistorische Anecdota* 1, 204-208. Caesar had noted that the Gauls, like others, believed that Minerva was the inventor of all skilled work (*De bello gallico* 6, 17).

[40] *Vita Eligii* (c.700-725), *MGH SRM* 4: 706-707. The first part of this is also found in an

"pray to," "make vows to," "immolate to," *etc.*, may mean that the name itself had apotropaic significance while the goddess herself had disappeared from the consciousness of working women. Some half dozen continental texts, from Martin of Braga's *Canons* of 572 to Burchard of Worms' *Decretum* (1008-1012), command women not "to practice any foolishness in their weaving, but to ask the help of God, Who gave them the skill to weave," but only the *Vita Eligii* hints that other deities have been involved.[41]

Minerva was the logical choice as patroness of skilled trades, but in a late 8th-century sermon, her name is rather surprisingly combined with Juno's as an insatiable and incestuous harlot, who prostituted herself to her father Jupiter and to Mars and Venus, her brothers–proof that the nature of the ancient divinities was now truly forgotten even by some of those who still remembered their names.[42]

Diana, Herodias, Holda, Friga:[43] The only explicit references to a cult of Diana made until the very end of our period are in St. Eligius' sermon and an anonymous 8th-century sermon, quoted above. The other mentions of her name reveal nothing about her role in folk religion: Martin of Braga had explained that the demons and evil spirits of the woods called themselves "Dianas" and an anonymous 8th/9th-century sermon identified *Deana* as a seeress or sorceress worshipped as goddess in Ephesus.[44]

But these texts may be clues to still active beliefs concerning Diana, since other sources show that she kept her hold on popular consciousness well into the Merovingian period. In the 6th-century life of Caesarius of Arles, a demon "whom the peasants called Diana" afflicted a servant girl with madness.[45] Gregory of Tours related that Diana was called upon to help women in labour and that within living memory her statue, mounted on a column, had been the object of a popular cult in the Ardennes.[46] In the *Life of St. Symphorian*, Diana was the midday demon who disturbed the minds of men as she haunted the woods and crossroads.

anonymous 8th-century sermon, Nürnberger, 44.

[41] *Canones ex orientalium patrum synodis* (572) 75, Barlow, 141. Minerva was sometimes identified with Sulis, a British goddess of healing springs. The distaff and spindle that are attributes of Celtic and Rhenish mother-goddesses probably represent their control over life not the homely arts; see Green, *Dictionary of Celtic Myth and Legend*, *s.v.v.* "Sulis," "Fates" and "Matronae Aufaniae."

[42] Levison, 311.

[43] See Grimm's *Teutonic Mythology*, 265-288 for this Friga.

[44] Martin of Braga, *De correctione rusticorum* (c. 572) 8, Barlow, 188; an anonymous sermon, Levison, 311. See Acts 19, 24.

[45] *Vita Caesarii* 2, 18, *MGH SRM* 3, 491.

[46] *De miraculis beati Andreae apostoli* 25, *MGH SRM* 1: 841; *HF* 8.33, *ibid.*, 401. For a discussion of the latter incident see Gauthier, *L'évangélisation des pays de la Moselle*, 240-242 and Hubert Collin, "Grégoire de Tours, Saint Walfroy le Stylite et la 'Dea Arduanna'. Un épisode de la christianisation des confins des diocèses de Reims et de Trêves au VIe siécle," in *La piété populaire au moyen âge,* 387-400.

Gregory of Tours testified repeatedly to her powers in this form.[47] Martin's use of the plural *Dianas* (if not a mistake) indicates that the goddess' name was applied to dryads, thus evoking her role as huntress but leaving her character as moon-goddess in abeyance.[48] In the anonymous sermon cited above, the euhemeristic explanation of her origins follows immediately upon the description of preternatural harvest-thieves, the *mavones*; it is possible that folklore made some connection between them and the goddess.[49]

Other beliefs attached to Diana's name emerged in the 10th and 11th centuries. In Regino of Prüm's *De synodalibus causis* and Burchard of Worms' *Decretum*, Diana appears as a goddess of the night, mistress of animals, and leader of a Wild Hunt, followed by a throng of women bound to her service:

> [C]ertain criminal women, who have turned back to Satan and are seduced by illusions of demons and by phantasms, believe and avow openly that during the night hours they ride on certain beasts together with Diana, the goddess of the pagans, and an uncounted host of women; that they pass over many lands in the silence of the dead of night; that they obey her orders as those of a mistress; and that on certain nights they are summoned to her service.[50]

This goddess' real identity is uncertain. The alternatives are the classical Diana, the Herodias of medieval legend (perhaps Herod's stepdaughter, not wife), the native German Holda or even the goddess Friga.[51] Regino cited Diana only. So did Burchard in his penitential,[52] but elsewhere he appears to be in doubt. In Book 10 of his *Decretum,* he introduces Herodias: the women ride "with Diana, goddess of the pagans or with Herodias." Herodias, Ratherius of Verona had revealed, was considered by some of his contemporaries as "a queen, in very truth, a goddess" to whom a third part of the world had been given as her price for John the Baptist's death.[53]

[47] *Vita S. Symphoriani* 10, *AASS* Aug 4: 497). The midday demon gives a woman the gift of prophecy (*HF* 8.33, *MGH SRM* 1: 348), lames a Poitevin priest and strikes dumb the wife of one of Gregory's men (*De virtutibus S. Martini* 3.9 and 4.36, *ibid.*, 634-635 and 658-659). See Roger Caillois, "Les démons de midi," *Revue de l'histoire des religions* 115 (1937) 141-173 and 116 (1937) 142-186, especially the latter, 163-165.

[48] *De correctione rusticorum* 8, Barlow, 188

[49] Levison, 311.

[50] Regino of Prüm, *De synodalibus causis* (c. 906) 2.371: 355; Burchard of Worms, Decretum (1008-1012) 10.1, *PL* 140: 831; *idem*, 19, 5.90, Schmitz 2: 429.

[51] Similar female figures were known by such names as Berchte, Perchte, Bona Socia or Dame Abundia (see Ginzburg, *Ecstasies. Deciphering the Witches' Sabbath*, trans. Raymond Rosenthal [New York, 1991], 89-110). For Diana's Ride, the Wild Hunt and medieval variations on that theme, see Claude Lecouteux, *Chasses fantastiques et cohortes de la nuit* (Paris, 1999).

[52] 19, 5.90, Schmitz 2: 429.

[53] *Praeloquia* (934-936) 1.10, *CCCM* 46A: 14. For Herodias in her role as leader of the witches, see Elliot Rose, *A Rasor for a Goat. A Discussion of Certain Problems in the History of Witchcraft and*

In another question, Burchard brought up a new name, Holda:

> Do you believe that there exists any woman who can do what certain women, duped by the devil, claim that they do out of necessity and by order? that is, that the witch/vampire [*striga*] whom common stupidity calls Holda must ride on fixed nights with a horde of demons transformed into the likeness of women on the backs of beasts, and that she is numbered among their company. If you participated in this false belief, you must do penance for one year on the appointed days.[54]

Jacob Grimm recognised in this passage the first appearance of a goddess of abundance and fertility, whose name is found in various forms (Holda, Holle, Hulle, Frau Holl, Huldr) in myth and folklore throughout Germanic territory. Nevertheless, some of Burchard's readers understood *Holda* not as a proper noun but as the epithet "generous," "propitious" or "lovely": the copyists of two recensions replaced it with *unholda*, feeling perhaps that *holda* was inappropriate to describe the baby-eating, man-eating *striga* of popular imagination. In another version, however, *striga* becomes *friga*, so that the goddess' name is given as Frigaholda (or, perhaps, *Friga holda*, "the generous Friga"). The similarity between *striga* and Friga makes confusion natural, particularly because of Holda's and the goddess' shared role as givers of abundance (although flight was associated with Freya and the Valkyrie rather than Friga).

Burchard placed these among questions meant for either men or women; therefore, the belief was common to both. Some women evidently thought that they participated personally in Diana's ride, but in the case of Holda's, Burchard gives the perhaps accidental impression that only Holda herself was thought to be human while her companions were demons in the guise of women (ghosts?).

The Parcae: Beliefs in deities who controlled human destiny were common in pre-Christian societies. In Greek myth, they were the *Moirae,* in Roman, the *Fatae* or *Parcae*, often represented as three goddesses spinning thread. They were sometimes identified with the triad of the Celtic *Matronae*. Representations in Burgundy and among the Treveri show the Mothers holding spindle, distaff and scroll, and in Britain, an inscription gives them the name *Parcae*. The Sagas record beliefs in an overmastering destiny (*wyrd*), sometimes represented as a woman with weaving implements, sometimes as the three Norns presiding over births.[55] Nevertheless

Diabolism (Toronto, 1962) 111-116. See also Lecouteux, *Chasses fantastiques,* 15-18, and Grimm, *Teutonic Mythology,* 283-288.

[54] 19, 5.70, Schmitz 2: 425.

[55] See Green, *Dictionary of Celtic Myth and Legend, s.v.* "Fates;" Davidson, *Lost Beliefs*, 107-121. See also Paul G. Bauschatz, *The Well and the Tree; World and Time in Early Germanic Culture* (Amherst, 1982), esp. 7-16. Isidore of Seville gave the classical description of the *Parcae*: "They claim that they are three in number: one who lays the warp of man's life, another who weaves it, the third who breaks it" (*Etymologiae* VIII, 11. 93, 734). But an 8th-century Irish text refers to the "seven daughters of the sea who are forming the thread of life" (Jacqueline Borsje, "Fate

our texts adduce no convincing evidence for such beliefs until the very end of our period. Earlier references to fate clearly refer to astrological beliefs. By contrast, Burchard of Worms' penitential represents destiny as "three sisters called *Parcae* by the common folk" but who had little in common with the remote, inexorable Greek and Latin goddesses of fate.[56]

Exceptionally, he gives an account of both beliefs and rites. Certain people believed that the *Parcae* presided over births, deciding the newborn's destiny and bestowing the power of shifting shape at will, "so that whenever that man chooses, he can be transformed into a wolf, called *Werewulff* in German, or into some other form." This was thought to be a change in nature as well as appearance: "If you believed that this has ever happened or could happen, that the divine likeness was changed into another appearance or kind by anyone except almighty God..." In another clause, the *Parcae* seem less ominous: "certain women" at "certain times of the year" (the New Year?) were accustomed to try to bribe "those three sisters whom ancient tradition and ancient stupidity named the *Parcae*." They set a table in their house with food, drink and three "little knives" for the sisters' refreshment, in the hope that if they came, they would help their hostess either at present or in the future: "thus they attribute to the devil the power that belongs to merciful God." Here the *Parcae* appear to be less figures of pitiless destiny than sprites, small ones at that, who can handle only "little" implements, and who make their way into the house but seldom (but in medieval Latin, the diminutive was often used to indicate contempt, not necessarily to refer to size).[57] The identification with the classical goddesses is made by Burchard, not the common people.[58]

Although these clauses are in the section supposedly dealing with specifically feminine beliefs and misdeeds, the first concerns a belief entertained by men as well as women. "Did you believe what certain [people] are accustomed to believe?" is directed at both sexes, since the masculine form is used. When it is really a matter affecting women only, Burchard employs a feminine pronoun or refers specifically to women. A masculine term is consistently used for the object of the *Parcae*'s attentions ("some man," that man," "him"), proof that while women may

in early Irish texts," *Peritia* 16 [2002] 214-231; here, 230).

[56] If Hutton (*The Pagan Religions*, 296) is correct that the concept of destiny was not native to the Germans but borrowed from the Greeks, Burchard's text is proof that by the beginning of the 11th century it was thoroughly absorbed into and transformed by German consciousness.

[57] 19, 5.151, Schmitz 2: 442; 19, 5.153, *ibid.*, 443. On the use of the diminutive in the early middle ages, see E. Coli, "Osservazioni sull'uso del diminutivo in Cesario d'Arles," *Giornale Italiano di Filologia*, n.s. 12 (33) (1981): 117-133.

[58] The *Parcae* appear as the "good women" in the 13th-century *Life of St. Germanus of Auxerre* of Jacob of Voragine in *The Golden Legend.* The saint was told that a table prepared after supper was meant for these nocturnal visitors (*Legenda aurea*, ed. G. P. Maggioni [Florence, 1998] 690-691). This episode is not found in St. Germanus' first *Vita,* written in the 5th century by Constantius of Lyons. I am indebted to Pierre Boglioni for this reference.

have been affected, it is certain that men were. The second clause, on the other hand, concerns women only: they are the ones who prepare the table, they are the ones who expect to benefit from the sisters' gratitude.

Burchard considered belief in the *Parcae*'s control of destiny to be a petty superstition, meriting only ten days of penance. Belief translated into action was a more serious matter: laying out knives, food and drink maybe constituted offerings, and as such called for a whole year of fasting.

A mortuary rite described in Burchard's penitential may contain a distant hint of the *Parcae*. The practice of striking carding combs together over a corpse before it was carried out of the house[59] perhaps symbolised the breaking of the thread of life spun by the goddesses of fate. It may be significant that carding combs and other weaving accessories are frequently found in Germanic ship-burials such as Sutton Hoo and in barrows and cemeteries.[60]

Satyri and pilosi: Classical myth and Isidore of Seville link satyrs and *pilosi* (the "hairy" ones) with nature and indiscriminate sexual activity, while the Vulgate associates *pilosi* with lawless abandon.[61] In Burchard of Worms, however, they are small house sprites frequenting the places where food is stored–not priapic and wild, but playful and needy, happy for shelter and willing to be bribed with gifts of toys and shoes to steal for their benefactors. "Did you make little children's bows and children's shoes and throw them into your pantry or storehouse for the *satyri* and *pilosi* to play with there, so that they would bring you other peoples' goods and enrich you as a result?"[62] Here the penance is a trivial ten days, implying that Burchard thought that such little creatures were no real danger to the faith. The word "throw" implies neither respect nor fear on the part of the householder.

Neptunus, Orcus, Geniscus: Neptune, Orcus and Geniscus appear together with Minerva and Diana "and other such nonsense" in St. Eligius' sermon, as deities who should not be believed (in) or invoked.[63] The words imply that the gods in

[59] 19, 5.95, Schmitz 2: 430.

[60] Bauschatz, *The Well and the Tree*, 37-39.

[61] Isidore of Seville, *Etymologiae* VIII, 11.103, 736. *Isaiah* 13, 21 prophesied that *pilosi* would dance in the ruins of Babylon.

[62] *Decretum* (1008-1012) 19, 5.103, Schmitz 2: 43). This agrees with Grimm's description of *pilosi* (the Germanic *scrat, scrato, schrätlein*) as being invariably male and small, hairy and lighthearted (*Teutonic Mythology*, 478-482). Nevertheless, they had a more somber side. In the Monk of St. Gall's account, a *pilosus* who stole beer or wine for a smith who allowed him to play at night with the hammers and anvils, was a "demon or ghost" and, as such, connected with the dead (*Gesta Karoli* 1, 23, *MGH SS* 2: 741). *Pilosi* may have been ghosts of children: miniature weapons, the size of toys, have been found in children's graves (A. Dierkens, "Cimitières mérovingiens et histoire du haut moyen âge," in *Histoire et méthode* 47 [Brussels, 1984]). See also Lecouteux, *Les nains et les elfes au moyen âge* (Paris, 1988), 101-102 and 182-188, and Arne Runeberg, *Witches, Demons and Fertility Magic* (Helsinki, 1947), 142-145.

[63] *Vita Eligii* 9(c.700-725), *MGH SRM* 4: 705-6. See also anonymous sermon, Nürnberger, 44.

question enjoyed an active cult in the northern parts of Merovingian territory during the 7th century, but no direct confirmatory evidence is provided. Hints about a cult to Neptune, however, are to be found elsewhere. Martin of Braga had named him as the demon who claimed to be god of the sea and left open the possibility that he was still worshipped or, at least, feared in the Galicia of his time.[64] In addition, the rites of a Celtic or Germanic water divinity may have been practiced in northeastern Gaul toward the end of the 8th century. The *Homilia de sacrilegiis* (originating in roughly the same area as St. Eligius' missionary activity) denounced *Neptunalia* observed by the sea or at the source of springs and brooks.[65] Perhaps significantly, the *Homilia* warns against praying by the sea or at the sources of water while Eligius' word "invoke" suggests prayer as well. As for the other figures, we can only speculate. Since Orcus was a god of the underworld, a reference to rituals in aid of the dead may have been intended.[66] Geniscus is surely a variant of Genius, the deity who, among the Romans, had protected the individual, the family or the place and who, in Celtic representations, was frequently shown as a hooded single or triple figure (the *Genius Cucullatus*).[67] In another contemporary sermon, the Geniscus is presented along with witches and spirits, creatures whom "certain rustics believe must exist." Here he is a minor local deity, not the important family or state Genius of Roman religion.[68]

[64] *De correctione rusticorum* (572) 8, Barlow, 188.

[65] 3, Caspari, 6. Green finds only "a small amount of evidence for some association between this deity and the indigenous cults of western Europe," although there is some epigraphic evidence that in Britain he was identified with Nodens (*Dictionary of Celtic Myth and Legend, s.v.* "Neptune"). There is no doubt of the importance of water deities, especially of those associated with healing springs. Minerva Sulis, Sequana, Apollo Grannus, Nehalennia are only few of many (*ibid., s.v.v.*). It will be remembered that the procession of the Germanic *Terra Mater* Nerthus ended with the ritual bath of the goddess in her lake and the drowning of the officiating slaves. Gregory of Tours recounts the cult of the lake god Helarius. His worship involved throwing offerings of food, wool and other objects into the lake (probably Saint-Andéol in the Aubrac Mountains of the Massif Central) (*In gloria confessorum* 2, *MGH SRM* 1: 749-750); see also Raymond van Dam, *Gregory of Tours: Glory of the Confessors* (Liverpool, 1988), 19 n.5. In the later Middle Ages, the term *neptunus* was applied to house sprites. Gervais of Tilbury (d. 1234) explains it as the French equivalent (*nuiton*) of the English *portunus*, a tiny creature, basically helpful and friendly, who nevertheless delights in misleading travellers (*Otia imperialia*, ed. Felix Liebrecht [Hanover, 1865] 29-30. This edition contains only the third part of the *Otia imperialia* and has been translated by Annie Duchesne as *Le livre des merveilles; Divertissement pour un empereur* [Paris, 1992].) See also Lecouteux, *Les nains et les elfes,* 93-94 and 174-178.

[66] I am aware of only one other appearance of *orcus* in our texts, in Wasserschleben's edition of the *Poen. Albeldense=Vigilanum* 84 (533), where it appears in the context of seasonal celebrations, and has been interpreted as either "ogre" or "bow" (if *orcum* is an error for *arcum*). See chapter 4, n. 113.

[67] The *Genius Cucullatus* was usually depicted with a fertility symbol. The triple figure was particularly common in Britain (see Green, *Dictionary of Celtic Myth and Legend, s.v.* "Genius Cucullatus").

[68] Levison, 310.

2.1.2 *Supernatural beings as objects of belief only*

Lamiae: In classical Latin, *lamiae* were witches or vampires who, according to Isidore of Seville, were reputed to steal children and tear them apart.[69] This seems to be their character in the first Synod of St. Patrick (c. 457), which chastised those who believed "that there is such a thing in the world as a *lamia*, which means witch/vampire [*striga*]" or who defamed "another being with that name."[70] For Martin of Braga, however, *lamiae* were, most unusually, river nymphs–demons inhabiting rivers.[71]

Mavones, Maones, Manes: The various names (*mavones, maones, manes, dusii hemaones, dusi manes* and "Magonians") given to mysterious beings who stole the produce of fields and orchards illustrate the ambiguities of terminology for popular beliefs and the difficulties faced by medieval authors who had to find Latin words for ideas rooted in alien traditions.

A passage in Gall Jecker's edition of Pirmin of Reichenau's *Dicta* is the earliest reference. Weather-magicians persuaded the gullible to pay them by telling stories about spirits, the *manus*, who had the power to make off with the crops. Another recension has *maones*, which Jecker took to be equivalent to *manes*.[72] They appear again at the end of the century in another sermon composed in roughly the same region: one should not believe in *mavonis* (*sic*) "as though the grain harvest and the grape harvest could be carried away."[73] The earliest *Life of St. Richarius,* probably from the second half of the same century, records a belief in the northwest among the Picards, that their crops were carried away by *maones* (in other recensions, *dusi hemaones* or *dusi manes*).[74] Similar beliefs existed farther south. Around early 9th-century Lyons, it was commonly believed that ships came through the clouds from a region called Magonia in order to take away the fruits or crops knocked down by hail or destroyed by storms. The aerial sailors supposedly paid malign magicians, the *tempestarii*, to produce the bad weather.[75] Four "Magonians" were

[69] *Etymologiae* VIII, 11.102: 736. To Gervais of Tilbury, *lamiae* were nocturnal female imps of small-scale mischief (they invaded houses, drained the barrels, pried into the stores, took babies out of their cradles, lit lights and sometimes made amorous advances to sleepers) as well as of major malice (*Otia imperialia* 86: 39-41). For Arne Runeberg they are wood spirits (*Witches, Demons and Fertility Magic*, 131-132). Since the original meaning of *strix* was owl, perhaps the terrifying Irish war-goddesses who adopted the shapes of birds on the battlefield lie behind the reference to *striga* in the Irish text; see Green, *Dictionary of Celtic Myth and Legend, s.v.* "Badbh, Morrigán."

[70] 16, Bieler, 56.

[71] *De correctione rusticorum* (c. 576) 8, Barlow, 188. Dowden connects *lamia* with the classical *lama* (slough, bog, fen) (*European Paganism*, 42).

[72] *Die Heimat des hl. Pirmin des Apostels der Alamannen* (Münster in Westphalia, 1927), 55. Missing from Caspari's edition.

[73] Levison, 311.

[74] *Vita Richarii* I, 2, *MGH SRM* 7, 445. See Levison, 310 n. 4.

[75] Agobard of Lyons, *De grandine et tonitruis* (815-817) 2, *CCCM* 52: 4. This episode is discussed

actually captured and after several days brought in chains to a local assembly, where Agobard the bishop of Lyons had some difficulty in saving them from the hands of the mob.

The similarity of names and habits establishes that the *manus, maones, mavones, dusi hemaones* and the inhabitants of Magonia are closely related. But there are differences: Magonians were foreigners; they appear to have been rather frail humans, unable either to steal the crops unassisted or to defend themselves–the Lyonnais were not afraid to seize and maltreat them. By contrast, the *mavones* and the others seem to have operated on home ground. They were probably spirits, since belief in them was seen as a remnant of pagan observances, and it took magicians to foil their raids on the harvest.

If Jecker was right that these spirits were properly known as *manes,* they have little in common with the shades of the dead, the infernal gods of Roman religion. However, Isidore of Seville's *Etymologiae* may provide a link. There the *manes* were gods of the dead but their power, instead of being located underground, lay between the moon and the earth, the very region of the clouds through which the aerial pirates came.[76] Thus, unknown peculiarities of Iberoceltic or Visigothic belief brought into Gaul by Spaniards fleeing the Moorish onslaught may lie behind such tales.

Silvaticae: Folk in Burchard of Worms' diocese believed in the existence of so-called *silvaticae*, "wild women" of the woods who "showed themselves at will to their lovers and, when they had taken their pleasure with them, hid and vanished, again at will."[77] Love encounters between deities and mortals are common in classical myth, but these sylvans are less goddesses than figures of folklore. Burchard placed this clause between the two dealing with the *Parcae*, thereby, Laurence Harf-Lancner observed, bringing together for the first time the concepts of eroticism and destiny, the essential characteristics of the medieval fairy.[78] Once again, Burchard refused to take belief in such supernatural beings seriously, and recommended only ten days of penance.

Dusiolus, aquaticae: *Dusiolus* and *aquaticae* are merely mentioned in Levison's anonymous sermon,[79] and are found in none of our other sources. The latter were probably water-nymphs, equivalent to Martin of Braga's *lamiae*. There are, however, indications about the nature of the former. The term *dusiolus* appears to be Gallic. Augustine mentioned *dusii* in the *City of God* and Isidore of Seville explained that "demons whom the Gauls call *dusii*" were known as *incubi* because they con-

further in chapter 7, section 1.

[76] *Etymologiae* VIII, 11.100: 736.

[77] 19, 5.152, Schmitz 2: 442.

[78] *Les fées au moyen âge. Morgane et Mélusine. La naissance des fées* (Geneva, 1984), 17-25.

[79] Levison, 310.

tinually inflicted their amorous attentions on women.[80] Since *dusiolus* appears between witches who molest babies and livestock on the one hand, and *aquaticae* and the *geniscus,* on the other, it is impossible to determine from this sermon whether he is a malign demon, an *incubus,* or a relatively innocuous local spirit, but the *dusi hemaones/dusi manes* of the *Life of St. Richarius* connects him with supernatural harvest-thieves.

Spiritus immundi: Unclean spirits are found in two passages of Burchard of Worms' *Decretum.* In one, Burchard followed Gregory I's *Moralia in Job* in declaring that "we know that unclean spirits who fell from the heavens wander about between the sky and the earth." This was in the context of theology. In his penitential, he shows the kinds of anxiety that they caused on a mundane level, and the means used to protect against them. Even in the case of necessity, some people were afraid to leave their houses (go outside the bounds of the settlement?) before dawn lest *spiritus immundi* harm them, until the cock's crow caused their power to wane and drove them off.[81] These may have been demons, minor imps of the woods, souls wandering away from their body (a belief hinted at in an anonymous 8th-century sermon[82]) or the ghosts of the unhallowed dead. Burchard did not object to belief in the spirits, but to the reliance on the cock's crow rather than on Christ and the sign of the cross.

2.1.3 *The origins of the gods*

Despite the scant evidence offered for dynamic cults or active beliefs concerning deities and other supernatural beings, clerical writers continued to cite them, usually to denigrate popular practices by linking them with the disgraceful memory of the ancient gods, or to explain the continuance of forbidden traditions by blaming the activity of demons. They followed the models presented by such eminent authorities as St. Augustine and Isidore of Seville to explain the gods as great or exceptionally wicked men and women to whom ignorant men had paid divine honours and/or whose identity had been taken over by demons to attract worship to themselves.[83]

When Caesarius of Arles dealt with this subject at length in his sermons, it was not to engage in polemics against still active cults as had Augustine, nor to

[80] Augustine, *De civitate Dei* 23; Isidore of Seville, *Etymologiae* VIII, 11.103, 736. See also Lecouteux, *Les nains et les elfes,* 169-174.

[81] *Decretum* 20.49, *PL* 140: 1031; *Decretum* 19, 5.150, Schmitz 2: 442. See also chapter 3, 2.2.

[82] "And there is another heresy which stupid men believe, that when the spirit leaves one man, it can enter another. But this, absolutely, can never happen, unless demons do so and speak through those men" (Levison, 312).

[83] Augustine, *De civitate Dei* 7.18 and 7.33; Isidore, *Etymologiae* 8.10. For a summary of Euhemerism, see Jean Seznec, *La survivance des dieux antiques* (1940; Paris, 1993) 21-48. See also David H. Johnson, "Euhemerisation and demonisation: The pagan gods and Aelfric's *De falsis diis*" in *Pagans and Christians,* 35-69.

present a scholarly exposition of the origin of the gods as would Isidore. Rather, he wished to discredit the customary celebrations of the New Year by holding up to horror and disgust the god who gave his name to January and its calends.[84] Similarly, the Council of Tours (567) pointed out that Janus had been a pagan man, a king indeed, but could not have been a god.[85] A 7th-century Spanish homily modestly relied on the authority of pagan literature to derive the name of the Calends from Janus, the human founder of a city, on whose account gluttons and drunkards celebrated that day in debauchery.[86] The late 8th-century *Homilia de sacrilegiis* adhered closely to Caesarius' text.[87] Another anonymous sermon proclaimed that it was "accursed Janus" who had taught the rites of the Calends.[88]

In a second sermon dedicated to the same feast, Caesarius extended the attack to include the deities commemorated in the names of the days of the week–names not drawn directly from the gods but from heavenly bodies. The sun and the moon were made by God to benefit man; instead of honouring them as gods, man should give thanks to their Maker. But the planets were named after wicked people who had lived in the time of the Jewish captivity in Egypt; men had sacrilegiously named days after them so that they would "seem more often to have in their mouth the names of those whose sacrilegious deeds they revered in their heart."[89] Caesarius' solicitude to explain the disreputable origin of the names of the Calends and the days of the week is acknowledgement that his flock did not consciously engage in the cult of the eponymous deities.

By contrast, the old gods were still believed to haunt the promontories and streams of Galicia a few decades after Caesarius' time. Martin of Braga's detailed description of the origins of idolatry was not meant to demonstrate human stupidity primarily, but the gods' demonic origin. After the Flood, men forgot God and began to offer worship to nature. The demons then began to show themselves to men and demand sacrifices on mountaintops and in groves. After adopting the names of certain wicked humans (Jupiter, a *magus* and incestuous adulterer; Mars, the sower of quarrels and discord; Mercury, the inventor of theft and fraud; the harlot Venus, whose numerous adulteries had included prostituting herself to her father Jupiter and her brother Mars), the demons persuaded "ignorant rustics" to build temples in their honour, with images and statues, and to offer animals and humans in sacrifice.[90]

Nature was teeming with still other demons honoured as divine: Neptune in

[84] *S.* 192.1, *CCSL* 104: 779-780.

[85] 23, *CCSL* 148A: 191.

[86] *Homiliare Toletanum* (2nd half, 7th century) Hom. 9, *PL Suppl* 4: 1941-1942.

[87] 23, Caspari, 12-13.

[88] Levison, 310.

[89] *S.* 193.4, *CCSL* 104: 785.

[90] *De correctione rusticorum* (572) 7-8, Barlow, 186-8. This appears to be the source of the reference in Levinson's anonymous sermon, to Juno-Minerva who committed fornication with her brothers, Mars and Venus.

the sea, *lamiae* in rivers, nymphs in springs and Dianas in the woods. Galician peasants would hardly have recognised all the names and personalities on which Martin, a highly educated foreigner surely more familiar with the classical tradition that with Iberian religions, expounded. Nevertheless, it seems clear that they believed in the existence of uncanny beings residing in nature and feared their power to do harm–justly so, for "they are all malign demons and vile spirits who molest and annoy unbelieving men who do not know how to arm themselves with the merest sign of the cross."[91] Significantly, while Martin described the cults practiced in temples in the past tense, he used the present in writing of the demons of the woods and waters.[92]

If demons still had power to cause harm, it was man's fault. God allowed them to do so because He was angry with those who wavered in their Christian faith and named the days after demons "who made no day but were criminals, the wickedest men among the Greeks."[93] Martin also mourned that *ignorantes* and *rustici* believed that the Calends of January were the beginning of the year–an idea patently false since scripture set the beginning of the year at the spring equinox, the "right division" between day and night established when God separated light and darkness.[94]

Atto of Vercelli used a Christmas sermon to denounce the festivities of the calends of January and of March, and to explain the origins of Mars and Janus. They had been particularly wicked and perverse men; their crimes had been so great that demons decided to inhabit their statues the better to fool people; under their names they claimed to be gods, and demanded celebrations in their honour at the beginning of months named after them. Thanks to custom, almost all misguided people had accepted this error, and this was why peasants continued to celebrate the calends to that day.[95] Atto likewise denounced public entertainments, especially during Lent and the Easter season, as inventions of the famous demons Liberus and Venus.[96] The persistence of these traditions, then, was due to the insidious influence of demons, not to a popular cult of the deities of long ago.

2.2 Effigies: *Idola* and *Simulacra*

Before turning to texts dealing directly with the worship of the represen-

[91] *De correctione rusticorum* (572) 8, Barlow, 188. Martin's nature goddesses differ significantly from Isidore of Seville's, for whom *Nymphae* (sometimes erroneously called Muses) are the goddesses of water; nymphs of mountains are *Oreades*, of woods *Dryades*, of springs *Hamadryades*, of fields *Naides*, and of the sea *Nereides* (*Etymologiae* 8, 11.96-97: 736).

[92] According to Gervais of Tilbury, the Catalonians of his time still believed that demons inhabited a mountain lake in the diocese of Gerona (*Otia imperialia* 66: 32).

[93] *De correctione rusticorum* 8, Barlow, 188-9.

[94] *De correctione rusticorum* 9, Barlow, 189-190.

[95] *S*. 3, *PL* 134: 836.

[96] *S*.9 *In die sanctae resurrectionis in albis*, *PL* 134: 845.

tations of idols *(idola, simulacra)*, it should be noted that the question of iconolatry, which aroused bitter controversy in the Byzantine Empire and was the target of polemical attacks in the *Libri carolini* (c.790) and Agobard of Lyons' *De picturis et imaginibus* (c.820), was of scant interest to pastors in the early medieval West. When the bishop of Marseilles had destroyed the images of saints in churches, he was reprimanded by Gregory I in 600: "We praised you whole-heartedly for forbidding that they be adored, but found it wrong that you smashed them ... For it is one thing to adore a picture, and another to learn what is to be adored from the story in the picture."[97] Two canonic collections recalled the prohibition of the Council of Elvira (300-306?) against having pictures in the church "lest what is painted on the wall be revered and adored."[98] Otherwise no evidence of contemporary anxiety about undue reverence for holy images is found even in the capitularies of Theodulph of Orleans, almost certainly the author of the *Libri Carolini*.[99]

Authoritative testimony concerning *idola* and *simulacra* is scarce in our documents. Maximus of Turin's sermons berating the local gentry for their indifference toward rural idolatry in the mid 5th-century are unmatched for their precision. They specify who worshipped idols (the peasants on great estates), what (stone statues), when (late at night or very early in the morning), where (in the fields and storehouses) and how (by bloody libations and burnt offerings). The landscape was permeated with peasant idolatry: "Wherever you turn, you see the altars of the devil or the profane auguries of the pagans or the heads of animals fixed along the boundary lines." And, in another sermon: "When you step into the store-room, you find tufts of yellowing turf and dead embers–an apt sacrifice to demons, when a dead divinity [*numen*] is worshipped by means of dead things. And if you go out into the fields, you notice wood altars and stone statues–a fitting ritual, by which inanimate gods [*dii*] are served on rotting altars."[100]

None of the early medieval accounts is so complete. Only documents from Francia and the missionary letters of popes provide clear references to man-made effigies; even they do not invariably inspire confidence that the authors based themselves on direct knowledge.

Merovingian Gaul: The religious situation of Merovingian Gaul in the mid 6th-century resembled that of Maximus' Turin, although Christianisation was perhaps somewhat less advanced there than it had been in Turin a hundred years earlier. Childebert, king of Paris, had to deal with a Christian populace which had in principle abandoned idolatry but which in practice still clung to its ubiquitous idols

[97] *MGH Ep* 2: 270. This was the pope's second letter to the same man on the same subject, see *ibid.*, 195.

[98] *Epítome hispánico* (c.598-610) 30.33: 169; Burchard of Worms, *Decretum* (1008-1012) 3.35, *PL* 140: 678-679. But in the next canon, Burchard quotes Gregory I, that pictures in the church serve to instruct the ignorant. See also the Council of Elvira 36, Vives, 8.

[99] See Anne Freeman, "Theodulf of Orleans," *Speculum* 32 (1957): 663-705 and P. Meyvaert, "The authorship of the *Libri Carolini*," *RB* 89 (1979): 29-57.

[100] *S*. 91.2, *CCSL* 23: 369; *S*. 107.2, *ibid.*, 420.

and traditional ways of life. The note of reportage is unmistakable in his *Edict of 554*:

> Because it is necessary that our authority be used to correct the common people (*plebs*) who do not observe the priests' teaching as they should, we order that this charter be sent out generally into every locality, commanding that those persons who were warned about their land and other places where statues were put up or man-made idols dedicated to a demon, and who did not immediately cast them down, or who forbade the priests to destroy them, should not be allowed to go anywhere after having provided guarantors, except for being brought into our presence. ... A report has reached us that many sacrileges occur among the population whereby God is injured and the people sink down into death through sin: night watches spent in drunkenness, obscenity and song even on the holy days of Easter, the Nativity and other feasts, or, when Sunday arrives [Saturday evening?], with dancing women [*bansactrices*] promenading through the villages. In no way do we permit the performance of any of these deeds whereby God is injured. We command that whoever dares to perpetrate these sacrileges after having been warned by the priest or by our edict shall receive a hundred lashes; but if he be a freeman or perhaps of higher status ...[101] [The rest of the text is lost.]

Here as in Turin, the idols were put up in fields and probably hidden away in storehouses. Two separate terms are used: *simulacra constructa* (effigies that have been set up or erected) and *idola daemoni dedicata ab hominibus factum* (man-made idols dedicated to demons). They suggest that the idols were of different types–one perhaps a minimally worked natural object (a pole, a rock, an animal head) and the other a more elaborate artifact.[102]

Childebert's ordinance casts a glimmer of light on the social and political process of Christianisation in Merovingian Gaul almost sixty years after Clovis and his warriors had accepted baptism at St. Remigius' hand. Despite social ambition and pressure from Church and king, some of the upper classes as well as the peasantry remained recalcitrant. Some landowners not only failed to carry out their duty to destroy the idols but even actively prevented the clergy from doing so. They may have been unwilling to rouse the peasants' resentment, or may themselves have been loyal to ancestral cults. But, though the people may have been reluctant to give up their idols, their enthusiastic participation (on their own

[101] *MGH CapRegFr* 1: 2-3. The ecclesiastical authorities at the Council of Macon (581) and the Council of Meaux (845) interpreted the latter part of this ordinance as being directed against the Jews (*Conc. Matisconense* 14, *CCSL* 148A: 226 and *Conc. Meldense* 73, *MGH Concilia* 3: 120). However, it is clear from Childebert's own text that he was concerned with the behaviour of Christian converts.

[102] In the considerably later *Vita S. Vigoris* (8th to the 11th century), Childebert authorises St. Vigor to destroy a stone idol in female form near Bayeux, despite the opposition of the local population (*AASS* Nov. I: 301-302).

terms) in Christian holy days proved that they were not opposed to Christianity. Social lines were blurred in this, too. Slaves, freemen and higher-ranking people joined in the vigils and drinking, singing, dancing and bawdiness that made genuine folk festivals of Christian feasts.

St. Columban's *Penitential* shows that around Luxeuil or Bobbio people frequented shrines (*fana*) in the second half of the 6th century. Some were still thought to be inhabited by deities or at least they still contained effigies of deities whose cult was celebrated in ritual meals. The effigies must have been of stone massive enough to resist attempts to overturn them–one thinks of the colossal statues of Jupiter dating from the Gallo-Roman period. The nuances of participation in such meals are clearly set out:

> But if any layman ate or drank in the vicinity of shrines out of ignorance, let him promise immediately never to do so again, and let him repent for forty days on bread and water; if, however, he did it for contempt after a priest preached to him that this was a sacrilege, and he communicated afterwards at the table of demons, and if he did it or repeated it only because of the vice of gluttony, let him repent for three *quadragesimae* on bread and water; if, in fact, he did this as a cult of demons or in honour of effigies, let him do penance for three years.[103]

The religious element remaining in these feasts was marginal: a Christian could be unaware that he was doing wrong to join in them, and he could disbelieve a priest's admonition that they were harmful or sacrilegious. What persisted was the continued attraction of traditional gathering places, festive dishes and (no doubt) good fellowship. The penance of three fasts of forty days implies doubt that such feasts were intended as pagan festivities. Nevertheless, the presence of effigies left open the possibility that acts of cult took place, perhaps the drinking of a toast or the pouring of a libation. But even this called for only three years of fasting, too light a penance for deliberate apostasy. This clause was included more or less in its entirety in nine other continental penitentials of the 8th and 9th century.[104]

The *Indiculus superstitionum* describes *simulacra* in Hainaut in the north as relatively small, light, portable objects. Some were made of "besprinkled flour," probably dough, some of rags, and some suitable for being carried around the

[103] *Paenitentiale Columbani* (c. 573) 24, Bieler, 104-5. Bieler's translation.

[104] In its entirety in the *Poen. Halitgari* (817-830) 42, Schmitz 1: 480; *Poen. Ps.-Theodori* (mid 9th century) 27.2, Wasserschleben, 596; *Anonymi liber poenitentialis* (9th century?) *PL* 105: 722. Three penitentials considered the cases only of those who sinned from ignorance and from cult: the *Poen. Merseburgense* (end of the 8th century) 49, *CCSL 156*: 141; *Poen. Valicellanum* I (beginning of the 9th century) 81, Schmitz 1: 305; *Poen. Vindobonense a* (late 9th century) 51, Schmitz 2: 354. The *Poen. XXXV Capitolorum* (late 8th century) 16.1, Wasserschleben, 516, and the *Poen. Martenianum* (9th century) 49.4, Wasserschleben, 292, 481, considered only the case of cult. The *Poen. Sangallense Simplex* (8th century, 2nd quarter) 26, *CCSL 156*: 121, considered only the case of ignorance. In the *Vindobonense* a, *fanassi* replaced *fana*.

fields.[105] The first two were evidently homemade and may have been used for divination,[106] as apotropaic or medicinal charms[107] or symbolic sacrifices (like the Roman *Argei*), rather than as objects of outright worship. The last is recognisably the avatar of a fertility divinity, like the Nerthus or Berecynthia whose processions around the countryside were described by Tacitus and Gregory of Tours.[108]

Unlike these, the *Ratio de catechizandis rudibus* carries little conviction in its description of effigies, despite the use of the present tense:

> Foolish, faithless and wretched men make idols for themselves with their own hands. They cast or sculpt gods for themselves in the image of man, some from gold, some from silver, some from bronze. Then they set them up and adore them. But others make themselves gods from wood and stone. Others also adore animals and worship them as gods. Still others give them the names of the dead who died badly in the midst of vices and sins, and whose souls now suffer eternal torments in hell.[109]

These are the idols and deities of the Old Testament and of Mediterranean antiquity, not of the northern territories worked by Carolingian missionaries. By contrast, when the same author had paraphrased God's commandments to Moses, he departed from the biblical text to condemn practices authenticated by other contemporary evidence: "Do not pay honour to idols ... do not use charms, do not read omens, do not make sacrifices to mountains, nor trees, nor at corners [at foundation stones?–*ad angulos*]." [110]

Papal correspondence: In papal letters, *idola* stands for statues of the gods. When writing about familiar practices in Sardinia and Sicily, Gregory I felt it unnecessary to elaborate on them or their rites. In 595, he complained to the empress Con-

[105] 26-28, *MGH CapRegFr* 1: 223. Lecouteux considers "rags" (*pannis*) to be a mistake for "bread" (*panis*), so that the clause deals with images made of bread (*Les nains et les elfes*, 178). However, it is doubtful that two titles of the *Indiculus* should concern effigies made of dough.

[106] Burchard of Worms describes a form of divination for the New Year using bread (*Decretum* [1008-1012] 19, 5.62, Schmitz 2: 423). Burchard categorises these and other methods as apostasy and idolatry.

[107] Like the anthropomorphic "idol made of dough" described by Pope Pelagius I, which were prepared in 6th-century Arles, and parts of which (eyes, ears, limbs) were distributed to the faithful, probably to ward off or heal diseases (*MGH Ep*.3,1: 445).

[108] Tacitus, *Germania* 40; Gregory of Tours, *In gloria confessorum* 76, *MGH SRM* 1: 794. When Martin of Tours met a funeral procession bearing a corpse covered in veils floating in the wind, his first thought was that it was the circumambulation of a deity, "because this was the custom of Gaulish rustics" (*Vita S. Martini* 12, *SC* 133: 278). A mother-goddess may have been carried around the fields in pagan Ireland, too, since Davidson reports that St. Brigit, whose feast coincided with Imbolc, the Irish spring festival, "was said to travel about the countryside on the feast of her festival and to bestow her blessing on the people and their animals" (*Myths and Symbols in Pagan Europe: Early Scandinavian and Celtic Religions* [Syracuse, N. Y., 1988], 39).

[109] Heer, 82.

[110] Heer, 81.

stantia that many Sardinians continued to offer sacrifices to idols "in the depraved manner of pagans" after paying a bribe to the local magistrate, without giving details. Writing to the bishop of Sardinia four years later, he recommended vigilance and harsh measures against those who venerated idols or soothsayers.[111] In his instructions the same year to the regional subdeacon concerning the misbehaviour of a priest of Reggio (Calabria), it was enough to say that Sisinnius was a "venerator and worshipper" of idols, who placed an idol in his own house. (The pope takes care to point out that he was, moreover, "tainted with the sin of homosexuality"–a "similar misdeed.")[112]

But the popes' ideas of Anglo-Saxon deities and *idola* were probably somewhat less accurate. The sacred groves, trees and springs of the northerners are notably absent from their letters. Gregory I wrote as if he believed that the Angles and Saxons thought that their gods were localised in effigies set up in elaborate shrines suitable for conversion into Christian churches. This appears from his urgent instruction to Mellitus to destroy only the *idola* in pagan shrines not the shrines themselves if they were solidly built. Once the idols were overturned, the Christian altar would take their place. The traditional sacrifices of many oxen (so similar to pagan Roman customs), followed by days of feasting, would become the mainstay of Christian rejoicing on dedicatory feasts or saints' days. Pagans would become Christians almost without noticing the transition.[113] The same concept of idols underlies his appeal to King Ethelbert of Kent: "Do not follow the cult of idols; overturn the structures used as shrines and edify your subjects by the great purity of your life and by exhortation, threats, persuasion, chastisement and good example."[114]

Boniface V's letter to Edwin King of Northumbria in 625 identifies idolatry with the worship of effigies, the cult of which entails ceremonies in shrines and rituals to obtain favourable omens:

> So we have undertaken in this letter to exhort your Majesty with all affection and deepest love, to hate idols and idol worship, to spurn their foolish shrines and the deceitful flatteries of their soothsaying, and to believe in God the Father Almighty and in his Son Jesus Christ and the Holy Spirit ... How can they have power to help anyone, when they are made from corruptible material by the hands of your own servants and subjects and, by means of such human art, you have provided them with the inanimate semblance of

[111] *Ep.* 5.38, *MGH Ep.* 1: 324; *Ep.* 9.204, *MGH Ep* 2: 192; Burchard of Worms included the latter in the *Decretum* (1008-1012) 10.3, *PL* 140: 833.

[112] *MGH Ep* 2: 238. The pope found Sisinnius' behaviour questionable in other respects as well–he was accused of having retained the property of his predecessor's children.

[113] See n. 7 above.

[114] Letter of Gregory I to Ethelbert, King of the Angles (601), Bede, *HE* 1.32: 112. Ethelbert was too cautious a prince to use fear and punishment to force conversion on his subjects, but an English prelate, the abbot Eanwulf, addressed the identical words, with rather more success, to Charlemagne (*Ep.* 87 [773], *MGH EpSel* 1: 199).

> the human form? They cannot walk unless you move them, but are like a stone fixed in one place, and, being so constructed, have no understanding, are utterly insensible, and so have no power to harm or help. We cannot understand in any way how you can be so deluded as to worship and follow those gods to whom you yourselves have given the likeness of the human form.[115]

Boniface's description of the idols, unlike Gregory I's, is specific enough to be based at least in part on precise information from Northumberland: they were manufactured locally, they were anthropomorphic, they were not made of stone (since they were compared to it), and it is possible that they either had movable parts or were carried around in procession.[116] The plural "servants and subjects" suggests that Edwin had at his disposal a significant number of skilled craftsmen, his own countrymen. The pope undoubtedly had the same type of effigy in mind when writing to Edwin's queen concerning her husband's subservience to "hateful idols."[117]

The popes seem unaware of the materials actually used by the Germans to make their effigies. Gregory II's exhortation to the Old Saxons to abandon the worship of idols made of various materials recalls the Old Testament:

> [L]et no one deceive you further by grandiose talk into seeking salvation in some sort of metal, adoring manufactured idols of gold, silver, bronze, stone or some other material. These lying *numina* that are known to be inhabited by demons were called, as it were, gods by the pagans of old.[118]

No hint here of the rags and dough of Hainaut–precious minerals were the first materials of which Gregory thought. If he thought of wood at all, he did not consider it worthy of special mention. In fact, the southerly Germanic tribes appear in general to have made their idols of wood (*e.g.*, the Saxon Irminsul destroyed by Charlemagne) or to have worshipped the tree itself (*e.g.*, the great oak at Geismar cut down by St. Boniface). Despite Adam of Bremen's glowing description of golden statues in the temple of Uppsala, Norsemen also were more likely to worship fairly primitive wooden effigies, such as those of the Volga Rus described by the Arab diplomat Ibn Fadlan (c. 922): "a long upright piece of wood that has a face like a man's and is surrounded by little figures," apparently representing the wives, daughters and sons of the main idol. The tendency of wood to rot, rather than the destructive fury of Christians (as Grimm suggested), is probably why so few Germanic idols survived except in bogs.[119]

[115] Letter of Boniface V to King Edwin of Northumbria (625), Bede, *HE* 2.10: 168-170. Colgrave and Mynors' translation.

[116] *Cf.* effigies in Hutton, *Pagan Religions*, 274.

[117] Bede, *HE* 2.11: 172.

[118] *Ep.* 21 (738-739), *MGH Ep Sel* 1: 35).

[119] In H. M. Smyser, "Ibn Fadlan's account of the Rus with some commentary and some allusions to *Beowulf*," in *Medieval and Linguistic Studies in Honour of Francis Peabody Magoun Jr.*, ed. J.

Gregory II, however, was apparently well informed on the conflict between the pagan Saxons and their Christian neighbours, the Franks. His warning against speeches promising rescue by idols illuminates the political turmoil in Saxony. The salvation sought was national survival not eternal salvation, and "grandiose talk" suggests inflammatory harangues urging the Saxons to put their trust in their native gods for help against Frankish encroachment.

2.3 IDOLATRY AND NATURAL PHENOMENA

In the passages above, "idols" (*idola*) are statues. Although rarely, the word was also used in a way that it might apply equally to the cult of natural phenomena or of deities immanent in nature. It is not always clear whether the author had in mind the natural objects, wood and stone, or the effigies made from these materials. This is the case when Gregory I wrote to the Bishop of Corsica about the reconversion of those who from neglect or need had slid back into the cult of idols, "for they must not worship wood and stones." He used the same words in a letter to the Bishop of Alexandria to describe the religion of the Angles "living at the ends of the world."[120] Material from southern Gaul and the Iberian Peninsula in the earliest part of our period show the same ambiguity.

2.3.1 *Gaul*

Idola in Caesarius of Arles' sermons on popular practices appears to mean trees and springs primarily, although it could also apply to immanent deities. One sermon upbraided the Christian converts who reconciled their old religion with their new and were prepared to fight to preserve the right to do so:

> If, dearly beloved, we rejoice indeed because we see you hasten faithfully to church, we are saddened and grieved because we know that some of you go off even more often to the ancient worship of idols, like godless pagans who lack the grace of baptism. We have heard that some of you pay their vows to trees, pray to springs ... In fact–what is even worse–there are unhappy wretches who not only do not want to destroy the shrines of pagans, but even do not fear nor blush to rebuild what was destroyed. And if someone who is mindful of God wants to burn sacred trees [*arbores fanatici*] or scatter and destroy diabolical altars, they go mad with rage and are overcome by great frenzy, so that they even dare to strike those who tried to overturn the sacrilegious idols for love of God, or perhaps meditate their death ... And why do such wretches come to church? and why did they accept the sacrament of baptism, if afterwards they are to return to the sacrilege of idols?[121]

Bessinger and R. P. Creed, (London, 1965), 92-110; here 97. See also Davidson, *Myths and Symbols,* 22-23 and Grimm, *Teutonic Mythology*, 112.

[120] 8.1 (597), *MGH Ep* 2: 1; 8.29 (527), *ibid.*, 30.

[121] *S.* 53.1, *CCSL* 103: 233-234. See also Burchard of Würzburg (8th century) *Hom.* 23, Eckhart, 842-843.

This is an ambiguous passage. It may be a matter of style that people are said to run (*currere*) to church but walk (*ambulare*) to the sites of the old idolatry, or it may imply that they felt some hesitation about the latter. A variety of terms describes the attacks on sacred objects: "destroy," "burn," "scatter, " "overturn." One verb each is enough for shrines, trees and idols, but two are needed for altars. Were there two sorts of altars–one a massive object, perhaps made of stone and hard to destroy, the other an easily-shattered composite? The cult of trees is plainly more important than that of springs, but were the trees the idols themselves or the shrines of the idols?

2.3.2 *Iberia*[122]

Iberian documents applied terms such "cult of idols" or "cult of the devil" without distinction to the cult of nature and the worship of anthropomorphic deities or their man-made representations. The Council of Braga (572), presided over by Martin of Braga, required bishops to teach the faithful "to flee the errors of idols, or other crimes."[123] These are probably the practices described in Martin's sermon as the cult of the devil: lighting candles to stones, trees and wells and by crossroads, celebrating the festivals of the pagan calendar, divining, invoking Minerva and observing lucky days.[124]

Martin hoped to eliminate such customs from Suevan Galicia with persuasion and education; neither he nor his council advocated the use of force. By contrast, Reccared's great unifying council, Toledo III (589), adopted aggressively coercive policies when faced with wide-spread idolatry "almost everywhere" in Visigothic Spain and Gaul.[125] Bishops and magistrates were ordered to destroy idols, and masters to forbid their slaves to practice idolatry. The council spelled out sanctions:

> That bishops and judges together are to destroy idols and masters to forbid their slaves to practice idolatry. Since the sacrilege of idolatry has developed throughout almost all of Spain and Gaul, the holy synod has decreed with

[122] Stephen McKenna's *Paganism and Pagan Survivals in Spain up to the Fall of the Visigothic Kingdom.* (Washington, 1938) is still the only thorough study of the evidence for paganism and superstition in Visigothic Spain. A general picture of peasant religion and culture is given in Salisbury, *Iberian Popular Religion.* See also Jocelyn N. Hillgarth, "Popular religion in Visigothic Spain," in Edward James, ed., *Visigothic Spain: New Approaches* (Oxford, 1980), 3-60.

[123] *Conc. Bracanense II*, 1, Vives, 81.

[124] *De correctione rusticorum* (c. 572) 16, Barlow, 198. This was repeated in the *Dicta Pirmini* (724-753) 22: 172-176 and in two anonymous sermons to the baptised (10th or 11th century and 12th century), Caspari, *Kirchenhistorische Anecdota* 1, 199-201, 204-208. The earlier of these sermons replaced "cult of the devil" with "works of the devil."

[125] Manuel Sotomayor observed that paganism in the Visigothic state was a political rather than a pastoral concern ("Penetracion de la Iglesia en los medios rurales de la España tardoromana y visigoda," in *Cristianizzazione ed organizzazione*, 639-670; here, 666).

> the consent of the most glorious prince that every priest, together with the district judge, is to examine carefully the sacrilege reported in his region, and to expel any that they find without delay. Moreover, they are to use force on all those implicated in such error by whatever punitive measures they can, always excepting the death penalty. If they fail to do this, both should know that they are subject to the danger of excommunication. Moreover, if any masters fail to uproot this evil from their estates and are unwilling to prohibit their households, they are to be banished by the bishop from the communion [of the faithful].[126]

The situation here was similar to that of Gaul a generation earlier. Idols were everywhere, their cults were practiced by the lowest in society, and the Visigothic lords, like their Merovingian counterparts, were either unable or unwilling to take energetic measures against them, be it from indifference, policy or sympathy for the old religions. The word "developed" indicates that the idolatry in question was a newly intensified development not just a remnant of past beliefs in the process of dying out, as it had been in Childebert's realm. Christianity had long been established in Spain, but ancient cults may well have experienced a resurgence during the long years of struggle between Arians and Catholics.

A picture of persistent and polymorphous Visigothic idolatry emerges from subsequent councils. In calling for new efforts, Toledo XII, summoned by Erwig in 681, and Toledo XVI, by Egica in 693, listed different kinds of "worshippers of idols," identified idols with demons and hinted at rituals and prestigious figures rivalling those authorised by the Church. They depict a heterogeneous society in which certain families or ethnic groups clung obstinately to traditional practices in defiance of king and Church, and provide a glimpse of the realities of power in late Visigothic society.

Toledo XII laid out a detailed programme for the guidance of officials and landowners, reconfirmed in 683 by the next Council. It admonished those who

> worship idols, venerate stones, light torches and honour sacred springs or trees, that they should know that they who are seen to sacrifice to the devil subject themselves to unforeseen death ... And, accordingly, as soon as they find them, the priests and civil authorities are to uproot all sacrilege of idolatry and the things against holy faith to which foolish men, entrapped by diabolic cults, devote themselves, and remove and destroy them. Moreover, they are to restrain with blows those who assemble for such vileness and hand them over, loaded down with iron, to their masters if, at least, their masters promise under oath to guard them so vigilantly that they will be unable to practice further wickedness. If their masters are unwilling to keep

[126] 16, Vives, 129-130. The *Edict of Reccared*, incorporated in the documents of this Council, ordered the religious and civil authorities tersely to seek out and destroy the cult of idolatry (16, *ibid.*, 135). The *Epítome hispánico* (c.598-610) contains merely the title (34.16: 178). On Reccared's motivation, see Biagio Saitta, "La conversione di Recaredo: Necessità politica o convinzione personale," in *Concilio III de Toledo. XIV centenario* (Toledo, 1991), 375-384.

> guilty persons of this sort in their charge, they are then to be brought before the king by those who had punished them, so that the prince's authority may exercise its free power to dispose of them. Nevertheless, let their masters, who have delayed in punishing the proclaimed faults of such slaves, be subject to the sentence of excommunication; let them also be aware that they have lost their power over the slave whom they refused to correct. If free-born persons are implicated in these faults, they are both to suffer the sentence of perpetual excommunication and to be punished with a particularly stringent exile. [127]

Such harsh measures were not successful. A few years later, Egica instructed the assembled ecclesiastic and civil authorities to destroy "idolatry or, what is the same thing [*vel*], the various errors of diabolic superstition" and to take the offerings made to idols "by peasants and any other persons" to the neighbouring churches.[128] The council obediently went to work. It attributed the persistence of idolatrous cults to proselytisation on the part of the devil who, working through his agents, used "malign persuasion" to entrap foolish folk into worshipping idols, stones, sacred springs and trees, lighting candles and becoming soothsayers and enchanters (*auguratores, praecantatores*) "and many other things too long to relate." Unlike Toledo XII, it put the entire burden of eradicating idolatry on the civil and religious authorities. They were to correct the culprits promptly, without regard to descent or rank. In punitive harshness, the council went beyond Egica's instructions, ordering that the confiscated offerings be exposed in the churches before the eyes of those who had meant them to be lasting gifts to their idols.[129]

These measures were necessitated by the continued inability of the Visigothic church and state to command the unconditional support of any social caste. The majority who fell into or persisted in idolatry may have been unfree peasants, but not all. Toledo XII, which still hoped to leave the responsibility for chastising backsliding slaves to their masters, had been forced to recognise that some could not be counted on to do so; Toledo XVI did not even make the attempt. Caught between the religious zeal of king and Church on the one hand and the recalcitrance of their peasants on the other, the landowners were unable to satisfy either. Toledo XVI shows that idolatry was far from moribund. Its practitioners could still marshal convincing arguments in its favour and it could still attract new recruits. The reference to descent and rank demonstrates that as the 7th century drew to a close, participation in idolatry, or at least sympathy and tolerance for it, persisted even among the elite of Visigothic society.

2.4 THE DEVIL AND DEMONS

Most texts use *idolum, diabolus* and *daemones* interchangeably. The worship of

[127] *Conc. Toletanum* XII (681) 11, Vives, 398-9. See also *Conc. Toletanum* XIII (683) 9, *ibid.*, 426.

[128] *Conc. Toletanum* XVI (693), Egica, *Charge to the Bishops* (693) Vives 485-486.

[129] *Conc. Toletanum* XVI (693) 2, Vives, 498-499.

idols was the worship of the devil or of demons, acts of idolatry were labelled diabolic or *daemonia* and sacrifices and offerings were made indiscriminately to the devil, demons and idols. Nevertheless, these terms are not always equivalent. In Iberia, the devil had a special function and, while idols were invariably passive objects of devotion, demons were presented in continental penitentials from the 8th century onward as responding actively to their votaries.

2.4.1 *The devil*

The dualist belief in the devil as a creative force, the principle of evil, not merely a fallen angel, appears to have remained deeply rooted in Spanish soil despite Priscillian's execution in 385. A 6th-century Galician council pronounced an anathema against those who believed that, "as Priscillian said," the devil had had some share in the creation of the world, or that he made bad weather (thunder, lightning, storms and droughts) on his own authority. Even four centuries later, a Spanish penitential anathematised those who believed that the devil controlled the weather.[130] It also cited the Priscillianist belief that the devil had never been a good angel created by God but had emerged from chaos.[131]

2.4.2 *Demons*

Demons and magic: Success in magic was sometimes ascribed to the intervention of demons. Sooth-saying (*divinationes*), amulets (*philacteria, characteres, ligaturae*) and enchantments or spells (*incantationes, carmina*) were routinely described as diabolical. Undoubtedly soothsayers and enchanters protested against the insult: the author of a Frankish penitential was obliged to insist that the divinations which the soothsayer believed that he performed by human skill were indeed the work of demons.[132] In some cases, however, the form of words used to describe a magical practice implies that the practitioners themselves thought that they achieved their results with the help of demons.

Fortune-telling and clairvoyance in particular were thought to be their province, and so were certain forms of black magic. Over a dozen continental penitentials from the early 8th to the 10th century warned that demons were the source of the divinations of *arioli*.[133] For Rabanus Maurus (d. 856), all the doings

[130] *Conc. Bracanense I* (561) 8, Vives, 68. See also *Collectio Hispana*, *Excerpta Canonum* 8. 6, *PL* 84: 83; *Decretales Pseudo-Isidorianae*, ed. Hinschius, 421-422; *Poen. Silense* (1060-1065) 207, *CCSL 156*A: 36. For Priscillian and his teachings, see Henry Chadwick, *Priscillian of Avila. The Occult and the Charismatic in the Early Christian Church* (Oxford, 1976). Priscillianism in Spanish popular religion is discussed in detail in Salisbury, *Iberian Popular Religion,* 191-226. See also Adhémar d'Alès, *Priscillien et l'Espagne chrétienne à la fin du IVe siècle* (Paris, 1936) and E.-Ch. Babut, *Priscillien et le priscillianisme* (Paris, 1909).

[131] *Poen. Silense* (1060-1065) 202-206, *CCSL 156*A: 36-37.

[132] *Poen. Oxoniense* I (8th century, 1st half) 21, *CCSL 156:* 90.

[133] *E.g., Poen. Remense* (early 8th century) 9.5, Asbach, 56.

of soothsayers were tainted by their association with demons.[134] In the late 8th-century *Homilia de sacrilegiis,* they gave answers through "seers or seeresses, that is, pythonesses."[135] People offered sacrifices to demons at tombs, funeral pyres and other places to gain knowledge of the future. Some, including priests and bishops, did so voluntarily, others, under force or intimidation.[136] Still others consulted an "invoker of demons" who apparently did not belong to the usual varieties of seer nor depend upon omens. However, the evidence of a late Spanish penitential suggests that professional help was not always necessary for sacrificing to demons or consulting them.[137]

Some people invoked "demons" when starting upon a piece of work, such as gathering healing herbs. This pertained to *ars magica*, wrote Regino of Prüm and Burchard of Worms, and they reminded the faithful that only the Creed and the Lord's Prayer were permitted.[138]

An 8th-century penitential states the general case for the intervention of demons in malign magic: "If anyone harms men by invoking demons, let him do five years of penance."[139] Others are specific. A score of continental penitentials and canonic collections from the 8th to the 11th century testify to the conviction that enchanters (usually *mathematici*) called on demons to disturb minds.[140] In Burchard of Worms' penitential, self-proclaimed weather magicians were believed to invoke demons in order to raise storms or alter men's minds.[141] Burchard prescribed only one year of penance, but the usual was five years, showing how seriously the clergy took this kind of magic.

Characteristics of demons: It was known that the spirits' skill in divination was due to their flitting through the air, which enabled them to see coming events and to report them to soothsayers.[142] In St. Eligius' sermon, "silly people" believed that they launched invasions from the moon.[143] Another contemporary sermon affirmed that they flew about in the air and had a thousand ways of doing harm.[144] Demons feared iron–Frankish demoniacs in the late 8th century wore iron rings

[134] *Hom.* 45, *PL* 110: 83.

[135] 5, Caspari, 6-7.

[136] *Poen. Arundel* (10th/11th century) 88, Schmitz 1: 461.

[137] Regino of Prüm, *De synodalibus causis* (c. 906) 2.361: 351; Burchard of Worms, Decretum (1008-1012) 10.30, *PL* 140: 837; *Poen. Silense* (1060-1065) 197, *CCSL* 156A: 36.

[138] Regino of Prüm, *De synodalibus causis* (c. 906) II, 5.52: 213; Burchard of Worms, *Decretum* (1008-1012) 1.94, *Interrogatio* 51, *PL* 140: 577.

[139] *Poen. Sangallense tripartitum* (8th century, 2nd half) 22, Schmitz 2: 181.

[140] *E.g., Poen. Remense* (early 8th century) 9.7, Asbach, 57.

[141] *Decretum* (1008-1012) 19, 5.68, Schmitz 2: 425). See also *idem,* 10.28, *PL* 140: 837.

[142] Ps.-Eligius, *De rectitudine catholicae conversationis.* 9, *PL* 40: 1175.

[143] *Vita Eligii* (c.700-725), *MGH SRM* 4: 707.

[144] Levison, 312. This notion is drawn from St. Augustine's *De divinatione daemonorum, CSEL* 41, 597-618.

and bracelets and put iron objects in their houses *ut demones timeant,* to scare demons away. They also feared the Sign of the Cross and were reputed to love blood.[145]

Finally, a phrase in Burchard of Worms' *Decretum* may suggest that demons were thought to participate in the Wild Hunt. In some passages, "hordes of demons in women's likeness" were believed to ride through the sky on the backs of beasts at fixed times. Another passage has women believing that together with other "limbs of the devil" they pass through locked gates to rise through the air to fight battles in the clouds. But "demons" is Burchard's word. In the popular mind the members of the throng were surely not demons or even ghosts, but living, flesh-and-blood women.[146]

2.5 THE CHRISTIAN TRADITION

In his analysis of the development and role of the cult of saints, Peter Brown showed the untenability of its equation with the cult of pagan gods.[147] Devotion to the saints and to angels, far from being condemned as a survival of paganism, became an integral part of medieval Christianity, encouraged by the hierarchy and embraced by the faithful. Churchmen, however, were anxious to keep control of the cults (objects, location, rituals) for fear of heterodox beliefs and syncretic practices. Caesarius of Arles deplored the custom (probably introduced by the Gothic masters of the Narbonnaise) of prolonging drinking bouts by toasts "in the names not only of living men but of angels and of other saints of antiquity as well" under the pretext of showing them honour.[148] The clergy deplored that the popular cult of St. Martin was celebrated with particularly improper carryings-on in 6th-century Auxerre: to a general condemnation of vigils for saints' feasts, a local Synod added a pressing injunction to the clergy to ban the vigils "observed in honour of the lord Martin." No doubt as the last great Christian feast before winter closed in, Martinmas (11 November) was the occasion of extravagant celebrations.[149] The

[145] *Homilia de sacrilegiis* 22, Caspari, 12; Martin of Braga, *De correction rusticorum* (572) 8, Barlow, 188; Isidore of Seville, *Etymologiae* VIII, 9.11: 714.

[146] Regino of Prüm, *De synodalibus causis* (c. 906) 2, 5.45: 212, and 2: 371: 354-356. See also Burchard of Worms, *Decretum* (1008-1012) 1.94, Interrogatio 44, *PL* 140: 576; *ibid.,* 10.1: 831-833 and 10.29: 837; *idem,* 19, 5.70, Schmitz 2: 425, and 5.90: 429. See chapter 7, section 2.1.

[147] *The Cult of the Saints. Its Rise and Function in Latin Christianity* (Chicago, 1981). For the older view that the saints were christianised versions of the gods, see Pierre Saintyves, *Les saints, successeurs aux dieux* (Paris, 1907). For saints as the heirs of classical heroes, see Jean-Claude Fredouille, "Le héros et le saint," in *Du héros païen au saint chrétien*, ed. Gérard Freyburger and Laurent Pernot (Paris, 1997) 11-25. See also Jean-Claude Schmitt's introduction to *Les saints et les stars. Le texte hagiographique dans la culture populaire. Études présentées à la Société d'ethnologie française, Musée des arts et traditions populaire* (Paris, 1979), 5-19, and Hippolyte Delehaye, *Les légendes hagiographiques* (Brussels, 1955), 140-201.

[148] S. 47.5, *CCSL* 103, 214.

[149] Syn. Autissiodorensis (561-605) 3 and 5, *CCSL* 148A: 265. For the cult of St. Martin, see

Epítome hispánico (c. 598-610) warned against praying to false martyrs and slighting the Church by adoring quasi-angels.[150]

This is not much in view of the well-documented tendency of the people to make saints for themselves. The spontaneous development of cults honouring the dead who were esteemed for their holiness, their power to perform miracles, or for some other reason, was a frequent medieval phenomenon.[151] Sometimes the cult was accepted by the Church or its local representatives, and its object was recognised as an authentic saint.[152] Occasionally, the cult remained suspect.[153] In neither case did it appear to be of more than local concern.

From the mid 8th century on, however, mentions of illicit cults of angels, saints, martyrs and confessors proliferated in continental sources from the Rhineland in the north to Piedmont in the south. Some Frankish condemnations were a response to rituals borrowed from traditional culture. Texts from Thuringia and Lombardy deal with questionable rituals or beliefs surrounding the figure of St. Michael the Archangel, whose cult was particularly popular among the Lombards. More numerous in continental documents are the references to beliefs in spurious angels and saints; these seem to have been prompted in the first instance by the activities of popular preachers, dismissed by the hierarchy as *pseudoprophetae*.

2.5.1 *Anomalous rituals in honour of the saints*

In the newly Christianised territories which the Pippinids were in the process of bringing under their rule, the rituals honouring saints owed at times more to the pagan past than to the Christian present. A sweeping Austrasian ordinance of 742

Walter Nigg, *Martin de Tour,* trans. Jacques Potin (Paris, 1978); Philippe Walter, *Mythologie chrétienne. Rites et mythes du moyen âge* (Paris, 1992), 65-80; Dominique-Marie Dauzet, *Saint Martin de Tours* (Paris, 1996), 267-310. E. Ewig discusses the popularity of St. Martin in the early Middle Ages as attested by the proliferation of churches dedicated to him and monasteries under his protection, in "Le culte de saint Martin à l'époque franque," in *Mémorial de l'année martinienne M.DCCCC.LX-M.DCCCC.LXI* (Paris, 1962), 1-18.

[150] 7.33: 122; 7.34: 122.

[151] *E.g.*, the cult of an unknown young woman named Criscentia developed in Paris because of the miracles occurring at her tomb (Gregory of Tours, *In gloria confessorum* 103, *MGH SRM* 1: 813-814). In the 12th century, Gilbert of Nogent related the story of a quite insignificant youth (*puer vulgaris*) who was honoured as a saint because he happened to have died on Good Friday (*De sanctis et eorum pignoribus*, *CCCM* 127: 97). For other such saints (either accepted or rejected by the hierarchy), see Pierre Delooz, "Toward a sociological study of canonised sainthood" in Stuart Wilson, ed., *Saints and Their Cults* (Cambridge and New York, 1983), 189-216, especially 199-201.

[152] Gregory of Tours recounts the story of a countryman whom two virgins instructed in a dream to build an oratory over their abandoned tombs; the then bishop of Tours eventually identified them as the hitherto unknown saints Maura and Britta (*In gloria confessorum* 18, *MGH SRM* 1: 757-758).

[153] St. Martin of Tours unmasked a supposed martyr as a brigand put to death for his crimes (*Vita Martini* 11, *SC* 133: 276).

against paganism and pagan customs (*paganiae, spurcitiae gentilitatis, ritus paganus, paganorum observationes*), restated in Charlemagne's *Capitulary* of 769, reveals that martyrs and confessors received sacrifices appropriate to gods or pagan heroes:

> We have decreed that each bishop, with the help of the *gravio*, who is the defender of the Church, should take care according to the canons that the people of God in his diocese do not perform pagan acts but cast off and spurn every filth of paganism, and that they should forbid sacrifices of the dead or sorcerers or soothsayers or amulets or omens or enchantments or the sacrificial victims which stupid men offer in the name of the blessed martyrs or confessors in the vicinity of churches, provoking God and his saints to anger, or those sacrilegious fires which they call *nied fyr*, and all those who love pagan observances.[154]

In addition, saints' feasts were occasions for drinking bouts. Charlemagne pressed his representatives to forbid the "evil of drunkenness and the societies to which they swear by St. Stephen or by ourselves or our children." St. Stephen's Day is celebrated at Yuletide (26th December); it is probable therefore that seasonal rites in honour of a Germanic deity were transferred to him.[155] The practice of drinking toasts in honour of the saints during commemorations of the dead was noted in the 9th century for the first time, as a feature of clerical assemblies.[156]

2.5.2 *St. Michael the Archangel*[157]

Although it is often stated that St. Michael's popularity in the early Middle Ages was due to his identification with Mercury or some other deity, the first reference to an improper cult to him appeared only in the 10th century. The Synod of Erfurt (932) reprimanded those who "irrationally sang and heard" (that

[154] *Conc. in Austrasia habitum* q. d. Germanicum (742) 5, *MGH Concilia* 2.1: 1, 3-4); *Karlomanni principis capitulare* (742) 5, *MGH CapRegFr* 1: 25. See also Charlemagne's *Capitulare primum* (769 vel paullo post), *ibid.*, 45.

[155] *Duplex legationis edictum* (789) 26, *MGH CapRegFr* 1: 64. Bächtold-Stäubli identifies this as the first recorded instance of a *Minnetrunk* (*s.v.v.* "Stephansminne" and "Karlsminne"), but Caesarius of Arles's description of forced drinking "in the names of angels and saints" seems to refer to the same custom.

[156] *Admonitio synodalis* (c. 813) 39bis, 51. See also Hincmar of Reims (852) *Capitulary* 1.14, *MGH CapEp* 2: 41; Burchard of Worms, *Decretum* (1008-1012) 2.161, *PL* 140: 652.

[157] The cult of St. Michael originated in the East. It became extremely popular in the Latin West, first in the 5th century and then especially during the 8th and 9th centuries. In the earliest Middle Ages, it was limited in the West to the areas with the most contact with the East (Italy, particularly among the Lombards, and southern France). Later it was introduced on the continent by Irish and Anglo-Saxon missionaries, St. Boniface himself actively promoting it. St. Michael was identified with Mercury by Thomas Barnes ("Michaelmas," *ERE* 8, 619-623) and with Jupiter in his role of conqueror of the titans by P. Saintyves in *Les saints, successeurs aux dieux*, 351. Olga Rojdestvensky, whose *Le culte de saint Michel et le moyen âge latin* (Paris, 1922) is the most detailed study available of this cult, disputes the traditional identification of St. Michael with either the Gallic or the German Mercury. See also Walter, *Mythologie chrétienne*, 243-253; Les

is, priests and laity) victory masses in honour of St. Michael (under whose banner they had gone into battle against the Hungarians). Among the innovations was lighting candles placed on the ground in the form of the cross.[158] In Lombardy, where he was very popular, the second day of the week was thought to be sacred to him. Going to mass then was considered particularly advantageous since he was thought to celebrate the mass himself that day. Around 963-968, Ratherius of Verona denounced this as folly,[159] explaining that there were neither days nor nights in heaven, and that St. Michael lacked the body parts (arms, hands, lips, lungs, teeth, *etc.*) essential for singing mass. This sermon led to misunderstandings and Ratherius took pains to make himself clear in a subsequent sermon:

> Bishop Ratherius does not say that those who go to St. Michael's church or who hear his mass do wrong, but Bishop Ratherius does say that he is a liar, who says that it is better to go to St. Michael's church or to hear his mass on Monday rather than on some other day. Bishop Ratherius says that he is a liar, who says that St. Michael sings the mass, because no creature except man alone can sing mass .... If you want to go to St. Michael's church, know, says Bishop Ratherius, that St. Michael does not welcome you on Monday more gladly than on another day. If you want to ask anything of him, know, says Bishop Ratherius, that St. Michael does not listen to you any more kindly on Monday than he would on Sunday, Tuesday, Wednesday, Thursday, Friday, Saturday.[160]

2.5.3 *Pseudoprophetae, spurious saints and angels, letters from heaven*

It is stretching the definition of idolatry to include under this rubric the cult of certain men who were venerated as prophets or saints, and who claimed to have direct contact with the word of God and His angels. They are viewed more correctly as the heretics and schismatics which contemporary authorities considered them. Nevertheless, elements in their doctrines, the places that they chose for their assemblies and the evocation of mysterious names appealed to modes of religiosity rooted in the pre-Christian past.

The concern of church and lay authorities about dubious claimants to sanctity was sparked by the activities of a certain Aldebert *natione generis Gallus*, and one Clement *genere Scottus* (both, therefore, presumably Celtic in origin, despite the

Pères Bénédictines de Paris, *Vies des saints et des bienheureux* (Paris, 1956) 9: 601-609; Michel Rouche, "Le combat des saints anges et des démons: La victoire de saint Michel," together with Pthe discussion following, in *Santi e demoni nell'alto medioevo occidentale (Secoli V-XI)* SSAM 36 (Spoleto 1989): 533-571.

[158] *Breviarum Canonum* 6, *MGH Concilia* 6.1: 112.

[159] *S. de Quadragesima* 35, *CCCM* 46: 83. For the cult of St. Michael among the Lombards, see Gasparri, *La cultura tradizionale dei Longobardi,* 155-161.

[160] *S. contra reprehensores sermonis eiusdem* 2-3, *CCCM* 46, 93-94.

former's Germanic name),[161] who enjoyed a considerable following in the Rhineland toward the mid 8th century. The ecclesiastical hierarchy labelled them pseudoprophets and heretics since they propounded unorthodox dogmas and laid claim to miraculous powers and private revelation from angels, saints and the word of God, the traditional sources of Christianity.[162] Thanks to St. Boniface's representations to the Council of Rome and Pope Zacharias' correspondence with him,[163] we know a good deal about their case, which came to a head in 744-745.[164]

Of Clement, the chief complaint (in the pope's opinion) was that, though a bishop, he was "so given to lechery that he had a concubine and had fathered two children on her"–hardly unusual among the lax clergy of the time. Nevertheless, Boniface had more serious charges against him. He accused Clement of rejecting the teachings of the Fathers of the Church, of defying the authority of the synods, of Judaising by permitting levirate marriage, and (most significantly perhaps) of teaching that Christ on his descent to the netherworld "freed all those who were imprisoned in hell, believers and unbelievers, adorers of God and worshippers of idols alike" and "many other horrible things concerning divine predestination."[165]

[161] The adoption of Germanic names by Gauls was not uncommon, but Hefele-Leclercq (*Histoire des conciles* 3: 875) and G. Kurth (*Vie de St. Boniface* [Paris, 1924], 89) considered Aldebert to be Frankish. Boniface may have called him a Gaul rather than a Frank out of a tactful regard for the feelings of his Frankish patrons, but he was usually careful about ethnic labels.

[162] Both Martin of Tours and Gregory of Tours were familiar with this type. Anatolius, a young hermit who established himself near Marmoutier, believed that he received visits and messages from angels, as well as a cloak from heaven (*Vita S. Martini* 23, *SC* 133: 302-306). In 587, a certain Desiderius claimed to have had communications from saints Peter and Paul, and performed healing miracles with the aid of crosses, ampulae filled with holy oil and a bagful of charms composed of various roots, mole teeth and mouse bones; Gregory suspected him of using the "black arts" (*HF* 9.6, 25, *MGH SRM* 1.1: 417-420). Another prophet wandered, dressed in skins, about the province of Arles in 591 during the outbreak of the plague, and then moved into Languedoc, performing healing miracles; he was accompanied by a woman called Maria, and by an unseemly band of followers (*HF* 10. 25, *ibid.,* 517-519). The episodes described by Gregory of Tours are analysed by Caterina Lavarra in terms of the religious dislocation and various natural and man-induced disasters suffered by the peasantry of 6th-century Gaul , in "'Pseudochristi' e 'pseudoprophetae' nella Gallia merovingia," *Quaderni medievali* 13 (1982): 6-43.

[163] The principal source of information is the *Concilium Romanum* (745), *MGH Conc.* 2.1: 37-44; for additional information, see in particular the pope's *Ep.* 57 (744) *MGH EpSel* 1: 104-105.

[164] The Council of Soissons (744) had made an attempt to deal with Aldebert by ordering that the crosses he had set up throughout the countryside be burned (7, *MGH Concilia* 2.1: 35). The clash between Boniface and these two trouble-makers is described in Lavarra, "'Pseudochristi' e "pseudoprophetae'," and in J. B. Russell, "Saint Boniface and the eccentrics," *Church History* 33 (1964): 235-247. The latter article is particularly useful for its analysis of the sources of some of Aldebert's beliefs. For a brief summary of the affair, see J. Laux, "Two early medieval heretics: An episode in the life of St. Boniface," *Catholic Historical Review* 21 (1935): 190-195. Russell set Aldebert and Clement's activities in the context of other outbursts of religious eccentricity in *Dissent and Reform in the Early Middle Ages* (Berkeley, 1965), 101-124, especially 101-107. See also Gurevich's illuminating remarks in *Medieval Popular Culture,* 62-73.

[165] *Conc. Romanum*, *MGH Concilia* 2.1: 40.

The Irishman Clement may in fact have shared in his own people's unwillingness to consign their ancient heroes to eternal damnation–in the Fenian Cycle, St. Patrick himself had rescued Finn McCool from hell, and was prepared to grant salvation to the pagan bards.[166] But Clement may also have been responding to the anxieties of newly converted Christians about the fate of their unbaptised kin. The Frisian king Radbod had refused baptism for this very reason: he did not wish to enter a heaven from which his pagan ancestors were barred.[167]

Aldebert's case was more complex and worrisome. A man of considerable and conspicuous asceticism, he claimed that his mother had had a prophetic dream before his birth, of a calf emerging from her right side, to foreshadow the grace with which he would be blessed.[168] He claimed also to have been visited as a child by an angel in human shape coming from the ends of the world, who brought him relics of "dubious holiness;" as a result, he could obtain from God anything for which he asked. Crowds of peasants accepted him as a man of apostolic sanctity and believed that he had performed many signs and marvels. His following evidently included members of the hierarchy, for he managed to have himself ordained by certain ignorant bishops, in violation of canon law.

At length, he began to fancy himself the apostles' equal. He dedicated chapels to himself instead of the apostles and saints, and questioned the worth of pilgrimages to Rome. He set up little crosses and shrines in the fields and at wells or springs and elsewhere, for the public celebration of mass; crowds, claiming that they would be helped by "St. Aldebert's" merits, flocked hither in neglect of their own bishops and churches. He distributed clippings of his nails and hair to be honoured and worn together with St. Peter's relics. Worst of all, when the people prostrated themselves at his feet to confess their sins, he sent them home absolved, assuring them that there was no need to confess since he already knew their secrets.[169]

Aldebert also claimed access to a letter fallen from heaven, which had been written by Jesus Christ Himself.[170] It had been found in Jerusalem by St. Michael the Archangel, then transmitted from hand to hand until it ended up by way of Mont St. Michel in Rome. The Pope and assembled Fathers were disposed to shrug this off as nonsense that only lunatics or silly women could believe, but they took a final charge against Aldebert more seriously, that he had composed a prayer ending with a sensational invocation of the names of eight "angels," to whom he prayed and whom he besought and–a strong word, not usual in prayers–con-

[166] *VSH* 1, cxxxi-cxxxii.

[167] *Vita Vulframmi* 9, *MGH SRM* 5; 668. Radbod's rejection of salvation was based not only on loyalty to his dead ancestors but on snobbery as well: a disinclination to fraternise with the poor in the kingdom of heaven.

[168] *Conc. Romanum*, *MGH Concilia* 2.1: 41.

[169] *Conc. Romanum*, *MGH Concilia* 2.1: 39-40.

[170] See Russell, "Saint Boniface and the eccentrics," 217, n. 45. A letter from heaven is mentioned by Licinian of Carthagena in the late 6th century (*PL* 72: 699-700).

jured.[171] The Council agreed as one man that the prayer should be burnt and the author anathematised: the names (except for Michael's) were not of angels but of demons and Aldebert, therefore, had called on demons for help, slipping them in under the guise of angels.[172]

These were assaults on several fronts against the Church that St. Boniface was attempting to reform or establish in the newly annexed Frankish possessions. He was struck by Clement's defiance of the Church's teaching authority and by Aldebert's arrogation of the Church's monopoly of the means of salvation: the right to control the sites and objects of devotion, administer the sacraments, validate relics, define dogma. To him, they were schismatics and the worst of heretics; to the pope, they were pseudoprophets or pseudochristians, and Aldebert, in particular, a madman. There are no accusations of paganism nor of sorcery.

Nonetheless, Boniface must have had suspicions that he did not care to express in direct formal charges against a bishop and an ordained priest. The *Indiculus superstitionum et paganiarum* was composed under his influence at the very period when Aldebert and Clement were becoming serious problems. The items that may be qualified as superstition rather than paganism cover issues raised at the Council of Rome: title 4 concerns "little huts, that is, shrines;" 9, "the sacrifice which is done for some saint;" 10, "amulets and ligatures;" 11, "the springs of sacrifices;" 18, "the doubtful places which they honour as holy" or "for holy purposes;"; and 25, "the fact that they make some of the dead into saints for themselves." These are not far removed from Aldebert's and Clement's teachings and practices.[173]

In the attachment to the dead, the mysterious charms that compelled God, the signs, prodigies and presages, the sites chosen for religious ritual, the amulets containing bits of nails and hair, the clairvoyance as to people's secret sins and, finally, the invocation and *conjuratio* of suspect spirits, there are elements which owe as much to the ancient patterns of belief and practice as to the new. So it was seen in a mid 9th-century penitential in a heading which bracketed the cult of angels with idolatry, superstition, sorcerers, soothsayers and forbidden practices.[174] Both this and the *Vetus Gallica* followed the Council of Laodicea (363) in proclaiming that those who abandoned the Church and went off to "name angels and set up congregations" (as had Aldebert) were engaged in secret idolatry.[175]

Although Aldebert vanished from history, among the people some memory of his teachings may have lingered well into the Carolingian era. The question of religious services in unsuitable surroundings continued to find a place in capitu-

[171] *MGH Concilia* 2.1: 42.

[172] *Conc. Romanum* 6, *MGH Concilia* 2.1: 43.

[173] *Indiculus superstitionum et paganiarum* (743?), *MGH CapRegFr* 1: 223.

[174] *Poen. Ps.-Theodori* (mid 9th century) 27, Wasserschleben, 595-596.

[175] *Poen. Ps.-Theodori* (mid 9th century) 27.7, Wasserschleben, 596. See also the *Collectio Vetus Gallica* (8th/9th century) 44.4c, Mordek, 524, and Council of Laodicea (380?) 35, Hefele-Leclercq, *Histoire des conciles* 1.2: 1017-1018.

laries and canonic collections. Prohibitions were repeated against the invention or use of names of unknown angels or of saints and martyrs of dubious memory (or honoured in memorial shrines of dubious authenticity); only Michael, Gabriel and Raphael were acceptable. Probably Aldebert's heresy lies behind some of these, although the authority cited was not the Council of Rome but of Laodicea which, however, had forbidden the cult of all angels.[176] The Council of Frankfurt forbade the erection of memorial shrines to new saints by roadsides, insisting that saints, chosen for their sufferings and virtuous life, be honoured only in church. Burchard of Worms' *Decretum* includes a 5th-century Carthaginian canon urging bishops to destroy, if possible, such structures beside fields and roads unless they were proven to contain the bodies or authentic relics of martyrs. If prevented by popular outrage, they were to warn their flocks against such places and dubious saints. To be reproved in every way were "the altars set up because of dreams or inane so-called revelations." This ancient ruling may have seemed relevant once more.[177] Finally, a very late recension of the council of Riesbach of 800 contained a clause that memorial shrines should not be built for unknown saints and martyrs.[178]

The letter from heaven also reappeared.[179] In his *Admonitio Generalis* of 789, Charlemagne commanded that "false writings and dubious accounts," especially a "supremely wicked and false letter wholly opposed to the Catholic faith" which had been circulated during the previous year, be neither believed nor read but burnt. Only canonic books and orthodox writings were to be read and transmitted.[180] An independent story of "letters written in gold by the hand of God" surfaced in 9th-century Northumberland, spread by a certain Pehtred. He reportedly claimed that these letters had been deposited on the tomb of St. Peter at the time of "Pope Florentius," that they foretold the precise time of the Day of Judgment and demanded suitable preparations for it, and that they revealed that God had created the Devil as devil not as angel of light.[181]

[176] *Admonitio Generalis* (789) 16, *MGH CapRegFr* 1: 55; *idem,* 42: 56; *Acta ad conc. Rispacense pertinentia* (799 vel 800) 7, *MGH Concilia* 2.1: 214; *Capitulare missorum generale item speciale* (802) 5, *MGH CapRegFr* 1: 102; *idem,* 21: 103. Haito of Basel, *Capitula* (806-823) 19, *MGH CapEp* 1: 216; Herard of Tours, *Capitula* (858) 3, *PL* 121: 764); *Ansegesis capitularium collectio* (1st half of 9th century) 1.41, *MGH CapRegFr* 1: 400; *idem,* 1.16, 1.66 and 1.78, *PL* 97: 711 and 712; Burchard of Worms, *Decretum* (1008-1012) 3.198, *PL* 140: 712. Burchard attributed this canon to a Council of Orleans.

[177] *Concilium Francofurtense* (794) 42, *MGH Concilia* 2.1: 170; Burchard of Worms, *Decretum* (1008-1012) 3.54, *PL* 140: 683. Burchard took his text from the Carthaginian Council of 401, see Dionysius Exiguus, *Registri Ecclesiae Carthaginensis Excerpta* 83, *CCSL* 149: 204-205.

[178] *Iordani Recensio Canonum Rispacensium* (c. 1550) *MGH Concilia* 2.1: 218.

[179] For letters from heaven, see E. Renoir, "Christ (lettre de) tombée du ciel" in *DACL* 3.1: 1534-46; W. R. Jones, "The heavenly letter in medieval England," in *Medieval Hagiography and Romance. Medievalia et Humanistica* (Cambridge, 1975), 163-178.

[180] 78, *MGH CapRegFr* 1: 60.

[181] Egred Bishop of Lindisfarne to Wulfsige Archbishop of York (830-837), Haddan and Stubbs 3: 615-616. This letter, written to protest Egred's innocence of Pehtred's heresies, is the

Such writings lent new relevance to ancient rulings against the use of noncanonical books. A 9th-century capitulary revived the ban of the Council of Elvira on reading "the famous [or infamous] books" in church.[182] Burchard of Worms' *Decretum* included both this and a canon based on the Council of Laodicea permitting the reading and chanting only of material sanctioned by divine authority and the Fathers.[183]

Over two hundred years after Boniface's struggle with Aldebert and Clement, Atto of Vercelli faced similar problems. He inserted in his capitulary, in the midst of the usual fulminations against tree-worshippers and clerics who frequented diviners, a condemnation of those who, "despising the teachings of the church, went over to falsely labelled prophets, angels and others of the holy dead and adopted their perverse doctrines."[184] He put this into a scriptural context with a quotation from *Matt* 24, 24: "For false Christs and false prophets will rise and mislead many." This, however, was not a general precaution prompted by the Bible, but the result of an actual threat to orthodoxy among his flock. In a heartfelt letter directed explicitly to the people of his diocese, he warned them of a "very grave matter," the imminent arrival of several pseudoprophets, who would "strive to turn many from the path of truth in order to lead their dupes to destruction." These men performed signs and wonders impressive enough and put forward arguments persuasive enough that many were quick to hail them as prophets. Atto feared that even priests would be "polluted" by their abominations.[185] "Polluted" suggests improper rites, not merely heretical ideas. Here again there is an undercurrent of something distinctly non-Christian.

2.6 Offerings and Rituals

In addition to the sacrifices and rituals already discussed, pastoral texts refer to immolations or sacrifices, offerings made to demons and idols–either tangible offerings and the accompanying rituals or behaviour thought inappropriate for Christians. Vows (*vota*) are mentioned but rarely in the context of the cult of deities or their effigies: the Council of Toledo XVI (693) explicitly stated that idolaters made "offerings of sacrilege" in fulfillment of a vow;[186] Arno of Salzburg instructed his priests to admonish the faithful to refrain from the "contagion of idols by not fulfilling pledges in the style of pagans" but idols might have been used figuratively in this case.[187] Prayer is never used in this connection although,

only source for them.

[182] *Capitula Angilramni* (c.847-852) 57, Cipriotti, 12. See also Burchard of Worms, *Decretum* (1008-1012) 3.199, *PL* 140: 712.This is based on *Council of Elvira* 52 (Vives, 10).

[183] *Decretum* (1008-1012) 3.198, *PL* 140: 712.

[184] *Capitulare* 48 , *PL* 134: 38.

[185] *Ep.* 3 (*ad Plebem Vercellensem*), *PL* 134: 104-105).

[186] 2, Vives, 498-499.

[187] *Conc. Rispacense* (798?) Instructio pastoralis 5, *MGH Concilia* 2.1: 198). See also an 8th-century Bavarian guide for pastoral care (Étaix, "Un manuel de pastorale de l'époque carolin-

as noted above, deities and demons were sometimes "invoked" for various reprehensible purposes.

2.6.1 *Tangible offerings*

Some offerings were made of durable materials, since Toledo XVI ordered that they be displayed perpetually in local churches. Votive offerings in the form of coins, replicas of animals or humans, miniature lamps, combs, tools and weapons, *etc.*, have been found in shrines, but our sources refer only to torches lighted during the ceremonies. The Council of Toledo of 681 as well as that of 693 included those who lit torches among idolaters.[188] Hints of human sacrifice are found in documents from the northern parts of Frankish territory but, in general, the offering of choice was food, usually meant to be shared between the deity and his devotees.

Human sacrifice: Pagans in the Rhineland continued to practice human sacrifices with the implied cooperation of certain Christians even during St. Boniface's time. Around 732, he reported to Gregory II that some of the faithful had actually sold their slaves to the pagans for sacrifice; the pope urged him to stop the trade and to treat the culprits as homicides.[189] These practices may well have been in Boniface's mind when he complained to an English friend of his dependence on the patronage of the Frankish princes to govern his flock, protect clerics and monastics "and suppress pagan rites and the sacrileges of idols in Germany."[190] It is not surprising, therefore, that Charlemagne believed that the pagan Saxons had recourse to human sacrifice. His terms for the capitulation of Saxony imposed the death penalty on anyone "who has sacrificed a man to the devil and offered him as a victim to demons in the manner of pagans."[191]

An anomalous passage in the early 8th-century *Poen. Sangallense Simplex* under the heading "About sacrifices" must be considered in this context: "If any cleric, layman or woman has crushed a child, three years (of penance)."[192] Despite anecdotal evidence for child sacrifice, this cannot be taken at face value: it must refer to the accidental overlying of a child. In fact, the title is misleading. Listed under it are losing the Eucharist, magic, self-mutilation, usury and using false weights and measures. The relatively light penance proposed rules out the possibility of

gienne," *RB* 91 [1981]: 118.

[188] 11, Vives, 398-399. In Gallic documents, lights were associated with nature cults only (*e.g. Conc. Arelatense* II [442-506] 21, *CCSL* 148: 119).

[189] Gregory II to Boniface, *Ep.* 28, *MGH EpSel* 1: 51. Fear of human sacrifice does not seem to have been a factor in Carolingian legislation against the sale of slaves outside the kingdom, since other valuables were also included in the ban. See, for example, the Capitulary of Mantua (781) 7, *MGH CapRegFr* 1: 190.

[190] Boniface to Daniel of Winchester (742-746) *Ep.* 63, *MGH EpSel* 1: 130.

[191] *Capitulatio de partibus Saxoniae* (775-790) 9, *MGH CapRegFr* 1: 69.

[192] 10, *CCSL 156*: 120.

human sacrifice.

Hints concerning other types of sacrifice involving the human body are of equally doubtful value. An 11th-century canonic collection explained that among different types of eunuchs were those who "are made effeminate for the sake of idols."[193] Our texts, however, give no reason to suppose that ritual castration was practiced in medieval Europe. Charlemagne and Pippin's capitulary for Italy treated it on a par with other forms of mutilation; Lothair I and his successors as rulers of Italy threatened to castrate anyone who dared castrate another person "according to a custom that has grown up."[194] By contrast, Regino of Prüm called for the death penalty on anyone, slave or free, who castrated a man either for "lechery or trade" or handed him over to be castrated.[195]

Two 8th-century continental penitentials included self-mutilation under suggestive headings. The *Sangallense Simplex,* as mentioned above, listed it under the title "About sacrifices;" the *Sangallense Tripartitum* put it under "About sorcery" (together with seven clauses dealing unambiguously with magic and three others more ambiguously with arson, the violation of tombs and abortion).[196] Self-mutilation had been practiced by the votaries of some of the eastern mystery religions in the Roman Empire as late as the mid 5th century, when Maximus of Turin drew a mocking picture of a shaggy, half-naked peasant priest anesthetised with wine, inflicting wounds on himself in honour of his goddess or god.[197] But here again, our documents provide no supporting evidence for ritual self-mutilation in the Carolingian epoch, except in mourning rituals.

[193] *Collectio canonum in V libris* (1014-1023) 3.193, *CCCM* 6: 404. The example of Origen was evidently in the forefront of the minds of some authors. Martin of Braga barred from the priesthood anyone who castrated himself in order to get rid of fleshly temptations (*Canones ex orientalium patrum synodis* [572] 21, Barlow, 130). See also *Epítome hispánico* (598-610), several canons, *e.g.,* 2.1: 105; *Capitula Dacheriana* (late 7th/8th century) 138, Wasserschleben, 157; *Collectio Vetus Gallica* (8th/9th century) 16.2, Mordek, 409); *Poen. Ps.-Theodori* (mid 9th century) 21.29 and 21.30, Wasserschleben, 589; *Poen. Silense* (1060-1065) 216, *CCSL* 156A: 38.

[194] *Capitulare italicum* (810) 5, *MGH CapRegFr* 1: 205; *Pactum Hlotharii* I (840) 33, *MGH CapRegFr* 2: 135). See also *Pactum Karoli* III (880) 33, *ibid.,* 141, *Pactum Berengarii* I (888) 33, *ibid.,* 146). For the trade in eunuchs between Italy and the Byzantine Empire during the Carolingian period, see Shaun F. Tougher, "Byzantine eunuchs: An overview, with special reference to their creation and origin," in Liz James, ed., *Women, Men and Eunuchs. Gender in Byzantium* (London and New York, 1997), 168-184.

[195] *De synodalibus causis* (c. 906) 2.87: 247.

[196] *Poen. Sangallense Simplex* (8th century, 1st half) 12, *CCSL 156*: 120; *Poen. Sangallense tripartitum* (8th century, 2nd half) 29, Schmitz 2: 378. This clause is included in more than twenty penitentials. Some of these belong to the Theodorian tradition, but the clause itself is not found in the *Penitential of Theodore.*

[197] S. 107, *CCSL* 23: 420-421. This passage is remarkable for the number of words for priest: *dianaticus, aruspex, sacerdos, pontifex, uatis.* Despite the word *dianaticus,* this man may have been one of the *Galli* in the service of the *Magna Mater.* The text serves as proof of the continuing appeal of the mystery cults for slaves even at this late date.

Animal sacrifices and food offerings: It has already been seen that Gregory I thought that the pagan Anglo-Saxons were accustomed to slaughter oxen in honour of their "demons," and then feast on them in the vicinity of shrines.[198] His suggestion that, once converted, they should be invited to continue the practice in the vicinity of churches and build themselves little huts of branches for the festivities, was possibly biblical in inspiration but, as to offerings of livestock, he may well have had direct information. A century or so later, Christian priests sacrificed bulls and he-goats to the gods of the Anglo-Saxons' Hessian and Thuringian kinsmen.[199] Pope Gregory II in 724 believed that the Thuringians of his day were still performing animal sacrifices, and he charged them to refrain from adoring idols and immolating flesh.[200] Animals were being sacrificed even within established Frankish territory: Carloman's so-called German Council of 742 and Charlemagne's *Capitulary* of 769 included sacrificial victims among "unclean pagan customs."[201]

The only sacrificial beasts identified explicitly are oxen and goats, but other animals were also likely to have been sacrificed. Condemnations of horseflesh found in pastoral letters leave little doubt that the writers at least thought that horses were being sacrificed. I emphasise "at least thought" because all the letters were written by outside observers, the accuracy of whose perceptions and interpretations is open to question.

A letter written in 751 by Pope Zacharias in response to a query from St. Boniface about his flock's dietary habits illustrates the difficulty. The pope condemned jackdaw, crow and stork as utterly unsuitable for consumption by Christians; hare, beaver and wild horse were worse yet. He followed with a revealing remark, "Nevertheless, dearest brother, you know well all the sacred scriptures."[202] The beaver, hare, stork, crow and jackdaw may be considered to be unclean animals on the basis of *Lev* 11, but not the horse–it is neither a ruminant nor does it have cloven hooves. It, then, was banned on other scriptural grounds, either as sacrificial food or as flesh containing blood (*Acts* 15, 29). St. Boniface was indeed familiar with the Scriptures. Moreover, he had consulted Gregory II some thirty years earlier about the lawfulness of sacrificial food once it had been blessed with the sign of the cross, and had been curtly referred to St. Paul: "If anyone says that this was food offered in sacrifice, do not eat of it for the sake of

[198] Gregory I to Mellitus (601), Bede, *HE* 1.30: 108.

[199] Zacharias to Boniface (748) *Ep.* 80, *MGH EpSel* 1: 174-175.

[200] *Ep.* 26 (726), *MGH EpSel* 1: 45. For evidence of sacrificial meals which, however, date from the 6th century, see Salin, *La civilisation mérovingienne* 4, 29-45.

[201] *Conc. in Austrasia habitum* q. d. Germanicum (742) 5, *MGH Concilia* 2.1, 1: 3-4. See also *Karlomanni principis capitulare* (742) 5, *MGH CapRegFr* 1: 25; *Capitulare primum* [769 vel paullo post] 7, *MGH CapRegFr* 1: 45; *Benedicti Capitularium Collectio* (mid 9th century) 1. 2, *PL* 97: 704.

[202] *Ep.* 87, *MGH EpSel* 1: 196. Merrifield found ritual deposits of wild birds (Great Northern Diver, raven, crow, buzzard and starling) in British sites dating from the Roman period, see *Archaeology of Ritual and Magic, passim*, and Hutton, *Pagan Religions*, 231.

him who said so and for the sake of conscience."[203] Boniface would hardly have questioned Zacharias on this matter if he had known, as he was in position to, that the horseflesh was a part of a sacrificial offering.

Curiously, this was not the first time that the issue of the consumption of horseflesh had arisen. Some years previously, he had "added" an observation to a report to Zacharias' predecessor Gregory III, that some Germans ate wild horse and the majority, domesticated horse. Gregory protested vigorously against this "filthy and loathsome" habit, and charged Boniface to curb it and impose due penance.[204] The word "added" is the pope's own. It gives the impression that Boniface passed on the information casually, that he himself had seen nothing wrong with the custom. Was it because it formed a part of the English diet, at least occasionally? We know that that was so at the end of the 8th century, when George and Theophylact, two papal legates, were scandalised to find that some of the English still ate horsemeat, "which none of the Christians of the East do!"[205]

It seems evident that the popes, and perhaps George and Theophylact as well, believed that horsemeat was sacrificial food. Impossible to know the source of their information: Gregory III was Syrian by origin and Zacharias was Calabrian-born, of Greek parentage; they could have had little direct knowledge of German culture. It is certain that the papal legates (by their names Greek also) had the benefit of personal observation, since they gave eye-witness descriptions of the "foul" English custom of maiming horses, slitting their nostrils, binding their ears together so as to make them deaf, and docking their tails.[206] But if all these customs were tolerated by the local clergy, it must have been because in their eyes such habits had no tinge of paganism.

It is instructive to compare these letters with the penitentials. The majority of the penitentials, if they mention it at all, repeat or paraphrase the *Penitential of Theodore* (668-756) in dismissing the whole issue with "It is not forbidden to eat horseflesh but it is not customary."[207] The English *Ps.-Egbertian Confessional* (c.950-c.1000) gives a curious variation: "Horseflesh is not forbidden although many households prefer not to buy it."[208] The only penitential texts to forbid it outright

[203] *Ep.* 26 (726), *MGH EpSel* 1: 46. The reference is to I *Cor* 10, 28. St. Boniface's biographer Willibald testifies to Boniface's knowledge of and devotion to Holy Writ (*Vita Bonifacii* 8-9, *PL* 89: 608-609).

[204] *Ep.* 28 (c. 732), *MGH EpSel* 1: 50).

[205] *Report of the Legates George and Theophylact of their proceedings in England* (787) 19, Haddan and Stubbs 3: 459.

[206] *Report of the Legates George and Theophylact of their proceedings in England* (787) 19, Haddan and Stubbs 3: 458.

[207] *Poen. Theodori* (668-756) II, 11.4, Schmitz 1: 545. This is found in about a dozen 7th/8th and 9th-century penitentials drawn from all parts of Christian Western Europe except Spain. One clause of the late 7th/early 8th-century *Collectio Canonum Hibernensis* balanced the opinion attributed to St Jerome against horseflesh with Theodore's ambivalence and that of an African synod (54.13, H. Wasserschleben, 218).

[208] 1.38, *PL* 89: 411.

are the mid 7th-century *Canones Hibernenses*, which treats it as very serious indeed, calling for four years of penance on bread and water, and the 8th-century *Old Irish Penitential* which associates it with magical acts: "Anyone who eats the flesh of a horse, or drinks the blood or urine of an animal, does penance for three years and a half."[209] Given its origin and early date, these form the most convincing testimony in pastoral literature of the ritual use of horseflesh, but they can be taken to apply only to the Irish.[210] Ethnographic evidence, however, suggests that such was the custom of the Germanic tribes as well and that the popes and papal legates had been right to be suspicious.[211]

Food offerings were not necessarily in the form of blood sacrifice. The Council of Orleans called for the expulsion of anyone who had dared to eat of food offered as part of the cult of idols.[212] A 9th-century penitential revived canon 40 of the Council of Elvira to forbid landowners to accept "that which had been given to an idol" (most probably produce) as part of their dues and threatened them with a five year penalty if they refused to obey.[213]

In some cases, the person who sacrificed to idols did so under duress. The *Ps.-Egbertian Confessional* advised priests to take this into consideration in prescribing penance to those who had been forced to offer food to demons, presumably while captive in the hands of pagans:[214]

Undefined offerings: More often, no indication is given as to the nature of the offering or idolatrous act. An 8th-century Italian or Frankish penitential condemned sacrifices to idols as "the crime and iniquity of all wickedness," and prescribed forty weeks of fasting; but, in another clause, it ordained much longer penances (five to fourteen years) for clerics who committed homicide, fornication

[209] 1.13, Wasserschleben, 137. See also *Old Irish Penitential* I, 2, Binchy in Bieler, 259.

[210] A late 12th-century continental penitential, the *Laurentianum*, also mentioned the consumption of the flesh of horses, asses and other forbidden beasts, but treated it far more leniently, prescribing only ten days of penance (36, Schmitz 1: 788).

[211] See Grimm, *Teutonic Mythology*, 47-49 and Próinseás Ní Chatháin, "Traces of the cult of the horse in early Irish sources," *Journal of Indo-European Studies* 19 (1991): 123-132. In the *Topographia hibernica,* Gerald of Cambrai described a horse-ritual including a sacred marriage and a sacrificial meal for the inauguration of a 12th-century king of Ulster (in Jeanne-Marie Boivin, *L'Irlande au moyen âge. Giraud de Barri et la Topographia Hibernica (1188)* [Paris and Geneva, 1993], 254). Ritual consumption of horseflesh was by no means limited to the Celts and Germans: according to a 14th-century account, almost the first move of the leader of the pagan reaction in Hungary after St. Stephen's death was to persuade the people to commit this abomination (*Chronicon Pictum*, 110). See Frederick J. Simmons, *Eat Not this Flesh. Food Avoidances from Prehistory to the Present* (2nd edition, Madison, Wisc., 1994), 180-188 for taboos on horseflesh. For the cult of the horse, see also Salin, *La civilisation mérovingienne* 4, 23-28.

[212] *Conc. Aurelianense* (533) 20, *CCSL* 148A: 102.

[213] *Poen. Ps.-Gregorii* (2nd quarter, 9th century) 10, Kerff, 174. For the Council of Elvira, see Vives, 8.

[214] *Confessionale Ps.-Egberti* (c.950-c.1000) 1.32, *PL* 89: 409.

or unspecified sins involving idols.[215] Benedict Levita paraphrased *Exod* 22, 20 in demanding death for anyone who sacrificed to the gods.[216] Over a dozen English and continental penitentials between the 7th and 9th centuries and two continental collections followed the *Penitential of Theodore* in their treatment of undefined sacrifices: "Those who sacrifice to demons in very trivial matters/ways are to do penance for one year, but those who do so in great matters/ways are to do penance for ten years."[217]

In two continental texts of the first half of the 9th century, "sacrifice" (*immolatio*) stood for the use of magic. The *Penitential of Ps.-Gregory* made a distinction between the relatively trivial matter of sacrificing to demons at wells and trees (meriting one year of penance) and the far more serious matter of sacrificing to them *in machinis*, surely a mistake for *in magnis* or *in maximis* (meriting up to ten years). The author went on to explain:

> In my opinion, to sacrifice to demons *in machinis* is to believe their extremely foul illusions, or to make use of divinatory science through what is called the false lots of the saints, or spells, symbols or whatever kind of pendants or bindings, in all of which is the skill of demons, coming from the pestiferous union of men and wicked angels.[218]

Rabanus Maurus also included magic under *immolatio* but interpreted Theodore of Canterbury's clause as dealing with magic and divination in general, making no mention of tangible offerings.[219]

2.6.2 *Sacrificial meals*

"Paganism is about eating," remarks Ken Dowden.[220] Repasts that included the use of idolothytes or "idol food" are recorded throughout our period and in texts from every part of western Christendom.[221] Prohibitions emphasise three

[215] *Poen. Oxoniense* II (8th century) 7 and 71, *CCSL 156*: 192 and 205.

[216] *Benedicti capitularium collectio* (mid 9th century) 2.27, *PL* 97: 756.

[217] *Poen. Theodori* (668-756) I, 15.1, Schmitz 1: 537. See also *Liber de remediis peccatorum* (721-731) 12, Albers, 407; *Canones Gregorii* (late 7th/8th century) 115, Schmitz 2: 535; *Poen. Remense* (early 8th century) 9.14, Asbach, 57; *Excarpsus Cummeani* (early 8th century) 7.13, Schmitz 1: 633; *Poen. Egberti* (before 766) 4.12, Schmitz 1: 577; *Poen. XXXV Capitolorum* (late 8th century) 16.4, Wasserschleben, 517; *Poen. Vindobonense* b (late 8th century) 7, Wasserschleben, 496-7, 482; *Double Penitential of Bede-Egbert* (9th century?) 29, Schmitz 2: 694; *Poen. Ps.-Theodori* (mid 9th century) 27.1, Wasserschleben, 596; Regino of Prüm, *De synodalibus causis* [c. 906] II, 367: 353; Burchard of Worms, *Decretum* (1008-1012) 10.12, *PL* 140: 835.

[218] *Poen. Ps.-Gregorii* (2nd quarter, 9th century) 27, Kerff, 183-184.

[219] *Poenitentiale ad Heribaldum* [c. 853] 20, *PL* 110: 491. This canon of Theodore's is first among five items listed here under the title *De cultura idolorum*; the next three deal with magic and the last with the consumption of sacrificial food.

[220] *European Paganism*, 159.

[221] Peter D. Good defines "idol food" as "any food associated with the rites of or participation in other cults" (*Dangerous Food: I Corinthians 8-10 in Context* [Waterloo, Ont., 1993], 54-55).

elements. In the first, a formal act of immolation which, though not described in these texts, is the essential factor. The food (*immolatum idolis, immolatitia, idolothytes)* was tainted by the act of sacrifice and the meal, therefore, was a communion between idols and the sharers in the feast. In the second, the sacrificial act is altogether absent. The focus is on the action of eating in dangerous places, by shrines or in "loathsome places." In the last, the critical factor is the company in which the meal takes place. Although the implication is that the food has been offered to idols, the sin lies not in eating it, but in doing so with the offerers of the sacrifice or other undesirables.

The texts are seldom explicit as to the nature of the food offered. Eating *immolata* was usually put on a par with eating blood, carrion, the flesh of suffocated or strangled animals and fornication, on the model of *Acts* 21, 25. Although Gregory II charged the Thuringians not to immolate flesh, meat was not necessarily the only food offered in sacrifice.[222] It was the principal offering, as desirable to demons and idols as to men, but no doubt grain, pulse, honey, oil and different fruits and vegetables were also given and shared in forbidden feasts here as elsewhere.[223]

Caesarius of Arles worried not only about the pagan backsliding implicit in food offerings, but also about his flock's propensity for drunkenness and gluttony. In one sermon, he urged the faithful to resist persuasion to worship idols or to drink of what has been sacrificed to them. The failure to mention food suggests that the heavy drinking of toasts, which he condemned elsewhere, was the main focus of his concern.[224] In another sermon, he berated those who "from naïveté or ignorance or surely–which is more believable–from greed, neither feared nor blushed to eat of sacrifices or of sacrilegious food prepared in the pagan fashion," and charged them to avoid the "devilish banquets held in the vicinity of a shrine or springs or of particular trees." Even if they had stayed away from the feast, they were not to eat of the food brought home from the shrines.[225] Under Caesarius' direction, the Council of Orleans (533) also tackled the question of the relationship between idolothytes and gluttony. It called for the exclusion from the Church of "those who had failed to keep intact the graces of their baptism, or who had partaken with reckless enjoyment of food immolated during the cults of idols."[226] Age or sickness kept Caesarius from the council that met in 541, once again at Orleans, but the uncompromising style of the clause concerning *immolata* is recognisably his: A baptised person who reverted to eating food offered to demons "as though to vomit" and who refused to give up this form of apostasy after having been warned by a priest was to be suspended from membership in the Catholic com-

[222] *MGH EpSel* 1: 45.
[223] *Cf.* Gooch, *Dangerous Food,* 30.
[224] *S.* 19.4, *CCSL* 103: 89.
[225] *S.* 54.6, *CCSL* 103: 239-240.
[226] 20, *CCSL* 148A: 102.

munity.[227] A 7th-century council held at Clichy again condemned those who consorted with pagans, imitated their practices or shared their "superstitious" food.[228]

The consumption of *immolata* or *idolothytes* was considered in about twenty penitentials, usually in conjunction with other forms of forbidden food. One imposed eleven weeks of fasting as penance for having eaten blood, carrion or food that had been offered to idols.[229] In another, two periods of forty days of penance were prescribed for eating food contaminated with human blood (a popular ingredient in magic potions), carrion or food offered to idols.[230] Still another, from the ninth century, differentiated between the various kinds of dangerous foods: a year of penance for eating what had been sacrificed to idols, but only two forty day periods for food contaminated with blood, carrion or from an animal seized by a wild beast.[231] An English penitential of the same century assigned twelve weeks of penance for eating carrion or food either consecrated or sacrificed to idols.[232]

Other texts upheld the ban on forbidden foods indirectly, by insisting that all other kinds were acceptable. The simplest formulation is found in the *Capitulary of Gregory II* (731): "Nothing about the eating of food should be considered unclean, unless it had been offered to idols."[233] A clause in an Irish collection with strong continental influences, also found among the canons of the Council of Aachen (816) and of Frankish canonic collections, anathematised anyone who condemned pious Catholics "as having no hope" (of salvation) because they had eaten meat, unless it contained blood or had been offered to idols or strangled.[234]

As we have seen, there were extenuating circumstances for eating forbidden foods. The *Penitential of Theodore* and half a dozen other penitentials advised confessors to take into consideration the individual–his age, upbringing and other relevant factors–in assigning penance.[235] It is noteworthy that only food offered to idols are mentioned in these texts, not the flesh of animals killed in forbidden

[227] 15, *CCSL* 148A: 136.

[228] *Concilium sub Sonnatio Episcopo Remensi Habitum* (627-630) 14, *MGH Concilia* 1: 204-205.

[229] *Poen. Sangallense tripartitum* (8th century, 2nd half) 35, Schmitz 2: 182.

[230] *Poen. Floriacense* (late 8th century) 41, *CCSL 156*: 100.

[231] *Poen. Hubertense* (early 9th century) 60 *CCCM* 156: 115.

[232] *Double Penitential of Bede-Egbert* (9th century?) 39.1, Schmitz 2: 696.

[233] 7, *Concilia Germaniae* 1: 37.

[234] *Hibernensis* (late 7th/early 8th century) 54.7, H. Wasserschleben, 216). See also *Collectio Vetus Gallica* (8th/9th century) 17. 12d, Mordek, 422; *Conc. Aquisgranense* (816) 65, *MGH Concilia* 2.1: 365; Regino of Prüm, *De synodalibus causis* (c. 906) Appendix I, 9: 395); *ibid.*, Appendix III, 11: 456. This canon is based on the Council of Gangra (mid 4th century) 2, Hefele-Leclercq, 1.2: 1033.

[235] *Poen. Theodori* (668-756) I, 15.5, Schmitz 1: 538). See also *Capitula Dacheriana* (late 7th/ 8th century) 90, Wasserschleben, 153; *Poen. Remense* (early 8th century) 9.18, Asbach, 58; *Excarpsus Cummeani* (early 8th century) 7.17, Schmitz 1: 634; *Poen. XXXV Capitolorum* (late 8th century) 16.3, Wasserschleben, 517; Rabanus Maurus, *Poenitentiale ad Heribaldum* (c. 853) 20, *PL* 110: 491; *Poen. Valicellanum* C.6 (10th/11th century) 60, Schmitz 1: 379.

ways. Ignorance and need were mitigating factors although they did not wholly exonerate the culprit. Those who had eaten idolothytes, carrion, blood or the blood of animals unknowingly were given lighter penances than those who had done so deliberately.[236] But even a child was responsible in some measure for the food that he had tasted in ignorance, whether it was an offering to idols, carrion or some other "abomination."[237]

Necessity in the form of threats of violence and perhaps hunger was tacitly accepted as a partial excuse. An early Spanish collection included clauses taken from the Council of Ancyra, concerning those who participated in idolatrous sacrifices and ate of immolated food. Several referred to a period when Christianity was still occasionally a persecuted religion, and were not pertinent to the actual situation in Spain c. 600 or to Western Europe in the centuries to follow.[238] Some, however, touched on issues that continued to disturb churchmen throughout the early Middle Ages. One dealt with the case of captives who ate immolated food, as much a problem in the 10th as in the 4th century. Forty days of penance was enough for those who ate polluted food after having been seized by pagans or betrayed by friends. If they proclaimed themselves to be Christians and were compelled by violence to eat, they were not to be deprived of Holy Communion. Nevertheless, they were to perform a lengthy penance even if they had truly done this with reluctance, without sharing in the rejoicings.[239] Four continental penitentials of the 8th and 9th centuries proposed fasts ranging from six to twelve weeks for those who had eaten food offered to idols, carrion or blood when "it was not necessary to do so."[240] Both ignorance and force are implied as partial excuses in a somewhat later penitential which imposed penances ranging from five to twelve years on those "who willingly slid back into the cult of demons, to making offerings to them or sacrifices to demons, or who knowingly ate idolothytes."[241]

In some cases, curiosity may have been a motive in eating forbidden food: Burchard of Worms' penitential imposed a mere ten days of penance for the sin

[236] *Poen. Oxoniense* II (8th century) 16, *CCSL 156*: 194. See also *Poen. Halitgari* (817-830), Schmitz 1: 486. *Poen. Merseburgense* a (late 8th century) 74, *CCCM* 156: 147. See also *Poen. Valicellanum* I (beginning of the 9th century) 97, *ibid.*, 320; *Poen. Vindobonense* a (late 9th century) 66, Schmitz 2: 355.

[237] *Poen. Oxoniense* II (8th century) 13, *CCSL* 156: 194; *Poen. Merseburgense* a (end of 8th century) 108, *ibid.,* 157; *Poen. Valicellanum* I (beginning of the 9th century) 82, Schmitz 1: 306; *Poen. Halitgari* (817-830) *ibid.*, 486.

[238] *E.g., Epítome hispánico* (c. 598-610) 2.43: 108, and 5.2 and 5.3: 116.

[239] *Epítome hispánico* (c. 598-610), *Ex ep. Papae Leonis ad Nicetam* 5: 210; *Poen. Casinense* (9th/10th century), Schmitz 1: 430. *Cf.* Council of Ancyra (314) 3 and 4 (Hefele-Leclercq, *Histoire des conciles* 1.1: 305-307).

[240] *Poen. Oxoniense* II (8th century) 4, *CCSL* 156: 191. See also *Poen. Sangallense tripartitum* (8th century, 2nd half) 35, Schmitz 2: 182; *Poen. Halitgari* (817-830) 44, Schmitz 1: 480; *Poen. Ps.-Theodori* (mid 9th century) 27.4, Wasserschleben, 596; *Anonymi liber poenitentialis* (9th century ?) *PL* 105: 722.

[241] *Poen. Arundel* (10th/11th century) 86, Schmitz 1: 460-461.

of eating "the food of Jews or other pagans [*sic*], which they prepared for themselves."[242]

In penitentials of the Columbanian school, the stress is not on the sacrificial character of the food but on the shrines near which it was eaten: "If indeed any layman ate or drank in the vicinity of shrines ..."[243] Columban's penitential makes no other reference to idolatry and gives no indication that the meal involved sacrificial rituals. Nevertheless, three 9th-century penitentials which include this clause follow it immediately with a reference to repeated acts of sacrifice, in one case under duress.[244]

Other texts situate the ritual meal in the countryside, by the trees, springs, stones, nooks or enclosures, corners or roadside crosses where people were accustomed to make and fulfill vows. About a dozen penitentials (none of Irish or Spanish origin) impose a year of penance for eating or drinking in such places.[245]

Some people held their repasts at such places for reasons that had, in their own minds, no religious component. Those who neither made offerings to idols themselves nor partook of sacrificial food wished nonetheless to participate in a communal feast with their non-Christian neighbours. A mid 9th-century continental penitential required two years of penance and exclusion from Holy Communion for those who "celebrate a feast at the same time in the abominable places of the pagans, and bring their own food, and eat together."[246] An English penitential emphasised that holding repasts in such places was sinful in itself, even if there was no question of offering a sacrifice: "If anyone dares to eat and drink in any other such place and does not offer a victim, nevertheless let him repent for a year on bread and water."[247]

Feasting in such disreputable places was, in general, a matter of strictly pastoral concern. When the secular authorities exceptionally took notice of this custom, we may doubt that the moral welfare of the participants was the chief incentive. The *Capitulatio de partibus Saxoniae* (775-790) fined those who made their vows at springs, trees and groves, made offerings "in the gentile manner" or ate "in honour of demons." The fines were heavy: sixty *solidi* for a noble, thirty for a free-

[242] *Decretum* (1008-1012) 19, 5.190, Schmitz 2: 451.

[243] *Paenitentiale Columbani* (c. 573) 20, Bieler, 104. See also *Poen. Merseburgense* (end of the 8th century) 49, *CCSL* 156: 141; *Poen. Valicellanum* I (beginning of the 9th century) 81, Schmitz 1: 305; *Poen. Halitgari* (817-830) 42, Schmitz 1: 480; *Poen. Ps.-Theodori* (mid 9th century) 27.2, Wasserschleben, 596; *Anonymi liber poenitentialis* (9th century?) *PL* 105: 722. The late 9th century *Poen. Vindobonense* a replaces the usual term for shrines, *fana*, with *fanassi* (51, Schmitz 2: 354).

[244] *Poen. Halitgari* (817-830) 43, Schmitz 1: 480; *Poen. Ps.-Theodori* (mid 9th century) 27.3, Wasserschleben, 596; *Anonymi liber poenitentialis*, *PL* 105: 722.

[245] *E.g., Poen. Remense* (early 8th century) 9.6, Asbach, 56.

[246] *Poen. Ps.-Theodori* (mid 9th century) 27.5, Wasserschleben, 596.

[247] *Confessionale Ps.-Egberti* (c.950-c.1000) 2.22, *PL* 89: 419.

man, and fifteen for an unfree tenant.[248] Given the Saxons' obstinate resistance to Frankish expansion, it is probable that Charlemagne worried that the Saxons might profit of the occasion of communal meals to foment further rebellion.

In times and places where Christianisation was not yet complete, opportunities for pagans and Christians to intermingle must have been frequent; no doubt even devout Christians felt a desire on festive occasions to continue their customary social contacts with their yet unbaptised neighbours. As we have seen, this was so in at least one case in the late 9th century; there is also evidence of shared feasts from earlier periods. It appears several times in the *Epítome hispánico,* which included canons imposing seven years of penance for celebrating pagans' feasts and sharing their meals, two years for rejoicing and banqueting with them, and thirteen for persuading others to join in. A canon credited to Pope Leo I takes a more nuanced position, requiring only a fast and the imposition of hands for those who only ate in company with immolaters, but (formal) penance for those who adored idols.[249] In Merovingian Gaul also, Christians willingly consorted with pagans. The Council of Clichy (626-627) deplored that "many a person eats food in company with pagans" and called for penalties for those who persisted in mixing with idolaters and immolaters.[250]

Canonic collections and penitentials continued to reiterate the rules against fraternising with pagans during their feasts, discriminating between those who merely "participated in a banquet held by pagans and ate food offered in sacrifice" and those who adored idols or were "tainted" with murder or fornication.[251] Rabanus Maurus cited the authority of the 4th-century Council of Laodicea in forbidding Christians to celebrate pagans' feasts with them or "to join in the depraved actions of the godless." A contemporary penitential treated this as a relatively minor offence, showing indulgence to those who, as prisoners, had shared in the pagans' food out of necessity, on condition that they rejected pagan customs and did penance when they were allowed to return (to Christian society).[252]

The authorities opposed communal feasts between Christians and non-Christians for social as well as religious reasons. When Pope Hadrian I complained to the Spanish bishops (c.785-791) about the many self-proclaimed Catholics who lived on friendly terms with Jews and non-baptised "pagans" (surely Moors), sharing their food, drink and "various errors," even contracting ties of marriage

[248] 21, *MGH CapRegFr* 1: 69.

[249] *Epítome hispánico* (598-610) 5.5, 116; 2.45: 108; *Ex ep. Papae Leonis ad Rusticum* 7: 208.

[250] 16, *CCSL* 148A: 294.

[251] *Benedicti capitularium collectio* (mid 9th century) 1.126 and 1.133, *PL* 97: 717 and 718. See also Rabanus Maurus, *Poenitentium Liber ad Otgarium* (842-843) 25, *PL* 112: 1418; Burchard of Worms, *Decretum* (1008-1012) 10.37: *PL* 839.

[252] *Poenitentium Liber ad Otgarium* (842-843) 24, *PL* 112: 1418; *Poen. Ps.-Theodori* (mid 9th century) 42.2, Wasserschleben, 611. See Council of Laodicea (380?) 39, Hefele-Leclercq, *Histoire des conciles,* 1.2: 1019.

with them, he was deploring friendly relations, not forbidden religious rituals.[253] The social element is in the forefront also in a 9th-century continental penitential which argued that there should be no contact with pagans since baptised Christians were not permitted to eat together nor to exchange the sign of peace with catechumens–"how much less then with pagans!" Similarly, an 11th-century Spanish penitential penalised Christians simply for sharing a meal with pagans.[254]

2.6.3 *Anomalies in Christian devotions*

The sometimes excessive cult paid to saints or angels and the inappropriate rituals in and around churches demonstrated how naturally new converts carried traditional attitudes and customs into their practice of the Christian religion. It is not surprising if they did so even in the observance of the sacraments.

Early Christian rulings against unsuitable offerings on the altar that were copied into western collections are of questionable relevance to the actual situation in medieval Western Europe. Some forbade the offering of anything except bread, wine and water, and specifically banned honey, milk or *sicera* (a fermented beverage), prepared dishes, poultry, livestock or vegetables.[255] Others permitted presentation of first fruits or incense for blessing at the Consecration but insisted on their distinction from the bread and wine mixed with water that were offered in sacrifice.[256] Since Western churchmen did not incorporate these into the decisions of their councils, it may be assumed that they did not think them apposite to the practice in their own dioceses.

Only two Western councils found it necessary to legislate about Mass offerings. The Council of Orleans (541) reminded clerics that the chalice could contain only wine mixed with water; anything else was sacrilegious.[257] The low-key tone of this clause suggests that the bishops considered that they were dealing with an error caused by ignorance, not with deliberate defiance. In late 7th-century Galicia, however, a considerable divergence in opinion arose about the Eucharist. Some priests offered milk at the Consecration, others distributed hosts dipped in wine (in milk?) to the faithful as the completion of the sacrament; still others consecrated unsqueezed grapes instead of wine, and distributed clusters of grapes at Communion. The Council of Braga (675) denounced these practices explicitly as superstitions practiced by persons in the grip of separatist tendencies (*schismatica ambitione*). While the Council did not actually label them schismatics, the words

[253] *MGH Ep* 3: 643.

[254] *Poen. Merseburgense* b (9th century, 1st half) 35, *CCSL* 156: 17); *Poen. Cordubense* (early 11th century) 175, *CCSL* 156A: 68).

[255] *Epítome hispánico* (598-610) 1.55: 102; *Collectio Vetus Gallica* (8th/9th century) 26.1, Mordek, 447-448. See also Regino of Prüm, *De synodalibus causis* (c. 906) 1.64: 53; Burchard of Worms, *Decretum* (1008-1012) 5.7, *PL* 140: 754. See *Breviarum Hipponense* 23, *CCSL* 149: 39-40.

[256] Regino of Prüm (d. 906) *De synodalibus causis* 1. 65: 54; Burchard of Worms, *Decretum* (1008-1012) 5.4, *PL* 140: 753). See also *ibid.,* 5.6 and 5.7: 754.

[257] 4, *CCSL* 148A: 133.

imply a deliberate and principled resistance to the authority of the hierarchy.[258]

Five continental penitentials dating from the 8th and 9th centuries describe the custom of performing, for pay, the fast imposed on someone else as penance. Sometimes this type of vicarious penance was evidently performed in good faith. The fullest discussion of this practice is found in the *Oxonian Penitential* II:

> Concerning those who fast for pay. We have heard of a habit of certain men and women, who think that it is right, when they hear that people received a penance for their sins and that fasting was imposed by the priests, to go to the ones on whom the fast was imposed and say: "Give me money and I shall fast for you." If any Christian man or woman be found who tried to do this, if he did it from ignorance, not expecting that this would be imputed to him as sin, let him be given a fast if he repents, so that he fast as much for himself as he had promised to fast for them and for which he had accepted money; and let the money he received be spent on the poor. And command everyone to whom you give penance: "Let no one fast for another man, because he both loses what he has given and remains in the state of sin himself."[259]

This seems to be an extension to the religious sphere of the custom of performing obligations (fighting battles, undergoing the ordeal) by proxy. It was found acceptable by some churchmen, such as the author of the 9th-century *Ps.-Theodorian Penitential*, who recommended it for those who were unable to perform the assigned penance personally.[260]

2.6.4 *Oaths*

In a society with limited technical means to establish truth and guilt or innocence, the oath, backed if necessary by kin or friends as conjurors, acquired paramount importance.[261] Under these circumstances, a Christian community could imagine no more powerful guarantor than God (specifically the Father and the Son) or His saints and angels. The authorities objected to oaths calling on them as irreverence or a violation of the biblical injunction against swearing, not as a form of paganism. Nevertheless these appear to be based on an attitude to

[258] 1, Vives, 372. The ban on the use of milk for consecration was repeated in Burchard of Worms' *Decretum* (5.1, *PL* 140: 751).

[259] 61, *CCSL* 156: 202. See also *Poen. Merseburgense* (late 8th century) 44, *ibid.*, 140; *Poen. Valicellanum* I (beginning of the 9th century) 110, Schmitz 1: 326; *Iudicium Clementi* (9th century) 3, Wasserschleben, 434; *Poen. Vindobonense* a (late 9th century) 48, Schmitz 2: 354.

[260] Wasserschleben, 622. See Vogel, *Le pécheur et la pénitence au moyen âge*, 30.

[261] Usually only when a person could not clear himself of a charge by oath either alone or with cojurors did he have recourse to some form of ordeal. For ordeals, see Robert Bartlett, *Trial by Fire and Water. The Medieval Judicial Ordeal* (Oxford, 1988; Dominique Barthelémy, "Diversité des ordalies médiévales," *Revue historique* 280 (1988): 3-25; Henri Platelle, "Pratiques pénitentielles et mentalités religieuses au moyen âge; la pénitence des parricides et l'esprit de l'ordalie," *Mélanges de Science Religieuse* (1983): 129-155; A. Michel, "Ordalies," *DTC* 11: 1139-1152.

God and supernatural beings that was rooted in traditional ways of thinking.

The traditional element is clearest in a text appearing twice in Burchard of Worms' *Decretum* which prohibits swearing by God's hair or head, blaspheming against God in other ways, and swearing by created objects or persons.[262] Swearing by the hair of the head was customary among various German peoples: among the Frisians, this was the weightiest and most solemn oath. Putting the hand on one's hair or heart, explained Hans Vordemfelde, symbolises the risking of the oath-taker's personality on the truthfulness of the oath.[263] The *Arundel Penitential* imposed penances for swearing (truthfully or not) by Christ's soul, body, or any part of His body, by God's saints or by creation. A Frankish capitulary of the 9th century added its own testimony, complaining that many miserable wretches, who were afraid to swear on the heavens, nonetheless did not fear to swear on the King of heaven and on all the celestial and angelical powers.[264]

[262] Burchard of Worms, *Decretum* (1008-1012) 12.15, *PL* 140: 879, and 19, 5.35, Schmitz 2: 417. In the first of these, a cleric guilty of blaspheming against God in any way is to be removed from his office, while a layman is to be anathemised. The second, which is in the penitential, merely imposes seven days of fasting.

[263] According to Vordemfelde, Bavarian and Swabian women put their hands on their plaits when swearing (*Die germanische Religion*, 120 n. 3).

[264] *Poen. Arundel* (10th/11th century) 37, Schmitz 1: 447-448; *Capitula Franciae occidentalis* (9th century, 1st half) 11, *MGH CapEp* 3: 46.

3
Nature

The cult of nature may be divided into three parts: (1) of the heavens (the sun, moon and firmament, including the stars and planets), (2) of humans and animals, and (3) of inanimate terrestrial nature (vegetation, especially trees, bodies of water, stones, fire and the earth itself).

3.1 THE CULT OF THE HEAVENS

The cult of heavenly bodies is found primarily in descriptions of rituals and magical practices originating in the belief that heavenly bodies are either divine or animated by souls.[1] Rituals and practices were based on two seemingly incompatible concepts of the relationship of man to heavenly bodies. On the one hand, there is a reciprocal relationship: the sun, moon, stars and heaven itself are sentient and respond to man; they demand worship, punish or reward, guarantee truth, depend on human help. On the other, their power is impersonal and pitiless, but predictable. Attempting to change their dictates is futile, but man may govern his own actions so as to take advantage of the most propitious disposition of the heavens and to unite himself to their annual cycle by participating in the great seasonal festivals. Cults of heavenly bodies are treated first in this section, then evidence for astrological beliefs; seasonal festivals are left to the next chapter.

3.1.1 *Heavenly bodies*

The sun: While there is archaeological and written evidence for solar cults in the late Empire and for the preeminence of such cults among the pagan Celts and Germans, and while solar symbols have been found in Christian tombs, pastoral literature makes little direct mention of them.[2] Sun worship formed part of the

[1] For sympathetic magic to exploit the forces of nature, see Raoul Manselli, "Simbolismo e magia nell'alto medioevo," and the discussion following, in *Simboli e simbologia nell'alto medioevo,* SSAM 23 (Spoleto, 1976), 293-348.

[2] European solar cults are summarised in Miranda Green, "Les dieux du soleil dans l'Europe ancienne," in Madanjeet Singh, ed., *Le soleil; Mythologie et représentations* (Paris, 1993), 294-309. Solar cults were not a part of traditional Roman religion, but had been introduced into the Empire from the East, *e.g.*, Mithra, particularly popular among soldiers, was originally a Persian sun-god (among other things); see Cumont, *Astrology and Religion among Greeks and Romans* (1912; New York, 1960), 73-76 and 89-102; Roger Beck, "Mithraism since Franz Cumont," in *Aufstieg und Niedergang* 17.4: 2002-2115; Vermaseren, "Hellenistic religions," in *Historia religionum* 1: 495-

official, public religion and culture of pre-Christian societies, and corresponded to the preoccupations of the aristocratic and military elite rather than of the peasantry. The fairly rapid assimilation of Christian concepts by the upper classes might explain the virtual disappearance of solar cults.

Neither Church Councils nor Capitularies mentioned sun-worship directly, and Caesarius of Arles' references to it are vague and unconvincing. The statement that "God placed the sun and the moon for us and for our benefit, not that we should honour those two shining lights as gods" was in the context of astrology rather than idolatry. Elsewhere, he asserted that "our sun of justice and truth is Christ and not that sun which is adored by pagans and Manichees."[3] Caesarius was usually explicit about his flock's shortcomings; thus it is unlikely that Christians in the Narbonnaise of the 6th century openly or consciously worshipped the sun.

Farther to the north, however, solar and lunar cults left traces in oaths to a much later period. St. Eligius' sermon forbids the faithful to "call the sun and moon lords or swear by them, for they are God's creatures and by His command serve men's needs."[4] This appears to reflect a contemporary practice containing elements of now-forgotten cults. Rhinelanders of the 11th century also swore by the sun, moon, heaven or earth and the rest of creation.[5]

The connection between the solstices and popular rituals is blurred by the discrepancy between the solstices and the two great seasonal celebrations, the feasts of St. John the Baptist and of the New Year. Nevertheless, St. Eligius' sermon interpreted the rites of St. John's Eve and other saints' days (no doubt around the solstice) as solar festivals: "Let no one perform solstice rites [*solestitia*]

532, esp. 512-515; John Helgeland, "Roman army religion," in *Aufstieg und Niedergang* 16.2: 1470-1505, esp. 1497-1498; G. H. Halsberghe, "Le culte de dieu Sol Invictus," *ibid.,* 17.4: 2181-2201). The lingering importance of solar cults in the 5th century is shown by the frequency of sermons emphasising the Christian interpretation of the solstices; see François Heim, "Solstice d'hiver, solstice d'été dans la prédication chrétienne du Ve siècle. Le dialogue des évêques avec le paganisme de Zénon de Vérone è saint Léon." *Latomus* 58 (1999): 640-660. Von Offele and G. Dottin questioned the existence of solar cults in Celtic religions (*ERE* 12: 48 and 74), but Green (*Dictionary of Celtic Myth and Legend*, 202, and *Symbol and Image,* 164-167) and Plummer (*VSH* 1, cxxix-clxxxviii) stressed its importance. August Nitschke found evidence of a Celtic solar cult in mainly Irish accounts of the birth and childhood of early medieval saints ("Kinder in Licht und Feuer–Ein keltischer Sonnenkult im frühen Mittelalter," *DA* 39 [1983]: 1-26). For the German cult of the sun, see Grimm, *Teutonic Mythology*, 700-706, and Y. Bonnefoy, *Mythologies*, trans. G. Honigsblum and ed. W. Doniger (Chicago and London, 1991), 1: 282-285. For solar and astral symbolism in epigraphy, see also H. Leclercq, "Astres" (*DACL* 1.2: 3005-3030). For the numerous solar symbols found in Merovingian tombs, see Salin, *Civilisation mérovingienne* 4: 121-133; for distribution of solar monuments see Green, *Symbol and Image,* 118, map 6.

[3] *S.* 193.4, *CCSL* 104: 785; S. 180.2, *ibid.,* 731.

[4] *Vita Eligii* (c. 700-725), *MGH SRM* 4: 707). See also anonymous homily (8th century), Nürnberger, 44..

[5] Burchard of Worms, *Decretum* (1008-1012) 19, 5.35, Schmitz 2: 417.

nor dances, leapings or devilish songs on the feast of St. John or some other solemnity of the saints."[6] The winter rites seem to have been particularly important in Italy. The Council of Rome (743) condemned celebrations of the winter solstice (*brumae*) as well as of the Calends of January. Two centuries later, Atto of Vercelli thought this clause relevant enough to include in his canonic collection. A capitulary negotiated between Charlemagne and his son Pippin and the Italian bishops (799-800?) denounced a yet-to-be explained ritual, the *brunaticus*, which involved lighting candles and offering bread.[7] It must have been brought to Charlemagne's attention by the Italian bishops; apparently it was unknown to the Frankish authorities elsewhere, for there is no hint of it in other Carolingian legislation. Such rituals formed a part of the complex of New Year's celebrations and will be studied in chapter 4.1 below.

This is all. There is no reference to any dawn or sunset ceremonies, such as the 5th-century Roman practice, condemned by Pope Leo I, of bowing to the rising sun on Christmas morning,[8] or to protective rituals in the case of a solar eclipse, a phenomenon far more terrifying than a lunar eclipse.

Nevertheless, certain magical practices appear to be connected with vestiges of sun worship. A belief in the healing power of the sun is implied in the practice of putting children on the roof (that is, exposing them to the sun) or placing them in an oven (which represents fire, the earthly image of the sun) to cure them of a fever. This was first described in the *Penitential of Theodore of Canterbury* and then in more than a score of English and continental penitentials throughout our period.[9] Theodore put it under the heading "Concerning the cult of idols" and the heavy penance of five years indicates that, in his view, the magic was consciously based on pagan cults (by contrast, he called for a maximum of only sixty days of fasting for consuming one's husband's blood as a remedy). Penances for this remained high, varying from five to ten years of fasting. Only at the end of our period did two penitentials reduce the penance to twenty and forty days respectively, showing that the practice was now seen as a minor, relatively innocuous magical practice, free of the taint of idolatry.[10]

The power of the sun was evoked by imitating or reversing its regular course through the sky. Imitating it by turning clockwise ("daesil") was powerful protective magic. Violating the natural order by reversing its motion was equally pow-

[6] *Vita Eligii* (c.700-725), *MGH SRM* 4: 705-706. This is the only reference given in Du Cange to solstice rituals, *s.v.* "Solstitium."

[7] *Conc. Romanum* 9, *MGH Concilia* 2.1: 15-16 (see also Atto of Vercelli, Capitularia 79, *PL* 134: 43); *Karoli Magni et Pippini filii capitula cum Italiae episcopis deliberata* (799-800?) 3, *MGH CapRegFr* 1: 202.

[8] *S.* 27.4, *PL* 54: 218-219.

[9] *Poen. Theodori* I, 15.2, Schmitz 1: 537.

[10] Burchard of Worms, *Decretum* (1008-1012) 19, 5.95, Schmitz 2: 430; *Poen. Arundel* (10th/11th century) 97, Schmitz 1: 464.

erful malign magic. Burchard of Worms' penitential describes a hostile practice based explicitly on the rotation of the sun. Women who wanted to deprive their husbands of health and strength fed them bread made of specially prepared flour. Naked and covered with honey, they rolled over grain sprinkled on a sheet, collected it carefully, and ground it by turning the mill against the sun.[11] The last step appears to have been the critical element, but the bewitching of the grain may already have included solar magic, if the woman rolled backwards and forwards over it in the direction of and against the sun.

The circumambulations described repeatedly in continental texts from the 8th century on were probably performed clockwise as well.[12] The Council of Mainz (813) categorically forbade the performance of "vile and lewd" songs around (*circa*) churches.[13] Somewhat earlier a capitulary from the Haute-Saône warned against inappropriate behaviour at church on Sundays or holy days, insisting that "dances and leapings and *circus* [a round dance or procession around an object?] and vile, lewd songs and diabolical pranks" were not to be performed "either in the roads or houses or in any place, because they are left over from pagan custom."[14] Rabanus Maurus altered a 6th-century Galician canon concerning New Year's practices, replacing "decking houses around with laurel and fresh branches" with "walking around (or surrounding) the house" with the greenery–if not a mistranscription, a slight but potentially significant shift in emphasis.[15]

Rituals were performed "around" dead bodies. Documents from 9th-century Trier condemned male and female dancers or acrobats who danced around corpses singing bawdy songs and performing games or pantomimes.[16] An anonymous Carolingian preacher denounced the "sacrifices of the dead around corpses in their tombs."[17] Konrad Haderlein suggests that the enigmatic title "concerning the sacrilege over the dead, *i.e. dadsisas*" of the *Indiculus superstitionum* deals with the

[11] *Decretum* (1008-1012) 19, 5.193, Schmitz 2: 451.

[12] A brief bibliography on the widespread custom of ritual circumambulation (*turnus sacralis*) is to be found in *History of Religions. Proceedings of the Thirteenth Congress of the International Association for the History of Religions,* ed. Michael Pye and Rita McKenzie (Lancaster, 1975), 120-121. See also Claude Lecouteux, *Démons et génies du terroir au moyen âge* (Paris, 1995), 108-123; Diana L. Eck, "Circumambulation," *Encyclopedia of Religion* 3: 508-511; Goblet D'Alviella, "Circumambulation," *ERE* 3: 657-659.

[13] 48, *MGH Concilia* 2.1: 272. See also *Benedicti Capitularium Collectio* (mid 9th century) Add. 39, *PL* 97: 876; Regino of Prüm, *De synodalibus causis* (c. 906) II, 5.87: 216; Burchard of Worms, *Decretum* (1008-1012) 1.94, *Interrogatio* 86, *PL* 140: 579.

[14] *Capitula Vesulensia* (c. 800) 22, *MGH CapEp* 3: 351. See also *Benedicti capitularium collectio* (mid 9th century] 2.196, *PL* 97: 771.

[15] *Poenitentium Liber ad Otgarium* (842-843) 24, *PL* 112: 1417. See also Martin of Braga, *Canones ex orientalium patrum synodis* (572) 73, Barlow, 141.

[16] *Capitula Treverensia* (before 818) 11, *MGH EpCap* 1: 56. See also anonymous sermon (c. 850-882) 11, Kyll, 11.

[17] Scherer, 439.

same practice, if *dadsisas* is interpreted as a scribal error for the unfamiliar Celtic word *daesil.*[18]

Another clause of the same *Indiculus* concerns "furrows around habitations."[19] Plowing (a symbolic ring-ditch?) in the direction of the sun's movement undoubtedly strengthened its protective magic against the evil spirits that lurked about outside the boundaries of human settlement.

The protective circle comes up again in the context of the Calends of January, at once the most propitious and the most ominous time of the year, when the night sky was full of ghosts. Burchard of Worms' penitential proposes a question concerning forbidden divinatory practices: "Did you sit on your roof after having drawn a circle around you with a sword so that you might see from there and understand what will happen to you in the coming year?"[20] A 10th-century penitential, probably from northern Italy, used a peculiar term, found nowhere else: "If anyone honours the Calends of January or other calends–except for saints' days–and goes about as a *cerenus, quod dicitur circerlus, aut in uecula*, let him do penance for four years ... because this is left over from pagan observances."[21] *Cerenus* may be a variation on *cervus* (stag) and *uecula* is without doubt a variation on *vetula/vitulo/vehicula* (old woman, calf, cart): masquerades of this sort are a staple of pastoral literature throughout the early Middle Ages. But *circerlus* must refer to someone, probably a masker, who walks or dances or is carried around the town or settlement.

Finally, a connection may exist between the eating of horseflesh, already discussed in chapter 2, and the remnants of solar cults. The horse, among both the pagan Celts and the Germans, was a solar animal; its flesh was eaten during ritual meals.[22] Condemnations of the consumption of horseflesh found in 7th and 8th-century texts concerning Ireland, the Rhineland and England indicate continued worries about such practices.

The moon: While the sun was particularly prominent in aristocratic and military religion, the moon, with its evident connection to growth and decay and the feminine cycle, played a correspondingly important role in agrarian popular

[18] *Indiculus superstitionum et paganiarum* (743?) 2, *MGH CapRegFr* 1: 223; Haderlein, "Celtic roots: Vernacular terminology and pagan ritual in Carlomann's Draft Capitulary of A. D. 743, Codex Vat. Pal. lat. 577," *Canadian Journal of Irish Studies* 18 (1992): 1-29; here, 15-20.

[19] 23, *MGH CapRegFr* 1: 223.

[20] *Decretum* (1008-1012) 19, 5.62, Schmitz 2: 423.

[21] *Poen. Oxoniense* (10th century) 29, *CCSL* 156: 91.

[22] Solar symbols and horses are found together in coins. The Celtic Jupiter, a sun-god, is frequently shown mounted on horseback on Jupiter-Giant columns. The connection of horses with the sun is strikingly demonstrated in Tacitus' account of the belief of the far northern Suiones that the sound and forms of horses could be discerned at sunrise (*Germania* 45. This comes from the Anderson edition; Grimm used an edition in which *equorum* was replaced with *deorum*, so that it was the gods who manifested themselves at dawn, see *Teutonic Mythology*, 700).

religion.[23] Continental pastoral literature offers an abundance of documentation about lunar cults from the 5th century onward. The descriptions of these cults are detailed and varied enough to vouch for their basis in actual social custom. They were manifested in spectacular rituals concerning eclipses or the darkening of the moon or its "labour," beliefs concerning the new moon, the relationship of the moon to women, the practice of fasting "in the honour of the moon," and an implied relationship between the weather and the moon. Missing from texts of our period is any explicit statement of the belief, noted in the mid 5th century by Maximus of Turin, that the growth and decline of living things were related to the phases of the moon.[24]

The protective rites described may have dealt with the phases of the moon and other phenomena as well as eclipses, the most terrifying lunar event. Eerie lunar phenomena were believed to be caused by enemies: magicians, demons, witches (*striae*) or monsters, who would have prevailed over the moon were it not for human intervention. From the valley of the Po to the valley of the Rhine, the most common means was noise (spells, cries, yells, trumpet blasts, drums, bells, anything that could be struck to make a din), but in the Rhineland, the sky itself was sometimes assaulted with weapons and fire.

The earliest Christian report about such practices in Western Europe comes from Cisalpine Gaul. In the mid 5th century, a number of Turinese believed that the moon was driven from the sky by sorcerers' spells, and that it could recover its light only with the help of din set up by the people. The noise disturbed Maximus of Turin in his pastoral duties so much that he devoted two sermons[25] to berating the faithful for their folly:

> Yesterday around nightfall, while I was chastising some of you for greed and avarice, such an outcry rose from the people that its blasphemy reached the sky. When I asked why the clamor, they replied that your outcry supported the struggling moon and that their clamor helped it in its need.[26]

Maximus leaves some doubt as to the lunar event that provoked the popular outburst. If his claim to ignorance as to the meaning of the noise was not merely a rhetorical device, the ritual was relatively unusual and, therefore, could not have been the regular waning of the moon. Yet in the first sermon, he refers to the

[23] See Bächtold-Stäubli 6: 477-534. Grimm gives a rapid overview of Indo-European moon beliefs in *Teutonic Mythology*, 700-720. These may be usefully compared to oriental lunar cults described in *La lune. Mythes et rites* (Paris, 1962). For Roman moon beliefs, see Sophie Lunais, *Les auteurs latins de la fin des guerres puniques à la fin du règne des Antonins* (Recherches sur la Lune 1, Leiden, 1979). Despite Pliny's story of sacrifices performed by druids on the 6th day of the moon, Green has found little evidence of lunar worship among the Celts (*Dictionary of Celtic Myth and Legend, s.v.* "Moon").

[24] *S.* 31.1, *CCSL* 23: 121.

[25] *S.* 30 and S. 31, *CCSL* 23: 117-123

[26] *S.* 30.1, *CCSL* 23: 117.

proverbial mutability of the moon, and, in the second, he relates the ritual to the tides, which seems to rule out an eclipse.[27] Although these sermons had no evident influence on early medieval pastoral literature except for a sermon by Rabanus Maurus, the same lack of precision is found in other documents concerning lunar rites.

Nearly a score of texts deal with the *obscuratio* or *defectio* of the moon.[28] Caesarius of Arles set the pattern. In a passage frequently echoed in English and continental penitentials and occasionally in continental sermons, he called on the faithful to reprimand those who raised a clamor when the moon was obscured. He warned them that they committed a grave sin in believing that they could "protect the moon with their uproar and sorcery when it is darkened *certis temporibus* by God's command."[29] Whether this applies to eclipses or to some other phenomenon depends on the interpretation of *certis temporibus*–"at fixed times" (perhaps as ordinary as the monthly waning of the moon), or "at certain times" (which his flock were unable to predict). If the latter, an eclipse is probable. Given favourable sky conditions, Caesarius' flock could have observed about twenty total or partial eclipses between 500 and 542; this afforded ample opportunity to develop protective rites without taking the edge off the fear caused by such an ominous event.

Maximus' sermons mentioned only cries and spells, but Caesarius' word *maleficia* (sorcery) shows that other forms of magic were also used. A second passage presents a clearer picture of the rituals, but creates even more doubt as to the celestial event in question:

> And what is this, when foolish men think that they have to run as if to the rescue of the moon in distress! When its fiery globe is covered [darkened? surrounded?] at fixed times, as a result of the natural movement of the air [mist? weather?], or is suffused with the burning colour of the setting sun nearby, they believe that it is as it were some assault of spells against the heavens, which they think they can defeat by a blast of the trumpet or the ridiculous jangle of ringing bells–using faulty pagan reasoning, they think that they win the moon's favour by their sacrilegious uproar.[30]

Here we are told that the moon was believed to be under attack from the spells of enemies (therefore, from malign magicians), and that the shrill sounds of trum-

[27] *S.* 30.3 and S. 31.1, *CCSL* 23: 118 and 121.

[28] In classical Latin, the usual terms for eclipse were *defectus* and *labores*, more rarely *obscuratio* (Lunais, *Les auteurs latins,* 208, 336-337).

[29] S. 13.5, *CCSL* 103: 67-68. See also *Liber de remediis peccatorum* (721-731) 3, Albers, 411; *Poen. Egbert* (before 7660 8.3, Schmitz 1: 581; *Double Penitential of Bede-Egbert* (9th century?) 30.3, Schmitz 2: 695; *Poen. Ps.-Gregorii* (2nd quarter, 9th century) 23, Kerff, 180; *Poen. Ps.-Theodori* (mid 9th century) 27.25, Wasserschleben, 598; Burchard of Worms, *Decretum* (1008-1012) 10.33, *PL* 140: 837 and in slightly different form, 19, 5.61, Schmitz 2: 423. Three 8th-century texts paraphrase Caesarius: *Dicta Pirmini*, 22, 176; *Vita Eligii*, *MGH SRM* 4: 707; anonymous 8th-century homily, Nürnberger, 44).

[30] S. 52.3, *CCSL* 103: 231.

pets and bells were a means of defence. But in this passage, Caesarius uses the term *certis temporibus* to describe two different phenomena: one owing to the atmosphere (this looks like a corona rather than an eclipse), the other to the setting sun. If "nearby" (*vicinum*) is taken literally, the red glow might refer to moonrise at sunset, especially spectacular at full moon. If, however, *vicinum* is used metaphorically, the description corresponds fairly well to the dull coppery glow of a moon in eclipse.[31]

Caesarius' explanation is sketchy in comparison to the wealth of myth found in many cultures to account for the eclipses.[32] More details emerged later. In northern Francia during the 8th century, people shouted "Triumph, O moon!" (*vince luna*) at the moon during an eclipse and "witlessly beat wooden and bronze vessels in the midst of the people's uproar" when the moon was darkened, thinking thus to bring back the moon into the sky "after it had been removed by witches" (the word used here is the feminine *stria* – not the masculine 'sorcerers' found in the 5th-century Turinese text).[33] These means suggest poor communities without the musical instruments used in Caesarius' Arles. A somewhat later anonymous sermon called for a harsh reprimand on those who thought that clangor, horn-blasts or cries were a defence against either the eclipse or storms.[34]

Rabanus Maurus borrowed freely from Maximus of Turin in his introduction to a sermon against those who "wore themselves out" by making an uproar during an eclipse: "While I was quietly at home yesterday, meditating how I might further your progress toward God, suddenly around evening and nightfall such an outcry arose from the people that its blasphemy reached the sky. When I asked why the clamor, they replied that your outcry supported the struggling moon and that by their efforts they supplied what it lacked."[35] But the following day, when he questioned visitors whether anything of the sort had come to their notice, he received a flood of information about Rhenish customs quite independent of the sorcerers' spells of the Turinese beliefs:

> on their part, they freely acknowledged that they had noticed similar things and even worse in the places where they had been staying. For one reported hearing blaring horns, as it were a summons to war; another, the grunting

[31] The colour may have given rise to the belief that the moon was wounded and bleeding. Popular Roman belief had held that this reddish hue was caused by the moon's angry reaction to incantations (Lunais, *Auteurs latins*, 234-238).

[32] For classical references see Caspari, *Homilia de sacrilegiis*, 30-32 and Lunais, *Les auteurs latins*, 9-14, 205-212 and 234-42. See also W. D. Wallis, "Prodigies and Portents–Eclipses," *ERE* 10: 368-369.

[33] *Indiculus superstitionum et paganiarum* (743?) 21, *MGH CapRegFr* 1: 223. See also *Poen. Merseburgense* a (late 8th century) 109, *CCCM* 156: 169, and *Poen. Vindobonense* a (late 9th century) 99, Wasserschleben, 422. *Homilia de sacrilegiis* (late 8th century) 16, Caspari, 10.

[34] Levison, 308.

[35] *PL* 110: 78.

> of swine. Truly, certain persons related that they had seen some people hurling lances and arrows against the moon, others scattering fire in the sky. And they affirmed that I know not what monsters were mauling the moon, and had they themselves not gone to its help, those same monsters would have devoured it completely. Others, moreover, to comply with the illusion of demons, split apart their own fences [hedges?–*sepes*] with weapons [tools?– *armis*], and broke the vessels that they had in the house, as though that would give the moon a great deal of help.[36]

It is not clear whether these are all protective measures. Some of the sounds the blaring of horns and the grunting (of pigs or of humans imitating pigs?) may have been made by the enemies of the moon, the monsters that were tearing it apart. But it is certain that the techniques described in this and other texts point to the very distant past indeed. The preference for bronze instruments and iron-tipped weapons (although wood, shells, stone and fire are also used) suggests that the origin of these rituals lies in the period when metal-using cultures were struggling to impose themselves on a neolithic society. The swinish grunting evokes the sacred boar of both Celts and Germans.[37] The spear hurled at the sky recalls the one reputedly hurled over the altar of the pagan Northumbrian gods by the high priest Coifi.[38] The symbolism of destroying vessels and especially hedges and fences (if *sepes* is correct and not a mistake for some other word such as seats, *sedes* –no easier to interpret) escapes me altogether; breaking down the barrier between the habitations and cultivated fields and the wild, where evil spirits are known to roam, seems dangerous, particularly when one is already engaged in a struggle against witches or monsters.

Special forces, distinct from the normal influence of the moon on growth, were thought to be active during the time of the new moon. In Burchard of Worms' penitential, the new moon was propitious for building a house or contracting a marriage.[39] Elsewhere it was a malign omen.[40] An anonymous 8th-

[36] *PL* 110: 79.

[37] Boar figurines and images are frequent in Iron Age artifacts. Green observes that they are "natural war symbols." Boar-crested trumpets and helmets are depicted on the Gundestrop cauldron, and Arduinna, like Demeter, was accompanied by a boar (*Dictionary of Celtic Myth and Legend, s.v.*). Tacitus noted that among the Germans, the Aestii, who were under the protection of the Mother Goddess, wore wild-boar amulets (*Germania* 45). Those who made a bad death, for instance, suicides, were also thought to return in the shape of swine (Lecouteux, *Chasses fantastiques*, 188). See also Bächtold-Stäubli 7: 1470-1509, *s.v.* "Schwein."

[38] Bede, *HE* 2.13: 182-186.

[39] 19, 5.61, Schmitz 2: 423. This is in agreement with Tacitus' observation that the Germans thought the new moon and the full moon the most auspicious times for beginning an enterprise (*Germania* 11).

[40] The fear of the new moon may also appear in Robert Grosseteste's description of a traditional rite practiced by those "who, when the new moon appears, bow down before it, make the sign of Christ's cross on themselves, say the Lord's Prayer, and then turn themselves

century sermon urged the faithful to undertake work fearlessly at this time.[41] St. Eligius connected the new moon with madness, as had Isidore of Seville; his sermon implies that people were afraid to commence some kinds of work then, in the belief that demons took that opportunity to launch an attack on them, using the moon as base:

> No one should be afraid to start on any work at the time of the new moon, because God made the moon for this purpose, that it should mark off time and temper the shades of night, not that it should prevent anyone's work or drive man mad, as think the fools who believe that they were invaded from the moon by demons.[42]

The new moon is sinister in the *Homilia de sacrilegiis* also. Christians forgetful of the sign of the cross "await vain things and name the new moon 'countermoon' [*contralunium*];" they believed themselves to be impeded by the moon from travelling, doing farm work (ploughing, transporting manure, pruning and cultivating vines, cutting wood in the forest), fencing or hedging in the house, or other work.[43] The countermoon is perhaps the faintly glowing darkened portion of the lunar disc sometimes visible at new moon–an ominous sign in the ballad "Sir Patrick Spens," where the sight of the new moon "wi' the auld moon in her arm" on the eve of the voyage presages the loss of the ship with all its crew. This may have been a common interpretation, making journeys seem particularly unsafe when it appeared. Reluctance to prune vines or cut branches off trees at the new moon, the time when the sap was thought most likely to run, is predictable. It is harder to understand why ploughing, manuring and cultivating would be considered equally dangerous, unless the *contralunium* augured badly for activities normally performed at the beginning of the month. Unwillingness to fence in the house appears to belong to the same pattern as the impulse to split hedges or fences during an eclipse.

The *Indiculus Superstitionum* explicitly connects women and the moon in a puzzling article, *de eo quod credunt quia femine lunam comendet, quod possint corda hominum tollere iuxta paganos.*[44] Homann interprets this as the belief that certain

around thrice in a circle; and they kiss the first man whom they meet, in the idea that thanks to this they are freed of all danger during the entire month" (*De decem mandatis*, ed. C. Dales and E. B. King [Oxford, 1987], 9). I am obliged to Pierre Boglioni for drawing my attention to this passage.

[41] Nürnberger, 44.

[42] *Vita Eligii* [c.700-725], *MGH SRM* 4: 707. According to Isidore of Seville, epileptics were commonly known as lunatics because they were trapped in the demons' toils " by the course of the moon" (*Etymologiae* IV, 7.7: 490).

[43] 13, Caspari, 8-9.

[44] 30, *MGH CapRegFr* 1: 223. See Homann, *Indiculus*, 137-141; Harmening, *Superstitio*, 250-252. This passage illustrates particularly well the difficulties in interpretation posed by ambiguities of grammar, spelling and vocabulary.

women could, like pagans, remove men's hearts because they had power over the moon, Harmening as the belief that since women had power over the moon, they could rob men of their senses.[45] The parallel between the lunar and the feminine cycle makes it easy to assume that women are in a special relationship with the moon,[46] but its exact nature is difficult to establish. It is not clear to me that the text states that women have power over the moon. The verb *comendet* is certainly a misspelling, but of what word? The possibilities are *commendent* (they commend, confide)–which implies the moon's power over women, not women's over the moon– and *comedent* (they eat, devour). *Commendere* may result in a more plausible text, but *comedere* has the merit of postulating fewer errors of transcription. *Quod possint* is also ambiguous. Does it mean "in order to be able" or, simply, "because they are able"? Finally, is *tollere* to be understood metaphorically as "to captivate" or literally as "to take" or "to take away"?

One translation, relatively benign and grammatically acceptable, is: "Concerning those who believe that women commend the moon because, according to the pagans, they are able to captivate the hearts of men." This version makes women grateful to the moon for influencing men favourably toward them, but it ignores the purposive weight of *possint.* Moreover, it suggests, implausibly, that only pagans believed in women's erotic attractions.

If one overlooks the absence of the reflexive *se*, and *commendere* is taken to mean "commend to" rather than "commend," one comes up with a more sinister possibility, that women surrender themselves to the moon to gain power over men: "Concerning those who believe that women confide themselves to the moon, like the pagans, so that they might take away the hearts of men." The Salic Law and Carolingian legislation attested to current popular beliefs in man-eating witches, while Burchard of Worms described the belief that in the dark of night women, seemingly asleep at home, flew off to fight battles in the sky, where they wounded or killed men, devoured them, then sent them back in the semblance of living men, but with straw and wood stuffed in place of their hearts.[47] The connection between these and moon beliefs is reinforced by the fact that the moon-goddess

[45] Homan, *Indiculus,* 137; Harmening, *Superstitio*, 251. Harmening connects this with the power of the moon to drive men mad in St. Eligius's sermon, see above.

[46] For a psychoanalytic perspective on myths connecting women and the moon, see M. Esther Harding, *Women's Mysteries Ancient and Modern* (New York, 1971). Although the connection between the moon and women is very common, it is not invariable. In Cambodia for example, the moon can be male as, according to Grimm, it was for the ancient Teutons (Eveline Porré-Maspéro et Solange Thierry, "La lune, croyances et rites du Cambodge," in *La lune*, 261-286, here 281; *Teutonic Mythology*, 103).

[47] *Pactus Legis Salicae* 64.3, *MGH Leg.* 1.4.1: 231; *Capitulatio de partibus Saxoniae* (775-790) 6, *MGH CapRegFr* 1: 68-9); Burchard of Worms, *Decretum* (1008-1012) 19, 5.170, Schmitz 2: 446. Such beliefs were not restricted to Germanic tribes or the medieval period. For example, Apuleius claimed that Thessalian witches cut the throat of their victims, pulled out the heart, then stuffed the wound with sponges.

Diana was the mistress of the aerial rides. The cavalcade through the night sky also evokes the hunt of Hecate and her following of ghosts on moonless nights. The *Indiculus* may then be interpreted to refer not to amorous women, but to female ghouls, the *striae*, and to ritual cannibalism.

Finally, we have seen that the eclipse or darkening of the moon was blamed on witches (*striae*).[48] If one identifies them with the women of Diana's cavalcade, and corrects *comendet* to *comedent*, a third translation may be hazarded: "Concerning those who believe that women devour the moon so that [or because?] they can remove the hearts of men." This implies a link between the witches' flight and eclipses: witches cause them to enable them to carry out their amorous or cannibalistic designs on men.

Fasting in honour of the moon for the sake of a cure is mentioned in a 9th-century continental penitential.[49] The meaning of the practice is obscure. Was it meant to heal the moon or the devotee? Was it done during the waning of the moon, in sympathy, or during its waxing, in the hopes of benefiting from its increased power, or on Monday, the day under its special protection? No particulars are given.

The texts imply that the weather and the moon were thought to be connected. Caesarius of Arles explained the darkening of the moon by the movement of the atmosphere. The countermoon apparently heralded a rising tempest. Three texts describe magic against storms in the same breath as similarly noisy rituals to drive off the moon's enemies: people hoped to avert hail by means of inscribed lead tablets or enchanted horns, and they beat drums and snail shells or drinking vessels and made other noise to drive off thunder.[50]

The firmament: Traces of the worship of the heavens and miscellaneous heavenly bodies lingered into the 8th century. St. Eligius' sermon spells out the existence of such cults: "Let no one believe that the sky or stars or earth or any created being at all is to be adored, but God alone."[51] The ecclesiastical authorities recognised the remnants of idolatrous pagan custom in the Bavarian ritual of *stapsaken*. If a person accused another of having robbed him and demanded restitution and monetary compensation, and the other denied the charge and refused to pay, the accuser would invite him to join in raising their right hands "toward the just judgment of God. And then both stretch their right hand toward the sky."[52] This

[48] Homilia de sacrilegiis (late 8th century) 16, Caspari, 10.

[49] *Poen. Ps.-Theodori* (mid 9th century) 27.26, Wasserschleben, 598.

[50] *Homilia de sacrilegiis* (late 8th century) 16, Caspari, 10; *Poen. Merseburgense* a (late 8th century) 109 (3rd recension), *CCSL* 156: 169; *Poen. Vindobonense* a (late 9th century) 97, Wasserschleben, 422. The two penitentials offer the wrath of God as their own explanation for eclipses and storms.

[51] *Vita Eligii* (c.700-725), *MGH SRM* 4: 708.

[52] *Conc. Neuchingense* (772) 6, *MGH Concilia* 2.1: 100-101. In Lothair's *Pavian Capitulary*, a conspiracy entered on "with the right hand" merited a lesser punishment than one formed by oath

rite shows every sign of originating in a sky (solar?) cult.

Less convincing are texts directed against custom, especially among clerics, of swearing by creation, something that by then was probably no more than a matter of habit, a formula devoid of meaning. The ruling of the *Statuta Ecclesiae Antiqua* (c. 475) to chastise or excommunicate clergy who swore *per creaturas* was incorporated into canonic collections and English and continental penitentials.[53] An Irish and a French text quoted *Matt* 5, 34-35 against swearing by heaven "which is the throne of God," or the earth "which is his footstool."[54] Pirmin of Reichenau referred more briefly to the same text, while Burchard imposed fifteen days of fasting on swearing "by heaven or by the earth."[55]

3.1.2. *The heavens and human affairs*

Priscillianism and the stars: The belief that the position of the sun and moon, the planets and the signs of the zodiac influenced human affairs was widespread. The most extreme form of this idea was expressed in Priscillianism. Priscillian (d. c. 385) reputedly taught that the different parts of the body were subject to signs of the zodiac and to the patriarchs (for example, the head to the Ram and to Reuben). His memory lingered the longest in Galicia–the shrine of St. James of Compostela may have been built over his tomb. Almost 200 years after his execution, the first Council of Braga (561) devoted two canons to condemning these doctrines, equating them with pagan beliefs: "If anyone believes that [human] souls and human bodies are bound by [or to] stars that determine their fate, as the pagans and Priscillianians claims, let him be anathema," and "If anyone believes that the twelve signs of the zodiac usually observed by astologers are distributed throughout the soul and the members of the body, and claims that they are assigned to the names of the patriarchs, let him be anathema."[56] These teachings

(6, *MGH CapRegFr* 2: 61). For a survey of the ritual significance of the hand in the ancient Mediterranean world and in medieval Christianity, see Karl Gross, *Menschenhand und Gotteshand in Antike und Christentum* (Stuttgart, 1985). See also Robert Hertz, "The pre-eminence of the right hand. A study in religious polarity," in *Right and Left: Essays on Dual Symbolic Classification*, trans. and ed. R. Needham (Chicago and London, 1973) 3-31.

[53] 74 [LXI], *CCSL* 148, 178. See also *Epitome hispánico* (c. 598-610) 19.74; *Collectio Frisigensis Secunda* (late 8th century) 72, Mordek 627; *Collectio Vetus Gallica* (8th-9th century); *Poen. Quadripartitus* (9th century, 2nd quarter) 118 and 191, Richter 16 and 22; *Benedicti Capitularium Collectio* (mid 9th century) add. 68, PL 97: 898; *Confessionale Ps.-Egbert* (c. 950-1000) 2.12, *PL* 89: 432; Burchard of Worms, *Decretum* (1008-1012) 12.15, *PL* 149L 879.

[54] *Hibernensis* (late 7th/early 8th century) 35. 3, H. Wasserschleben, 125. See also *Capitula Franciae occidentalis* (9th century, 1st half) 11, *MGH CapEp* 3: 46.

[55] *Dicta Pirmini* (724-753) 18: 168; Burchard of Worms, *Decretum* (1008-1012) 19, 5.35, Schmitz 2: 417.

[56] 9 and 10, Vives, 68. Jean Daniélou traced the symbolic connection of the signs of the zodiac with the patriarchs to concepts prevailing among the hellenised Jews of Philo's Alexandria ("Les douze apôtres et le zodiaque," *Vigiliae Christianae* 13 [1959]: 14-21).

reappear without ascription in the *Epitome hispánico* (c. 598-610), in canons against belief in fate, the relationship betwen parts of the body[57] and "signs" (constellations) or stars, or in the control of stars over the human body. The affirmation, set in the midst of these canons, that "God made the soul"must be meant to counter once more the belief that the stars were responsible for the creation of souls.[58]

Horoscopes and astrologers–nascentia, divinationes temporum, mathematici:[59] The frequent condemnations of the observation of heavenly bodies were more concerned with attempts to harmonise human activities with the heavens than with attempts to tell the future. Horoscopes (*nascentia*) are mentioned only thrice. St. Eligius' sermon forbids anyone to "lay out for himself the fate or, rather, the fortune or *genesis* commonly called a horoscope, so as to say that things will come about as the horoscope predicted."[60] This sermon was supposedly preached to a very recently converted audience in northeastern Gaul and its author may have had little knowledge of local practice. By contrast, Atto of Vercelli (d. 961) was completely familiar with the customs of his flock. Having discussed the legitimate observation of the stars (for telling time, navigating, predicting rain), he continued: "For Almighty God set the stars in heaven to serve man on earth ... The Creator of all gave them this law at the beginning and beyond that they have no power either to help or harm, despite the astrologers (*mathematici*) who teach that the stars rule horoscopes or govern marriages and that they should be taken into consideration by architects."[61] He poured scorn on such claims, pointing out that countless married couples who proved barren had followed the stars carefully, and that attention to the stars had not saved buildings from destruction in a recent fire in Pavia.

This is an important document for the social anthropology of 10th-century Piedmont, and not only because it highlights current preoccupations–fear of childlessness, fear of fire.[62] Most of the texts dealing with observation of the heavens relate it to agricultural work. But Atto provides evidence here that townsmen (professional builders and their clients) relied on celestial guidance as well as countryfolk. So when, a few sentences later, Atto explains how the "enemy of mankind" causes the "wound of stupid and rustic false belief" to persist and "the

[57] 11 and 12: 216.

[58] 13: 216.

[59] See Theodore Otto Wedel, *The Medieval Attitude toward Astrology, Particularly in England* (1920; Yale Studies in English, 1968), esp. 24-48; M. L. W. Laistner, "The western Church and astrology during the early Middle Ages," *Harvard Theological Review* 34 (1941): 251-275; Flint, *Rise of Magic*, 92-101 and *passim*.

[60] *Vita Eligii*, *MGH SRM* 4: 707. See also anonymous sermon, Nürnberger, 44. A resemblance too close to be coincidental is found between this passage and canon 61 of the *Conc. Quinisextum* (692) (Mansi 11: 970-971).

[61] *S.* 3, *PL* 134: 837.

[62] Particularly interesting since our texts dealing with procreative techniques concern contraception and abortion almost exclusively; magic to conceive is very rare.

poison of superstitions to spread," defends old customs and stiffens obduracy, he is not talking of the folly and coarseness, the superstition and hidebound resistance of the peasantry alone, despite the epithet *rusticus*, but of city people as well.[63] His remarks in letters, too, were addressed to all segments of the population. In a "letter to all the faithful of his diocese" he included consultation of heavenly signs among the forms of divination or magic which many in the region practiced. In another letter to the people of Vercelli, he warned against pseudoprophets who misled the ignorant and unsophisticated by performing prodigies and "signs" (interpreting horoscopes or heavenly phenomena?) and preaching false dogma.[64]

Other texts may contain implicit references to horoscopes. The *Homilia de sacrilegiis* condemned the belief that "evil or good fate existed in men." This probably refers to the fatal influence of heavenly bodies, although it must be remembered that the *Parcae* also governed destiny. English Christians were urged not to practice vain divinations, as did the pagans. These divinations are explained as faith in the sun, moon and course of the stars and, significantly, in setting up "divinations of the times" in order to arrange one's affairs–surely by casting a horoscope.[65]

Atto's word *mathematicus* had originally meant mathematician, but by the dawn of the Middle Ages, its meaning was well-established as astrologer.[66] Caesarius of Arles used it in this sense when he challenged the hope of a long life held out by the *mathematicus,* and when he insisted that the devil spoke through *mathematici* and Manicheans when they assured man that he had no need to confess his sins because his actions were fated by the stars.[67] *Mathematici* were astrologers to Isidore of Seville, as well.[68] In penitential literature, however, the word had lost its connection with astrology. Instead, it meant a malign enchanter who inebriated or befuddled men.

The dearth of texts concerning astrology and astrologers was explained by Theodore Wedel as the result of the "general decline of learning which overtook Western Europe during the first medieval centuries." M.L.W. Laistner blamed the lack during this period of handbooks of scientific astrology. But Valerie Flint is surely right that vaguer words such as *haruspex, augurium* and *divinatio* did service at times for astrologer and astrology.[69] It is possible, too, that pastoral texts mention the belief in heavenly phenomena as presages seldom because such beliefs

[63] *S.* 3, *PL* 134: 837. From this sermon, we learn also that the constellation of the Bear was known to *nostri rustici* as the Wain (*plaustrum*), the head of Taurus as the Hen (*gallina*) and the Belt of Orion as the Sickle (*falx sector*).

[64] *Ep.* 2, *PL* 134: 104; Ep. 3, *ibid.*

[65] 3, Caspari, 6; *Confessionale Ps.-Egberti* (c.950-c.1000) 2.23, *PL* 89: 419.

[66] Augustine, *De diversis quaestionibus* 45, *CCSL* 44: 67-69; Gregory the Great, Moralia in Job 33.19, *CCSL* 143B: 1689, and *In evangelia homeliae* 10.5, *CCSL* 141: 69.

[67] *S.* 18.4, *CCSL* 103: 85; S. 59.2, *ibid.*, 259-260.

[68] *Etymologia* III, 27.2: 456.

[69] Wedel, *Medieval Attitude toward Astrology,* 14; Laistner, "The western Church and astrology;" Flint, *Rise of Magic*, 95.

were so common even among the most respectable clerical and intellectual circles that it was not a matter for condemnation. In fact, heavenly portents abound in early medieval literature. According to Paul the Deacon, to cite but one author, bloody signs in the sky announced Clothair's downfall and a comet or eclipse foretold pestilence.[70] A fund of common knowledge evidently persisted despite the lack of handbooks.

The calendar. The observance of propitious days disturbed churchmen in part because the days of the week were identified with the names of the sun, moon and planets. For Caesarius of Arles and Martin of Braga, this usage permitted "wretched and untaught men" to pay tribute to the deities or demons whose names the planets bore. This effectively gave the gods credit for having created the days. The correct names were those that God had established: the first, second, third days, and so on.[71] Their sermons do not imply that the faithful consciously–let alone deliberately–paid cult to the pagan deities, but, on the contrary, that they needed to be reminded of the origin of the planetary names. By the late 8th century, even some clerics themselves had forgotten it. "He who observes the days which the misguided pagans called suns, moons, Marses, Mercuries, Jupiters, Venusus, Saturns ..."–the man who wrote these words knew that the names of the days were objectionable, but he no longer knew why.[72]

Such sermons had little impact on popular behaviour. Throughout Western Europe except in Portugal, the days continued to be named after the sun, moon, and either Roman or Germanic deities. The ecclesiastical authorities may have thought that fighting a losing battle on this score was not worth their while, since the issue is absent from the documents of church councils and penitentials.

By contrast, churchmen persisted in their attempts to prevent the faithful from accommodating their work to the course of the heavens, especially the phases of the moon and its position throughout the year. The *Calendar of 354* provides an insight into the indications followed at the dawn of the Christian era. No doubt details varied according to time and place, but the essence of the observances must have remained reasonably constant throughout our period.

According to this *Calendar*, while the moon was in Aries, Cancer, Libra and Capricorn, one profitably lent or borrowed money, made a will, set up the web for weaving, washed wool, gelded beasts, contracted agreements, or traveled. Taurus, Leo, Scorpio and Aquarius were suitable for asking favours, addressing the powerful, submitting accounts, making tools, sending children to school, laying foundations, grafting trees, propagating plants, layering grape-vines, turning the soil, sowing and reaping. Under the signs of Gemini, Virgo, Sagittarius and Pisces, it

[70] *Historia Longobardorum* 4.15, 5.31 and 6.5, *MGH Scriptorum Rerum Longobardicarum et Italicarum*, 121, 154 and 166.

[71] Caesarius of Arles, *S.* 193.4, *CCSL* 104: 785; Martin of Braga, *De correctione rusticorum* 9, Barlow, 189.

[72] *Homilia de sacrilegiis* 12, Caspari, 8.

was advantageous to set out on a voyage, launch a ship, harvest olives, gather the vintage, prepare charcoal and lime, rack wine and groom one's hair and beard.[73]

Within these general guidelines, there were specific indications for each day, predictable from the character of the tutelary god. Saturday was bad all around: everything would be confused and difficult, those born that day would be in danger, what was lost or stolen would not be found, the sick would get weaker. Tuesday, Mars' day, was better only in being apt for such warlike activities as enlisting in the army or readying weapons. The other days were more propitious in every way: Wednesday, ruled by Mercury, was good for business deals; Monday, sacred to the moon, for manuring and building water tanks; Sunday, for seafaring and shipbuilding; Thursday, Jupiter's day, for politicking; and Friday, under Venus' guidance, for family affairs: marrying and disposing of one's children. The general prognosis was good for all five of these days: newborn babies would be sturdy, lost or stolen property would be found, the sick would get better. In addition, each of the twenty-four hours had its own ruling planet, with Saturn and Mars again being unfavourable, Jupiter and Venus propitious, and the rest neutral.[74]

Life was regulated by the calendar, especially the phases of the moon, among the Celts and Germans as well.[75] The Coligny calendar shows the Celtic month divided into two parts, before and after the full moon. According to Bede, the ancient Angles, like the Hebrews and Greeks, based their months on the moon "inasmuch as among them the moon is called *mona* and the month *monath*." The name of each month reflected the special characteristics or appropriate activities of that month. The calendar imposed by Charlemagne was similar.[76]

Without being as encyclopedic as the *Calendar of 354* in describing the relationship between the heavenly bodies and work, our texts show that such ideas persisted. Caesarius of Arles found "no small proof of the devil" in foolish men's belief "that days and calends, the sun and moon are to be revered." This meant keeping the fifth day,"Jupiter's day," as a day of rest, and paying attention to the day on which one left home or returned, a habit of the laity as well as, "worse yet," some clerics.[77] We have seen that this observance came up some dozen times in subsequent pastoral literature. Martin of Braga implied that some other day ("Jupiter's day or that of some demon") was honoured this way as well, but per-

[73] *Chronographus Anni CCCLIIII,* ed. T. Mommsen, *MGH AA* 9.1: 47. For a discussion, see Henri Stern, *Le calendrier de 354* (Paris, 1953).

[74] *Chronographus A. CCCLIIII*, 42-45.

[75] See J. A. MacCulloch, "Calendar (Celtic)," and H. Munro Chadwick, "Calendar (Teutonic)," in *ERE* 3: 78-82 and 138-141. For the observation of the phases of the moon to determine activities, see also Harding, *Women's Mysteries*, 26-27.

[76] Bede, *De temporibus*, ed. Charles W. Jones (Cambridge, Mass., 1943) 211-213; Einhard, *Vita Caroli Magni* 29, *MGH Scriptores Rerum Germanicarum* 25: 33. For an evaluation of *De temporibus,* see Page, "Anglo-Saxon paganism," 124-127.

[77] *S.* 52.2, *CCSL* 103: 230-231; *S.* 54.1, *ibid.*, 236; *S.* 193.4, *CCSL* 104: 785; *S.* 1.12, *CCSL* 103: 8-9.

haps only by pagans.[78] Waiting for a propitious day to start out on or return from a journey appeared again in Martin's writings, the *Hibernensis* and other continental documents down to the 11th century and beyond.[79] Given the dangers of the road, bad weather, wild beasts, brigands, it is not surprising that many thought it foolhardy to ignore the heavens or any other omen–an overheard sneeze, the caw of a crow or the flight of an owl–when embarking on a trip.

The *Homilia de sacrilegiis* echoed Caesarius but added new material as well:

> he who observes days ... and who believes that journeys should be undertaken and commerce carried out depending on those days, or who thinks that they can be of either help or hindrance in any other need, or who honours that very day which they call Jupiter's for the sake of Jupiter and does no work that day–he is not a Christian but a pagan.[80]

The day thought suitable for commerce may have been Wednesday, as in the mid 4th century. If so, the *Homilia* casts a light on the article of the *Indiculus superstitionum* concerning "the feasts/days they make for Jupiter and Mercury" and fills the gap left in the *Indiculus* by the absence of any reference to the familiar "fifth day."[81]

Friday was still preferred for marriages in 6th-century Galicia. According to Martin of Braga, this, like waiting for a propitious day for travel, was tantamount to devil-worship. Pirmin of Reichenau mentioned Venus' or "some other" day as being chosen, but two considerably later sermons based on his *Dicta* referred only to Friday.[82] At least some parish priests tolerated this custom–a pastoral letter of the Carolingian period penalised them for giving Communion to people who persisted in the observance of Thursday and Friday "on account of pagan custom."[83] One of Atto of Vercelli's sermons denounced a new Piedmontese custom of treating Friday as a day of rest, excesses of banqueting and convivial merrymaking.[84] This may have been an outgrowth of the practice of marrying on Friday,

[78] *De correctione rusticorum* (572) 18, Barlow, 202-203.

[79] *E.g.,* Martin of Braga, *De correctione rusticorum* (2nd half, 6th century) 16, Barlow, 198; *Hibernensis* (late 7th/early 8th century) 64.8 , H. Wasserschleben, 232. See also *Vita Eligii* (c.700-725), *MGH SRM* 4: 705-706; *Dicta Pirmini* (724-753) 22: 173; *Homilia de sacrilegiis* (late 8th century) 12 and 27, Caspari, 8 and 16; Rabanus Maurus (d. 856), *Hom.* 43: *Contra paganicos errores, quos aliqui de rudibus Christianis sequuntur, PL* 110: 81; two anonymous sermons to the baptised (10th or 11th century and 12th century), Caspari, *Kirchenhistorische Anecdota* 1: 201 and 205.

[80] 12, Caspari, 8.

[81] 20, *MGH CapRegFr* 1: 223. See also the reference to the *feriati dies* of Jupiter and Mercury in the Anonymous sermon (9th century, 1st half), Scherer, 439.

[82] Martin of Braga, *De correctione rusticorum* (572) 16, Barlow, 198; *Dicta Pirmini* (724-753) 22: 173. See also two anonymous sermons to the baptised (10th or 11th century and 12th century), Caspari, *Kirchenhistorische Anecdota* 1: 200 and 205.

[83] *Epistola Canonica sub Carolo Magno* (c.814) 5, Mansi 13: 1095.

[84] *Ep.* 4 *Ad omnes fideles suae diocesis, PL* 134: 105. Germana Gandino notes that Atto does not see Friday merrymaking as a form of idolatry but as a misapprehension of the religious meaning of the day ("Cultura dotta e cultura folklorica a Vercelli nel X secolo." *Bollettino storico-bibliografico*

since numerous texts testify that feasting and shows unsuitable for clerics were an integral part of wedding celebrations.

The observance of one day in particular was not related in any obvious way to the course of the heavens. Between the 6th and 11th centuries, over two dozen texts from every part of Western Europe, including Ireland, condemned fasting on Sunday for any reason whatsoever: religion, carelessness, contempt of the day, restraint, obstinacy, divination, zeal. Texts linked this practice variously to Manichean, Priscillianian and Judaising tendencies, perhaps to excesses of ascetism, or to magic.[85]

Martin of Braga's *Canons* forbade Christians "to abide by pagan traditions and observe or honour the elements or the moon or the course of the stars or the empty falsehood of signs [constellations?–*signa*] for building a house, planting corn or trees or contracting a marriage." From the late 8th century or 9th century onward, churchmen showed considerable interest in this canon, repeating it in collections and in penitentials.[86] Its influence is also evident in the denunciation by the 10th-century *Confessional of Ps.-Egbert* of attempts to "arrange affairs" according to the sun and moon and course of the stars.[87]

Other authors were inspired by St. Paul's complaint to the Galatians concerning their observance of days, months, times and years.[88] A 9th-century continental penitential borrowed St. Ambrose's words to interpret this for the benefit of the faithful:

> they observe days if they say, as may happen, "I think one should not set out tomorrow but, on the following day, one should not begin anything." And this is how they are usually deceived by *magi*. But the ones who honour months are those who study the course of the moon, perhaps saying, "Grain should not be prepared [harvested? threshed?–*frumentum confici*] on the seventh day of the moon." Or again, "It is not desirable to bring a purchased slave to the house on the ninth day of the moon." It is usually easier for evil

subalpino [1992]: 253-279; here, 263).

[85] *E.g.*, Martin of Braga, *Canones ex orientalium patrum synodis* (572) 57, Barlow, 138; *Poen. Theodori* (668-756) I, 11.2 and 11.3, Schmitz 1: 534; *Conventus et Synodus Erfordiensis, Gesta Synodalia* (932) 5, *MGH Leg* 6.1: 109). Priscillianians probably fasted on Sundays during the twenty-one days before Epiphany and during Lent (Chadwick, *Priscillian of Avila*, 14).

[86] Martin of Braga, *Canones ex orientalium patrum synodis* (572) 72, Barlow, 141. See also *Collectio Vetus Gallica* (8th/9th century) 44.4b, Mordek 523-524; Rabanus Maurus, *Poenitentium Liber ad Otgarium* (842-843) 24, *PL* 112: 1417-1418; *Poen. Quadripartitus* (9th century, 2nd quarter) 144, Richter, 18; Halitgar of Cambrai, *De Poenitentia* (9th century, 1st half) 4.26, *PL* 105: 685; of Radulph of Bourges, *Capitulary* (between 853 and 866) 38, *MGH EpCap* 1: 262-263; Regino of Prüm, *De synodalibus causis* (c. 906) II, 373: 357; Burchard of Worms, *Decretum* (1008-1012) 10.13, *PL* 140: 835.

[87] 2.23, *PL* 89: 419.

[88] *Gal* 4, 10-11. For the observance of days, see Harmening, *Superstitio,* 162-172, and Lynn Thorndike, *A History of Magic and Experimental Science,* 1, *The First Thirteen Centuries* (1923; New York, 1953). 672-691.

> spirits to flourish through such things. Certainly, people observe times thus, when they say, "Today is the beginning of spring; it is a feast. After tomorrow come the Vulcanalia and the like." They add again, "Tomorrow one should not leave the house." Thus those who honour years say, "The New Year is at the calends of January"....[89]

This text appears with modifications in the collections of Regino of Prüm and Burchard of Worms. The latter dropped the reference to the Vulcanalia, showing yet again his concern for local relevance. Both replaced *frumentum* with *strumenta* (*instrumenta*), that is, instruments or lessons.[90]

Regino of Prüm recorded a ruling by a 7th-century Council of Rouen concerning those who observed days, the moon and months, and hoped that the "effective power of hours" could be manipulated to make matters better or worse.[91] Ghärbald of Lüttich demanded that the observers of months and times be haled before the bishop's court for a hearing along with other sorcerers or miscreants (*malefici*),[92] while the *Penitential of Arundel* imposed a year for the sacrilege of observing days, times, or the phases of the moon in the performance of work.[93] Waiting for the right day and right phase of the moon had been condemned in St. Eligius' sermon as well,[94] but a West Frankish synod (c. 900) qualified the ban on this kind of practice. It forbade Christians to observe

> the day or the moon for marrying a husband or wife, initiating a change, or for doing any kind of work; the moon only may be observed (but not the day) for the purpose of blood-letting or cutting down or planting trees, because the humours and sap of the trees either are in harmony with the movement of the moon's atmosphere or follow its nature.[95]

These texts demonstrate the importance of the heavens, of the moon above all, for determining the tempo of peasant life. The *Homilia de sacrilegiis* in particular provides precious insights into the time constraints on agricultural work. We have seen its testimony that the "countermoon" of the new moon was thought to be

[89] *Poen. Quadripartitus* (9th century, 2nd quarter) 140, Richter, 18. See St. Ambrose, *PL* 17: 359-360.

[90] *De synodalibus causis* 2.372: 356-357; *Decretum* 10.11, *PL* 140: 835. Some texts simply refer to those who observed months and years without elaborating further, *e.g.*, St. Boniface's letter to Pope Zacharias, Ep. 50 (742) *MGH EpSel* 1: 84, and *Capitula Treverensia* (before 818) 5, *MGH EpCap* 1: 55.

[91] *Conc. Rothomagense* (650) 13: Mansi 10: 1202; Regino of Prüm, *De synodalibus causis* (c. 906) II, 5.51: 213.

[92] *Capitulary 2 of Ghärbald of Lüttich* (between 802 and 809)] 10, *MGH CapEp* 1: 29).

[93] *Poen. Arundel* (10th/11th century) 95, Schmitz 1: 463.

[94] *Vita Eligii* (c.700-725), *MGH SRM* 4: 705.

[95] 6, Hartmann, 44-45. For appropriate times for blood-letting, see Ps.-Bede, *De minutione sanguinis sive de phlebotomia*, in *The Complete Works of Venerable Bede, 6, Scientific Tracts*, ed. J. A. Giles (London, 1843), 349-352.

particularly unsuitable for plowing, manuring, pruning and cultivating vines, cutting down trees, and putting up fences.[96] In addition, it reprimanded those who watched the signs of the heavens and the stars for the time to plow, and for sending the bulls, rams and he-goats in among the flocks (for breeding).[97]

In treating the daily concerns of peasants so directly and in such detail, the *Homilia de sacrilegiis* casts doubt on the validity of the opinion that descriptions of astrological practices and beliefs during the early Middle Ages "belong to the learned literature of the day, and trace their origin to foreign, not to native sources. In the course of centuries, this learned superstition became the common property of the uncultured and the stock in trade of the makers of almanacs."[98] The question of the mutual interdependence of learned and folkloric cultures is beyond the scope of this study, but it is evident that the author of the *Homilia* was describing practices that at the very least had been adopted, adapted and preserved by the "uncultured" to meet their needs without regard to learned culture. They were a part of oral tradition, the store of knowledge passed from one generation to the next, perceived as essential for the well-being of the community, and not an added on, alien system of thought. Burchard of Worms underlined this in a recapitulation all the sinful practices associated with heavenly cults:

> did you observe the traditions of pagans that even in these days fathers invariably pass on to their sons, as it were by a hereditary right instituted by the devil? That is, did you worship the elements, namely, the moon and the sun, and the course of the stars, the new moon, or its eclipse, in order to succeed in restoring her to her glory by your noise or help, or so that these elements could help you or you them, or did you observe the new moon for building a house or contracting a marriage? If you did, you shall do penance for two years on the appointed days for it is written: "Whatever you do by word or deed, do it all in the name of our Lord Jesus Christ."[99]

3.2 The Cult of Men and Animals

3.2.1 *Men*[100]

Traces of a cult of living human beings appear in oath rituals. The concept of divine kingship probably underlies the banned Carolingian custom of swearing by the king and his heirs, in particular at the feast of St. Stephen at Yuletide.[101] About

[96] 13, Caspari, 8-9.

[97] 10, Caspari, 8.

[98] Wedel, *Medieval Attitude toward Astrology*, 45.

[99] *Decretum* (1008-1012) 19, 5.61, Schmitz 2: 423. Note that here, forbidden knowledge is transmitted by men to their sons – the only example in our texts.

[100] For the cult of humans, see Vordemfelde, *Die germanische Religion*, 112-127.

[101] *Duplex legationis edictum* (789) 26, *MGH CapRegFr* 1: 64; *Capitulare missorum* [803] 22, *ibid.*, 116. See also *Ansegesis capitularium collectio* (1st half of 9th century) 3.942, *ibid.*, 430; *Benedicti capitularium collectio* (mid 9th century) 2.241, *PL* 97: 775.

a dozen penitentials, from both the continent and England, followed the *Penitential of Theodore* in dismissing the practice of swearing on a man's hand as meaningless according to the Greek tradition. A single penitential added "among Latins, it is a sin."[102] The particular importance of the right hand in swearing oaths appears in the *Staptaken* ritual described above.

Curative powers apparently resided in the human head or skull. The works of Rabanus Maurus, Regino of Prüm and Burchard of Worms included the use of a burnt human head as a tonic among women's magical practices.[103]

The multi-faceted magical efficacy of female nudity appears in Burchard of Worms' penitential.[104] The ritual of preparing grain by rolling around naked in it has already been described. This was malign magic, intended to cause a husband to grow weak or sick. However, when a woman wanted him to "burn more ardently with love" for her, she fed him bread which she had had kneaded on her bare buttocks. A form of rain magic was worked through the instrumentality of a naked little girl. The element of nudity is particularly important here, since Burchard mentioned it four times.[105]

Other references to nudity are harder to evaluate. Maximus of Turin described the typical 5th-century peasant priest in the service of a goddess (*dianaticus*, *haruspex*) as part naked: bare chest, half-clad legs, but with his bushy head covered with carefully-dressed locks (or perhaps with false hair); he is drunk and bleeding from self-inflicted wounds. Maximus' description is mocking, meant to to contrast the drunken peasant's appearance to the seemly bearing of a Christian priest. But since the man's choice of clothing appears deliberate (he took pains with his hair) and since he had just finished his early-morning devotions, this may be a case of ritual nudity.[106] Toward the end of the 8th century, the papal legates George and Theo-

[102] *Poen. Theodori* [668-756] I, 6.3, Schmitz 1: 530. See *Poen. Valicellanum* E. 62 (9th/10th century) 40 Wasserschleben, 563. None of the texts mention the common practice of swearing on weapons; see Jean-Luc Chassel, "Le serment par les armes. Fin d'antiquité – haut moyen âge," in *Le serment. Recueil d'études anthropologiques, historiques et juridiques* (Paris, 1989) 79-110.

[103] Rabanus Maurus, *Poenitentiale ad Heribaldum* (c. 853) 20, *PL* 110: 491. See also Regino of Prüm, *De synodalibus causis* 2. 369: 354; Burchard of Worms, *Decretum* (1008-1012) 19, 5.177, Schmitz 2: 448. The word translated here as "head" is *testam* in the text. Vogel's translation of it as "testicles" ("Pratiques superstitieuses," 760) is questionable on both grammatical and ethnological grounds. The Celts among others believed that the head is the source of semen, see Claude Sterckx, *La tête et les seins. La mutilation rituelle des ennemis et le concept de l'âme* (Saarbrücken, 1981), 115, and Campanile, "Aspects du sacré," 170: a burnt human skull might have been a remedy for infertility. In Pliny's *Natural History* 28.2, the skull of a hanged man provides a remedy for the bites of a rabid dog. For the magical power of heads, see J. A. MacCulloch, "Heads," *ERE*, 6: 537.

[104] For ritual nudity, see M. J. Heckenbach, *De nuditate sacra sacrisque vinculis. Religionsgeschichtliche Versuche und Vorarbeiten* IX, 1911.

[105] *Decretum* (1008-1012) 19, 5.193, Schmitz 2: 451; 19, 5.173, *ibid.*, 447; 19, 5.194, *ibid.*, 452. These magical rituals are discussed in more detail in chapters 7 and 8.

[106] *S.* 107, *CCSL* 23: 420-421.

phylact protested against the practice of some of the English clergy of approaching the altar bare-legged to celebrate mass, their tunics so short that they ran the risk of displaying themselves indecently. If "bare-legged" may be taken to include bare feet, there may be an element of the fear of standing shod in holy places, but it is more likely that the practice was a result of the priests' poverty or boorishness.[107] At roughly the same time, on the other side of the Channel, officials were urged to take steps against groups of people who wandered around naked and in irons, proclaiming that they had inflicted this on themselves as penance. Since they were not accused of obscenity, it is not clear whether they were in fact nude, or dressed in rags.[108]

3.2.2 *Animals*[109]

Although animals played a significant role in condemned forms of popular beliefs and practices, the texts give little direct evidence of animal-worship.[110] Throughout our period, mumming in the guise of stags and yearlings, calves or lambs at the New Year, and of stags and bears at commemorations of the dead was a widespread ritual. Omens were sought in the flight and calls of birds and in the behaviour of other animals. Stag horns and "serpents' tongues" were used as amulets. The flesh of certain animals was eaten in defiance of often-repeated prohibitions. But any element of animal worship in these practices was so attenuated that it escaped the attention of our authors, and they are considered later, in the study of New Year's rituals, divination, healing magic, commemorations of the dead and alimentary restrictions. Nevertheless, about half a dozen passages demonstrate that traces of animal cults (in which domesticated animals appear more important than wild beasts) still existed. Persuasive evidence for acts of worship, such as prayers or sacrifice, is absent even here. The single text that mentions the cult of animal or therianthropic divinities does so as recalling now-

[107] *Legatine Synods–Report of the Legates George and Theophylact of their proceedings in England* (787) 10, Haddan and Stubbs 3, 451).

[108] *Admonitio Generalis* (789) 79, *MGH CapRegFr* 1: 60-1. See also *Capitulare missorum generale item speciale* (802) 45, *ibid.*, 104; *Ansegisis collectio* appendix (1st half of 8th century) 34, *ibid.*, 447.

[109] There is ample archaeological and written evidence for the importance of animals in the religious life of Celts and Germans. See Green, *Symbol and Image in Celtic Religious Art*, 131-151, and *Animals in Celtic Life and Myth* (London and New York, 1992); Vordemfelde, *Die germanische Religion*, 100-111. Archaeological evidence for animal sacrifices and ritual use is examined in Patrice Méniel, *Les sacrifices d'animaux chez les Gaulois* (Paris, 1992).

[110] Nevertheless, hints of the cultic importance of animals can be found in hagiography, particularly in scenes where powerful animals are forced to perform saints' bidding, as in the case of the bear compelled by St. Corbinian to take the place of a pack animal which it had killed (*Vita Corbiniani retracta* 2, 10, *MGH SRM* 6: 609). For the importance of animals in hagiography in general, see Pierre Boglioni, "Il santo e gli animali nell'alto medioevo" in *L'uomo di fronte al mondo animale nell'alto medioevo*, SSAM 31 (Spoleto, 1985), 935-1002; in Irish *Lives*, see Sister Mary Donatus, *Beasts and Birds in the Lives of the Early Irish Saints* (Philadelphia, 1934).

forgotten customs.

In 5th-century Turin, the heads of livestock had been staked around the bounds to safeguard property. Something of the sort occurred among the Franks as well. Gregory I exhorted Queen Brunhilde to take steps to prevent people from making offerings to idols, worshipping trees, and exhibiting "sacrilegious sacrifices of animal heads." The people who indulged in these practices were not pagans, but ostensibly Christians, as Gregory makes clear: "many Christians hurry to church and at the same time–horrible to relate!–do not abandon the cult of demons."[111] Such practices probably had less to do with animal worship than with the cult of heads, widely-practiced by the Celts and to a lesser degree by the Germans.[112] A related custom appears in one recension of Ps.-Eligius' *De rectitudine catholicae conversationis,* that of placing the likenesses of livestock (*pecudum similitudines*) at crossroads.[113] Testimony of this sort is reinforced by both archaeological and written evidence.[114]

The Council of Orleans (541) condemned certain Christians' custom of swearing on the heads of wild beasts or cattle while invoking pagan names.[115] The names presumably belonged to the deities to whom the animals were sacred (for example, the bear to Arduinna or Diana, the stag to Cernunnos, the wolf to Woden).

Martin of Braga stated explicitly that his Galicians venerated mice and moths "as god" on certain days (around the New Year), when cloth and bread were put into a box for them to eat, but this was hyperbole. His own words show that the rags and crumbs were not offerings but merely omens (*praefigurationes*) for the New Year.[116] This must have been a strictly local practice; its perfunctory mention in St.

[111] Maximus of Turin, *S.* 91.2; *CCSL* 23: 369; Gregory I, *Ep.* 8.4 (597) *MGH Ep* 2: 7.

[112] For the cult of human heads, see Claude Sterckx, *La tête et les seins*; of animal heads, Bächtold-Stäubli, *s.v.* "Tierköpfe."

[113] *De rectitudine conversationis catholicae* 5, *PL* 40: 1173. In other recensions of the document, *pecudum* is replaced by *pedum* (feet)*;* if *pedum* is correct, the practice of placing votive offerings at crossroads is meant.

[114] Animal and human skulls with enlarged occipital openings, such as those found by Méniel at Gournay (Picardy), point to ritual use and could easily have been placed on spikes (*Les sacrifices d'animaux*, 36-37); Salin believed that animal teeth found in Anglo-Saxon tombs were probably remnants of skulls fixed on posts over tombs (*La civilisation mérovingienne* 4: 28). According to Ibn Fadlan, Volga Rus merchants displayed the heads of sacrificed cattle and sheep on stakes around the representations of their deities (Smyser, "Ibn Fadlan's account of the Rus," 97). A 9th-century traveller to Denmark described seeing sacrificed animals fastened to a pole outside each man's house as proof that he had performed his duties (Davidson, *Myths and Symbols,* 36).

[115] 16, *CCSL* 148A: 136. Vordemfelde (*Die germanische Religion*, 110-111) considered that the *Lex Frisionum* referred to this practice in a clause concerning oaths sworn "on clothing or cattle" (12.1, *MGH Leges* 3: 666).

[116] *De correctione rusticorum* (2nd half, 6th century) 11, Barlow, 190-191. Michel Meslin sees this as sympathetic magic, "un rite de bon départ, un *augurium* dont la fonction magique consiste à préserver le futur, à le rendre favorable à l'homme" ("Persistences païennes en Galice, vers la

Eligius' sermon is evidence only of the hagiographer's familiarity with *De correctione rusticorum*.[117] It is noteworthy that Pirmin of Reichenau, who relied heavily on the same work in compiling his *Dicta*, omitted to mention this rite.

According to Burchard of Worms' penitential, people trusted in the protective power of the cock's crow to quell and drive off evil nocturnal spirits milling around the homestead.[118] The meaning of sitting on the hide of a bull or ox at crossroads to foretell the future on New Year's Eve is more difficult to decipher.[119] Clearly a protective power resides in the skin, but contact with it may also give the man the ability to see more or differently. There may be a link between it and the animals on which the women were mounted during their night rides through the skies.

The controversial article *De cerebro animalium* ("concerning the brains of animals") in the *Indiculus superstitionum* may refer to a cult of animal brains, although no other reference to a Germanic cult of brains (as opposed to heads or skulls) has been found.[120] Animal brains were, however, used in medication and charms, and perhaps this is what is intended here.[121]

Finally, a letter by Aldhelm of Sherborne (d. 709) contains an explicit reference to a former cult of animal- or therianthropic divinities. He wrote of shrines in which *ermuli cervulique* had been "grossly and awkwardly worshipped" but which had now been converted to Christian uses.[122] Clearly, *cervulus* is a stag-deity, but what the *ermulus* may be is unknown. Rudolph Arbesmann showed that it was not one of the popular masks used at the New Year, as Du Cange had believed.[123] It

fin du VIe siècle," in *Hommages à Marcel Renard* 2 [Brussels, 1969], 512-524; here, 519).

[117] *Vita Eligii* (c.700-725), *MGH SRM* 4: 706.

[118] *Decretum* 19, 5.150, Schmitz 2: 442. Burchard makes it obvious that *cantus galli*, a conventional term for dawn, should be taken literally. The power of the cock's crow to chase away demons was known to Prudentius (Hymnus ad galli cantum 10, in *Carmina*, ed. Michael Cunningham, *CCSL* 126: 4). For additional texts, see Boglioni, "Il santo e gli animali," 972-973. See also Bächtold-Stäubli 5, 1340-4, *s.v.* "Hahnekrähen."

[119] 19, 5.62, Schmitz 2: 423. See Wilfrid Bonser, "Animal skins in magic and medicine," *Folklore* 73 (1962): 128-129 for further references.

[120] 16, *MGH CapRegFr*, 1, 223. Homann casts doubts on this interpretation, as also on the suggestion that this refers to a form of divination (*Der Indiculus* 98-100). Haderlein cites Albin Saupe (*Der Indiculus superstitionum et paganiarum, ein Verzeichnis heidnischer und abergläublischer Gebräuche und Meinungen aus der Zeit Karls des Grossen, aus zumeist gleichzeitigen Schriften erlaäutert* [Leipzig, 1891], 21) on the absence of documentary evidence for the use of animal brains in divination among the pagan Germans ("Celtic roots," 5).

[121] An external application of boar's brains mixed with honey was proposed as a remedy for boils on the genital organs and as a protection against snake bite (*Medicina antiqua* 1: 123, lines 23-25; 2: 121).

[122] *Ep. ad Ehfrid*, *MGH AA* 15: 489; *PL* 89: 93. I have used Arbesmann's translation, see next note.

[123] Du Cange, *s.v.* "Ermulus." Arbesmann proved that this passage refers to idols in stag- and other animal-forms in pagan shrines ("The 'cervuli' and 'anniculae' in Caesarius of Arles," *Traditio* 35 (1979), 89-119; 103). It is tempting to identify *ermulus* with the ancient Celtic deity

may have been the image of an animal or hybrid god, but by Aldhelm's day, a century after Gregory I launched his mission to the Anglo-Saxons, such cults had disappeared.

3.3 Cult of inanimate objects: trees, springs, stones, fire and the earth

The extent to which inanimate natural objects were venerated in themselves is difficult to determine. The usual formulation of the type "do not pray, make vows, *etc.*, *ad arbores et fontes et lapides*" suggests in most cases that prayers, vows and offerings were made near trees or in groves or by springs and stones, rather than to them; such passages are studied in chapter 5. But at times texts seem to refer not merely to location but to objects of devotion. *Ad* followed by the accusative may take the place of the dative, as in an early 9th-century sermon; *ad arbores vel ad fontes aut alicubi, nisi ad deum et sanctos eius et sanctam matrem ecclesiam dei auxilia querere.*[124] Here it appears that supplicatory rituals were directed to, not merely performed in the vicinity of, trees, springs and other places, as they were to God, His Saints, and Holy Mother Church.

3.3.1 *Trees, springs and stones*

The veneration of trees is well-attested among the pagan Celts and Germans by archaeological, iconographic and onomastic data,[125] as well as the writings of classical authors, medieval chroniclers and hagiographers.[126] The authors of our

whose name "[C]ernunnos" is inscribed over a horned god on a stone in Paris, and whose cult was widespread throughout Celtic territory. Images of horned gods have been found in England (especially in the north, among the Brigantes), although none is identified as Cernunnos who, however, lived on in medieval legend as Herne the Hunter. See Green, *Symbol and image*, 86-96, and *Dictionary of Celtic Myth and Legend, s.v.* "Cernunnos."

[124] Arno of Salzburg, *Synodal Sermon* (c. 806), Pokorny, 393-394.

[125] For references, see Green, *Dictionary of Celtic Myth and Legend, s.v.* "Tree." See also *idem, Symbol and Image*, 151-155; Bächtold-Stäubli 1: 954-958, *s.v.* "Baum;" Vordemfelde, *Die germanische Religion*, 79-84; Mannhardt, *Wald- und* Feldkulte; Grimm, *Teutonic Mythology,* 66-87 and 115-119. See also Bauschatz, *The Well and the Tree.* John Scheid, however, argued that the popular notion of a sacred tree or grove as an object of worship in itself was a result of the German romantic movement; in classical literature, sacredness came from the presence of a deity ("*Lucus, nemus*: Qu'est-ce qu'un bois sacré?" in *Les bois sacrés,* 13-20). See also Jean-Louis Brunaux, "Les bois sacrés des Celtes et des Germains" (*ibid.,* 57-65). The same case can be made for other natural objects or forces associated with religious belief.

[126] There are records of bloody sacrifices to or on trees. According to Tacitus, Germanicus found heads tied to the trees of the Teutoburgian forest where Arminius had annihilated Varus and his legions (*Annales* 1.61). St. Germanus, a great hunter in his youth, used to hang the heads of the animals he had slain on a great pine outside Auxerre (*Vita Amatoris* 24, *AASS* May 1: 58). The heads of men, horses and dogs were hung on trees outside the temple of Uppsala (Adam of Bremen, *Gesta* IV, *MGH SS* 7: 380). Missionaries were determined to eradicate tree cults–St. Boniface, for example, cut down the great oak of Jupiter at Geismar in Hesse (*Vita Bonifatii*

texts were very conscious of tree cults, and wrote repeatedly of them, but provided singularly few details, failing even to name the specific varieties singled out by devotees.[127] Only Martin of Braga and those who copied him identified the laurels and young branches that decorated houses at the New Year. The rest used the generic *arbor* or, rarely, *lignum* (wood). This reflects poorly on the authors' interest in a matter of such importance to their flocks.[128]

Adjectives such as *profanae, fanatici, sacrivi, consecratae* testify to the status of trees as objects of worship. Churchmen saw the reluctance to use the branches of such trees as proof of idolatrous awe. Those knowing of the existence in their neighbourhood of "profane trees" which were the object of vows should make every effort to destroy the tree, said Caesarius of Arles. He urged the faithful to burn "fanatical trees" (perhaps tree shrines) and to chop "sacrilegious trees" down to the very roots.[129] Clearly, in Caesarius' opinion, the devotees believed either that the tree was divine itself or that the *numen* residing in the tree would die when it died.[130] Even the dead tree was thought too sacred to be converted to domestic uses, despite massive deforestation in the region. He remarked bitterly on the folly of those who did not fear eternal fire as much as they feared using the branches for fuel:

> What a thing is that, that when those trees to which people make vows fall, no one carries wood from them home to use on the hearth! Behold the wretchedness and stupidity of mankind: they show honour to a dead tree and despise the commands of the living God; they do not dare to put the branches of a tree into the fire and by an act of sacrilege throw themselves headlong into hell.[131]

Trees dedicated to a religious purpose were *sacrivi* to the Synod of Auxerre (561-605) and St. Eligius, and *consecratae* to the Council of Nantes (7th or 9th century). The Synod forbade fulfilling vows to sacred trees, while Eligius ordered the destruction of fountains and "the trees that they call sacred," berating his flock likewise for fearing to take the dead wood home to use in the fire.[132]

auctore Willibaldo 8, *MGH SS* 2: 342-344).

[127] *E.g.,* the Celtic image of the sky god was an oak tree and Jupiter columns were sometimes decorated with oak leaves and acorns, while the German world tree Yggdrasil was an ash (Davidson, *Myth and Symbol,* 22-23).

[128] *Canones ex orientalium patrum synodis* (572) 73 and *De correctione rusticorum* (572) 16, Barlow, 141 and 198.

[129] *S.* 14.4, *CCSL* 103: 72; S. 53.1., *CCSL* 103: 233 and 234.

[130] The Roman general Gaius Suetonius Paulinus also believed that destroying the sacred groves on the island of Mona, where the Druids had reputedly been accustomed to offer human sacrifices, would put an end to such practices (Tacitus, *Annales* 14.30).

[131] *S.* 54.5, *CCSL* 103: 239.

[132] *Syn. Autissiodorensis, CCSL* 148A: 265; *Vita Eligii* (700-725), *MGH SRM* 4: 707-708. Auxerre was paraphrased by the *Epítome hispánico* (c.598-610) 43.4, 190. In Eligius's sermon, the choice

Interestingly, these fathers ignored the cult of stones although the second Council of Arles (442-506) testified to its existence. This Council warned bishops that they were guilty of sacrilege if they did not take action against unbelievers in their territory who "lit torches or honoured trees, wells and stones."[133] Caesarius may already have been in residence in Arles while this council was in session, and it is difficult to imagine that he was not aware of its deliberations. The omission from his sermons of any mention of this cult could hardly have been an oversight.

The *Vetus Gallica* included the Arlesian canon. Benedict Levita and Burchard of Worms dropped the term "unbelievers" and put the responsibility on the priest for his own parish rather than on the bishop for the diocese.[134] The difference may be insignificant. But possibly they included this canon with its minor modification because if Christians still practiced such cults privately, the parish priest was in a better position to know of them and to prevent them than the bishop in his see. Burchard's introduction of the clause, "Concerning those who lit torches at trees and springs," shows that he interpreted the canon as referring to devotions in a natural setting, not the outright worship of trees and wells.

An illuminating clause of the Council of Nantes, copied into 10th and 11th-century collections, condemned the cult of trees consecrated to demons and of stones found among ruins and in wild places in the woods.[135] "Demons' tricks" (perhaps eerie phenomena like the will-o'-the-wisps of marshlands) attracted people to such sites. The stones (and trees?) were believed to have healing power; people who sought a cure for themselves or for others brought lights and gifts to fulfill vows made there. The trees, so sacred that no one dared to cut branches or twigs from them, were to be uprooted and burnt. The stones were to be dug up completely and hidden from their votaries–these might have been large slabs of rock partly buried in the ground rather than stone statues.[136] To the Council, these objects were idols.

Other texts speak explicitly of the worship or worshippers of inanimate objects. Martin of Braga condemned the practice of placing bread in a spring (probably at the New Year). This may be an offering, similar to the offerings of food

of the verb "cut down" implies that *sacrivos* applied only to trees not to springs. *Sacriuus* is used in the Salic Law to designate certain barrows, presumably those meant for sacrifice (*Lex Salica* 2, 6, *MGH Leg* 1, 4.2: 30). For the relevant canon of the Council of Nantes, see below, n. 135.

[133] 21, *CCSL* 148: 119. See also *Collectio Vetus Gallica* (8th/9th century) 44.12, Mordek, 522.

[134] *Benedicti Capitularium Collectio* (mid 9th century), 3.316, *PL* 97: 839. See also Burchard of Worms, *Decretum* (1008-1012) 10.21, *PL* 140: 836.

[135] *Conc. Namnetense* 18, Aupest-Conduché, 54=20, Sirmond 3: 607. See also Regino of Prüm, *De synodalibus causis* (c. 906) 2.366: 352-353; Burchard of Worms, *Decretum* (1008-1012) 10.10, *PL* 140: 834.

[136] Perhaps like a stone idol described in the earliest *Life of St. Samson of Dol.* Samson was witness to rites in honour of a Cornish deity represented by "an abominable *simulacrum* on top of a mountain. " This was not a statue but a menhir, since we are told that the saint scratched a cross on "the standing stone" (*Vita Samsonis* 6.48, *AASS* Iul 6: 584).

and clothing which Gregory of Tours described as having been made to Lake Helarius.[137] The Toledan councils of 681 and 693 put "worshippers of stone, lighters of torches and devotees of the rites of springs and trees" under the heading of idolaters (*cultores idolorum*).[138] Burchard included a letter from Gregory I (598) deploring the actions of those who honoured trees and who committed "many other illicit acts" against true religion.[139] Nature cults were not dead even in regions which had long been Christianised. Atto of Vercelli claimed that he had taken pains to deal with those "who worshipped trees and approached springs thus piously"–this, in the middle of the 10th century.[140]

A couple of instances of magic from Frankish territory hint at the nature of the power immanent in trees. Around the end of the 8th century, country folk drove livestock through hollow trees and holes in the ground. Some fifty years later, some women who "despised" their husbands' seed put the semen into a rotting tree to thwart their desire to beget children.[141] One testifies to the belief in the life force of an old but living tree to protect animals from disease, the other to the sense that the dying or dead wood was a still active force for the negation of life.

3.3.2 *Fire and the hearth*[142]

Rituals over the hearth are related to the cults of the household fire and of the ancestors. Martin of Braga set the custom of pouring food and wine on the log in the hearth among New Year's rituals although this custom, intrinsic to Roman family ritual, was not traditionally limited to the New Year. Pirmin's *Dicta* echoed Martin's text, but in a 12th-century sermon, it was water that was poured on the

[137] Martin of Braga, *De correctione rusticorum* (572) 16, Barlow, 198; *Dicta Pirmini* (724-753) 22: 172; Gregory of Tours, *In gloria confessorum*, *MGH SRM* 1: 749-750.

[138] *Conc. Toletanum* XII (681) 11, Vives, 398-9; *Conc. Toletanum* XVI (693) 2, *ibid.*, 498-499.

[139] *MGH Ep* 2: 21. See also Burchard of Worms, Decretum (1008-1012) 10.2, *PL* 140: 833.

[140] *Capitula* 48, *PL* 134: 38.

[141] *Vita Eligii* (c.700-725), *MGH SRM* 4: 706; Poen. Ps.-Theodori (mid 9th century) 16.30, Wasserschleben, 576. There may a hint of magic or ritual in a practice described by Burchard of Worms in the midst of a list of perversions: "Did you commit fornication as some are accustomed to do, by putting your virile member in perforated wood, or something similar, so that by its movement and [resultant] pleasure you ejaculated semen?" (*Decretum* 19, 5.124, Schmitz 2: 436). Burchard saw this as a sexual sin only but it may, unsuspected by him, have contained elements of a fertility rite.

[142] According to Miranda Green, fire in pre-Christian religions "was perceived as the terrestrial element which corresponded to the sun in the heavens" with both positive and negative features. Fire festivals were particularly important in the more northerly parts of Europe, which enjoyed less sunshine than the Mediterranean regions (*Dictionary of Celtic Myth and Legend, s.v.* "Fire"). See also Grimm, *Teutonic Mythology,* 600-630. For European fire-festivals, see Fraser, *The Golden Bough*, 610-647. For cults of fire and of new fire in particular, see also M. Edsman, "Fire," (trans. D. M. and M. Paul in *ERE* 5: 340-346); for the role of the hearth in religion, A. E. Crawley *et al.,* "Hearth gods," (*ibid.,* 559-565).

hearth, while grain and wine were placed on the log.[143]

Other rituals are connected with ancestors more directly. Burning grain after a death for the well-being or health of the house or of the living first appeared in the *Penitential of Theodore of Canterbury*, then in about a dozen penitentials from every part of Europe except Ireland. This might have been either a propitiatory offering to the resentful dead or a means to drive the ghost away. Once again, the penalty of five years indicates that the practice was seen as seriously idolatrous until, at the end of our period, Burchard of Worms dismissed it as a trivial piece of magic by reducing the penance to twenty days.[144] Here the relationship between the living and the dead is distinctly hostile, but Burchard described a divinatory ritual that suggests a friendlier relationship. Grain was tossed on the freshly swept hearth; if the seeds skipped about on the still hot surface, they were warning of danger.[145] In this case, the hearth was the meeting-ground of the family, where the living turned to the dead for counsel and the dead responded as best they could.

Other aspects of fire cults may be discerned in magic involving fire, the oven, ashes and the use of smoke. In his New Year's sermons, Caesarius of Arles upbraided his congregation for refusing to share the household fire (and other goods) with outsiders, whether neighbours or strangers. This was an *augurium*–to share the first fire, bearing the symbolic importance of all new beginnings at the beginning of the year, meant to surrender a part of the prosperity of the family to others.[146] The same refusal to share fire appears in the *Homilia de sacrilegiis*.[147] Since the *Homilia* shows many signs of Caesarius' influence, its say-so would not in itself be convincing evidence for the prevalence of identical customs in two such different geographical and cultural settings divided by some two and a half centuries. However, the somewhat earlier *Indiculus superstitionum*, which originated in a very similar cultural milieu and which unquestionably reflects actual practices, provides some support in an article on "the pagan observance at the hearth or at the beginning of any business." Observances at beginnings surely refer to signs and omens when setting out on a voyage or starting a piece of work, but quite possibly the "observance of the hearth" refers to unwillingness to share one's hearth-fire at the New Year.[148]

Again, the *Indiculus superstitionum* draws attention to "the fire made by rubbing wood together, that is, *nodfyr*."[149] The German council of 742 also condemned

[143] Martin of Braga, *De correctione rusticorum* (572) 16, 198; *Dicta Pirmini* (724-753) 22: 172; Anonymous sermon to the baptised (12th century), Caspari, *Kirchenhistorische Anecdota* 1: 204.

[144] Poen. Theodori (668-756) I, 15.3, Schmitz 1: 537; Burchard of Worms, *Decretum* (1008-1012) 19, 5.95, Schmitz 2: 430.

[145] *Decretum* (1008-1012) 19, 5.101, Schmitz 2: 431.

[146] *S.* 192.3, *CCSL* 104: 781; S. 193.3, *ibid.*, 784. Meslin traces this refusal to give fire to an ancient taboo against violating the integrity of the family hearth ("Persistences païennes," 522).

[147] *Homilia de sacrilegiis* (late 8th century) 25, Caspari, 14.

[148] 17, *MGH CapRegFr* 1: 223. See Homann, *Der Indiculus*, 100-106.

[149] 15, *MGH CapRegFr* 1: 223. See Homann, *Der Indiculus*, 96-98.

"those sacrilegious fires which are called the *nied fyr*" as pagan observances.[150] The needfire–this, too, a fire newly made–had an apotropaic or therapeutic function in protecting or curing livestock driven through or around it. Bonfires were used in this way much earlier and much farther to the south as well. Peter Chrysologus (d. 450) had condemned the people of Ravenna who, at the New Year, "drive their animals [*bestiola*] around the place where they light a fire, and trust in such idiocy. They think that they are doing themselves good, and thus commit a serious sin."[151]

Belief in the therapeutic effects of smoke and fire was wide-spread. Caesarius of Arles preached against consulting healers specialising in the art of fumigation. In the *Homilia de sacrilegiis,* fumigation was used to drive the devil out of demoniacs.[152] Fumigators or perfume-burners *(suffitores)* appear among a list of pagan practices in a Carolingian pastoral letter.[153] It has already been seen that parents tried to exploit the healing power of fire by putting their feverish children in the oven, a practice containing perhaps elements of a ritual of rebirth. The art, known to Rhenish women, of burning human heads and mixing the ashes in a potion for their husbands, relied on the healing effects of fire as well.[154]

3.3.3 *The earth*[155]

Only St. Eligius' sermon suggests that the earth was adored along with the rest of creation.[156] We have seen, however, that a handful of texts which condemned swearing by natural objects specifically mentioned the earth, quoting *Matt* 5, 35, "because it is God's footstool." One text stands out. According to a 9th-century west Frankish capitulary, many people who obeyed the biblical injunction against swearing by the sky, nevertheless continued to swear by God and the heavenly hosts and by the earth, not just by the earth covering the saints' bodies (which, it appears, would have been acceptable).[157] There is no indication of the prayers and

[150] *Conc. in Austrasia habitum q. d. Germanicum* 5, *MGH Concilia* 2.1: 1, 3-4. See also *Karlomanni principis capitulare* (742) 5, *MGH CapRegFr* 1: 25. *Nedfratres* replaces *nodfyr* in Benedict Levita's collection (1. 2, *PL* 97: 704).

[151] *Homilia de pythonibus*, *PG* 65: 27-28.

[152] 22, Caspari, 12.

[153] *Epistola Canonica sub Carolo Magno* (c.814) 5, Mansi 13: 1095). For *suffitores*, see 6.3.1.

[154] Burchard of Worms, *Decretum* (1008-1012) 19, 5.177, Schmitz 2: 448.

[155] For a survey of earth cults, see Dieterich, *Mutter Erde.* In Celtic and Germanic paganism, the earth was usually personified by various mother goddesses, such as the *Terra Mater* Nerthus and Berecynthia, whose cults were described by Tacitus and Gregory of Tours, or the numerous *Matronae,* whose single or triple images are found widely scattered in England and on the continent, and who are depicted with signs of fertility such as babies, eggs, fruit or bread. In Ireland, fertility was guaranteed by the marriage of the king of Tara to the land in the form an earth goddess (*e.g.*, Ériu). See Green, *Dictionary of Celtic Myth and Legend, s.v.v..* "Mother goddesses," "Sacral kingship."

[156] *Vita Eligii* (c.700-725), *MGH SRM* 4: 708.

[157] *Capitula Franciae occidentalis* (9th century, 1st half) 11, *MGH CapEp* 3: 46. Also condemning

offerings that are outward signs of an active cult.

Nevertheless, a belief in the life-giving and life-sustaining power of the earth survived. Both continental and English texts present evidence of recourse to the earth for help in times of sickness of humans or animals. St. Eligius preached against driving cattle through openings in the ground. In 10th-century England, women indulged in the "greatly pagan" practice of taking a baby to the crossroads and dragging it along the ground or, more probably, through an opening in it (*per terram*). On the continent at about the same period, people "led" sick children through holes in the ground to cure them.[158] In Burchard of Worms' penitential, women used a related method to treat colicky babies:

> Did you do as certain women are accustomed to do? Those, I say, who have crying babies: they dig out the earth, and make a hole through a part, and they pull the baby through that hole, and they say that thus the crying baby will stop crying. If you did this or agreed to it, you shall do penance for five days on bread and water.[159]

A continental text introduced a variation on this theme. In the *Penitential of Ps.-Theodore*, it was not the baby (specifically, a "little son") who was drawn through the hole in the ground. Instead, another person (surely the father or mother) passed through the opening for the sake of the child, and then blocked the opening with thorns, presumably to keep the sickness or evil spirits causing it from following.[160] In all these cases, the apparent intention is to secure the welfare of a much-valued possession by entrusting it to the protection of the powerful forces residing in the earth.

The matter itself of the earth, the soil, had medicinal value. In the *Penitential of Ps.-Theodore*, it figures in a list together with skin, scabs, worms and body wastes, substances that other sources show were used as medicine.[161]

3.3.4 *The baptism of nature*

Although nature cults withstood the attempts of churchmen to destroy them, they were not necessarily static. A curious passage in Atto of Vercelli's St. John's Day sermon demonstrates the persistent sense among countryfolk in 10th-century Piedmont that nature and natural objects were personal forces; it also demonstrates strikingly the profound impact on popular consciousness of the Church's

swearing by creation are the *Hibernensis* (late 7th/early 8th century) 35. 3, H. Wasserschleben, 125; *Dicta Pirmini* (724-753) 18: 168; *Poen. Arundel* (10th/11th century) 37, Schmitz 1: 447-448; Burchard of Worms, *Decretum* (1008-1012) 19, 5.35, Schmitz 2: 417.

[158] *Confessionale Ps.-Egberti* (c.950-c.1000) 4. -, *PL* 89: 426; *Poen. Arundel* (10th/11th century) 97, Schmitz 1: 464).

[159] *Decretum* 19, 5.179, Schmitz 2: 448.

[160] *Poen. Ps.-Theodori* (mid 9th century) 27.16, Wasserschleben, 597.

[161] 31.32, Wasserschleben, 604.

emphasis on the sacraments. Atto protested with dismay against the innovative practice of women (he used the very derogatory word *meretriculae*) who, on the eve of St. John's, "baptised" clumps of grass and leafy boughs so that they might call them (the earth and trees? the turves and branches? the persons participating in the rite?) *compatres* and *commatres*, that is, godparents or intimate friends. Afterwards they strove "as though for the sake of piety" to preserve the greenery, which they hung in their houses.[162]

The teachings of the Church on marriage and incest made kinship by baptism equivalent to kinship by blood. The obligations of mutual support among kindred were felt strongly at a period when the power of the state to protect or control its subjects was unreliable at best. This gives rise to two possible explanations for this rite. It may have been a way, as Agnès Fine declares, of creating and reinforcing ties between the members of the community.[163] But it could also have been an attempt to bind earth and trees with ties of mutual, family obligation.[164] If so, we have here a wholly new way of looking at nature. It was still personified, still powerful, but no longer terrifying. There is no sense of the trembling awe which kept 6th-century Provençal peasants from using fallen branches as firewood: the sacramental magic of baptism had given the countryman or woman the means to harness the powers of Nature for his or her own benefit.

[162] *S.* 13 (*In annuntiatione beati praecursoris et martyris domini nostri Jesu Christi, Joannis Baptistae*), *PL* 134: 850-851. This passage is discussed further in chapter 4, 1.4.

[163] She considers this to be a way of creating liens of intimate friendship (the "compérage de Saint-Jean") between the persons who participate in this ritual, of which she quotes examples from Sardinia and Corsica as recently as the 1960's and 1980's (*Parrains, marraines. La parenté spirituelle en Europe* [Paris, 1994], 127-163, esp.149-158). See also John Lynch, *Godparents and Kinship in Early Medieval Europe* [Princeton, 1986] and J-P. Bouhot, "Le baptême et sa signification" in *Segni e riti nella chiesa altomedievale occidentale,* SSAM 33 (Spoleto, 1987), 251-267.

[164] Fine notes that the person who performed the baptism was also considered to be a godparent with special authority over the child (*Parrains, marraines*, 213-214). On the spiritual relationship between baptiser and baptised, see also John Lynch, "Spiritual kinship and sexual prohibitions in early medieval Europe" in S. Kuttner and K. Pennington, eds., *Proceedings of the Sixth International Congress of Medieval Canon Law.* Monumenta iuris canonici. Series C: Subsidia 7 (Vatican, 1985), 271-288.

4
Time

Among the pre-Christian peoples of Western Europe as elsewhere, the most important festivals of the year were associated with the lunar and solar calendars, with the cycle of the birth, growth, death and rebirth of nature.[1] Fertility and good fortune were guaranteed by rites designed to propitiate the forces of nature at the beginning and end of the growing and campaigning season, at the beginning of open pasturage, at planting, harvest and storage, at the solstices and, above all, at the New Year. These rites included sacrifices, expiatory offerings, observation of presages, sympathetic magic and feasting.

Both the cyclical concept of time and the very character of the rituals were deeply offensive to the Church. In the Christian view, time and nature are made and their course determined by God alone. Time is the medium in which God's

[1] Arnold Van Gennep called such ceremonies "rites of passage" (*Rites of Passage*, trans. Monika B. Vizedom and Gabrielle L. Caffee [Chicago, 1960]). Eliot D. Chapple and Carleton S. Coon prefer the term "rites of intensification;" the function of these is to restore "equilibrium for the group after a disturbance affecting all or most of its members" (*Principles of Anthropology* [New York, 1942], 507-528). A brief essay by Victor Turner and Edith Turner ("Religious celebrations," in *Celebration. Studies in Festivity and Ritual* [Washington, 1982], ed. Victor Turner, 201-219) offers valuable insights into the function and meaning of festive ritual in general. Jean-Jacques Wunenburger examined the origin, functions and essential elements of feasts in archaic religion in *La fête, le jeu et le sacré* (Paris, 1977), 23-108. See also Károly Kerényi, *The Religion of the Greeks and Romans*, trans. Christopher Holme (London, 1962). For festivals in the Middle Ages, see E. K. Chambers, *The Medieval Stage*, 2 vols. London 1903), especially vol. 1; J. Verdon, "Fêtes et divertissements en Occident durant le haut moyen âge," *Journal of Medieval History* 5 (1979): 303-314; Salin, *La civilisation mérovingienne* 3: 437-455. See Mikhail Bakhtin, *Rabelais and His World*, trans. Hélène Iswolsky (Bloomington, Indiana, 1984), on the regenerative aspects of ribaldry and the grotesque in medieval festivals. For the purificatory, protective and fertility-ensuring significance of seasonal festivals, see Arne Runeberg, *Witches, Demons and Fertility Magic*, 177-239. For the role of play in ritual, see Johan Huizinga, *Homo Ludens* (reprint, London, 1971), esp. 19-46, and for the criticism of Huizinga's equation of play and the sacred, Roger Caillois in *L'homme et le sacré* (3rd ed., Paris, 1950), 199-213. E. O. James's *Seasonal Fasts and Festivals* (London, 1961) is a useful handbook of feasts around the world; see especially 199-319. See also J. A. MacCulloch and H. Munro Chadwick, "Calendar (Celtic)" and "Calendar (Teutonic)," *ERE* 3: 78-82 and 138-141; MacCulloch and B. S. Philpotts, "Festivals and fasts (Celtic)" and "Festivals and fasts (Teutonic)," *ibid.*, 5: 838-843 and 890-891. For liturgical feasts, see Mario Righetti, *Manuale di storia liturgica*, 4 vols. (Milan, 1969) and for the pagan origins of Christian feasts, F. X. Weiser, *Handbook of Christian Feasts and Customs. The Year of the Lord in Liturgy and Folklore* (New York, 1958).

will is fulfilled. It is linear, beginning with the Creation (declared on the basis of *Gen* 1, 4 to have occurred at the spring equinox) and marching irreversibly toward the Last Judgment. Its central events are the Incarnation, Passion, Resurrection and the Last Judgment. Such an exalted view of cosmic time leaves little room for the mundane concerns of natural time in which man's immediate needs have to be met. To pre-Christian agrarian mentality, however, time was the measure of nature and determined natural processes; like nature itself, it was to be worshipped.[2] No wonder then that the celebration of seasonal rites constituted idolatry in the eyes of the Church. Even apart from the explicitly religious aspect, the festivals were accompanied by sexual license, excess in food and drink, masquerades and suggestive songs and dances, all of which the Church judged to be sinful. Festivals of this sort, wrote Károly Kerényi, reveal "the meaning of workaday existence, the essence of the things by which men are surrounded and of the forces which operate in their life." The urge to participate in the natural cycle by communal celebrations–a part of Kerényi's "festive sense," the awareness of the special sacred festive quality of time[3]–was not sublimated, still less suppressed, by conversion to Christianity.

The Church employed two principal strategies against these festivals. The first was to attempt–with very limited success–to ban them altogether. The faithful were to concentrate on the ultimate goal of personal salvation, and exhibit a Christian indifference to the secular and quotidian. The second was to establish liturgies to replace pagan rites, to obliterate the old natural calendar with a Christian calendar that substituted the events of sacred history for the turning points of the year. The liturgy of the Rogations was meant to fulfill the functions of the Ambervalia, while the Circumcision, the feast of St. John the Baptist, All Souls' Day and Christmas were superimposed on the Calends of January, Midsummer, the Celtic feast of the dead at Samhain, and the birthday of the Unconquered Sun respectively.

The results were mixed. Although the new Christian rituals and feasts were accepted, the traditional ones were not wholly abandoned, but often continued to be celebrated simultaneously. At the same time, Christian feasts and sacred places became the occasion and site of customs highly objectionable to pious churchmen. They found to their great distress that many of the faithful who had participated

[2] H. Hubert and M. Mauss, "Introduction à l'analyse de quelques phénomènes religieux," in M. Mauss, *Oeuvres* (ed. V. Karady; Paris, 1968), 30.

[3] "The festive quality which distinguishes certain demarcations of time attaches to all things within the compass of the feast and for those caught up within this compass and breathing this atmosphere–the festive persons themselves–is a wholly valid spiritual reality. Among spiritual realities this festive quality is a thing on its own, never to be confused with anything else. It can be confidently distinguished from all other feelings and is itself an absolute distinguishing mark." This quality, notes Kerényi, exists even apart from the biological needs which man expects nature to fill (*The Religion of the Greeks and Romans*, 53).

in Church services immediately afterwards indulged in obscene songs[4] and language, dancing,[5] mumming and questionable games, attended obscene shows, profaned the precincts of the Church with disgraceful behaviour, and spent the eve of Christian feasts in drunkenness and debauchery. Terms for such practices abound in our texts: *cantica turpia (luxuriosa, amatoria), cantationes, cantare; verba turpia proferre; ballationes/vallationes, saltationes* (evidently a kind of acrobatic dance composed of forward or backward vaults, bounds, maybe somersaults), *saltare, choros ducere* (round or choral dance), *circus*; *mimarcia; cervulum et anniculam facere, faciem transformare; lusa, ludi, iocationes, ioci, iottici, inebriare, lascivire.*

In this chapter, we consider the rituals of three types of festivals: (1) the natural, seasonal cycle (2) the liturgical cycle and (3) celebrations of family and communal life.

4.1 Seasonal cycle

4.1.1 *New Year–the Calends of January*[6]

The New Year was the main focus of the Church's efforts to Christianise seasonal celebrations. Here she failed utterly. Not surprisingly, the feast of the Circumcision failed to capture the popular imagination. Despite earnest attempts to persuade the faithful to observe it with prayer, fasting and penitence, the very pagan celebrations of the Calends of January (in Italy combined with the *brumae*, the festival of the winter solstice) continued to be practiced and even to evolve

[4] For medieval churchmen's attitude to music see T. Gérold, *Les Pères de l'Église et la musique* (Paris, 1932), 88-105. See also J.A. MacCulloch, "Music (primitive and savage)," *ERE* 9: 5-10. The bawdy song described in the Icelandic "Tale of the Völsi" (Boyer and Lot-Falck, *Les religions de l'Europe du Nord*, 79-84) may be an example of the kind of songs condemned by churchmen.

[5] For the origins and ritual significance of the dance, see C. Sachs, *World History of the Dance*, trans. B. Schönberg (Paris, 1930); R. Foatelli, *Les danses religieuses dans le christianisme* (Paris, 1939); L. Spence, *Myth and Ritual in Dance, Game and Rhyme* (London, 1947); E. L. Backman, *Religious Dance in the Christian Church and in Popular Medicine*, trans. E. Classen (London, 1952); M. Brillant, *Problèmes de la danse* (Paris, 1953); P. de Felice, *L'enchantement des danses et la magie du verbe; Essai sur quelques formes inférieures de la mystique* (Paris, 1957). For dances in churches, see Jacques Chailly, "La danse religieuse au moyen âge," in *Arts libéraux et philosophie* (Paris and Montreal, 1969), 357-380, and Louis Gougaud, "Les danses dans les églises," *Revue d'histoire ecclésiastique* (1914): 1-22 and 229-245. The magico-religious significance (for fertility, protection and victory in war) of classical hopping- or vaulting-dances, probably somewhat like the *saltationes* so frequently mentioned in early medieval texts, are discussed by Hubert Petersmann, "Springende und tanzende Götter beim antiken Fest" in Jan Assmann and Theo Sundermeier, eds., *Das Fest und das Heilige. Religiöse Kontrapunkte zur Alltagswelt* (Gütersloh, 1991), 69-87. See also H. Leclercq, "Danse, *DACL* 4.1: 248-258, and Louis Séchan, "Saltatio, " in Daremberg and Saglio 4.2: 1025-54.

[6] The celebrations of the Calends of January combined rituals dating from the earliest period of Roman history with others drawn from customs of the different peoples within the orbit of Roman influence. A detailed discussion of the origins, evolution and meanings of this feast is found in Michel Meslin, *La fête des kalendes*. See also Fedor Schneider, "Über Kalendas Ianuariae und Martiae im Mittelalter," *Archiv für Religionswissenschaft* (1920-1921): 82-134.

throughout the entire early Middle Ages. Sermons are the source of the most detailed descriptions, but New Year's rituals are also to be found in continental and English penitentials, the acts of a few continental councils, capitularies, and the correspondence of St. Boniface. They are wholly absent from Irish sources, and largely so from Iberian.[7]

Churchmen objected strongly to the Calends on both theological and moral grounds. In their eyes, the very practice of celebrating the New Year in January paid implicit honour to Janus by giving him credit for the creation of time. In the mid 5th century, Maximus of Turin reminded his flock that Janus had not been a god, but the human founder of a town (Genoa); whoever observed the Calends of January sinned by paying divine honours to a dead man.[8] A century later, the god had been forgotten. By the time of Caesarius of Arles, there was in fact no cult of Janus himself; the connection between the god and the season was made not by the people but by preachers, to stigmatise the festivities as much as possible. As churchmen gave up the struggle against the pagan names of the months, this theological polemic against New Year's festivities was gradually abandoned.

On the other hand, the opposition to seasonal magico-religious practices continued as vehemently as before, with little apparent success. Comparison of the earliest descriptions with those of later writers shows that although organised official celebrations (civic processions, circus games and theatrical performances) had disappeared, they were replaced by spontaneous popular manifestations (perambulations of town and countryside, pranks). Magical rituals were reinterpreted in terms of local culture and became more and more elaborate throughout the period. Mumming, sexual license, scandalous dances, songs and jokes, bouts of drunkenness and gluttony, giving gifts and refusing to give alms, taking auguries, were all intended, "according to pagan custom," to guarantee a prosperous New Year or, at the least, to gain a foreknowledge of what was to come. These customs are described, or at least mentioned, in over sixty of our texts, as well as in other types of documents.[9]

They were so well-known that many of our authors found it enough to refer to them only in the most general terms: "those who honour the Calends of

[7] A handful of references are found in Iberian documents. Calends of January customs are described in Martin of Braga's *De correctione rusticorum* (572), the *Epítome hispánico* (c. 598-610 43.2: 190 (and probably 1.73: 104) and a sermon of the second half of the 7th century, found in a *Toledan Homiliary* (*Hom.* 9, *PL Suppl* 4: 1941-1942). Customs similar to those of the Calends of January appear in two Spanish penitentials, but are not ascribed specifically to the beginning of the year.

[8] *S.*63.2, *CCSL* 23: 266.

[9] Isidore of Seville was particularly concerned about the confusion of the sexes: while some men disguised themselves as animals or monsters, others put on women's faces and imitate women's gestures; worse than all the other excesses was the mingling of men and women in wild dances (*De ecclesiasticis officiis* I, 41.2, *PL* 83: 775).

January," "if anyone does anything during the Calends of January which was invented by the pagans;" "whoever observes the Calends of January with enchantments and sorcery rather than religious piety ...."[10] Since most condemnations depended on two of Caesarius of Arles' sermons, it is worth considering at length what he had to say about the rituals (he admitted that he did not give an exhaustive description):

> During these days wretched men and, worse yet, some even who are baptised, don false appearances, monstrous disguises, in which I know not whether they are primarily laughingstocks or rather objects of sorrow. What sensible person indeed could believe that he would find sane people who deliberately transform themselves into the state of wild beasts while playing the stag [*cervulum facientes*]. Some are clothed in the hide of beasts, others don animal headdresses, rejoicing and exulting if thus they have changed themselves into the likeness of beasts so as not to appear to be men ....
>
> Now truly, what is this! how vile! that those who are born men dress in women's clothing and, by the vilest perversion, sap their manly strength to resemble girls, not blushing to clothe their soldier's muscles in women's gowns: they flaunt their bearded faces, and they aim to look like women .... [T]here are those who observe omens during the Calends of January, by refusing to give fire from their house or any other goods to anyone, no matter who asks; yet they accept diabolical gifts from others, and give them to others themselves. That night, moreover, some yokels [*rustici*] arrange little tables with the many things necessary for eating; they intend that the tables remain arranged like this throughout the night, for they believe that the Calends of January can do this for them, that throughout the entire year they will continue to hold their feasts in like plenty. And ... command your household to get rid of these and other practices like them, which would take too long to describe, which are thought by ignorant people to be trifling sins, or none at all; and command your household to observe these Calends as they do the Calends of the other months ...
>
> And, therefore, the saintly fathers of ancient times, considering how most of mankind spent those days in gluttony and lechery, going mad with drunkenness and sacrilegious dancing, ordained throughout the whole world that all the churches should proclaim a public fast, so that wretched men might know that the evil that they brought upon themselves was so great that all the churches are obliged to fast for their sins ... In fact, let no one doubt that anyone who shows any kindness to stupid men lewdly indulging in amusements during those Calends is himself a sharer in their sins ...[11]

In the second sermon, Caesarius amplified on the "demented" customs of

[10] *Conc. Turonense* (567) 23, *CCSL* 148A: 191-2; *Conc. Rothomagense* (650) 13, Mansi 10: 1202; *Poen. Arundel* (10th/11th century) 93, Schmitz 1: 462.

[11] *S.* 192.2-4, *CCSL* 104: 780-782. This sermon was repeated in all essential points in the 8th century by Burchard of Würzburg, *Hom.* 3, Eckhart, 837-839.

those who, for the sake of foolish gaiety, observed the Calends or "the folly of other superstitions" with license to get drunk and to indulge in "obscene chanting at games." He emphasised the indecent flaunting of men in women's clothing, make-up grotesque enough "to make the demons themselves blanch," bawdy songs in praise of vice sung with shameless gusto and accompanied by disjointed gestures, and the mumming in the likeness of she-goats and stags. "The inventor of evil makes his entry through these in order to master souls ensnared by the appearance of play."[12]

He called on the sober and upright members of his congregation to reprimand their neighbours and subordinates, to forbid them to use indecent language or sing bawdy songs, and to deny alms to those who "by sacrilegious custom" were carried away by insanity rather than by playfulness. "And, therefore, if you do not want to share in their guilt, do not allow a little stag or a little yearling or monstrosities of any other sort to appear before your houses, but rather chastise and punish them and, if you can, even tie them up tightly.... Admonish your household not to follow the sacrilegious customs of the unhappy pagans."[13]

Worse yet, insisted Caesarius, was the Arlesian custom of refusing to share the household fire with neighbours or with travellers who were in need, while, at the same time, exchanging gifts with others. He called once more on the faithful to join in a fast to shame "the fleshly and lewd joy of the pagans" and to pray for those who "by sacrilegious custom celebrated those calends for the sake of gluttony and drunkenness."[14]

Mumming:[15] Of all New Year's practices, mumming is the one most frequently mentioned by medieval pastors, in terms that clearly reflect Caesarius of Arles' influence. Nevertheless, they did not copy Caesarius blindly. They cited some of the masks that he described, ignored others and hinted at some that he had not mentioned.

Caesarius had described three different types of disguises: as animals, both domesticated and wild–specifically, as a stag (*cervus/cervulus*), female yearling (? *annicula*) or she-goat; as women; and as some other, undefined monstrosities (*species monstruosae, portenta*). What these monstrosities were we may guess from Peter Chrysologus' condemnation of the processions of New Year's maskers in fifth-century Ravenna. Here, too, men had dressed in the likeness of women, cattle or beasts of burden, but also in the masks of gods whose shameless behaviour they mimed; some of them blackened their faces with charcoal, daubed their bodies with dung and straw, and clothed themselves in rags and pelts.[16] That god-masks

[12] *S*. 193.1, *CCSL* 104: 783.

[13] S. 193.2, *CCSL* 104: 784.

[14] *S*. 193.3, *CCSL* 104: 784.

[15] See Roger Caillois, *Les jeux et les hommes* (Paris, 1958), 136-154.

[16] *Homilia de Pythonibus*, *PG* 65: 27; S. 155.6, *CCSL* 24B: 965; S. 155.2, *ibid.*, 962); S. 155 B.2, *ibid.*, 968. Colour plates of contemporary New Year's costumes accompanying Meslin's "Du

played a part in Arlesian mumming as well is suggested in Caesarius' reference to "those monstrous portents, that is, Mars, Mercury, Jove, Venus and Saturn" in his condemnation of the pagan names of weekdays.[17]

Such portents, whether gods or monsters, virtually disappear from subsequent texts. A sermon formerly attributed to Maximus of Turin shows unmistakable signs of Caesarius' influence in denouncing all the forms of mummery that Caesarius had mentioned: as women, beasts of burden, wild animals, and *portenta.*[18] In the middle of the 9th century, Herard of Tours wanted to ban and punish those who fabricated different kinds of portents, but this may refer to a form of divination. Two Spanish penitentials also mentioned those who "feigned monstrosities" (*qui ... monstruose fingunt*); this, however, is probably related to the festivities of May, not of January; moreover, the exact meaning of *monstruose* is in doubt.[19] The masks of the gods were abandoned, perhaps because the gods themselves had sunk into oblivion.

Animal masks, however, remained immensely popular, at least in pastoral literature. In the beginning, a number of different types of animal headgear or costume are cited, but in later texts mention is usually made of two only, the stag and some version of the female animal (yearling or she-goat). The difficulty of using pastoral literature to form a precise idea of early medieval culture is illustrated in the history of these two masks, which has been traced and analysed by Rudolph Arbesmann.[20] The first time that these creatures were paired was in one of Caesarius of Arles' sermons where, without referring to the Calends of January, he decried the "unhappy custom left over from the profane observances of pagans," that is, "the utterly foul wickedness of the *annicula* or *cervulus*."[21] The repeated references to them in the New Year sermons show that they were identified particularly–although perhaps not exclusively–with that feast.

Reports of the stag mask (*cervulus* or *cervus*) predated Caesarius of Arles' sermons. In the 4th century, the bishop of Barcelona Pacianus had written a book about this mask and St. Ambrose had also made a passing reference to it in one of his sermons.[22] However, the other creature, whose name Germain Morin, Caesarius' editor, gives as *annicula*, and which had been transcribed previously as *hinnicula, iuvenca, agnicula* (young female animal) or *anula* (old woman), underwent drastic changes.[23]

paganisme aux traditions populaires" give an idea of what of such costumes.

[17] *S.* 193.4, *CCSL* 104: 785.

[18] *S.* 16, *PL* 57: 257.

[19] Herard of Tours, *Capitula* (858) 3, *PL* 121: 764; *Poen. Albeldense* (9th century, 2nd half) 102, *CCSL* 156A: 12; *Poen. Silense* (1060-1065) 194, *ibid.*, 36. For the two latter, see May festivities below.

[20] "The 'cervuli' and 'anniculae' in Caesarius of Arles," *Traditio* 35 (1979) 89-119.

[21] *S.* 13.5, *CCSL* 103: 67.

[22] *Paranesis, PL* 13: 1081; *De interpellatione Iob et David* 4 (2), 1.5, *CSEL* 32.2., 271.

[23] Arbesmann, "'Cervuli' and 'anniculae'," 93.

Annicula itself is found only in one or two other texts. A peculiar phrase in a Carolingian letter concerns those who act or play the doe (*cervola*) and *anniculi* "that is, *suffitores*"[24] (the gender of the animals is reversed here). Either words are missing before *suffitores* (fumigators or perfumers) or, conceivably, *suffitores* were shamanistic healers who wore animal masks, used smoke in their rituals and put spells on horns. The *Indiculus superstitionum* may also contain a reference. Haderlein postulates that the clause *de cerebro animalium* ("concerning the brains of animals") is a mistranscription for *de cervulo [sive] annicula[m]* ("concerning the little stag or female yearling"). This entirely plausible reading, if correct, fills a puzzling gap in the *Indiculus*, which otherwise takes no notice of the celebration of the Calends.[25]

Except for these two, subsequent literature adopted some variant of the formula of a synod held at Auxerre (561-605): "It is not permitted to play the *uetolus aut ceruolus* (calf or stag) during the Calends of January."[26] *Vetolus* underwent numerous transformations, from *vetula* (old woman) through (among others) *vitulus, vecula, uicola, vegula* and *vehicula* to the wholly incomprehensible *feclus* of the *Homilia de sacrilegiis.*[27]

Such variations and those in the accompanying verb ("to act" or "play," "to observe," "to walk" or, most often, "to go") show a profound confusion as to the nature of the ritual involved, but it is clear that it entailed miming the animals' movements and habits in some way, not merely dressing in pelts or donning an animal headdress.[28] A scribe felt it necessary to gloss an early ninth-century penitential with the remark that "stags and *vetula* is what they do in the pagan style. It is done in jest, because men clothe themselves to resemble wild beasts, or put on the false image of beasts." Another penitential explained "to walk in [the guise of] a stag or *vetula*" in like manner: "they change into the habit of wild beasts, and put on the pelts of cattle and animal headdresses; in this way they transform themselves into animal-kind."[29]

[24] *Epistola Canonica sub Carolo Magno* (c.814) 5, Mansi 13: 1095.

[25] "Celtic roots," 5-8. It is quite possible, however, that the omission of references to New Year's customs was deliberate. St. Boniface, under whose influence the *Indiculus* must have been compiled, and who was outspoken about the failings of his flock (and the flocks of others), did not mention that his Germans practiced New Year's rituals. In his letter of 742 to Pope Zachary, he complained only of the bad example given to pilgrims by the public celebrations of the Romans (*Ep.* 50, *MGH Ep Sel* 1: 84).

[26] 1, *CCSL* 148A: 265.

[27] *E.g.*, respectively in *Epítome hispánico* (c.598-610) 43. 2, 190; *Poen. XXXV Capitolorum* (late 8th century) 18, Wasserschleben, 517; *Poen. Bobbiense I* (early 8th century) 30, *CCSL* 156: 70; *Double Penitential of Bede-Egbert* (9th century?) 33, Schmitz 2: 682; *Anonymi liber poenitentialis*, *PL* 105: 722; *Homilia de sacrilegiis* (late 8th century) 17, Caspari, 10-11.

[28] *E.g.*, respectively in *Vita Eligii* (c.700-725), *MGH SRM* 4: 705; *Poen. Hubertense* (8th century, 1st half) 35, *CCSL* 156: 112; *Poen. Sangallense tripartitum* (8th century, 2nd half) 21, Schmitz 2: 181; *Poen. Floriacense* (late 8th century) 31, *CCSL* 156: 100.

[29] *Poen. Valicellanum I* (beginning of the 9th century) 88, Schmitz 1: 311; *Poen. Ps.-Theodori* (mid 9th century) 27.19, Wasserschleben, 597. By referring only to animal masks, these glosses mili-

The original texts mentioning animal masks, Arbesmann noted, came from northern Italy, Gaul, Spain (but not, it should be noted, from Galicia)–all inhabited by Celtic populations, among whom there is ample evidence for a pre-Christian deer cult. Stag rituals in these regions can, therefore, be safely accepted. But the penitentials, in which the mention of these masks becomes increasingly perfunctory, were written mainly in Frankish territory, where there is no such evidence.[30] The pertinence of these texts to actual practice is therefore open to question.

It does not follow, however, that "the tradition of the Kalends masks on Frankish soil is purely literary."[31] Not every practice described and attacked by the early writers, however great their authority, was enshrined in medieval literary tradition. The custom of mumming in women's attire, as vehemently condemned by Caesarius of Arles as the stag-mask, is mentioned only some half dozen times in comparison with the almost thirty references to animal costumes. Even the *Ps.-Theodorian Penitential*, which drew directly on Caesarius (except for *annicula*), failed to note this practice. If the authors mentioned one mask often and the other seldom, we must assume that one was common, the other rare. The possibly counterproductive effects of penitentials and preaching must also be taken into account. An attentive congregation might pick up promising suggestions from admonitions to avoid certain types of behaviour. Pacianus of Barcelona had been dismayed to learn that the very attempt to condemn a stag rite gave it publicity and currency: "Unhappy me! What wickedness did I let loose! I think they would not have known how to play the stag had I not by my rebuke shown them how."[32] Burchard of Worms may have had this in mind when warning confessors of the need to use the utmost discretion in questioning penitents and distinguishing between public and private sins.[33]

Moreover, the variations in spelling may at times point to descriptions of different practices in different areas: mumming as an old woman in one region, as a heifer in another, perhaps with the maskers riding on a cart in a third. It is from this point of view that we should consider an unusual passage in an eighth-century penitential:

> If anyone honours the Calends of January or any other calends, unless they are feasts of the saints, and goes about as a *cerenus,* as the *circerlus* is called, or in/as a *uecula,* let him do penance for four years, because this truly is a remnant of pagan customs.[34]

tate against the contention of Du Cange (*s.v.* "Vetula") and Meslin (*La fête des kalendes,* 83) that *vetula* and its variations signify "old woman" rather than "heifer."

[30] Grimm, *Teutonic Mythology,* 755.

[31] Arbesmann, "'Cervuli' and 'anniculae'," 105.

[32] *Parenesis* 1, *PL* 13: 1081.

[33] *Decretum* 19.5, between questions 152 and 153, Schmitz 2: 442.

[34] *Poen. Oxoniense I* (8th century, 1st half) 29, *CCSL* 156: 91.

It is possible but unlikely that *cerenus* and *circerlus* are both a corruption of *cervulus*. The words are sufficiently far from the familiar original that one can expect something different–for instance, a candle-bearer circumambulating the town or village on foot or in a cart, if *uecula* can be understood as *uehicula*.[35]

In all probability, the practices described in garbled forms of the words of Caesarius of Arles and the Synod of Auxerre varied widely. Arbesmann is undoubtedly right in insisting that "though leading churchmen carried on a never-ceasing struggle against the beast-mimicry of New Year's Day, they had no real knowledge of its origin and initial significance. The same holds true for those who were reproved for taking part in the masquerade."[36] But, if the original forms of the ritual were lost, others evidently took their place; if the initial meaning was forgotten, undoubtedly new meanings arose.

Transvestism as part of New Year celebrations appears to have been less common, appearing in only a handful of texts. Ps.-Maximus denounced the kind of man who, "softening" his manly vigour, "deteriorated into woman" and "by his walk and behaviour acted as though sorry to have been born a man."[37] The *Homilia de sacrilegiis* followed Caesarius: "And, how vile! men who want to look like women put on women's dresses!"[38] Burchard of Würzburg simply copied Caesarius.

Other texts do not restrict such mumming to the New Year. Pirmin of Reichenau warned against going about in the likeness of stags or little old women at the Calends or any other time, and wearing the clothes of the other sex either then or during other fooleries (*lusa*).[39] Two continental penitentials mention mumming in the guise of the opposite sex.[40] Finally, two Spanish penitentials describe a ritual, of which the context is unclear, in which men dressed as women and mimed their gestures.[41] None link soldiers to cross-dressing.

Feasting: gluttony, drunkenness; singing, dancing and pranks: A century before Caesarius of Arles, St. Maximus had scathing words for the gluttony, drunkenness, indecent dancing and sexual licence with which Turinese, who had just received the Eucharist, celebrated a "banquet of superstition" at the New Year: "Befuddling the mind with wine, distending the belly with food, twisting the limbs in

[35] For carts used in New Year's processions, see Ginzburg, *Ecstasies*, 182-183.

[36] "'Cervuli' and 'anniculae'," 111.

[37] *S.* 16, *PL* 57: 257.

[38] 24, Caspari, 14.

[39] *Dicta Pirmini* (724-753) 22: 175. *Lusa* appears to stand for a variety of amusements which might include the hunt, secular songs, determined and excessive merriment, and the music of lyre and pipe–these being forbidden to the clergy, but permissible to "worldly men and the princes of the earth" (*Council of Friuli* [796 or 787] 6, *MGH Concilia* 2.1: 191).

[40] *Poen. Hubertense* (8th century, 1st half) 42, *CCSL* 156: 112. See also *Poen. Merseburgense b* (c.774-c.850) 29, *ibid.*, 176.

[41] *Poen. Albeldense* (c. 850) 102, *CCSL* 156A: 12. See also *Poen. Silense* (1060-1065) 194, *ibid.*, 36. These texts are discussed below, under the heading of May celebrations.

dances and engaging in depraved acts so that you are forced to forget what is God" is paying dues to an idol, he warned.[42] These feasts broke down the barriers of self-control even for good, habitually temperate Christians. Maximus understood the sick reaction and depression that followed such debauches, but also the psychological forces leading to them:

> What sensible person who understands the sacrament of the Lord's birth does not condemn the Saturnalia nor reject the lechery of the Calends? ... For there are many who still carry on with the superstitious old customs of the foolishness of the Calends. They celebrate this day as the highest feast. Where they look thus for happiness they find, rather, sorrow. They wallow in wine and sicken themselves on feasting so that he who is chaste and moderate all year gets drunk and pollutes himself; and if he does not does so, he thinks that he has been deprived of the feast.[43]

Caesarius, too, was outraged by the gluttony, drunkenness and debauchery to which celebrants gave free rein, by the leaping dances (*saltationes*) special to seasonal feasts, and the contortions and bawdy songs praising vice performed by soldiers in women's clothes.[44] His successors in general played down such excesses. Only an anonymous seventh-century Toledan sermon treated them in comparable depth, adding new details to elements borrowed from Caesarius and, perhaps, Maximus. Those who had duly celebrated Christ's birth should avoid the feasts of pagans and customs of fools: bacchanalian songs, loud outbursts of raucous laughter, participation in foolery, contorted leaps or dances resulting at times in broken limbs, the discordant sound of clapping. Janus' feast, it claimed, gave gluttons and drunks the occasion "to serve under Venus' banner" and led to fighting.[45] The term *cacchinnatio,* wild laughter, suggests the possibility that hallucinogens were used, with the hypnotic effect intensified by rhythmic clapping.[46]

Among later texts, only St. Eligius' sermon put inordinate drinking among other forbidden seasonal antics.[47] The rest ignored the gourmandising and drunkenness of the New Year, although penitentials in general had a great deal to say about excess of this sort. The reticence may have been due to the fact that among more northerly peoples festive drunkenness was acceptable, even normal, if not indeed compulsory. It is significant that from the beginning of the 8th to the end

[42] *S.* 63.1, *CCSL* 23: 266.

[43] *S.* 98.1, *CCSL* 23: 390.

[44] *S.* 192.4 and *S.* 193.1, *CCSL* 104: 781 and 783.

[45] *Hom.* 9, *PL Suppl* 4: 1941-1942.

[46] William A. Emboden noted that that Democritus spoke of the "immoderate laughter" of those who drank a mixture of cannabis, wine and myrrh ("Ritual use of *cannabis sativa*: A historical-ethnographic study," in *Flesh of the Gods*; *The Ritual Use of Hallucinogens*, ed. Peter T. Furst [New York and Washington, 1972], 214-236). Bakhtin maintained that immoderate, irreverent laughter is an essential aspect of the carnivalesque (*Rabelais and His World*).

[47] *Vita Eligii* (c.700-725), *MGH SRM* 4: 705).

of the 9th century, penitentials of Irish, English and continental origin followed the example of Theodore's in acquitting monks, priests and deacons of guilt for getting drunk to the point of vomiting if (among other extenuating circumstances) they had done so out of obedience to their superiors, for joy at the Nativity, Easter or a saint's feast. Moreover, they found it acceptable for a superior to order his subordinates to drink to this extent as long as he did not do so himself.[48]

Similarly, the singing or chanting, dancing and leaping, bawdy talk and pranks or games that distressed Caesarius are missing from most of the later accounts of Calends of January rituals. Yet such practices continued. We have the evidence of the Toledan homily and still more of St. Boniface. The latter wrote in barely respectful terms to Pope Zacharias in 742, upbraiding him for New Year's celebrations that ran day and night in the neighbourhood of St. Peter's itself, and scandalised pilgrims, the visiting *carnales homines idiote, Alamanni vel Baioarii vel Franci* of Boniface's own flock. These had been eye-witnesses to round dances performed in the roads, pagan outcries and sacrilegious songs, tables set day or night with food, determined refusal to give charity, and the spectacle of women flaunting amulets and ligatures on arms and legs, and openly offering them for sale.[49]

Significantly, the pope did not dispute the accuracy of Boniface's charges, but protested that he had done what he could to prevent such behaviour. The following year, the Roman Council attempted yet again to ban some of the same customs. It directed a canon (repeated by Atto of Vercelli) against the "pagan" way of observing the Calends of January and the *bromae*: setting tables with dishes of food, singing and dancing through the villages and along roads.[50] If the pope himself was powerless to prevent such goings-on in the heart of Latin Christendom, the outlook was poor indeed for pastors in more remote areas.

There is no wholly convincing evidence, even in the earliest of our documents, of frequentation of the New Year's spectacles (theatrical presentations, games and races) which had formed an integral part of the celebrations of the calends of January under the Roman Empire.[51] Caesarius condemned spectacles in other sermons but there is only a slight indication that this belonged to the festivities of the season. In the introduction to a New Year's sermon, he listed the passion for shows as one of the sins which become pride if persisted in knowingly.

[48] Poen. Theodori (668-756) 1.4, Schmitz 1: 525. In general, Theodore took clerical drunkenness very seriously, giving it pride of place in his penitential.

[49] *Ep.* 50, *MGH EpSel* 1: 84. See also the pope's reply (Ep. 51, *ibid.*, 90).

[50] *Conc. Romanum* (743) 9, *MGH Concilia* 2.1: 15-16. See also Atto of Vercelli (d. 961) *Capitulare* 79, *PL* 134: 43; Burchard of Worms repeated the injunctions against singing and dancing, but omitted *brumae*, which might have been unknown in the Rhineland (Decretum [1008-1012] 10.16, *PL* 140: 835 and 19, 5.62, Schmitz 2: 423).

[51] See Meslin, *Les fêtes des kalendes,* 66-70, for a discussion of the public games held during the New Year. Verdon suggests that Merovingians and Carolingians preferred activities in which they could participate (songs, dances, races, sports) to spectacles, where they were merely onlookers ("Fêtes et divertissements en Occident," 308).

"This is what happens when for the sake of silly merriment, the calends or other superstitious follies are observed with licence for drunkenness and the degenerate chant [song?–*cantus*] of games [*ludi*]."[52] But these *ludi* were probably pranks and masquerades rather than the formal games of the circus: in the same sermon, Caesarius urged his flock to show no indulgence to those "who run mad rather than play" in the sacrilegious ritual of mumming.[53] A reference to "spectacles of animal games" appears in the Toledan homily, but the expression should probably be understood as part of an extended figure of speech, not the description of an actual event.[54]

Spontaneous or ritualised forms of play took the place of formal games. St. Eligius charged the faithful not to perform "fooleries" at the New Year, and the *Homilia de sacrilegiis* included "other wretched things or games" among the rituals of the New Year.[55] Children no doubt participated in these as much as adults. When St. Samson of Dol came across a group of youngsters running loose through the countryside "on account of that abominable day," he sent them off with a gentle scolding and a penny each, thus showing a spirit of indulgence that would have pained Caesarius of Arles.[56]

Arrangement of favourable omens: Caesarius of Arles described two seemingly contradictory practices meant to augur prosperity during the New Year. On the one hand, one jealously guarded the goods of one's house, and refused hospitality or even the loan of fire to needy neighbours or travellers, for fear of subtracting from one's own wealth; on the other, one accepted and gave "diabolical" gifts (*strenae*).[57] At the root of the refusal of charity was a kind of mercantilistic theory of good fortune, the intuition that each person prospered at the expense of his neighbour, and that the only way to safeguard one's luck was to refuse to share it with others. The exchange of gifts was equally devoid of generosity, being meant to ingratiate oneself with those who could be of use.

To Maximus of Turin a century earlier, the hypocrisy and abjection of the one emphasised the heartlessness of the other. His exasperated rage is palpable in his description of the obsequious gestures (proferred lips, hand insinuated into hand) accompanying this ritual; for once, his habitual rough jocularity gives way to complete, almost physical disgust. He emphasised the injustice of such ruinously

[52] *S*. 193.1, *CCSL* 104: 783.

[53] *S*. 193.2, *CCSL* 104: 784. Circus games were held in Arles during Caesarius's lifetime. The town had a fine arena graced with an obelisk imported from Egypt (J. H. Humphrey, *Roman Circuses: Arenas for Chariot Racing* [Berkeley and Los Angeles, 1986], 398). Caesarius does not use the word *ludi* in reference to spectacles or the circus, but he refers to gambling as *tabulam ludere* (*e.g.*, *S*. 61.3, *CCSL* 103: 269).

[54] *Hom*. 9, *PL Suppl* 4: 1942.

[55] *Vita Eligii* (c.700-725), *MGH SRM* 4: 705; Homilia de sacrilegiis (late 8th century) 17, Caspari, 10-11.

[56] *Vita Samsonis* 2.13, *AASS* Jul 6: 590.

[57] *S*. 192.3, *CCSL* 104: 781. See also *S*. 193.3, *ibid.*, 784.

extravagant gifts: the inferior deprived his own children of necessities to give to his superior in the hopes of winning his favour while the rich man gave generously to the well-off and refused charity to beggars.[58]

The practice of exchanging gifts and denying alms and hospitality at this time of the year either died away or was no longer considered to be particularly harmful. References to it are scarce. The Synod of Auxerre and St. Eligius mentioned it only in passing, without comment, and the *Homilia de sacrilegiis* merely paraphrased Caesarius.[59] But St. Boniface testified that even in the mid 8th century, Romans refused to give objects made of iron as well as fire and other goods during the New Year.[60] Only in one of Atto of Vercelli's sermons is there a serious attack on the custom. Although he used terms such as "sorcery," "traditions associated with Janus" and "rites associated with pagans," his focus was not on magical beliefs or paganism as such, but on greed, selfishness and social injustice. "Certain false Christians" of his diocese refused entry to anyone who came to the door without gifts, refused to practice hospitality and utterly refused to give or lend anything from the house. "It were better for you," he wrote, "if you received Christ or the poor into your house as guests, rather than the peasants–whom you and yours oppress–when they come bearing platters of food and containers of wine, as they were told to do."[61]

Insofar as our texts allow us to judge, the refusal to give to the poor at this time of the year kept the same meaning throughout our period. The same cannot be said of the custom of loading a table with food. In the Narbonnaise and Galicia of the 6th century, this was, Meslin noted, expression of typically Roman religious mentality, which considered the table to be a sacred object, with a bare table being a symbol of famine at any time, but particularly so at the beginning of the year.[62] But when a similar ritual was described later and farther to the north, it had changed both in detail and in purpose.

Caesarius of Arles denounced the rustic custom of setting little tables on New Year's Eve with all the foods "necessary for eating," in the belief that this would ensure abundance for the coming year.[63] Martin of Braga warned his flock that garnishing the table was the practice of a diabolical cult.[64] In the eighth-century *Homilia de sacrilegiis,* "garnishing" meant preparing tables at night with loaves of bread and other foods, and "treating them with reverence" on the day itself. A

[58] *S*. 98.2, *CCSL* 23: 390.

[59] *Syn. Autissiodorensis* (561-605) 1, *CCSL* 148A: 265; *Vita Eligii* (c.700-725), *MGH SRM* 4: 705; *Homilia de sacrilegiis* (late 8th century) 25, Caspari, 14-15. The clause from the Synod of Auxerre was copied into *the Epítome hispánico* (c.598-610) 43.2: 190.

[60] *Ep*. 50, *MGH EpSel* 1: 84.

[61] *S*. 3 (*In festo octavae Domini*), *PL* 134: 835-836.

[62] "Persistences païennes," 521.

[63] S. 192.3, *CCSL* 104: 781.

[64] *De correctione rusticorum* (572) 16, Barlow, 198. See also *Anonymous sermon to the baptised* (12th century), Caspari, *Kirchenhistorische Anecdota* 1: 204.

second paragraph is close to Caesarius' own text: tables were spread with many different things in the belief that this would guarantee plenty for all the repasts of the year.[65] Another continental sermon of same period likewise condemned those who set tables during the Calends.[66] Despite the efforts of the pope and the Roman Council, Romans in the mid 8th century heaped their tables with dishes of food (*dapibus*) on New Year's day or night.[67] Some two centuries later, Atto of Vercelli included this Council's canon in his capitulary with a modification: *dapibus* becomes *lampadibus* –with torches, not with dishes of food. Burchard of Worms also paraphrased the same Council with alterations, anathematising those who prepared "tables in their house with stones (*lapibus*) and lavish foods." One may conjecture reasonably that *lapidibus* is a mistake for either *dapibus* or *lampadibus*.[68]

Here Burchard followed the tradition inherited from Caesarius of Arles and Martin of Braga. But in another clause, he gave a completely new meaning to the custom. Women laid the tables with food and drink and three "little" knives for the refreshment of the *Parcae*, the three sisters who could help those who treated them well.[69] The ritual took place "at certain times of the year;" the New Year is not specified but undoubtedly it was the most important. The only other penitential to mention the *Parcae's* table does not limit the practice to women: "He [*qui*] who prepares a table in servitude to the *Parcae* is to do penance for two years."[70] In these, the table no longer bore a symbolic meal but food and drink meant to be consumed by other-worldly visitors. It is not Roman or Gallo-Roman magic but, despite the classical reference, a Germanic propitiatory rite.[71]

The branches decking the houses of Turinese in the mid 5th century and Galicians in the late 6th were no token ornaments. Maximus of Turin described how his flock returned from the countryside on the second day of January, brandishing twigs, almost staggering under the weight of the burden, in the belief that thus "they measured out for themselves the prosperity or grief of the whole year, not realising, the wretches! that they were carrying a load of guilt rather than a sackful of gifts."[72] Galicians, wrote Martin of Braga, "wasted time in the pagan fashion" and wreathed their houses in laurel and green branches,[73] or "placed

[65] 17 and 25, Caspari, 10-11 and 14-15.

[66] Levison, 310.

[67] Boniface to Zachary, *Ep.* 50, *MGH EpSel* 1: 84; *Conc. Romanum* 9, *MGH Concilia* 2.1: 15-16.

[68] *Decretum* (1008-1012) 10.16, *PL* 140: 835. See also *idem,* 19, 5.62, Schmitz 2: 423.

[69] *Decretum* (1008-1012) 19, 5.153, Schmitz 2: 443. See chapter 2, n. 59.

[70] *Poen. Arundel* (10th/11th century) 83, Schmitz 1: 460.

[71] McKenna believes that this was already the meaning of Martin of Braga's decorated table: "They thought that the demons would eat this food and in return would grant them abundance of everything during the rest of the year" (*Paganism and Pagan Survivals*, 98). Germanic influences are entirely plausible in an area that had been settled by the Suevi during the 5th century.

[72] *S.* 98.3, *CCSL* 23: 391-392.

[73] *Canones ex orientalium patrum synodis* (572) 73, Barlow, 141; *De correctione rusticorum* 16, *ibid.,* 198.

pieces of laurel" around.[74] This condemnation was repeated by clerics who were unlikely to have observed the use of laurel in their more northerly dioceses.[75]

Another attempt to assure success during the year, the practice of starting all kinds of work on New Year's Eve is described in Burchard's penitential. The question is addressed to both sexes (the interrogative pronoun is masculine), but the activities mentioned are associated with women rather men: spinning, weaving and sewing together and "every other kind of work."[76] The *Indiculus superstitionum*'s clause concerning pagan observances when beginning some work and Regino of Prüm's question, "Did you do or say anything in any activity, or did you begin anything with sorcery or with the art of magic?" may also apply to New Year's practices.[77]

Divinatory and other rites: The term *auguria observare* in Caesarius of Arles' Calends sermons applies to the arrangement of favourable omens, what may be called the "observance" rather than "observation" of auguries. The rituals considered until now have been of this type. Caesarius' failure to mention actual attempts to read the future at this crucial turning point of the year is all the more astonishing because divination played an important role in his own time and because of the rich variety of techniques appearing in subsequent literature. Some texts only hinted at such practices either specifically at the New Year or at beginnings in general,[78] others gave concrete details.

Some methods involved the observation of animal behaviour. In Galicia, claimed Martin of Braga, a "day of moths or mice" (*dies tiniarum vel murorum*) was observed, during which bread and rags were placed in a box to see how the moths and mice dealt with them.[79] This form of augury had limited appeal either in

See also Rabanus Maurus, *Poenitentium Liber ad Otgarium* (842-843) 24, *PL* 112: 1417; Burchard of Worms, *Decretum* (1008-1012) 10.15, *PL* 140: 835.

[74] *De correctione rusticorum* 16, Barlow, 198. See also *Dicta Pirmini* (724-753) 22: 172; *Anonymous sermon to the baptised* (12th century], Caspari, *Kirchenhistorische Anecdota* 1: 204.

[75] The *laurus nobilis* symbolised distinction and victory in Roman society and, as an evergreen, especially one associated with Apollo, it was a natural symbol for life and prosperity; it may also have been used as a mood-altering substance. For the cultic use and psychotropic effects of the laurel, see Rätsch, *Dictionary of Sacred and Magical Plants*, 57-58.

[76] *Decretum* (1008-1012) 19.5, 104, Schmitz 2: 432). McNeill and Gamer translate *filare* ("to spin") as "to wind magic skeins" (*Medieval Handbooks of Penance*, 355). Du Cange also gives *filare* (*s.v.*) a magical connotation. Sewing together as well may have a magical nuance of connecting beginning and end. The text, however, may simply refer to protective magic for ordinary spinning and needle-work.

[77] *Indiculus superstitionum et paganiarum* (743?) 17, *MGH CapRegF* 1: 223; Regino of Prüm, *De synodalibus causis* (c. 906) 1.304: 145.

[78] *E.g.*, *Homilia de sacrilegiis* (late 8th century) 17, Caspari, 10-11.

[79] *De correctione rusticorum* (2nd half, 6th century) 11, Barlow, 190-191. It is tempting to speculate on how Martin of Braga, a high ecclesiastic and foreigner, acquired knowledge of what is obviously a very private peasant practice. Martin, a Pannonian by origin, might have been particularly sensitive to magical practices involving insects: jewellery in the form of insects is very rare, but out of a total of sixty-four samples of ornaments in that shape (flies, crickets, or

practice or to clerical authors, for only St. Eligius mentioned it in passing, and without tying it to the Calends of January.[80] Ps. Maximus records the attempt to "peer into the coming year by means of random signs given by birds and wild beasts."[81] Auguring from birds and beasts also appeared in two ninth-century documents from the Moselle region.[82]

Bread was used for divination. Martin of Braga (and, after him, Pirmin of Reichenau and an anonymous twlefth-century preacher) mentioned the custom of throwing bread into a spring or well. This is placed in the midst of other divinatory practices, and might be an attempt to foretell the future from the way the crust was carried off by the current.[83] In the Rhineland, according to Burchard of Worms: loaves were prepared "in one's name;" a firm, dense, well-risen dough augured well for the year.[84] This use of bread leads once again to a Carolingian text dealing with Italian customs. The canon describes a practice apparently associated with the winter solstice (*bruma*), a prelude to the New Year's festivities:

> Concerning those wicked men who honour [or keep–*colunt*] the *brunaticus*, and who burn candles beneath the *maida,* and make vows concerning their own men: that each should indeed remove them from such iniquity. But if they want to offer bread to the church, let them do so in a natural fashion, not with an admixture of any of that wickedness.[85]

The text is unclear. Is the *brunaticus* the winter solstice itself or a solstice ritual? As to *maida*, Du Cange was surely wrong in his guess that they were related to the ancient Gaulish custom of placing wood or wax models of human limbs at the crossroads.[86] Were he right, placing a candle under them would have been a form

possibly moths), none came from Iberia, twenty-four came from Hungary and twenty-one from southern Russia (Salin, *Civilisation mérovingienne* 4: 180-186). German belief associated rodents with night and death (*ibid.*, 67).

[80] *Vita Eligii* (c.700-725) 16, *MGH SRM* 4: 706.

[81] *Hom.* 16, *PL* 57: 258.

[82] *Capitula Treverensia* (before 818) 5, *MGH EpCap* 1: 55; *Anonymous sermon* (c. 850-882) 5, Kyll, 10.

[83] Martin of Braga, *De correctione rusticorum* (572) 16, Barlow, 198; *Dicta Pirmini* (724-753) 22: 172; *Anonymous sermon to the baptised* (12th century), Caspari, *Kirchenhistorische Anecdota,* 1: 204. Meslin considers this to be divinatory ("Persistences païennes," 521), but McKenna thought that it was a fertility rite (*Paganism and Pagan Survivals in Spain*, 103).

[84] *Decretum* (1008-1012) 19, 5.62, Schmitz 2: 423).

[85] *Karoli Magni et Pippini filii capitula cum Italiae episcopis deliberata* (799-800?) 3, *MGH CapRegFr* 1: 202.

[86] Du Cange (*s.v.*) apparently arrived at this definition on the basis of *maidanum* or *maydanum=platea* or *campus*, a usage for which he gives 14th-century examples (*cf.* Urdu and Arabic *maydan*). A more plausible explanation is suggested by Chambers (*The Medieval Stage* 2: 304). He accepted Friedrich Diez's translation as breadtrough (*backtrog*) from the Latin *magis,-idis* (*Etymologisches Wörterbuch der romanischen Sprachen* [Bonn, 1853] no. 4980, *s.v.*). *Cf.* modern Italian *madia* ("kneading-trough," "bread-bin").

of black magic used against one's followers. The editor of this text, A. Boretius, however, pointed out that the last part of this ordinance indicates that *maida* had to do with magically-prepared bread. The magical process must have consisted of singeing or fumigating the loaves or breadtrough, or marking them with the smoke of candles held below. Moreover, the fulfillment of the magic required that the breads be presented in church, which argues against its being malign in intent. Also unclear is who or what are the "they" (*eos*) that should be "removed from such iniquity"–the men for whom the offering was made (which seems unlikely), the loaves of bread or the trough (but *maida* is neuter plural), or the participants (in which case one would expect to find the reflexive pronoun)?

Prognostication by "observing the foot" is mentioned by Martin of Braga and Pirmin of Reichenau.[87] This may mean taking care in the morning to set the auspicious foot on the ground first, to enter or leave with the right foot leading, or simply to avoid stumbling, always a dire omen.[88] There may also be a connection between this practice and the custom, forbidden by the Council of Saragossa in 380, of walking barefoot during the three weeks preceding Epiphany.[89]

Forms of divination involving enigmatic figures or objects, the roof and the crossroads are juxtaposed in the texts and are possibly associated with each other and with the New Year. *Inpurae* or *inpuriae* appear in three eighth-century sermons from roughly the same region of Frankish territory. In the *Vita Eligii* sermon, the statement that "no Christian should believe *inpurae* nor sit *in cantu* [in a corner? within a wheel?] for that is a diabolic deed" comes between Calends of January practices and solstice rituals at the feast of St. John the Baptist.[90] This is fairly straightforward, but the syntax of the others is obscure. In Pirmin of Reichenau's *Dicta*, *inpuriae* are placed between weather sorceresses and the masks of the Calends: "Do not believe weather-sorceresses, nor give them anything for that reason, nor *inpuriae* who/which, they say, put men on the roof [whom/which, they say, men put on the roof?–*inpurias, que dicunt homines super tectus mittere*] so that they can tell them the future, whatever of good or evil is coming to them. Do not believe in them, because only God knows the future. Do not go about as stags or old women at the Calends or some other time."[91] An all but incomprehensible sentence in Levison's *Anonymous Sermon* puts them together with other superstitions, but clearly associates them with New Year's customs.[92] It combines the Calends of January "which accursed Janus taught," the *inpurae*, setting tables, awaiting omens,

[87] *De correctione rusticorum* (572) 16, Barlow, 198; *Dicta Pirmini* (724-753) 22: 172.

[88] *Cf.* the habit of 6th-century Arlesians of making the sign of the Cross as a counterspell when they tripped (Caesarius of Arles [502-542] *S.* 13.1, *CCSL* 103: 64-65, and *S.* 134.1, *CCSL* 104: 550).

[89] 4, Vives, 17. This canon was directed against the Priscillianists. For the mythical significance of bare feet, see Ginzburg, *Ecstasy*, 226-295, and Heckenbach, *De nuditate sacra,* 1-68.

[90] *Vita Eligii* (c.700-725), *MGH SRM* 4: 705.

[91] *Dicta Pirmini* 22: 173-174.

[92] Levinson, 310-311.

observing birdsong, sneezes and signs (constellations?) and "the *mavonis* [*sic*] as though they were able to carry off the crops and the grape-harvest ...." We may only guess at the relationship between all these elements.

What the *inpurae* are is also a matter of guesswork. Blaise (using Pirmin's text as a source) defined *impuriae* as "filth" ("ordures"), thereby ignoring an impressive body of German scholarship. Caspari, Jecker and Boudriot believed that the word came from the Greek *empuros* and referred to a burnt offering or to divination from fire or, rather, smoke.[93] If they were right, the custom is perhaps remotely related to the refusal to give fire and, more directly, to the divinatory observances involving the hearth. The German scholars theorised that St. Eligius' biographer and Pirmin used a source which depended on Caesarius and Martin, whose texts, they believed, dealt with Greek not German customs: among the Germans, fire was used for healing, not for divination. Both Caesarius and Martin had been exposed to Greek culture, the former from the important Greek colony in Arles, the latter from his travels in the East.

Interpreting *inpuriae* as fire is not altogether satisfactory. The readership envisioned for the *Vita Eligii* may have been clerical, but the *Dicta Pirmini* and Levinson's *Anonymous Sermon* were meant as guides to preaching; an unfamiliar Greek term would have been out of place.[94] Moreover, putting fire on a thatched roof to observe the smoke would be risky, a way of courting evil omens for the coming year. But interpreting *inpuriae* as filth does not lead to any very plausible form of augury either. Divination from excrement was a well-known technique, mentioned in the *Indiculus superstitionum,*[95] but our authors had no hesitation in using the word *stercora* if that is what they had in mind; furthermore, it is difficult to understand the purpose of putting the droppings on the roof.

Alternative interpretations are possible. *Inpuriae* may be the latinisation of a vernacular word of unknown origin and meaning. But if we can assume that, despite the importance of the concept of purity in Christian thought, it was possible for scribes to mistake so familiar a word as *impurae*, *inpuriae* may also be understood to mean women. Both ancient and medieval literature bear witness to seeresses' important divinatory role in German society; clerics might well have vilified them as impure. Putting a human being on the roof made sense. It was a good vantage point for seeing and consulting the Furious Host, the army of the dead which rode out on New Year's Eve. Almost three centuries after Pirmin, Burchard of Worms' penitential provided independent evidence of related divinatory practices:

[93] Blaise, *s.v.* "Impuriae;" *Dicta Pirmini,* 174 n.; Jecker, *Die Heimat des Hl. Pirmin*, 141-143; Boudriot, *Die altgermanische Religion*, 33. See also Boese, *Superstitiones Arelatenses,* 75-76, Homann, *Die Indiculus*, 102-103, and Harmening, *Superstitio*, 129-130.

[94] It must be admitted that the not particularly well-educated author of Levison's sermon wandered from topics likely to be relevant to his hearers, and also that he had a smattering of the classics, enough to discourse on the incestuous amours of Juno-Minerva.

[95] 13, *MGH CapRegFr* 1: 223.

> Did you observe the Calends of January in the pagan fashion ... by sitting on the roof of your house in the middle of a circle that you traced with your sword, so as to see and learn from there what would happen to you in the coming year, or did you sit at the crossroads on the hide of a bull, so that you would discover your future?[96]

Militating for this interpretation is Levinson's *Anonymous Sermon.* The first two parts of the sentence dealing with the Calends take exactly the same form: "Et Kalendas Ianuarias, quod maledictus Ianus docuit, hoc custodiunt vel inpuras, quod mensas conponunt ...." If (despite the generally unorthodox syntax of the sermon) the grammar is consistent in this case and the second *quod* functions, like the first, as a feminine plural relative pronoun, then the translation is, "And here they observe the Calends of January, which accursed Janus taught, or the *inpurae* who arrange tables ..."–just as, in Burchard's day, women were the ones who laid the tables for the *Parcae.* In addition, Pirmin's text is more plausible if we take *inpuriae* to mean not fire, but women who put men on the roof, or who were themselves put on the roof by men, to observe the future.

Cantus is ambiguous also. Du Cange (*s.v.* "cantus") defines *in cantu sedere* as "sitting on the side." Niermeyer defines *cantus* (*s.v.*) as corner or cornerstone, and Ernout and Meillet (*Dict. étymologique, s.v.* "cant[h]us") as the rim of a wheel. These point in two directions. If the term refers to "corner," it is related to the practice of frequenting crossroads or corners (*anguli*).[97] If the *cantus* is a wheel, then we are once again in the realm of the magic circle, particularly potent if the rim was made of iron, which demons were known to fear.

Finally, the *Homilia de sacrilegiis* includes a show of weapons or tools (*arma*) in the fields among forbidden New Year's practices.[98] This may be part of a divinatory ritual, perhaps to consult the Furious Host, or a game or dance of the sort that had been condemned from Caesarius of Arles' time onwards. This and Burchard of Worms' description of using a sword to trace a protective circle are among the very few references in our sources to weapons–a dearth surprising given the prestige of weapons in Germanic societies.[99]

4.1.2 *February rituals–dies spurci, spurcalia*

Two February festivals appear in our texts. In the 6th century, some Gaulo-Romans still marked February 22, originally the ancient Roman *Caristia* (commemoration of the dead) by making food offerings to the dead and partaking of

[96] *Decretum* (1008-1012) 19, 5.62, Schmitz 2: 423.

[97] *E.g., Poen. Remense* (early 8th century) 9.6, Asbach, 56.

[98] 17, Caspari, 10-11.

[99] A bow is used in a ritual described by Spanish penitentials (see May rituals below), and toy bows are given to *pilosi* and *satyri* to play with, in Burchard's penitential. For the cult of weapons, see Vordemfelde, *Die germanische Religion*, 28-47. For weapons at this period, see Wolfgang Hübener, "Waffen und Bewaffnung," in Claus Ahrens, ed., *Sachsen und Angelsachsen* (Hamburg, 1978), 463-471.

a sacrificial meal.[100] About the other feast, we know virtually nothing except the approximate name (*spurcalia, dies spurci)* and its purpose (according to the *Homilia de sacrilegiis*), namely, to drive out winter. The *Indiculus superstitionum* merely names it and locates it in February.[101]

Despite the incongruence, pointed out by Caspari, of attempting to drive out winter in February in a northern climate, Caspari and Homann were agreed that the Spurcalia were unmistakably Germanic, of either Anglo-Saxon or Frankish origin. Holleman however suggested that they were a survivor of the Roman Lupercalia.[102] The word first appears in Aldhelm's *De laudibus virginitatis*, with the meaning of a ritual of lustration.[103] (It may be significant that like Aldhelm, St. Boniface, the guiding influence on the *Indiculus,* was an Englishman.). The derivations proposed are *spurcus* (unclean, obscene) and *porcus* (swine), which also has the connotation of uncleanness. The name implies obscene practices, particularly in view of the fact that the excesses of Carnival usually take place in February. Other possibilities are an expiatory offering of pigs or, in view of that animal's well-known oracular character, a divinatory rite.[104]

These are the only indications. The term *spurcitiae gentilitatis* found in eighth- and ninth-century legislation,[105] should be understood to mean "filthy pagan practices" in general (sacrifices to the dead, consultation of soothsayers, charms, dancing in front of churches, and the *Needfire*). It does not apply specifically to February rituals.

4.1.3 *May rituals*[106]

Romans, insular Celts and Germans all observed the beginning of May as a festive period. The Romans celebrated the *Floralia* from April 28th to May 1st. May 1st was also the date of the Irish *Beltene* and the German *Walpurgisnacht.* Popular custom marked this feast with bonfires, fiery wheels, Maypoles, singing, dancing, burnt sacrifices. Our texts offer only a scattering of data about such

[100] *Conc. Turonense* (567) 23, *CCSL* 148A: 191-192.

[101] *Homilia de sacrilegiis* (late 8th century) 17, Caspari, 10-11; *Indiculus superstitionum et paganiarum* (743) 3, *MGH CapRegFr* 1: 223. For a discussion of these texts, see Caspari, *Homilia de sacrilegiis*, 35-37; Homann, *Indiculus*, 34-39; Harmening, *Superstitio*, 148-154. For rituals to expel winter and death, see Grimm, *Teutonic Mythology*, 764-774.

[102] According to Grimm, Germans and Slavs celebrated such rites between March and May (*Teutonic Mythology*, 764-784); Holleman, *Pope Gelasius I and the Lupercalia*, 185-189 and *passim.*

[103] *PL* 89: 122.

[104] See Bächtold-Stäubli 7: 1482-1485.

[105] *E.g., Conc. in Austrasia habitum q. d. Germanicum* (742) 5, *MGH Concilia* 2.1, 1: 3-4=*Karlomanni principis capitulare* (742) 5, *MGH CapRegFr* 1: 25. See also *Concilium in Francia habitum* (747 inuente), *MGH Concilia* 2.1: 47; *Council of Clovesho* (747), Haddan and Stubbs 3, 377-378; *Karoli Magni capitulare primum* (769 vel paullo post) 7, *MGH CapRegFr* 1: 45; *Hludiwici Pii capitulare missorum* (819) 10, *ibid.,* 1: 290.

[106] Green, *Dictionary of Celtic Myth and Legend,* s.v. "Beltene;" Grimm, *Teutonic Mythology,* 759-762 and 1050.

customs. In the early 8th century, the people of Noyon observed one or more days in May as an unauthorised holiday from work.[107] According to two continental penitentials from the late eight- and late ninth-centuries, "illicit acts" were performed during the *Maiae,* among them the preparation of a potion to be consumed either on or before the day itself (May 1st?).[108]

Two Spanish penitentials appear to refer, at least in part, to May festivities. A difficult passage in Francis Bezler's new edition of the mid ninth-century *Penitential of Vigila* (*Poen Albeldense*) and of the *Penitential of Silo* reads: "Qui in saltatione femineum habitum gestiunt, et monstruose fingunt, et maias, et arcum et palam, et his similia exercent, I annum peniteant." This differs from Wasserschleben's version of the former penitential, which reads *orcum et pelam* (which he interpreted to mean "ogre and child") instead of *arcum et palam.*[109]

Bartolomé Bennasser translated the passage to describe a single festival: "For those who, during dances, dress like women or who disguise themselves monstrously as *majas* [after the goddess of the month of May] or ogres [monsters subject to Orcus, that is, Pluto], or who dress themselves in the hide of beasts, a year of penance."[110] According to Bezler, however, two separate feasts are being de-

[107] *Vita Eligii* (c.700-725), *MGH SRM* 4: 706.

[108] *Poen. Merseburgense a* (end of 8th century) 32, *CCSL* 156: 135; *Poen. Vindobonense a* (late 9th century) 35, Schmitz 2: 353. Grimm noted that in his day a May drink composed of wine and certain herbs, especially woodruff, was drunk in the Lower Rhine and Westphalia (*Teutonic Mythology*, 778 n).

[109] *Poen. Albeldense* 102, *CCSL* 156A: 12; *Poen. Silense* 194, *ibid.,* 36. Wasserschleben, 533. Gonzalez Rivas's edition of the *Albeldense* has *menstruose fingunt* rather than *monstruose fingunt.* His edition of the *Silense,* like Bezler's, gives *monstruose,* but in both editions *monstruose* is the spelling used when, clearly, the meaning is "menstruous": *e.g.* "A *monstrouse* woman may eat only the flesh of the blessed lamb" on Easter Sunday (*CCSL* 156A: 20; Gonzalez Rivas, *La penitencia en la primitiva Iglesia española*, 174). This raises the possibility that the rite was a form of protective magic relying on the well-known magical efficacy of menstrual blood. According to Pliny, menstrual blood is a charm against hail, whirlwind and stormy weather; naked menstruous women walking through a field of grain kill noxious insects and other vermin; menstrual blood smeared on doorposts nullify the arts of wizards (*Historia naturalis* 28.23, Loeb. ed. 58-60). Among the Sioux, a menstruating woman walking through the field at dawn was thought to cause the seed to sprout more rapidly (Alice Fletcher, "The foundation of the Indians' faith in the efficacy of the totem," *American Association for the Advancement of Knowledge* (1897) 326, cited in Károly Bartha *et al.,* eds., *A Magyarság Néprajza* (*Ethnography of the Hungarian People*; Budapest, 1941-1943) 2: 386. In a necessarily rapid survey of anthropological literature, this is the only evidence that I have been able to find to support the often-repeated assertion that menstrual blood was used to fertilise the crops. For the use of menstrual blood to drive off demons, see Heckenbach, *De sacra nuditate*, 53-54. See also Thomas Buckley and Alma Gottlieb, "A critical appraisal of theories of menstrual symbolism," in *Blood Magic: The Anthropology of Menstruation* (Berkeley, 1988) 5-53. Bruno Bettelheim interpreted a number of rites practiced by men in tribal societies as mimicking menstruation (*Symbolic Wounds. Puberty Rites and the Envious Male* [revised ed., New York, 1962], 99-108).

[110] "Pour ceux qui, dans les danses, s'habillent en femmes ou se déguisent, de façon monstrueuse, en *majas* [*maia*: à l'origine, divinité du mois de mai] ou en ogres [*orcos*: monstres qui obéis-

scribed.[111] The first is the Calends of January, with men aping women and monsters, very much in line with the usual masquerades of the season, and recalling Isidore of Seville's words: "The faithful transform themselves into monsters by assuming animal garb, while others use feminine gestures to make their manly features womanish."[112] The second part of the text concerns a May celebration during which an agrarian fertility rite is performed with the aid of a bow and a spade (*arcum et palam*). Bezler rejects Wasserschleben's transcription of *orcum et pelam* as an unwarranted attempt to make sense of a difficult text.[113]

4.1.4 *Midsummer–the feast of St. John the Baptist*

Midsummer festivities appear in four sermons, only in the context of the feast of St. John the Baptist, which does not in fact coincide exactly with the solstice. Nowhere else in the literature do we find so complete a fusion between pagan and Christian concepts as here. If the substance of the rituals is solidly pagan, the form has taken on a largely Christian cast.

Two of the texts come from Caesarius of Arles' sermons. The influence of the gospel image of John the Baptist is unmistakable in the description of ritual bathing in springs, marshes and rivers during the night and early hours, although they were accompanied with "vile or lewd" love songs.[114] Bawdiness was an essential feature. Caesarius urged his hearers to keep the members of their household from obscene behaviour and indecent speech so that "the sacred solemnities" not be polluted by bawdy songs. His warning that the saint would listen to their prayers "if he knows that we celebrate his feast peacefully, chastely and soberly" implies that the actual celebration tended to degenerate into an orgy.[115] St. Eligius' sermon condemned "solstices [*solestitia*], dances or vaults, or devilish songs on the feast of St. John or other solemnities of the saints."[116] These rites appear to be rain-making and fertility magic; they show no obvious characteristics of a fire festival, unless it can be assumed that the leaps and dances were performed over and around the fire in rites of the sort described in later literature.[117]

saient à Orcus, c'est-à-dire Pluton] ou se vêtent de peaux de bêtes, un an de penitence" (*Histoire des Espagnes* [Paris, 1985] 1: 35).

[111] *Les Pénitentiels espagnols,* 287-293. See also *idem,* "¿El ogro y el niño o el arco y la pala?" *Revista de literatura medieval* 4 (1992): 43-46.

[112] *De officis ecclesiae* 1.41, *PL* 83: 775.

[113] Wasserschleben related this rite to a Spanish custom of reenacting the story of St. Christopher and the Christ Child on the feast of Corpus Christi, which sometimes coincided with the beginning of May (Wassershleben, 71). For a similar custom, see Chambers, *The Medieval Stage* 1: 120 and n. 3.

[114] *S.* 33.4, *CCSL* 103: 146.

[115] *S.* 216.4, *CCSL* 104: 860-861.

[116] *Vita Eligii* (c.700-725), *MGH SRM* 4: 705-706.

[117] Descriptions of midsummer festivities were given by John Beleth and William of Auvergne (*Rationale divinorum Officiorum*, *CCCM* 41: 263-268; *De legibus* 1: 82). See also Chambers, *The Medieval Stage* 1: 116-159; Grimm, *Teutonic Mythology*, 588-590.

A northern Italian rite already discussed in the context of the cult of the earth gives a striking example of syncretism.[118] Atto of Vercelli denounced a custom that had

> developed in numerous places concerning so glorious a solemnity. Certain little trollops [*meretriculae*] abandon the churches and the divine offices; they pass the whole night any which where, in the streets and crossroads, by springs and in the countryside; they form round dances, compose songs, draw lots and pretend that people's prospects are to be predicted from things of this sort. Their superstition has given rise to madness to the point that they presume to baptise grass and leafy boughs, and hence they dare to call [the turf and trees?] godparents [co-parents?–*compatres vel commatres*)]. And for a long while afterwards, they strive to keep them hung up in their houses, as though for the sake of piety.[119]

Here Atto adds divination to the usual rites of singing and dancing in the countryside and in the vicinity of springs, but he restricts such practices to women. The opprobrious *meretriculae* is probably meant merely to denigrate them, but may also hint at obscene behaviour. Atto, however, was most perturbed by the baptism of nature, which, as we have seen, has two possible, mutually compatible, explanations: the creation and reinforcement of ties among members of the community, and the attempt to bind earth and trees with bonds of mutual, family obligation.[120] The godparent was *compater* or *commater* to the infant's parents. The relationship entailed the mutual obligation of support between godparent and godchild, and godparents and parents. Quite apart from its religious and folkloric aspects, this practice is of great sociological interest in its reflection of the importance of the bonds of kinship in tenth-century Piedmont.

Oblique hints of a special baptismal ritual held on the Feast of John the Baptist appear in passages concerning co-parents of St. John. John Lynch has ident-

[118] I assume that Atto of Vercelli's sermon 13, for the "annunciation" of St. John the Baptist, is actually for June 24th, which celebrates the saint's birth. (In fact, the next sermon is for his nativity.) The feast of the conception of St. John the Baptist is recorded in early Byzantine liturgies but was not adopted in the West. The practices described in this sermon do not seem appropriate for the feast of his decapitation, which falls on 29 August. For the feasts of St. John, see Weiser, *Handbook of Christian Feasts and Customs*, 320-331.

[119] *S*.13, *PL* 134: 850-851.

[120] See chapter 3, nn. 164 and 165. A British text published in 1607 hints at a like relationship between the Irish and wolves (see Plummer, *VSH* I, cxlii n. 8). For the relationship between co-parents, see Joseph H. Lynch, *Godparents and Kinship*, 192-201, and *idem*, "*Spirituale vinculum*; The vocabulary of spiritual kinship in early medieval Europe," in *Religion, Culture and Society in the Early Middle Ages. Studies in Honour of Richard E. Sullivan*, ed. T. F. X. Noble and J. J. Contreni (Kalamazoo, 1987) 181-204. The importance of such ties is exemplified by their use as an instrument for political domination, see Angenendt, "The conversion of the Anglo-Saxons considered against the background of the early medieval mission," in *Angli e Sassoni*, 747-781. See also Julian Pitt-Rivers, "Pseudo-kinship," *International Encyclopedia of the Social Sciences* 8: 408-413.

ified three such passages.[121] An eighth-century sermon, probably from Alemannia, urged the faithful to "honour your *comparatus* of St. John well because you have agreed to mutual loyalty and love, and whoever violates this loyalty and love commits a serious sin."[122] This seems to have nothing to do with the usual rites of baptism, for the very next sentence concerns the obligations towards the children "whom you received at the font." An eighth or ninth-century penitential and one of a slightly later period mention incest or attempted marriage with a *cummater de sancto Joh.* or *commater sentiana.*[123] To these may be added the similar testimony of a late twelfth-century penitential.[124] Clearly, by the 8th century a ritual existed in different parts of Europe, performed on the eve or on the feast of St. John, which was not associated with the normal administration of baptism, but which created ties of spiritual kinship that were recognised by some churchmen at least. The baptism of turf and branches, however, may have been a form of this rite peculiar to Vercelli, or newly introduced there in Atto's time.

4.1.5 *Other seasonal festivals*

Passing mention is made of the rituals of the Calends of March, the *Vulcanalia* and the *Neptunalia.*

During his sermon for the octave of the Nativity (*i.e.*, for the Calends of January), Atto of Vercelli referred in an aside to excesses practiced at the beginning of March, affirming that it was through the rites of these two calends that the devil exercised his power over the entire year.[125] The Calends of March, the beginning of the year in the primitive Roman calendar, had been celebrated by agrarian and military rituals. Schneider and Meslin maintain that under the Empire, the Calends of March and of January were in effect combined into one great feast, but this does not appear to be the case here, since Atto took pains to differentiate between the two.[126] The beginning of March may also have been the "other calends" mentioned in an eighth-century penitential.[127] Martin of Braga and two imitators mentioned the *Vulcanalia.*[128] Originally, this had been a Roman harvest and storage festival held in late August; no doubt the same term could have been applied to harvest festivals held elsewhere throughout the Middle Ages and quite possibly

[121] *Godparents and Kinship,* 215-218.

[122] Morin, 517.

[123] *Poen. Merseburgense c*, 13, Wasserschleben 436; Poen. *Valicellanum E*, 62, 18, Wasserschleben, 559.

[124] *Poen. Laurentianum* 18, Schmitz 1: 787.

[125] Atto of Vercelli (d. 961) *S.* 3, *PL* 134: 836.

[126] See Meslin, *La fête des kalendes,* esp. 9-14, and Schneider, "Über Kalendas Ianuariae und Martiae im Mittelalter."

[127] *Poen. Oxoniense I*, 29, *CCSL* 156: 91.

[128] Martin of Braga, *De correctione rusticorum* (2nd half, 6th century) 16, Barlow, 198; *Dicta Pirmini* (724-753) 22: 172; *Anonymous 12th-century sermon*, Caspari, *Kirchenhistorische Anecdota*, 1: 204.

celebrated with rites objectionable to conscientious churchmen.[129] The *Vulcanalia* also appears in a ninth-century continental penitential, in a context with no apparent reference to actual practices.[130] Finally, the *Homilia de sacrilegiis* warns that whoever "observes the *Neptunalia* in the sea or at the source of a spring or rivulet" has forfeited his faith and baptism.[131] It is safe to assume that water-rites celebrated in eighth-century Francia had but little in common with the classical Neptunalia.[132]

4.2 Liturgical Cycle

The populace appears to have embraced many of the feasts of the Church (the great feasts of Christmas, Easter, and the Ascension, saints' days, dedications of basilicas,[133] rogations and *letaniae*) with enthusiasm, gladly participating in the liturgy, prayers and processions. Nonetheless, churchmen and the civil authorities were not wholly satisfied with the popular reaction. In some areas, the people tried to pick and choose: Louis the Pious and the Pavian bishops issued an edict requiring the faithful to observe the feasts preached by the clerical authorities, and to refrain from celebrating "with inane superstition feasts that should never be observed."[134] Among the forbidden customs found in a mid ninth-century capitulary from the diocese of Milan are the observance of unauthorised saints' feasts, "unusual" litanies and other "superfluous" practices.[135] Regino of Prüm instructed bishops to investigate the feasts kept in their dioceses.[136] These texts point to a proliferation of local cults that had developed spontaneously or had escaped ecclesiastical and official control.

[129] The Roman *Vulcanalia* was celebrated in the Circus Flaminius with games and sacrifices: fish from the Tiber were flung on the fire to appease the dead, and clothes were hung out into the sun (Daremberg and Saglio 5: 1002-1003). But Martin, a Pannonian by birth, may have had some other elements in mind as well: Vulcan had been the Roman name of the chief god of the Pannonians; he had been a solar deity whose rites were celebrated with some savagery and obscenity, see A. von Domaszewski, "Volcanalia," *Archiv für Religionswissenschaft* 20 (1920-1921) 79-81.

[130] *Poen. Quadripartitus*, 140, Richter, 18.

[131] 3, Caspari, 6.

[132] Neptune had originally been the god of streams, fountains, lakes, *etc.*, and therefore was associated with fertility. In Roman times, the Neptunalia (July 23) had been celebrated by building little huts of leafy branches; see Daremberg and Saglio 4: 698b and *passim*.

[133] The Christian custom of celebrating the dedication of basilicas was adopted in imitation of pagan Roman customs; see P. de Puniet, "Dedicace des églises," *DACL* 4.1: 375.

[134] *Capitula episcoporum Papiae edita* (845-850) 17, *MGH CapRegFr* 2: 83. See Silvana Zanolli, "Le sagre paesane" (in Paolini, *La terra e il sacro*, 79-110) for the importance of saints' festivals, especially of the local patron saint, in the affirmation of collective identity by rural communities.

[135] *Capitula Eporediensia* (c. 850) 10, *MGH CapEp* 3: 242).

[136] *De synodalibus causis* (c. 906) 2: 5.73: 215.

4.2.1 *Adaptations of pagan rituals*

Concerns about behaviour during Christian festivals were more common than worries about the choice of festivities. They probably lie behind Charlemagne's and Charles II's edicts ordering the people to celebrate the principal feasts only in the appointed places, in towns or villages, under the eyes of the authorities, not secretly at home or elsewhere.[137] These concerns centred on traditional rituals practiced by the faithful as they processed through the countryside on their way to church and assembled in front of it: singing, dancing, mumming and pranks, with concomitant drunkenness and licence, behaviour viewed by the Church as left over from the pagan customs which it had hoped to eliminate by suppressing pagan feasts.[138] A well-known episode in the *Life of St. Eligius* is illustrative. The saint was accustomed to preach against games, leaping dances and "meaningless superstitions." One feast of St. Peter (June 29th), he met a crowd near Noyon, on its way to celebrate as usual, and began to upbraid it vehemently in front of the basilica. He was interrupted by angry voices from the crowd: "No matter how often you talk, Roman, you will never be able to uproot our customs, but we will continue always and forever to carry on our feasts as we have hitherto. No man will ever be able to forbid us the ancient pastimes that are so dear to us."[139]

On this issue as on others, Caesarius of Arles set the pattern. In one sermon tried to impress on priests and bishops the necessity, however inapt they felt, of preaching against drunkenness, indecent songs, dancing and vaulting at any time but especially during the solemnities of the saints.[140] Another time, he attacked his congregation directly for the indecent talk, round dances and bawdy songs in which they indulged at every occasion, at home, on the road, at a banquet or assembly:

> There are unhappy wretches who neither fear nor blush to perform dances and vaults before the basilicas of the saints themselves and who, if they are Christians when they come to church, are pagans when they leave it, for this custom of dancing is left over from pagan custom. And now see what kind of Christian he is, who had come to church to pray, forgets prayer and does not blush to mouth the sacrileges of pagans. But just consider, brothers,

[137] *Capitula excerpta de canone* (806 vel post?) 21, *MGH CapRegFr* 1: 133; *Capitulare Papiense* (876) 7, *MGH CapRegFr* 2: 102.

[138] James has emphasised the social importance of the popular celebrations during the feasts of the Church: "The major festivals, including the feast-day of the local saint, attracted crowds of countrymen into the towns–crowds who came, perhaps, as much for the markets, the dancing and the drinking as for the ceremonies. It must have been on such occasions that the sense of belonging to the *civitas*-community was at its strongest" (*The Origins of France: From Clovis to the Capetians, 500-1000* [London and Basingstoke, 1982] 53).

[139] *Vita Eligii* (c.700-725) 20, *MGH SRM* 4: 711-712.

[140] *S.* 1.12, *CCSL* 103: 9. Caesarius might have had in mind either the sung narrative of a racy legend or bawdy love song. For coarse and abusive language, oaths, obscenity, *etc.*, as a manifestation of folk culture and social intimacy, see Bakhtin, *Rabelais and His World*, 145-167.

whether it is right that lewd song should flow like poison from lips which Christ's body entered.[141]

Caesarius was particularly dismayed that men and women whom conscience could not lead to learn a few words had no difficulty in learning indecent songs, and that some came to church less for love of Christ and to pay honour to the saints than to seize the opportunity to drink and revel, quarrel and complain, and indulge in lewd speech.[142]

The secular powers threw their weight on the side of the Church. Childebert's *Edict of 554* tried to put a stop to nights spent in drunkenness, obscenity and singing, with dancing women (*bansatrices*) disporting themselves through the villages during the octaves of Easter, Christmas, at other feasts and before the Lord's Day.[143] Reccared's great council of 589 followed with an ordinance banning the populace's impious customs on saints' feasts, namely, *saltationes* and obscene songs loud enough to drown out the liturgy.[144] This appeared in abbreviated form in an early collection but, in a ninth-century pentitential, the title given to the same ordinance was "Against dancing and dirges" (*vallamatica et plancatica*). The compiler evidently interpreted such customs as inappropriate mourning rituals.[145]

The clergy were even more concerned about such conduct than the civil authorities. Unlike the latter, they were particularly scandalised at the role of women. The Council of Chalon-sur-Saône (647-653) called on priests to prevent obscene and wicked songs and the indecencies performed by women dancing at the dedications of basilicas and the feasts of martyrs, by denying them access to the sacred precincts.[146] The Roman councils of 826 and 853 singled out women as the chief culprits: "Certain people, especially women" were happy to attend holy days and the feasts of saints, not for the right reasons, but to dance, sing indecent verses, participate in round dances and generally behave like pagans, "so that even if they come to church with minor sins, they leave with major ones."[147] Pope St. Leo IV and a tenth-century Saxon capitulary urged the clergy to forbid women to sing and dance in the churchyard.[148] But when St. Eligius had preached against dancing, leaping and diabolical songs during the solemnities of saints, he did so

[141] *S.* 13.5, *CCSL* 103: 67.

[142] *S.* 19.3, *CCSL* 103: 89; *S.* 55A, *ibid.,* 244-245. See also *S.* 55.2, *ibid.,* 242, and *S.*225.5, *CCSL* 104: 891.

[143] *MGH CapRegFr* 1: 2-3. On the celebration of saints' feasts in Merovingian Gaul, see Hen, *Culture and Religion,* 84-89.

[144] *Conc. Toletanum* III (589) 23, Vives, 133. See also Regino of Prüm, *De synodalibus causis* (c. 906) I, 392: 178.

[145] *Epítome hispánico* (c.598-610) 34.23: 178; *Poen. Quadripartitus* (9th century, 2nd quarter) 292, Richter, 33.

[146] *Conc. Cabilonense* (647-653) 19, *CCSL* 148A: 307.

[147] *Conc. Romanum* (826) 35, *MGH Concilia* 2.2: 581). See also *Conc. Romanum* (853) 35, *MGH Concilia* 3: 328.

[148] Leo IV (d. 855) *Homilia,* 39, *PL* 115: 681; *Capitula Helmstadiensia* (c. 964) 7, Pokorny, 506.

without ascribing it to either sex in particular.[149]

Texts in a similar vein are to be found enshrined in continental legal texts, sermons and, especially, penitentials, from the 8th to the 11th centuries. For example, the Council of Mainz (813) strictly forbade the performance of "indecent, bawdy songs around churches."[150]

Clearly, it was the Christian feasts that drew the populace to the churches, since they were quite able to make merry elsewhere if they wanted.[151] This made the bawdiness and lack of respect particularly offensive to the clergy. Regino of Prüm's question concerning those who "dared to sing wicked songs which provoke laughter around a church" implies that some of the songs were mocking or, at any rate, humorous.[152] Rabanus Maurus and Burchard of Worms thought it worth their while to include a fourth-century African canon which had noted that behaviour during martyrs' feasts was outrageous enough to offend "the virtue and modesty of matrons and countless other women."[153] An anonymous eighth-century sermon is unique in condemning the practice of performing dances, leaping and diabolical without tying it either to Christian feasts or to the precincts of the church.[154]

Three continental penitentials called for the excommunication of those who danced, vaulted or sang amorous songs outside the church during feasts.[155] Herard of Tours also forbade "indecent and bawdy songs, vaults and diabolical fooleries on holy days."[156] This may have involved masquerades–two other penitentials proposed long penances for dancing in front of the church and changing one's appearance by wearing the clothing of the other sex or by putting on an animal disguise.[157] Pirmin of Reichenau brought up the topic twice: "Flee dancing, vaulting, and indecent and bawdy songs as you would the devil's arrow; do not dare to perform them either by the church nor in your houses, nor in the roads nor in any another place, because they are remnants of pagan custom." In a second passage,

[149] *Vita Eligii* (c.700-725), *MGH SRM* 4: 705-706.

[150] 418, *MGH Concilia* 2.1: 272. See also *Benedicti capitularium collectio* (mid 9th century) Add. 39, *PL* 97: 879; Regino of Prüm, *De synodalibus causis* (c. 906) 1.393: 179; Burchard of Worms, *Decretum* (1008-1012) 3.87, *PL* 140: 691.

[151] E.g., *Poen. Ps.-Theodori* (mid 9th century) 38.9, Wasserschleben, 607.

[152] *De synodalibus causis* (c. 906) II, 5.87, 216. See also Burchard of Worms, *Decretum* (1008-1012) 1, Interrogatio 86, *PL* 140: 579.

[153] *Poenitentium Liber ad Otgarium* (842-843) 25, *PL* 112: 1418. See also Burchard of Worms, *Decretum* (1008-1012) 10.36, *PL* 140: 838.

[154] Nürnberger, 44.

[155] *Poen. Merseburgense a*, 2nd recension (late 8th century) 119, *CCSL* 156: 168; *Poen. Vindobonense a* (late 9th century) 94, Schmitz 2: 356. See also the *Iudicium Clementis* (9th century) 20, Wasserschleben, 435.

[156] *Capitula* (858) 114, *PL* 121: 772.

[157] *Poen. Hubertense* (8th century, 1st half) 42, *CCSL* 156: 112. See also *Poen. Merseburgense b* (c. 774-c.850) 29, *ibid.,* 176; Burchard of Worms, *Decretum* (1008-1012) 10.39, *PL* 140: 839.

he added jokes, diabolical games and the gestures or words of mimes and prostitutes (dramatic performances?–*iocus, lusa diabolica* and *mimarciae*) to the list of forbidden merry-making activities.[158] A couple of anonymous sermons of the same period merely reiterated Caesarius' words[159] but in the following century two documents introduced a relatively new element, the *circus* (which was probably ritual circumambulation or simply a round dance, not some sort of spectacle).[160]

Of particular interest is a canon included by Benedict Levita, which forbade the faithful to stand at the crossroads and in the highways on Sundays, spending the time "as usual" in vain tales and speeches or singing and vaulting. This does not appear to refer merely to a pleasant gossip after mass but rather the recounting in song of the old tales, perhaps acted out with dance steps.[161]

4.2.2 *In church*

In the earliest part of our period, the questionable behaviour was carried into the church itself.[162] The Council of Orléans (533) denounced the fulfillment of vows in church with singing, drinking and lewdness "because God is angered rather than pleased by such devotions."[163] An ordinance against eating in church was included in the *Epítome hispánico* (c. 598-610).[164] The Synod of Auxerrre dealt with the performance, in church, of round dances by the laity or songs by girls, or ritual meals, all traditional observances in pagan shrines.[165] In the mid 7th century, the Council of Chalon-sur-Saône was still trying to keep women from dancing and singing not only in the enclosed space in front of the basilicas, but within the basilicas themselves.[166] Clearly, some of the faithful saw songs, dance, drink, banqueting and perhaps some kind of fertility ritual as normal parts of religious ceremonies.[167]

[158] *Dicta Pirmini* (724-753) 28: 188-189; *ibid.,* 22: 176.

[159] *Anonymous sermon* (8th century), Morin, 518; *Anonymous sermon* (late 8th/9th century), Levison, 309.

[160] *Capitula Vesulensia* (c. 800) 22, *MGH CapEp* 3: 351. See also the *Poen. Ps.-Theodori* (mid 9th century) 38.9, Wasserschleben, 607. Benedict Levita's version omits the reference to the circus (*PL* 97: 771).

[161] *Benedicti capitularium collectio* 2.205, *PL* 97: 772.

[162] The ninth-century *Statutes of St. Boniface* (21, *MGH CapEp* 3: 363) repeated the admonition of the Council of Auxerre against permitting singing, dancing and feasting in church (see below), but this is the only suggestion that such practices were carried into the Carolingian era.

[163] *Conc. Aurelianense* 12, *CCSL* 148A: 100.

[164] 7.27: 122.

[165] *Syn. Autissiodorensis* (561-605) 9, *CCSL* 148A: 266. See also *Epítome hispánico* (c. 598-610) 43.7: 190.

[166] 19, *CCSL* 148A: 307.

[167] Religious dancing had been a part of early Christian celebrations, especially in the East; even in the West, St. Ambrose saw certain kinds of church dances as playing an important role in religious ritual. Foatelli describes a number of religious ceremonies throughout the Middle Ages in which dancing had clerical blessing (*Les danses religieuses,* 31 *ff*).

In addition, the Council of Aachen (816), a ninth-century penitential and several collections incorporated rulings from early councils concerning communal feasts (*agapae*) and eating and reclining within the church.[168] The *Vetus Gallica* also forbade bishops and clerics to hold feasts in the churches except for the refreshment of transient clerics.[169] These may have been included in the penitential and collections for historical reasons, but the fact that a contemporary council also saw fit to reiterate the ruling suggests that feasting did occur on church premises. But by the 9th century, this was merely a symptom of a lack of respect for the church and had nothing in common with pagan ritual.

4.2.3 *Easter season*

Attending spectacles may have been a part of the festivities following the great liturgical feasts. A fourth-century African decree against frequenting the theater and circuses on holy days, especially during the first week of Easter, was relevant enough in tenth-century Piedmont for Atto of Vercelli to copy it into his capitulary. On White Sunday, he preached against attendance at theatres, wedding and other songs, vaulting dances and the "circus" (circumambulation, animal shows, races?) during the Octave of Easter. He emphasised the moral damage to young and old caused by beguiling performances of titillating stories about debauched maidens and lustful harlots–inventions, he claimed, of those notorious demons, Liberus and Venus.[170]

Rejoicing at the end of the grey penitential days of Lent tended to slide into excess elsewhere, too. In the 7th century, the *Toledan Homiliary* offered an Easter sermon inviting the faithful to celebrate the feast "not in feasts and drunkenness nor in lechery and vain talk" but piously, and calling on them to chastise "even your servants" for the sanctification of the entire household.[171] In the 9th century, an anonymous pre-Easter sermon called on Christians to avoid urging or forcing specially-invited"neophytes" to drink to excess at banquets.[172] Rabanus Maurus warned against the temptation to relax into games, drunkenness, lechery and vice

[168] *Conc. Aquisgranense* (816) 80, *MGH Concilia* 2.1: 367. See also *Poen. Quadripartitus* (9th century, 2nd quarter) 311, Richter; Atto of *Vercelli (d. 961) Capitularia 22, PL 134: 33; Burchard of Worms, Decretum (1008-1012) 3.82, PL 140: 690; Collectio Frisingensis Secunda* (late 8th century) 37, ed. Mordek, 624; *Collectio Vetus Gallica* (8th/9th century) 22.3: *ibid.*, 437. See the *Council of Laodicea* (380?) 28, Hefele-Leclercq, *Histoire des conciles* 1.2: 1015.

[169] Mordek, 22.3a, 437. This is taken from the early fifth-century *Breviarum Hipponense* 29, *CCSL* 149: 41.

[170] *Capitularia* 78, *PL* 134: 43; S. 9 (In die sanctae resurrectionis in albis), *PL* 134: 844-845. Gandino maintains that Atto's classical terminology describes actual behaviour at fairs, the existence of which is attested for Vercelli at the beginning of the 10th century ("Cultura dotta e cultura folklorica," 261).

[171] *Homiliare Toletanum* (2nd half, 7th century) *Hom.* 31, *PL Suppl* 4: 1959].

[172] *Anonymous sermon* (9th/10th century?) 2, Caspari, "Eine Ermahnung zu würdiger Feier des bevorstehenden Osterfestes," 200. The neophytes in question were evidently adults, who were to be baptised during the Holy Saturday liturgy. For forced drunkenness, see 4.3.1 below.

at this time (however, even Lent had been enlivened by unfitting pastimes: business, gaming, litigation and, most of all, the hunt).[173]

4.2.4 *Rogation days*

Churchmen denounced some of the popular manifestations accompanying the processions of Rogation days (*rogationes, laetaniae*).[174] In general, English authorities seem to have taken Gregory I's advice to heart and permitted–or, at least, failed to denounce–the continuance of traditional pagan rituals in the vicinity of churches. The behaviour during the *Dies laetanorium*, however, exceeded the limits of tolerance, and the Council of Clovesho (747) reprimanded those "careless or ignorant people" who passed these days in "vanities," that is, games, horse races and exceptionally lavish banquets.[175]

The Anglo-Saxons' Rhenish cousins were given to similar amusements. Participants had to be admonished to behave properly and not to dress sumptuously, nor sing lewd songs nor indulge in worldly games while they followed the procession; they should sing the *Kyrie Eleison* "not uncouthly as hitherto, but better."[176] Rabanus Maurus' flock treated the Rogations as an opportunity to ride out on bravely caparisoned horses and gallop across fields roaring with laughter, encouraging each other with word and gesture to race. They ended the day's doings with their friends and neighbours in an all-night feast, during which they engaged in drinking contests to the accompaniment of musicians, and interspersed their "particularly large" drinks with "songs invented by the skill of demons"–surely heroic lays of the pagan past.[177] At the beginning of the 10th century, people still decked themselves out in costly garb at such times, and indulged in drunkenness and communal feasting, while "trifling women" (*mulierculae*) performed round dances. Regino of Prüm exhorted the faithful to wear sackcloth and ashes, go on foot and not horseback (to prevent racing?), and sing prayers in unison to beseech

[173] *Hom.* 18, *PL* 110: 36; *Hom.* 11, *ibid.,* 24-25. The latter provides a valuable glimpse of a battue and of the relative value of dogs and servants.

[174] The Rogation Day of 25 April (*litaniae majores*) was introduced in the 5th century by Pope Gelasius as a Christianisation of the pagan *Robigalia*; the Minor Rogations (the three days before Ascension Thursday) were introduced by Bishop Mamertus of Vienne to intercede against natural catastrophes (*The Oxford Dictionary of the Christian Church, s.v.* "Litany," "Rogation Days"). See also H. Leclercq, "Rogations," *DACL* 14.2: 2459-2461; D. de Bruyne, "L'origine de la Chandeleur et des Rogations," *RB* 34 (1922), 14-26; Righetti, *Manuale di storia liturgica* 2: 296-301.

[175] 16, Haddan and Stubbs 3, 368. The clergy were no more immune to the attraction of such pastimes than the laity. Some fifty years later, Alcuin wrote severely to Eanbald II, Archbishop of York, advising him to insist that young clerics riding with him behave modestly and sing psalms in dulcet tones, instead of chivvying foxes and hallooing after them across the fields (*MGH Ep* 4: 168).

[176] *Statuta Rhispacensia Frisingensia Salisburgensia* (799, 800) 34, *MGH CapRegFr* 1: 229.

[177] *Hom. 19: In Litaniis*, *PL* 110: 38. *Cf. Is* 5 12 for the musical instruments mentioned here.

God for his mercy and protection.[178]

References to horse races and games during the Rogations may explain the clause of the *Indiculus superstitionum* "concerning the pagan race called *yrias* [conducted by runners, horsemen, charioteers? dressed] in torn rags and sandals" (*de pagano cursu quem yrias nominant scisis pannis vel calciamentis*), although torn rags and sandals hardly accord with the elaborate clothing described in the later texts. Haderlein interprets *scisis* (translated here as "torn") as *scilicet* ("that is"), and suggests the translation, "about the pagan race commonly referred to as *yrias*; that is, head-band and sandals [only]"–that being the "proper habit in chariot races, Celtic or Gallo-Roman."[179]

4.3 Communal celebrations

Condemnations of popular rituals during seasonal and liturgical feasts were often based explicitly on their pagan origins. This is not the case with the rituals practiced by private, more-or-less restricted groups gathered together for a banquet or commemorative feast or to settle conflicts. Nevertheless, some of the rituals were based on concepts of the pre-Christian past, and provide an insight into the folklorisation of pagan customs.

With the exception of Caesarius of Arles, our authors generally ignored the celebratory assemblies of lay people. They concentrated on clerical assemblies at commemorative banquets and regular diocesan meetings. Clerics' customs as to eating and drinking, offering of toasts and entertainment allow us to form a picture of lay gatherings by inference. The frequent admonitions to the clergy reflect the higher standards to which they were held, although some attempts were made in the 9th century to enforce the same on the laity.[180]

Among these customs, ritual drunkenness was the most important. Gorging and heavy drinking on festive occasions were not foreign to either Celtic or Mediterranean societies, but were traditionally a matter of social and religious obligation among Germanic peoples in particular.[181] The origin of the great German feasts

[178] *De synodalibus causis* (c. 906) 1.280: 132.

[179] 24, *MGH CapRegFr* 1: 223; Haderlein, "Celtic roots," 2. A chariot race ending in a near-fatal accident is described as part of the rituals in honour of a Breton mountain deity in the *Vita Samsonis* 48, *AASS* Jul. 6, 584).

[180] *Conc. Aquisgranense* (816) 83, *MGH Concilia* 2.1: 368. See also Rabanus Maurus, *Poenitentium Liber ad Otgarium* (842-843) 27, *PL* 112: 1419. Canon 64 of the Council of Laodicea (380?) required clerics to leave wedding banquets before the games began (Hefele-Leclercq, *Histoire des conciles,* 1.2: 1023).

[181] For banqueting and conviviality in the Roman Empire, see Paul Veyne, "The Roman Empire," in *A History of Private Life,* 1: *From Pagan Rome to Byzantium*, trans. Arthur Goldhammer (Cambridge, Mass., and London, 1987), 186-194. For Roman drinking contests, see John H. D'Arms, "Heavy drinking and drunkenness in the Roman world: Four questions for historians," in *In Vino Veritas*, ed. Oswyn Murray and Manuela Tecusan (Oxford, 1995), 304-317. For Celtic banquets, in which competition for the choicest cut of meat was an important feature, see

had been the sacrificial communion meal (usually at midsummer, midwinter and at funeral feasts), in which the living ate and drank in honour of the gods and the dead. Excess was essential to induce the ecstasy that united celebrants to the gods and to one's fellows. Partaking only moderately of food and especially drink was an insult to gods and men, both living and dead. This tradition had a great influence on early medieval social customs. Drinking and drunkenness were communal affairs, taking place at convivial gatherings.[182] Private drunkenness appears to have been unknown–at least it found no place in our documents.

4.3.1 *Feasting and ritual drunkenness*[183]

In the first half of the 6th century, Caesarius of Arles, already deeply concerned about the gluttony and drunkenness endemic in his flock (he dedicated two sermons wholly to the topic), took note of the extreme social pressure to drink to excess and of rituals which had probably been introduced by the Visigothic and Ostrogothic occupiers of his diocese. In feasts lasting until dawn, people were forced, sometimes by social superiors, to drink unseemly amounts ("A powerful person forced me to take too much, and I could not act otherwise at his feast or at the king's"). Men ridiculed those who were unable to hold their own in drinking; some gorged themselves to the point of vomiting and drank "measure beyond measure."[184] The abstemious person was held to be less than a man, and was mocked for his virtue to such an extent that, in Caesarius' eyes, his sufferings were tantamount to the pains of martyrdom. The clergy were no better than the

Green, *Dictionary of Celtic Myth and Legend, s.v.v.* "Feast" and "Mac Da Thó." Maurice Cahen's *Études sur le vocabulaire religieux du vieux-scandinave. La libation* (Paris, 1921) is the invaluable source for German drinking traditions. See also Grimm, *Teutonic Mythology*, 59-63; and Derolez, *Les dieux et la religion des Germains*, 188-190. For ritual feasting, see Karl Hauck, "Rituelle Speisegemeinschaft im 10. und 11. Jahrhundert," *Studium Generale* 3 (1950) 611-612. See Salin, *La civilisation mérovingienne* 4: 45-49, for archaeological evidence of drinking feasts.

[182] Supporting testimony is found in the *Lives* of missionary saints. For example, St. Columban happened upon a crowd of Suevi who were preparing a sacrifice in honour of Woden near Bregenz; the central feature was a large vessel, "called *cupa* in the vernacular," filled with beer (*Vita Columbani* auctore Iona 27, *MGH SRM* 4: 102). St. Vaast was invited to a banquet where jugs of beer were set out "according to pagan rite," some prepared for Christians, others for pagans (*Vita Vedastis* 7, *MGH SRM* 3: 410-411).

[183] This section by no means exhausts the subject of drunkenness which, as a social and moral evil, played a large role both in penitentials and church law. I have considered only those passages in which drinking and drunkenness appear to have a strong ritual element.

[184] *S.* 1.12, *CCSL* 103: 8; *S.* 47.2, *ibid.*, 212; *S.* 46.3, *ibid.,* 206. In his discussion of the cultural aspects of drinking, Wolfgang Schivelbusch underlines the persistent nature of the etiquette of drunkenness: "Au XVIe siècle encore, toute beuverie débouche obligatoirement sur l'ivresse de chacun des participants; ce serait une offense inouïe à l'égard des commensaux si l'un des beuveurs s'interrompait. Il est interdit à la fois de ne pas accepter une boisson offerte et de ne pas rendre la pareille" (*Histoire des stimulants*, trans. Eric Blondel *et al.* [Paris 1991], 81).

laity, drinking too much themselves and forcing others to do so.[185]

Drinking contests and toasting were enforced rituals. Special goblets, larger than usual, were provided for competitions conducted according to "fixed rules," with praise as the winner's reward. Sometimes, by "foul custom," a "relic of paganism," three men volunteered or were compelled to vie in this contest.[186] At the end of the feast, when the celebrants were already sated, they were obliged to start drinking again, this time "to various names, not only of living men but also of the angels and other saints of antiquity," with savouries being served to increase thirst.[187] The participants evidently found ready substitutes in the angels and saints for the deities and ancestors to whom they had traditionally drunk.

Heavy, prolonged, ritualised drinking was also a feature of rural gatherings in Caesarius' diocese. Certain peasants (or uncouth persons – *rustici*) were reputed to invite their kin or neighbours to drinking bouts lasting several days, until all the drinkables were gone, thus using up in "four or five days of lamentable and shameful drinking" a supply adequate for two or three months of normal family use. This occurred when they had either wine or some other drink–surely beer or mead–available in large enough quantities "as though for a wedding feast." Caesarius, however, was not describing weddings. Although his words imply that the guests were summoned to celebrate the availability of liquor (that is, when the new wine or beer was ready for drinking), the grim, single-minded determination with which the guests drank themselves into unconsciousness does not suggest rejoicing but rather a commemoration of the dead.[188]

By the 7th century, drinking toasts was no longer perceived to be a survival of pagan practice. It continued unabated at commemorative banquets even (our texts create the impression, mainly) in clerical circles (see chapter 8). Now indeed it was accepted as a legitimate excuse for the extremes of drunkenness. We have seen that penitentials in the Theodorian tradition found it acceptable for monks, priests and other clerics to get drunk to the point of vomiting under certain circumstances, namely, sickness, prolonged abstinence from food and drink, or drunkenness in obedience to superiors "for joy at the Nativity or Easter, or in commemoration of one of the saints." A superior incurred no guilt for this, unless he got equally drunk himself.[189] Surely among the saints honoured so vigorously had been holy members of the order or the chapter.

Drinking toasts to saints was enshrined among Scandinavian Christians as an

[185] *S.* 55A.3, *CCSL* 103: 246.

[186] *S.* 46.3, *CCSL* 103: 206; *S.* 46.8, *ibid.,* 210-211.

[187] *S.* 47.5, *CCSL* 103: 214. See also *S.* 55.4 and *S.* 55A.3, *ibid.,* 243 and 246 for other references to drinking "to names."

[188] *S.* 47.7, *CCSL* 103: 114.

[189] *Poen. Theodori* (668-756) I, 1.4, Schmitz 1: 525. Among the others are two Irish penitentials, the late eighth-late ninth-century *Bigotianum* (I,1.1, Bieler, 214) and the *Old Irish Penitential* (c. 800) I, 15, Binchy, *ibid.*, 261. See Jean Imbert, "L'ébriété dans les pénitentiels," in *Life, Law and Letters. Historical Studies in Honour of Antonio García y García*, ed. Peter Linehan (Rome, 1998), 475-487.

obligatory religious ritual at fixed periods of the year, for example, in twelfth-century Norway, at Christmas, Easter, Michaelmas and perhaps the feast of St. John the Baptist.[190] The more southerly Germans also continued to drink toasts. Charlemagne forbade drinking feasts where associations were sworn in St. Stephen's name or in that of the emperor or his children.[191] Since it is known that toasts were drunk to the Virgin Mary, and if we postulate that *petendum* (entreaty?) is a mistake for *potandum* (drink), the clause "concerning the *petendum* which good people call 'of the Virgin Mary'" of the *Indiculus superstitionum* may deal with a libation in her honour.[192]

Many great clerics, imbued with ambient ideals of *noblesse oblige*, perpetuated these customs. Hospitable bishops encouraged their guests to get drunk, and few authorities took a severe view of this mistaken generosity. St. Boniface was an exception. In a letter written in 747 to Archbishop Cuthbert of Canterbury, he criticised the habitual drunkenness of the English and particularly the penchant of English bishops to urge ever greater amounts on drinkers in order to get them drunk. He claimed that this was a habit peculiar to pagans and the English ("our people"), rather surprisingly insisting that Franks, Gauls, Lombards, Romans and Greeks were all exempt from this failing.[193] Despite his favourable opinion of the Lombards, a mid ninth-century Council of Pavia also felt obliged to instruct bishops to be satisfied with modest meals, to refrain from urging their guests to eat and drink, and to present a constant example of sobriety, avoiding unsuitable entertainments at their gatherings–all indecent topics, exhibitions of play-acting, storytellers' nonsense (heroic tales?), stupid jokes, and performances of buffoons.[194]

Toasting and the concomitant drunkenness was a feature at rowdy banquets held for a variety of worthy causes. In a canon "concerning the gatherings or brotherhoods which they call *consortia*," the Council of Nantes (7th or 9th century) commanded priests and laymen to attend none but those pertaining to religious obligations, that is, to Mass, lights, mutual prayers, funerals, charities and other pious works. It banned attendance at "meals and banquets forbidden by Divine Authority where, as we know from experience, difficulties and excessive demands

[190] Lucien Musset, "La pénétration chrétienne dans l'Europe du Nord et son influence sur la civilisation scandinave," in *La conversione al cristianesimo,* 301.

[191] *Duplex legationis edictum* (789) 26, *MGH CapRegFr* 1: 64. The feast of St. Stephen is celebrated on 26 January. Cahen notes that "Yule" means both Christmas and the libation made at that time (*Études sur le vocabulaire religieux*, 54).

[192] 19, *MGH CapRegFr* 1: 223. Homann points out that since *Minnentrinken* in honour of saints such as Gertrude and John the Baptist were sanctioned, the objections to toasts to the Virgin could not have been to the practice itself, but to the excesses that it entailed (see *Der Indiculus*, 108-110). The word *boni* ("good people") is interesting. If it implies that people who were not "good" drank this toast to other, more questionable figures, we may have here another indication as to drinking rituals.

[193] *Ep*. 78, *MGH EpSel* 1: 170-171.

[194] 3, *MGH Concilia* 3: 221. See also Burchard of Worms, *Decretum* (1008-1012) 14.6, *PL* 140: 891.

[compulsory toasts?], indecent and meaningless merriment, and brawls often leading to homicides, hatred and dissension" might arise. Priests and clerics who did so were to be deprived of their rank, and laymen and laywomen were to be separated from the Church until they had done penance.[195] The mention of women in this context is noteworthy.

Forced drunkenness (as opposed to the drinking of compulsory toasts at institutionalised banquets) remained an important concern in approximately two dozen penitentials. One of the earliest, the sixth- or seventh-century *Ambrosian Penitential*, of Irish or English origin, made a distinction between the different motives for forcing others to get drunk: if from a wrong-headed idea of good nature, the guilty man had do penance as for his own excesses; if, however, from malice or sensuality, to befuddle them or make fools of them, he was the murderer of their souls and was to do penance accordingly.[196] Most penitentials continued to make explicit or implicit distinctions between base motives (wickedness or malice) and more innocent ones (good-fellowship or kindliness). The motive might also be self-assertive pride, to satisfy which some men habitually compelled their associates to drink too much.[197] A recension of the eighth-century *Merseburg Penitential* takes a kinder approach to laymen's failings in this respect than to clerics': if the latter got others drunk from a kindly impulse, they were to do forty days of penance, if from ill-will, they were to be judged guilty of homicide, but laymen were to do only seven days of penance.[198]

Drinking contests reappeared only in Burchard of Worms' penitential, and there by implication: "Did you ever get drunk by way of bragging, that is to say, so as to boast that you were able to outdo others in drinking and thus led yourself and others into drunkenness by your vanity? If you did so, do penance for thirty days on bread and water."[199]

Our authors did not recognise any element of pagan tradition in these drinking rituals. Nevertheless, an eighth-century sermon put the usual adjuration against drunkenness in the midst of beliefs and practices traceable to the non-Christian past.[200] Rabanus Maurus thought Caesarius of Arles' description of a drunkard relevant enough to include in one of his own homilies: "If he finds a chance, he drinks until he vomits and, once he gets himself drunk, he jumps up like a

[195] 15, Sirmond 3: 605. See also Regino of Prüm, *De synodalibus causis* [c. 906] II, 441: 386-387. Aupest-Conduché rejects the attribution of this canon to the Council of Nantes (*Bulletin philologique et historique* [1973] 38).

[196] 1.5 and 1.6, Körntgen, ed., *Studien Zu den Quellen*, 259. Schivelbusch remarks on the ambivalent nature of social drinking: the fraternal bond formed between the drinkers consists in reciprocal control, obligations and rivalries "qui ne sauraient être considerés comme des actes amicaux" (*Histoire des stimulants*, 81).

[197] Burchard of Worms, *Decretum* (1008-1012) 14.15, *PL* 140: 892.

[198] 52, 3rd recension [57], *CCSL* 156: 142.

[199] *Decretum* (1008-1012) 19, 5.85, Schmitz 2: 428).

[200] *Anonymous homily* (8th century), Nürnberger, 43.

madman in a frenzy to dance diabolically, leap about and sing filthy, amorous and lewd verses."[201] This is ecstasy deliberately sought, not the accidental result of drunkenness.[202]

What kind of alcohol was consumed during these drinking bouts? Despite the fact that beer and mead were more readily available in many areas than wine, the nearest reference to them in this context is Caesarius of Arles' reference to the "other drinks" which *rustici* prepared for their feasts. Since he put wine in first place, that was presumably, as one might expect in Provence, the usual choice. The *Poen. Casinense* also gives us to understand that the liquor at these gatherings was wine: "If anyone forces a man to drink more than is fitting, and he gets drunk, and if he causes ill-will, quarreling or a brawl or something unlawful because of the wine, he is to be judged as a homicide."[203]

4.3.2 *Other feasts*

Caesarius of Arles noted that rustics accumulated vast amounts of wine and other potables in preparation for weddings. This was apparently accepted as normal by other authors, for it was not remarked upon again.[204] A fifth-century Gallic council included a description of the entertainment at wedding banquets which included bawdy love songs and suggestive dances and vaultings, rituals originating in fertility magic. Priests and clerics forbidden to marry were instructed to avoid such festivities "lest ears and eyes dedicated to the sacred mysteries be polluted by the contagion of vile spectacles and words." This was based on canons originating with the fourth-century Council of Laodicea, and may have had little relevance to Gallic customs, but it was copied by the sixth-century Council of Agde and into collections, especially after the 8th century.[205] No attempt was made to prevent the laity from being exposed to such celebrations.[206]

[201] Caesarius of Arles (502-542) *S.* 16.3, *CCSL* 103: 78; Rabanus Maurus (d. 856) *Hom.* 44, *PL* 110: 82-83.

[202] In another sermon, Rabanus Maurus noted that drunkenness and gluttony were not considered to be serious sins but, on the contrary, praiseworthy by members of all classes, both men and women, young and old, including the clergy (*Hom.* 53, *PL* 110: 119).

[203] 44, Schmitz 1: 410. The *Vita Columbani* and the *Vita Vedastis*, however, describe beer-drinking (see n. 182 above). The latter adds the suggestion that special rituals went into the preparation of beer meant to be used in ceremonial drinking.

[204] *S.* 47.7, *CCSL* 103: 215.

[205] *Conc. Veneticum* (461-491) 11, *CCSL* 148: 15). See also *Conc. Agathense* (506) 39, *ibid.,* 209-210; *Epítome hispánico* (598-610); Ch*rodegangi Regula Canonicorum* (762) 68, *Concilia Germaniae* 1: 117; *Collectio Vetus Gallica* (8th/9th century) 40.10, Mordek, 499-500; Regino of Prüm, *De synodalibus causis* (c. 906) 1.335: 158; Burchard of Worms, *Decretum* (1008-1012) 2.134, *PL* 140: 648. It is found in modified form in Martin of Braga's *Canones ex orientalium patrum synodis* (572) 60, Barlow, 138. This canon is based on canons 53 and 54 of the fourth-century Council of Laodicea (Hefele-Leclercq, *Histoire des conciles*, 1.2: 1023).

[206] This can be contrasted with an injunction to Christian laymen and laywomen to avoid pagan celebrations, dancing and bawdy language at weddings, expressed in an anonymous sermon, probably of the first half of the 5th century and of unknown provenance, apparently

Other canons specifically targeting the clergy also come from texts based on earlier Church legislation. The Council of Mainz (852) revived a fourth-century canon forbidding clerics to attend presentations of plays and instructed them to withdraw from banquets and weddings before the actors appeared.[207] Sometimes the clergy were forbidden altogether to attend such celebrations so as to avoid the temptation to the pleasures of secular life.[208] Radulph of Bourges undoubtedly drew on actual experience when he targeted the rural clergy in particular, forbidding them to attend weddings, and painted a lively picture of the pleasures denied to them: "fleshly desire, earthly business, the empty delights of this world, shameless gatherings, vain shows, death-dealing worldly pomps, obscene language of jokes, useless laughter."[209]

Other texts, also drawn from the council of Laodicea, forbade all Christians "to clap and leap" at weddings, charging them to feast and dine in a seemly manner.[210] Herard of Tours singled out vaulting at weddings.[211] These were seen as magical rituals–the condemnation of acrobatic dances is a constant feature in our literature and, while clapping is mentioned much more seldom, it was noted as an element at New Year's celebrations and at commemorative banquets.[212]

4.3.3 *Games*

In early medieval pastoral literature, *ludi*, games, do not have any obvious connection with the organised *ludi* of the Roman Empire. They were condemned usually as part of illicit celebrations marking the change of seasons, liturgical festivals or other communal gatherings. Most other references give no indication of anything intrinsically sinful in games, even in Caesarius of Arles and Rabanus Maurus' condemnations of gambling[213] or in canons forbidding them to the clergy.[214]

There are two exceptions. St. Eligius urged his hearers to forbid the per-

intended for an urban audience (Jean Leclercq, "Sermon ancien sur les danses déshonnêtes," *RB* 59 [1949] 196-201).

[207] 23, *MGH Concilia* 3: 252. See also Regino of Prüm, *De syondalibus causis* (c. 906) 1.337: 159; Atto of Vercelli (d. 961) *Capitularia*, 42, *PL* 134: 37.

[208] *Poen. Quadripartitus* (9th century, 2nd quarter) 74, Richter, 12. See also Burchard of Worms, *Decretum* (1008-1012) 3.108, *PL 140:* 695.

[209] *Capitulary of Radulph of Bourges* (between 853 and 866) 19, *MGH EpCap* 1, 248.

[210] *Collectio Vetus Gallica* (8th/9th century) 40.2i, Mordek, 496. See also Rabanus Maurus, *Poenitentium Liber ad Otgarium* (842-843) 27, *PL* 112: 1419; Burchard of Worms, *Decretum* (1008-1012) 9.10, *PL* 140*:* 817.

[211] *Capitula* (858) 112, *PL* 121, 772.

[212] *Homiliare Toletanum* (2nd half, 7th century) *Hom.* 9, *PL Suppl* 4: 1941; *Admonitio synodalis* (c.813) 39bis, ed. Amiet, 51.

[213] Caesarius of Arles (502-542) *S.* 61.3 and *S.* 81.5, *CCSL* 103: 269 and 368; Rabanus Maurus (d. 856) *Hom.* 11: *In Dominica III Quadragesimae*, *PL* 110: 24.

[214] *E.g., Conc. Romanum* (679), Haddan and Stubbs 3, 133); See also Pope Zachary to Frankish notables (748) *Ep.* 83, *MGH EpSel* 1: 186); *Conc. Romanum* [826] 11, *MGH Concilia* 2.2: 572).

formance of "diabolical games and the dances or songs of pagans" because "any Christian who participated in them would become a pagan thereby." A canon of the Council of Tribur (895) ruled that there should be no mass or prayers offered for a cleric killed in the course of a battle, brawl or "the games of pagans," "but the judge's hand strikes;" he was not, however, to be deprived of burial.[215] It is impossible to speculate about the nature of the games that offended the bishop of Noyon, but the ones targeted by Tribur were evidently tests of strength, skill or daring. Both contained ritual elements perceived not merely as immoral or reckless but as wholly incompatible with Christianity.

[215] Vit*a Eligii*, *MGH SRM* 4: 707; *Conc. Triburiensis*, *Canones extravagantes* 10, *MGH CapRegFr* 2: 248.

5
Space

Pastoral writers were preoccupied by devotions in unauthorised spaces–trees, springs and stones, roads and crossroads, graves, corners and "other places." These were, in the words of the sixth-century Council of Tours, "the chosen places of the pagans" or, as in a ninth-century continental penitential, the "loathsome locations" where prayers and gifts were offered, help sought and banquets celebrated in defiance of the Church.[1] The texts treat shrines and altars (*fana, arae*) in the same way as natural sites, giving the impression that they were frequented because they were inhabited by nameless powers like the others, not because they were thought to be sacred to a clearly defined deity. Mills, basilicas and even the family house also had a mystical quality. In other words, all topographical features, man-made or natural, could become suspect in the eyes of churchmen if they were the site of ritual or magical activity that they did not initiate or explicitly permit.

All places were potentially sacral, appropriate for the performance of rites that had, or at some time had had, religious elements. Not that all spaces were invariably "sacred" to the participants in the various rites. Most normally were not. Although some, such as crossroads, were always uncanny, usually it was some particular quality that invested certain trees or groves, springs, or blocks of stone with a religious significance that others did not have. Ghärbald of Lüttich implied as much when he asked whether there was any place in his diocese where "stupid people made and observed their vows by springs, trees and stones for reasons of health or some kind of devotion," the names of the people involved and the nature of their offerings.[2]

The Church opposed the frequentation of such sites on two counts. First, because of their pagan associations. At least during the early phases of conversion, some of the faithful continued to believe that the powers permeating the pagan countryside had in some way survived the Christian onslaught and kept their places in their traditional haunts. So explained Martin of Braga: "Many of the demons driven from heaven preside over the sea, rivers, springs or the woods; men who do not know God honour them and sacrifice to them as though they were gods."[3] But, by the 10th century, the pagan element in folk pieties seems to

[1] *Conc. Turonense* (567) 23, *CCSL* 148A: 191-192; *Poen. Vindobonense a* (late 9th century) 50, Schmitz 2: 354.

[2] *Capitulary 2* (between 802 and 809) 12, *MGH CapEp* 1: 30.

[3] *De correctione rusticorum* (572) 8, Barlow, 188.

have been forgotten. Regino of Prüm thought it necessary to spell out the connection for the clergy: people made their devotions at trees, springs and stones as though at an altar, and brought candles and offerings, "as if some divinity [*numen*] were to be found there who could bring a benefit or work them a mischief."[4]

The distinction between sacred trees, springs and stones on the one hand and numinous localities on the other has already been noted. Trees modified with epithets such as *fanatici, sacrivi* or *sacrilegi,* or rocks or stones to which gifts were brought in offering appear to have been objects of worship in themselves. Their cult could be destroyed or obliterated: the trees cut down, uprooted or burnt, the stones dug up and smashed. In other cases, the object itself does not appear to have been sacred. When phrases such as *ad arbores, ad fontes, ad lapides* (to/at trees, springs, stones) are followed by *ad cancellos, ad angulos, ad trivia, etc.,*–terms that refer to locations of devotion, not its objects–then it can usually be assumed that the trees, springs and stones also have to be seen as locations merely, not as idols or representations of deities. Pirmin of Reichenau made this distinction when he exhorted the faithful not to adore idols *nor* to adore or fulfill vows at stones, trees, *anguli* (corners? cornerstones?), springs or crossroads.[5]

Second, the places themselves, even without pagan associations, were condemned because they were outside the control of the Church, which tried to draw a rigid line between the sacred and the profane. There was an economic side to be considered also, since the fulfillment of vows included offerings (*conpensus, oblationes, munera*). As early as the 6th century, the Synod of Auxerre expressed concern that people made offerings at home and passed vigils there or among bramble bushes (?–*sentius*), and by sacred trees and springs–if they had a vow to fulfill, they should keep vigil in the church, and give the offerings to the charitable fund of the church or to the poor.[6] But it was only in the 8th century that churchmen began repeatedly to insist in penitentials that the faithful should not perform any religious acts whatsoever except in church.[7] In particular, the frequent admonition that mass should be said only in the church, not in houses or other locations, speaks of the determination to make the parish or other duly authorised church the exclusive focus of religious activity.[8] The devotional acts were not necessarily wrong in themselves. In fact they were often the ordinary devotional practices of Christians, similar to those performed in front of the altar.[9]

[4] *De synodalibus causis* (c. 906) 2, 5.43: 212. See also Burchard of Worms, *Decretum* (1008-1012) 1.94: Interrogatio 42 and 10.32, *PL* 140: 576 and 837.

[5] *Dicta Pirmini,* 22: 172. See also *ibid.,* 28: 188-189.

[6] *Syn. Autissiodorensis* (561-605) 3, *CCSL* 148A: 265. I postulate that *sentius* is an error for *sentibus.*

[7] *E.g., Liber de remediis peccatorum* (721-731) 1, Albers, 410; *Poen. Burgundense* (early 8th century) 29, *CCSL* 156: 65.

[8] *E.g., Admonitio synodalis* (c. 813) 30, Amiet, 48.

[9] For the increased centralisation of religious life in the hands of the clergy and in churches, see Yves Congar, *L'ecclésiologie du haut moyen âge* (Paris, 1968), 61-127.

The performance of acts appropriate to church in an inappropriate place was worse in the eyes of churchmen than the performance, in the same place, of acts objectionable in themselves. Continental and English penitentials typically imposed three years of penance for making and fulfilling vows in forbidden locations, but only one year for eating and drinking in the same place, a practice which may have entailed the consumption of *idolothytes*, *i.e.*, sacrificial offerings.[10] In a sermon to the diocesan clergy, Arno of Salzburg seems to suggest that it was actually wrong to seek God's help in time of need "whether at trees or springs or anywhere, except in front of God and His saints and in holy mother church."[11]

St. Boniface's struggle with the renegade bishop Aldebert is illustrative. Unquestionably Aldebert was an unsavoury character, but the miraculous signs at his birth, his visions and mystical experiences contained little which hagiographers did not claim for many unexceptionable saints (although his habit of distributing nail and hair clippings as relics would have raised eyebrows under any circumstances). He was dangerous in part because of his unorthodox doctrines, but in part also because of his practice of setting up little crucifixes and oratories and holding religious services in the fields, by fountains and elsewhere. The crowds that abandoned legitimately constituted clergy and churches to flock to him were no doubt drawn by his charisma but surely also by the attraction of sites so familiar to them.[12]

5.1 Loci abhominati

5.1.1 *Natural sites*

The popular sites for unauthorised cult were trees, groves, woods and shrubs (*arbor, lucus, silva, locus silvestris, sentius*), bodies of water such as a spring, marsh, river, stream, sea, current or well (*fons, palus, flumen, rivus, mare, torrens, puteus*), stones (*lapis, saxum, petra*) and mountains (*montes*). Trees and water appear to have been the most popular. They maintained their importance, at least in the thoughts of clerical writers, from Caesarius of Arles to Burchard of Worms, in all parts of Western Europe except Ireland.[13] Separately or together, they appear in some form in about eighty-five percent of the nearly one hundred passages dealing with cultic practices performed in the vicinity of a natural object. Stones come far behind them, appearing in about twenty percent.

By far the most common term for tree used in our texts is *arbor,* generally appearing in phrases such as Caesarius' "No should make vows near/to [*ad*] a tree," or "A Christian ought not to make vows near/to trees nor adore near

[10] *E.g., Poen. Remense* (early 8th century) 9.6, Asbach, 56.

[11] *Synodal Sermon* (c. 806), Pokorny, 393-394.

[12] *Conc. Romanum* (745) 6, *MGH Concilia* 2.1: 39-41.

[13] Although nature cults or ritual acts that took place near trees, springs and wells are missing from the *Penitential of Theodore*, they appear regularly in the majority of the Theodorian penitentials.

springs."[14] The overwhelming majority of later texts paired trees and springs, often combined with other sacred places, and used some variation of the formula "to make" or "to fulfill" vows.[15]

The word *arbores* may refer either to clumps of trees or to individual trees remarkable for their size or shape or because they stood alone in a field. In the 8th and 9th centuries, however, a handful of texts dealt explicitly with groves or woods (*luci, silvae*) as well as trees. When Gregory III wrote to the Hessian nobility and commoners in 738 to exhort them to abandon pagan customs, he included practices at groves and springs.[16] The same word, *lucus*, found its way into three Carolingian texts. The conquered and forcibly baptised Saxons were penalised if they made "vows to [or near–*ad*] springs, trees or groves, or if they brought anything there according to the customs of pagans and ate in honour of demons." A short time later, the Council of Frankfurt called for the destruction of trees and groves. The *Homilia de sacrilegiis*, dating from roughly the same period, also spoke of *luci:* "whoever goes to the ancient altars, groves, trees, stones or to any other place, or offers animals or some other thing there, or holds a feast in that place ..."[17]

Religious rites occurring in woods appear in two unrelated sources. The seventh or ninth-century Council of Nantes called for the destruction of stones which ("by the tricks of the demons" or, alternatively, "for their amusement") were "venerated in ruinous and woody areas where people make vows and bring [offerings]."[18] The image evoked is of a ruined and abandoned shrine reclaimed by the forest. The *Indiculus superstitionum* likewise cited "the rites [or sacred places, objects or sacrifices–*sacra*] of the woods which are called *nimidas*."[19]

As sites of devotion, for making and fulfilling vows, bodies of water appear in our texts almost as often as trees, and usually in conjunction with them. They appear alone once only, in the clause of the *Indiculus superstitionum* "concerning the springs of sacrifice."[20] Caesarius of Arles' word *fontes* (spring or well) is the most common term in our documents, but a few other terms also appear. Caesarius spoke of the St. John's Eve custom of bathing in the marshes and rivers as well

[14] *S.* 1.12 and *S.* 54.5, *CCSL* 103: 8 and 239.

[15] *E.g., Poen. Hubertense* (early 9th century) 24: *CCSL* 156: 110; *Poen. Remense* (early 8th century) 9.6, Asbach, 56; Regino of Prüm, *De synodalibus causis* (c. 906) 2.365: 352; *Anonymi liber poenitentialis*, *PL* 105: 722.

[16] *Ep.*43, *MGH EpSel* 1: 69.

[17] *Capitulatio de partibus Saxoniae* (775-790) 21, *MGH CapRegFr* 1: 69; *Conc. Francofurtense* (794) 43, *MGH Concilia* 2.1: 170; *Homilia de sacrilegiis* (late 8th century) 2, Caspari, 6. Such an elegant word as *lucus* is to be expected from members of the Carolingian elite, but is surprising when coming from the rather crude pen of the author of the *Homilia.*

[18] *Conc. Namnetense* 18, Aupest-Conduché, 54 = 20, Sirmond 3: 607. See also Regino of Prüm, *De synodalibus causis* [c. 906] 2.366: 352-353; Burchard of Worms, *Decretum* [1008-1012] 10.10, *PL* 140: 834.

[19] 6, *MGH CapRegFr* 1: 223. Homann considers that *nimidas* is the vernacular for grove (*Der Indiculus*, 51-53); Haderlein derives the word from the Celtic *nemetos* ("Celtic roots," 10).

[20] 11, *MGH CapRegFr* 1: 223.

as springs.[21] The *Ratio de cathecizandis rudibus* forbade sacrifices to/near springs and rivers.[22] In the early 11th century, Rhenish women practiced a rain-making ritual which included leading a naked little girl into the nearest river and sprinkling her with the water, in the hope of making rain.[23] This, however, was a form of sympathetic magic rather than a pious practice. The procession went to the river as the most convenient source of water, not to carry out their devotions in its vicinity. The *Homilia de sacrilegiis* condemned "Neptunalia" performed in the sea or at the source of a brook or spring.[24] The word *torrens* (rapids) was twice substituted for the familiar *fontes* in the eleventh-century Latin translation of the tenth-century Anglo-Saxon *Ps.-Egbertian Confessional.*[25] Finally, three continental penitentials described a ritual used by women to exploit the mystical power of domesticated water by suspending sick children over a well (*puteus*).[26]

Stones or rock formations (*petrae, lapides* and *saxa*) as sites of cult appear in fewer than two dozen texts.[27] They play an insignificant role in penitentials, being mentioned in four only.[28] On the other hand, considering that Caesarius of Arles ignored them altogether, they are relatively well represented in sermons or guides for sermons, appearing in seven.[29] The rest of the references are drawn from canons, ordinances and collections. With the exception of one each from the Iberian peninsula and England, all come from the territory covered by the Frankish state or from the Rhineland.

Petrae is used in just over half these texts, *lapides* in nine and *saxa* in one only. No doubt these words had different connotations to authors from different regions and backgrounds, ranging from large massive objects (boulders and menhirs, even hills or crags) to small movable pieces of undressed or dressed stone (milestones, tombstones, boundary markers or stone figures). In classical usage,

[21] *S.* 33.4: *CCSL* 103: 146.

[22] Heer, 81.

[23] Burchard of Worms, *Decretum* (1008-1012) 19, 5.194: Schmitz 2: 452.

[24] 3, Caspari, 6.

[25] 2.22, *PL* 89: 419; 4. -, *ibid.*, 426.

[26] *Poen. Merseburgense* a (late 8th century) recension 2, 88, *CCSL* 156: 155. See also *Poen. Vindobonense a* (late 9th century) 79, Schmitz 2: 355; *Poen. Casinense* (9th/10th century) 57, Schmitz 1: 412.

[27] For cults of stones and rock formations, see Toutain, *Les cultes païens* 3.1: 169-178 for Spain, 3.2, 357-364: 293-295 for Gaul.

[28] *Poen. Ps.-Theodori* (mid 9th century) 27.18, *Confessionale Ps. Egbert* (c. 950-1000) 2.22, *PL* 89: 419; Wasserschleben, 597; *Poen. Arundel* (10th/11th century) 92, Schmitz 1: 462; Burchard of Worms, *Decretum* (1008-1012) 19, 5.66 and 5.94: Schmitz 2: 424 and 430. In penitentials, *petrae* are amulets, probably gems, used legitimately by people possessed by demons (*e.g., Poen. Theodori* 2, 10.5, Schmitz 1: 544).

[29] Martin of Braga, *De correctione rusticorum* (572) 16, Barlow, 198; *Vita Eligii* (c.700-725), *MGH SRM* 4: 706; *Dicta Pirmini* (724-753) 22: 172; *Ps.-Boniface*, S. 61, *PL* 89: 855; *Homilia de sacrilegiis* (late 8th century) 2, Caspari, 6; Rabanus Maurus (d.856), *Hom.*67, *PL* 110: 127; *Anonymous Sermon* (10th or 11th century), Caspari, *Kirchenhistorische Anecdota* 1: 199-201.

saxum meant rock or large piece of stone; *lapis* was a general term meaning stone and any object made of stone. These were gradually replaced by *petra*, a word of Greek origin, brought in perhaps by sailors, and made popular among Christians probably by the play on *Petrus* in the Vulgate.[30] No two of these are ever combined in the same passage except in Martin of Braga's model sermon and generally we are given no context to explain why one was chosen rather than another.[31]

The Council of Tours (567) was the first of the sources to mention stones (*petrae*) among the "designated places of pagans" that were the site of forbidden devotions. The canon was prompted by the custom of celebrating February 22nd, the feast of the Chair of St. Peter, with repasts in honour of the dead, and may hint at the connection of such places with the underworld.[32] Like the churchmen at Tours, Martin of Braga gave *petrae* priority among the sites of illicit religious practices: "For example, what is lighting candles at stones, trees, springs and crossroads but the cult of the devil?"[33]

In general, the documents do not differentiate among the practices associated with *petrae*, *arbores* and *fontes*.[34] There are two exceptions. The *Indiculus superstitionum* refers to "that which they do over stones"[35] whereas it spoke of the "rites" and "sacrifices" of woods and springs. A ninth-century sermon specifically mentions sacrifices immolated over stones or by springs or trees.[36] Here plainly the stones function as altars.

Burchard of Worms used *lapides* twice in the same sentence to mean two different types of stone–one a fairly large object, solid and permanent enough to identify the site of a forbidden repast, the other small enough to be carried about, and used in building a cairn:

> Did you eat any part of an idolothyte, that is, of the offerings made in some places at the graves of the dead, or near springs, trees or stones, or at crossroads, or did you carry stones to a cairn, or crown crucifixes at crossroads with wreaths? If you did so or consented to any of these, you should do penance for thirty days on bread and water.[37]

In addition, *lapides* appears in two unusual contexts in Burchard's penitential. The forbidden customs of the New Year include preparation of a table in the house *cum lapidibus vel epulis*, surely a mistake for *dapibus* or *lampadibus* (dishes or

[30] *Dict. étymologique, s.v.v.*

[31] Martin of Braga reported that people threw *lapides* on mounds of stone (*acervi petrarum*) in honour of Mercury (*De correctione rusticorum* [572] 7, Barlow, 18).

[32] 23, *CCSL* 148A, 191-192.

[33] *De correctione rusticorum* (572) 16, Barlow, 198.

[34] *E.g., Vita Eligii* (c.700-725), *MGH SRM* 4: 706; *Poen. Arundel* (10th/11th century) 92, Schmitz 1: 462.

[35] 7, *MGH CapRegFr* 1: 223.

[36] Ps.-Boniface, *S.* 6.1, *PL* 89: 855.

[37] *Decretum* (1008-1012) 19, 5.94: Schmitz 2: 430.

lamps).[38] The other was a divinatory rite meant to discover the prognosis for recovery from an illness:

> Did you do what some people do when they visit a sick person? If they find nothing [no other indication?–*nihil*] when they approach the house where he lies sick, they find some stone lying nearby, they turn it over and they look in the place where it had been lying to see if anything is alive beneath it; and if they find an earthworm, ant or fly or anything else moving about there, they claim that the sick person is convalescing. If, however, they find nothing moving about there, they say that he is about to die. If you did so, or believed it, you are to repent for twenty days on bread and water.[39]

As with the choice of the unusual *lucus,* the author of the *Homilia de sacrilegiis* preferred to use an unconventional although impeccably correct word, *saxa,* for stones. Since *lucus* has a sense quite distinct from *arbores*, *saxa* may also have had a precise technical meaning for him that differentiated it from the more common *petrae* or *lapides.*

The only explicit reference to hills or mountains is in the late eighth-century *Ratio de cathecizandis rudibus*, which puts *montes* as the first in the list of forbidden locations for offering sacrifices.[40] The absence otherwise of any other mention of the cult of mountains or rocky promontories is astonishing in view of the well-known attraction of such sites for pagan cults, attested in both archaeological and written records.[41]

[38] 19, 5.62, Schmitz 2: 423. Wasserschleben's edition also gives *lapidibus*, while another recension gave *lampadibus* (53a, 643). For *dapibus*, see *Concilium Romanum* (743) 9, *MGH Concilia* 2.1: 15-16; for *lampadibus*, Atto of Vercelli (d. 961) *Capitularia*, 79, *PL* 134: 43.

[39] *Decretum* 19, 5.102, Schmitz 2: 432. See also Wasserschleben, 90, 650. There appears to be a word missing after *nihil*–probably *ibi* ("there").

[40] Heer, 81.

[41] Many of the divinities of the continental Celts favoured high mountains, see Green, *Dictionary of Celtic Myth and Legend,, s.v.* "Mountains." See also José Maria Blásquez Martínez, *Religiones primitivas de Hispania* (Rome, 1962) 1: 9-10 and 37-38. Among the Germans, dwarfs and the dead were thought to inhabit mountains but, according to Grimm, there is no record of Tannhauser's Venusberg before the end of the Middle Ages (*Teutonic Mythology*, 454-455). See also E. Sidney Hartland, "Stones" in *ERE* 11: 864-869, and R. A. S. MacAlister, "Stone Monuments," *ibid.*, 877-881. Contemporary Christian sources also testify to pagan mountain cults. The Breton cult of an "abominable" stone effigy set on a mountain peak (*Vita Samsonis* 48, *AASS* Jul 6: 584) has already been mentioned, as has the god Helarius who inhabited a mountain-lake in the Massif Central (Gregory of Tours, *In gloria confessorum* 2, *MGH SRM* 1: 749-750). Valerio of Bierzo was incensed by the rituals practiced on the peaks of high mountains by the pagan mountaineers of Astorga during the first half of the 7th century (*Replicatio sermonum a prima conversione* 1, Aherne, 115) although, like so many other hermits, he himself was drawn to the harshness of rocks and mountains, and keenly alive to the opportunities they offered for spiritual development, see *ibid.,* 71. Farther to the east, St. Gall chivvied the Alpine demon-gods from their lakes and mountaintops (*Vita Galli* auctore Wettino 1.7, *MGH SRM* 4: 261). The late 10th-century *Miracula S. Geraldi* contains the description of a large cliff with a gaping hole in it

5.1.2 *Man-made sites*

Shrines and altars (fana, arae): It has been seen that Gregory I and Boniface V had thought that Anglo-Saxon and Germanic shrines, like Roman temples, housed idols. On one occasion, Caesarius of Arles himself discussed the refusal to destroy shrines and altars and *arbores fanatici* as part of the "ancient cult of idols," which may have implied that he, too, had in mind shrines dedicated to deities.[42] More usually, he treated *fana* and *arae* as he did the trees and springs, as places thought to be sacred in and of themselves, which should be destroyed wherever they were found. These were all sites of "diabolical feasts" where sacrificial food was eaten, but, remarked Caesarius scornfully, the participants indulged in such feasts less out of idolatry than of naivety, ignorance or greed. Nevertheless, there were apparently leftovers, for Caesarius insisted that this food was not to be eaten even at home.[43]

St. Columban's penitential (c. 573) and others modelled on it considered the possibility that some persons may have eaten and drunk in the vicinity of shrines for idolatrous reasons, either the cult of demons or in honour of statues, but they also considered other motives likelier –ignorance, contempt (of priests' teachings) or gluttony.[44] Columban's penitential was written in territory which had been christianised for generations, where probably the majority frequented shrines without explicitly religious intent. In time, many forgot altogether what the shrines had originally signified–a ninth-century scribe found it necessary to give an explanation, "We call shrines the temples of idols."[45]

In St. Eligius' sermon, lights were set up and vows made at *fana*, which the *Indiculus superstitionum* identified as "little huts."[46] Both documents put them ahead of stones, springs and trees, signifying perhaps that their authors considered rites at shrines to be more thoroughly tainted with paganism or superstition than rites elsewhere.

Caesarius of Arles called on his hearers to destroy *arae* in three sermons.[47]

by the roadside near St. Gerald's shrine at Aurillac. Wayfarers thought it a great shame to pass by without throwing pebbles into the opening (Anne-Marie Bultot-Verleysen, "Des *Miracula* inédits de saint Géraud d'Aurillac (†909). Étude, édition critique et traduction française," *Analecta Bollandiana* 118 [2000] 47-141; here, 90-92).

[42] *S*. 53.1, *CCSL* 103: 233.

[43] *S*. 54.6, *CCSL* 103: 239- 240.

[44] 20, Bieler, 104.

[45] *Poen. Valicellanum I*, 81, Schmitz 1: 30).

[46] *Vita Eligii* (700-725) *MGH SRM* 4: 706; *Indiculus superstitionum et paganiarum* 4: MGH *Cap RegFr* 1: 223.

[47] *S*. 14.4: *S*. 53.1, *S*. 53.2, *S*. 54.5, *CCSL* 103: 72, 233, 234 and 239. This word was reserved for pagan altars, Christian altars, even when illegitimate, being called *altaria*, as in a clause in Burchard's *Decretum* concerning the "altars" erected in fields or along the roads as a memorial to martyrs, and which contained neither a body nor relics (3.54: *PL* 140: 683). See also the nice distinction that Bede, following St. Cyprian, drew between the *altare* that the West Saxon king Redwald had "for the sacrifice of Christ," and his *arula* "for the (sacrificial) victims of demons"

Such altars were readily accessible, as openly displayed as sacred trees and shrines next to houses or in the fields that belonged to the estate or nearby.[48] The variety of verbs that he chose (destroy, shatter, scatter, crush) suggests that they were made of different substances, some of stone or metal, some of more ephemeral materials like the yellowing turves and rotting wood described by Maximus of Turin. After Caesarius, the word virtually disappears from pastoral literature. Only in the *Homilia de sacrilegiis* are altars put among the sites of sacrifices or feasts. Here they are obviously made of stone, for they are "antique."[49] Finally, a ninth-century penitential borrows from Isidore of Seville to derive the origin of *ariolus* from *ara.*[50]

The absence of the word altar does not mean that *de facto* sacrificial altars had disappeared. Other texts speak of sacrifices performed over stones and over tombs. Nevertheless, it appears that most churchmen were no longer worried that their charges continued to offer formal sacrifices even when they left food and other objects in numinous places.

Crossroads and forks (bivium, trivium, quadrivium, compitum): Crossroads and forks are quintessential transition point with special significance in the beliefs of many societies.[51] It is there where the living can encounter the dead and other spirits, good or evil, where they can look into the future and find healing for man and beast, where they can take malign spirits in order to confuse and lose them.[52] It is also suitable for making malign magic. In myth, Hecate and Diana were to be met at the crossroads. Records of devotional or magical practices at *bivia, trivia, quadrivia* and *compita* are found in over a dozen continental and English documents between the 6th and 12th centuries. The choice of words appears to be random, but *compitum* (appearing twice only) has the added meaning of "the point at which the boundaries of the fields converged;" rites performed there were probably fertility rites.[53]

Sometimes, crossroads were treated together with and in the same way as other suspect locations, as in Martin of Braga's injunction against lighting candles near stones, trees and springs or by *trivia*, Pirmin of Reichenau's against leaving offerings of wooden limbs at *trivia,* trees or "anything else," or Atto of Vercelli's denunciation of the women who spent St. John's Eve dancing, singing, casting lots

(*HE* 2.15: 190).

[48] *S.* 14.4 and *S.* 54.5, *CCSL* 103: 72 and 239.

[49] 2, Caspari, 6.

[50] *Poen. Ps.-Gregorii* 16, Kerff, 177.

[51] See Bächtold-Stäubli, *s. v.* "Kreuzweg;" J. A. MacCulloch and R. Wünsch, "Cross-roads," in *ERE* 4, 330-336; Grimm, *Teutonic Mythology*, 1074 and 1113-1115.

[52] According to Bächtold-Stäubli, Germans assembled at the crossroads to read the future on the eves of St. Silvester's, St. Andrew's and St. Thomas' feast days (December 31st, November 30th and December 21st respectively), as well as those of Christmas and the New Year.

[53] *ERE* 4: 336.

"all over the place, through roads and *compita*, springs and countryside."[54]

Various structures were sometimes placed at the crossroads and these, rather than the crossing itself, became the focus of devotional practices. In the case of Martin of Braga's Galicia, the structure was a cairn made of stones or pebbles flung by "greedy men" as they passed through the point where four roads met. Martin claimed that this was done as a sacrifice to Mercury "as to the god of wealth" but, as noted previously, this was probably an offering to him in his role as guide to the dead.[55] In the Rhineland of the early 11th century, people brought headwreaths for the crucifixes (and perhaps stones for cairns) at crossroads.[56] The heretic Aldebert, too, set up crosses which the Council of Soissons ordered to be burned.[57] They were probably at intersections, too, since we have additional testimony from two penitentials concerning feasts held next to crucifixes placed at the junction of four roads (*ad cruces in quadrubio*).[58] In two eighth-century Burgundian or Frankish penitentials, it appears that in certain places little enclosures or shrines (*cancelli*) were built there. Another goes very far in implying that all revered sites derived their sacred character from their location at the crossroads (*quadrubius*).[59]

Cancelli, anguli, fori: Trees, springs, stones and crossroads are easily identified, and their ritual importance is readily understood. However, two other sites appearing in conjunction with them present a puzzle. Over a dozen documents of the 8th and 9th century included *cancelli* or *cancellae* among the sites of ritual activity.[60] With the exception of St. Eligius' sermon, all are penitentials, one English, the rest continental. The usual Christian meaning, "chancel" must be ruled out. Possibly here it has the relatively rare meaning of "enclosure." If so, it may be a corruption of *sacellus*, *i.e.*, a small shrine, probably in memory of the dead–Gregory of Tours mentioned a *cancellus* built over a tumulus.[61] The *Hubertian*

[54] *De correctione rusticorum*, 16, Barlow, 198); *Dicta Pirmini* (724-753) 22: 175; Atto of Vercelli (d. 961) *S.* 13 (In annuntiatione beati praecursoris et martyris domini nostri Jesu Christi, Joannis Baptistae), *PL* 134: 850. See also *Epítome hispánico* (c. 598-610) 43.4: 190; *Dicta Pirmini* 22: 172 and 176; Burchard of Worms, *Decretum* (1008-1012) 19, 5.66, Schmitz 2: 424; *Anonymous sermon to the baptised* (12th century), Caspari, *Kirchenhistorische Anecdota* 1: 204.

[55] 7, Barlow, 189. In Burchard of Worms' penitential, cairns are built in fields not at crossroads.

[56] Burchard of Worms, *Decretum* 19, 5.94: Schmitz 2: 430.

[57] *Conc. Suessionense* (744) 7, *MGH Concilia* 2.1: 35.

[58] *Poen. Merseburgense a* (end of 8th century) 3rd recension, 54: *CCSL* 156: 141; *Poen. Vindobonense a* (late 9th century) 50, Schmitz 2: 354. C. Tolley suggests that such crosses may have served the same functions as sacred trees ("Oswald's Tree," in Hofstra *et al.*, eds., *Pagans and Christians,* 149-173).

[59] *Poen. Hubertense* (8th century, 1st half) 24: *CCSL* 156: 110; *Poen. Oxoniense* I, 24: *CCSL* 156: 90.

[60] *E.g., Poen. Parisiense Simplex* (8th century, second quarter) 21, *CCSL* 156: 76.

[61] *HF* 6.10, *MGH SRM* 1: 255.The word *tumulus* was applied to Christian as well as pagan graves. During the 8th-10th centuries, before the general establishment of parishes with church-

and *Oxonian I* penitentials explicitly situated these structures at crossroads, the place suitable for the burial of the unhallowed dead.[62]

Angulus is mentioned in four continental penitentials, the *Dicta Pirmini* and the *Ratio de catechizandis rudibus*, all dating from the 8th century.[63] Since Pirmin puts it in a list that already included a crossroad (*trivium*), it probably did not refer to the corner where two roads meet; it may have been a turn in the road. Or if it was a nook or a secret place, it was perhaps akin to *cancellus*. *Cancellus* and *angulus* invariably appear as part of a list, although never together, but it cannot be assumed that the same thing is being meant. A possible interpretation of *angulus* is cornerstone, since the practice of making foundation offerings when building a house is well-known and confirmed by the archaeological record.[64]

Fori appear in a single Frankish or Burgundian penitential, dating from the first half of the 8th century. They are associated with trees, *cancelli* and crossroads as a site for unsuitable devotions.[65] The word is in the accusative, *foros*, and it is conceivably a mistranscription of *fontes*, however strange it is to find an error in the copying of so familiar a word. Otherwise it may refer either to wine-vats or to assembly places for trade or legal hearings.

Unidentified sites: Some twenty passages in documents dating from the early 8th century onward hint at the existence of other, unnamed places of cult. Expressions such as "anywhere except in church," "elsewhere," "nor in any other place," "at no matter what devilish deceit"[66] and, especially, the "dubious places which they honour for the sake of the saints" of the *Indiculus superstitionum*[67] may refer to places that are mentioned in some other text, but leave open the possibility of the

yards suitable for burial grounds, Christians were often buried along roadsides, in abandoned ditches, or along ditches marking boundaries; see Cécile Treffort, "L'inhumation chrétienne au haut moyen âge (VIIe-début XIe siècles)," *L'information historique* 59 (1997) 63-66. In the *Vita Boniti* 37, a paralytic nun spends the night at the *cancellus* over his tomb (*MGH SRM* 6, 137).

[62] But memorials were built at roadside for the honoured dead also. The Council of Frankfurt (794) forbade the erection along the roads of such structures for "new saints" (42, *MGH Concilia* 2.1: 170). Burchard copied an African canon into his *Decretum,* which was directed against altars set up in for unauthorised martyrs, see n. 47, above.

[63] *E.g., Poen. Remense* (early 8th century) 9.6, Asbach, 56. One of the recensions of this penitential uses the spelling *angelus* rather than *angulus*. Blaise, *s.v.*, gives its primary meanings as corner or cornerstone; Lewis (*Latin Dictionary, s.v.*) adds "a retired or secret place, a nook, corner, lurking-place" as metonymic meanings.

[64] For a description of human and animal remains thought to be foundation deposits made in the British Isles during the Saxon period and later, see Merrifield, *The Archaeology of Ritual and Magic,* 116-119.

[65] *Poen. Oxoniense* I, 24: *CCSL* 156: 90.

[66] *E.g.,* respectively, *Poen. Remense* (early 8th century) 9.6, Asbach, 56; Arno of Salzburg, *Synodal Sermon* (c. 806), Pokorny, 394; *Dicta Pirmini* (724-753) 28: 188; *Anonymous sermon* (c. 850-882) 2, Kyll, 10.

[67] *MGH CapRegFr* 1: 223.

existence of still others. The *Homilia de sacrilegiis* testifies to the existence of *sarandae* (maybe a mistake for *sacrandae*–"venerable"), ancient monuments known as the "greater ones," where the possessed were taken to be rid of demons.[68] Undoubtedly there were others of which our authors took no note.

5.2 Cult

Night was the usual time for visits to unauthorised sites for religious or magical purposes. In the 6th century, Arlesians bathed in springs, marshes and streams during the night or early morning hours of St. John's Day and, four centuries later, the women of Vercelli spent the entire night out in the countryside.[69] The Synod of Auxerre (561-605) forbade the practice of passing the eve of saints' feasts in all-night vigils anywhere except in church, banning such vigils at Martinmas in particular.[70] A tenth-century English penitential imposed three years of fasting on anyone "who keeps night watches by rapids or any other created thing except at church."[71] In addition, the repeated references to different kinds of *luminaria* must mean that the hours after sunset were favoured for these rites. Such lights probably functioned both as an offering to friendly powers and as protection from the malign spirits roaming about in the dark.

It is impossible to judge how many persons actually frequented forbidden places at any given period or in any given region. Since Caesarius of Arles returned to the subject in seven or eight sermons, we might assume that this was a common practice in the Arles of his day. Nevertheless, he spoke only of "some people." This does not suggest large numbers. Other texts give even less of an idea. Such vagueness may be contrasted with statements, admittedly rare, concerning other practices or beliefs. Caesarius himself was unequivocal that drunkenness was a generalised vice among the men of his diocese; according to the sixth-century Council of Narbonne, "many" kept Thursday as a holiday; Agobard of Lyons said that practically everyone in his diocese irrespective of condition, origin and age believed in the existence of weather-making magicians; the *Canon Episcopi* affirmed that many people were implicated in witchcraft beliefs. It may well be that, by contrast, only few people took the risk of going to uncanny places at uncanny hours to perform rites so persistently condemned by all the authorities of church and state.

The need for help in time of sickness, either of men or of the herds, emerges from our texts as the principal motive for frequenting such places. Caesarius of Arles exhorted his flock to go to church for bodily health and forgiveness for their

[68] 22, Caspari, 12.

[69] Caesarius of Arles (502-542) *S.* 33.4: *CCSL* 103: 146; Atto of Vercelli (d. 961) *S.* 13, *PL* 134: 850.

[70] 3 and 5, *CCSL* 148A: 265. The first of these was copied into the *Epítome hispánico* (c.598-610) 43 4: 190.

[71] *Confessionale Ps.-Egberti* (c.950-c.1000) 4. -, *PL* 89: 426.

sins instead of turning to trees and wells, sorcerers and magic.[72] The Council of Nantes took steps against persons who, looking for health, made vows at forbidden places and brought candles or gifts. Burchard of Worms included this twice in the *Decretum*. His penitential specifically instructed confessors to question penitents whether they had asked for "health of body or mind" at such places.[73] During an outbreak of disease among animals and the human population and other misfortunes at the beginning of the 9th century, the people of Salzburg preferred to turn to cunning men, magic and to the powers residing in or around trees, springs and elsewhere rather than to the Church and legitimate physicians.[74]

Practices varied. Around seventh- or eighth-century Noyon, amulets were used at chosen sites, although we are not told how. Perhaps they were sought in such places or left there, dipped in the water or pulled through the trees or the crossroads, in order to benefit from the numinous power of the place.[75] Somewhat earlier, Neustrian herdsmen and hunters left bewitched bread, herbs or *ligamenta* in trees or at crossroads as charms to ward off or to inflict disease.[76] Those to be protected from illness were sometimes taken to the site to exploit its powers directly. We have seen from St. Eligius' sermon that cattle were driven through hollows in trees or openings in the ground and, from an English penitential, that mothers took babies to the crossroads and dragged them along the ground (or through a ditch). Madmen were brought to ancient monuments for exorcism.[77] An eleventh-century Spanish penitential recorded the practice of putting a sick man in the road or at the gate.[78]

Immersion in water or exposure to it had magical benefits also. Health was probably part of the reason for bathing in Provençal marshes, springs and rivers on Midsummer Eve. Elsewhere on the continent, as we have seen, women turned to the underground sources of springs and hung their babies in slings over wells to bring them good health.[79] In Spain, washing oneself or bathing "inversely"

[72] *S.* 13.3, *CCSL* 103: 66-67.

[73] *Conc. Namnetense* (7th century? 9th century?) 18, Aupest-Conduché, 54. See also Regino of Prüm, *De synodalibus causis* (c. 906) II, 366: 352-353; Burchard of Worms, *Decretum* (1008-1012) 10.10, *PL* 140: 834. *Idem,* 19, 5.66, Schmitz 2: 424.

[74] Arno of Salzburg, *Synodal Sermon* (c. 806), Pokorny, 393-394.

[75] *Vita Eligii* (c.700-725), *MGH SRM* 4: 70. Practices apparently based on similar concepts have already been noted: bringing home branches or twigs was mentioned by Maximus of Turin (*S.* 98.3, *CCSL* 23: 392), wreathing or circumambulating the house with laurel and greenery by Martin of Braga (*Canones ex orientalium patrum synodis* 73, Barlow, 141), and hanging the house with "baptised" grasses and branches by Atto of Vercelli (*S.* 13, *PL* 134: 850-851).

[76] *Conc. Rothomagense* (650) 4: Mansi 10: 1200. See also Regino of Prüm, *De synodalibus causis* (c. 906) II, 5.44: 212; Burchard of Worms, *Decretum* (1008-1012) 19, 5.63, Schmitz 2: 423-424; *Poen. Arundel* (10th/11th century) 94: Schmitz 1: 463.

[77] *Vita Eligii* (c.700-725), *MGH SRM* 4: 706. *Confessionale Ps.-Egberti* (950-1000) 4. -, *PL* 89: 426; *Homilia de sacrilegiis* (late 8th century) 22, Caspari, 12.

[78] *Poen. Silense (*1060-1065) 89, *CCSL* 156A: 25.

[79] Caesarius of Arles (502-542) *S.* 33.4: *CCSL* 103: 146; *Poen. Merseburgense a* (late 8th century)

(*inuersum*), either with or without some kind of spell (*per aliquam incantationem*), was magic, the water downstream from a mill being efficacious specifically in case of illness. The ninth-century *Albeldensian Penitential* treated bathing *inversum* with a spell as considerably more serious than trying to cure a sickness by merely bathing below the dam, without using a spell. A somewhat later penitential, which implied that *incantatio* was not necessarily a part of the ritual, did not differentiate between the two practices.[80] *Inuersum* is unclear: it may mean walking into the stream backward, swimming on one's back, turning against the current or against the apparent course of the sun (surely black magic), or reversing the usual order of washing. *Per incantationem* is also ambiguous, since it may mean "by way of enchantment," "with the help of magic" or "while reciting a spell."

Divination may have been a secondary motive for resorting to forbidden places. In general, the texts are ambiguous on this point. It is not clear from Caesarius of Arles' "We have heard that some of you make/fulfill vows at trees, pray at springs, observe diabolical omens" whether he meant to condemn auguries practiced near trees and springs, or if he was talking of two separate practices.[81] The same is true of a passage in Gregory III's letter to the Hessians and Thuringians (c. 738), in which he urged them to foreswear their traditional loyalties and customs:

> Turn your minds wholly to God, utterly rejecting and casting off soothsayers and lot-casters, the sacrifices of the dead or of groves and springs, auguries [*or* the sacrifices of the dead or auguries of groves and springs], amulets, enchanters and sorcerers, that is to say, evil magicians, and the sacrilegious observances that are habitually performed in your territory.[82]

Some continental documents, however, strongly suggest a connection. The eighth-century *Hubertian Penitential* proposed five years of penance for whoever had committed sacrilege, "that is, what they call *aruspices* [practiced divination? acted like soothsayers?], if he made his devotions at springs, *cancelli* at crossroads or at trees, or offered a sacrifice or questioned diviners in any matter or augured from birds or with any other wicked device."[83] A somewhat later penitential also associated divination with cults practiced in woods and by springs. It condemned "dia-

recensions 2 and 3, 88, *CCSL* 156: 155). See also *Poen. Vindobonense a* (late 9th century) 79, Schmitz 2: 355; *Poen. Casinense* (9th/10th century) 57, Schmitz 1: 412.

[80] *Poen. Albeldense* (c. 850) 99 and 100, *CCSL* 156A: 12; *Poen. Silense* (1060-1065) 192 and 193, *ibid.,* 36. See Bezler, *Les pénitentiels espagnols,* 287-288.

[81] *S*. 53.1, *CCSL* 103: 233.

[82] Ep.43, *MGH EpSel* 1: 69. According to Tacitus, sacred horses used by the Germans for divination were kept in woods and groves (*Germania*, 10); Plutarch claimed that German women used to "foretell the future by observing the eddies of rivers, and taking signs from the windings and noise of streams" (*The Lives of the Noble Grecians and Romans* ["the Dryden translation; " Chicago, London, Toronto, 1952] 585).

[83] 24, *CCSL* 156: 110.

bolical enchantments or sacrifice to demons be it by springs or trees or ... devotions anywhere except at church or [reading] the future in any way" and eating or drinking "near springs and trees by way of divination." Atto of Vercelli complained that Piedmontese women cast lots and looked for omens as they sang and danced their way through the countryside, by roads, crossroads and springs at midsummer.[84] The clearest testimony is found in Burchard of Worms' question concerning the practice of sitting on an ox-hide at the crossroads in order to foresee the future during the Calends of January.[85]

Caesarius of Arles distinguished between the pious acts performed near trees and those performed near springs. One made or discharged vows by trees, but one adored or prayed by springs, as well as made vows there.[86] (Caesarius, therefore, did not consider making vows to be prayer.[87]) Atto of Vercelli's capitulary also seemingly differentiated between the cult of trees and springs in a clause concerning "the matter of those who honour trees and approach springs thus in a religious manner."[88] "Pray" is used in this context in only two other documents. In the *Homilia de sacrilegiis,* one prayed by the sea or at springs or the source of streams; in Burchard of Worms' penitential, prayer was offered indiscriminately at springs, stones, trees and crossroads, as well as at the lawfully appointed places.[89]

Making and/or fulfilling vows in these places is mentioned at least once in more than two dozen sermons, penitentials, decisions of church councils and canonic collections. There are few indications as to the nature of the votive offerings. Some were very simple–pebbles thrown by passers-by on cairns or wreaths crowning roadside crosses.[90] Others were more elaborate. Most frequently men-

[84] *Poen. Parisiense I* (late 9th century or later) 14: Schmitz 1: 683; *ibid.,* 23: 684; Atto of Vercelli (d. 961) *S.* 13, *PL* 134: 850-851.

[85] 19, 5.62, Schmitz 2: 423. This appears to be related to the Scandinavian "utiseta," a form of divination practiced by incubation at crossroads; see R. L. M. Derolez, "La divination chez les Germains," in A. Caquot and M. Leibovici, eds., *La divination* (2 vols.; Paris, 1968) 278-279.

[86] *S.* 13.5, S. 14.4, S. 53.1 and S. 54.5, *CCSL* 103: 68; 71, 233 and 239; S. 229.4: *CCSL* 104: 909*; Pascahal sermon*, para. 7, R. Étaix, *RB* 75 (1965) 209.

[87] He may have considered the bargaining *do ut des* aspect of vow-making to be incompatible with prayer. The concept of prayer is discussed by Samuel D. Gill, who pointed out that there was as yet no adequate study of prayer as a religious phenomenon (*Encyclopedia of Religion* 11: 489-494); but *Quand les hommes parlent aux dieux. Histoire de la prière dans les civilisations*, ed. Michel Meslin (Paris, 2003) goes a long way to fill this gap. Although dated, Friedrich Heiler's 1932 analysis of primitive prayer in *Prayer: A Study in the History and Psychology of Religion* (trans. and ed. Samuel McComb with J. Edgar Park; New York, 1958) 1-64 is still useful. See also "Prayer," *ERE* 10: 154-205, especially, Jesse Benedict Carter, "Prayer (Roman)" and Enid Welsford, "Prayer (Teutonic)," 199-202.

[88] 48, *PL* 134: 38.

[89] *Homilia de sacrilegiis* (late 8th century) 3, Caspari, 6; Burchard of Worms, *Decretum* (1008-1012) 19, 5.66, Schmitz 2: 424.

[90] Martin of Braga, *De correctione rusticorum* (572) 7, Barlow, 18; Burchard of Worms, *Decretum*

tioned are lights, models of the human body, and food.[91]

Lights are found in about a dozen texts. They were first mentioned in a Galician document, but all subsequent references to them (wax or tallow candles, lamps, torches) come from the northern parts of Frankish territory or the Rhineland.[92]

Votives in the form of carvings or models of the human body or members of the body were given as a pledge in case of illness or, more commonly, in thanksgiving for a cure. The Synod of Auxerre (561-605) forbade them categorically, banning altogether the making of carvings or wooden models of feet or mannikins. In fact, such offerings were common at Christian shrines, and were considered a praiseworthy expression of popular piety.[93] The case was different when they were placed, not in a church, but in other locations.[94] Models of feet (of a flammable material such as wood or rags) were deposited at the crossroads in Merovingian Noyon.[95] The title "concerning wooden feet and hands made in the pagan fashion" in the *Indiculus superstitionum* undoubtedly refers to the same practice in Saxony and Thuringia.[96] Pirmin of Reichenau assured the faithful that

19, 5.94: Schmitz 2: 430.

[91] See Merrifield, *The Archaeology of Ritual and Magic*, 22-57 and 88-93 for the archaeological record of both the pagan and Christian periods for deposits of votive offerings (*e.g.*, bones and skulls, weapons, tools, ornaments, plaques, pottery, food) found in bodies of waters, on land and under houses.

[92] *E.g.*, Martin of Braga, *De correctione rusticorum* 16, Barlow, 198; *Vita Eligii* (c.700-725), *MGH SRM* 4: 706; *Conc. Namnetense* (7th or 9th century) 18, Aupest-Conduché, 54; Burchard of Worms, *Decretum* 19, 5.66, Schmitz 2: 424. The practice of lighting torches was condemned by the second Council of Arles (442-506) 21 (*CCSL* 148: 119) and two Iberian councils, *Toletanum XII* (681) 11 and *XVI* (693) 2 (Vives, 398-9 and 498) in the context of outright nature worship.

[93] 3, *CCSL* 148A: 265. See Anne-Marie Bautier, "Typologie des ex-voto mentionnés dans des textes antérieurs à 1200," in *La piété populaire au moyen âge. Actes du 99e Congrès National des Sociétés savantes, Besançon, 1974* (Paris, 1977), 237-282. For votives in the form of people, animals and body parts, see Merrifield, *Archaeology of Ritual and Magic*, 89-90. These come from England from the 14th and 15th century but must be very close in spirit to the offerings made on the continent at an earlier period as well.

[94] An insight into such practices and the reactions of the authorities is offered in Ruth Harris' discussion of the popular cult that grew up around the grotto of Lourdes almost immediately upon the apparitions–vigils, processions, a wide range of offerings, including flowers, lights, statuettes of the Virgin, money, food, jewellery and pieces of cloth. The practices of simple devotees can for once be seen, in the scrupulous documentation of events by civil and clerical authorities and contemporary observers. The authorities, especially the clergy, did not question the good faith and sincerely Christian piety of the crowds in attendance, but were worried about the disruption of order, the unorthodox venue and some of the more questionable attributes of the apparition (*Lourdes. Body and Spirit in the Secular Age* [London, 1999], especially 84-91). It may be assumed that in the 11th century as well, the faithful were acting out of genuinely Christian piety, and that many of the clergy were well aware of the fact.

[95] *Vita Eligii* (c.700-725), *MGH SRM* 4: 708.

[96] 29, *MGH CapRegFr* 1: 223.

putting wooden models of limbs at crossroads or by trees and in other places was of no use in bringing about a cure.[97]

Possibly the bread that sixth-century Galicians placed in the springs at the New Year and even the enchanted bread hidden in trees or thrown into crossroads by hunters and herdsmen should be considered as food offerings of a sort. More usually, however, food was offered as part of a ritual meal. Burchard of Worms asked penitents if they had taken bread or other offerings to springs, stones, trees or crossroads, or had eaten there. Here he seems to differentiate between food taken to the site as an offering and food consumed there, but another question makes clear that a part of the food offered was consumed by the votaries.[98] The *Homilia de sacrilegiis* bears witness that in Christian Francia, animal and other sacrifices were still being offered by "ancient altars and in groves, at trees, stones and elsewhere" preparatory to a communal meal, even at the end of 8th century.[99] The same may have been true of tenth-century England as well, for a contemporary penitential imposed a year of penance for sacrificing victims at natural sites.[100]

Occasionally, one may form an idea of the nature of unidentified offerings. When the Synod of Auxerre and the *Epítome hispánico* urged the faithful not to leave *vota* among the shrubs or by trees, springs or crossroads, but to bring them to the church to be distributed to the poor, the offerings in question must have been food, clothing, or money.[101] On the other hand, it is evident that carvings or models of some sort were the likeliest target of Charlemagne's order to take away and destroy the "observances that stupid people make at trees, rocks and springs."[102]

In other texts, no indication is given as to the kind of offering or sacrifice. This is the case with the titles concerning "the rites of the forests called *nimidas*," "that which they do over stones,"and "the springs of sacrifices" of the *Indiculus superstitionum,* and with the objects immolated over stones or by springs and trees in Ps. Boniface's sermon, the *sacrificium* of the *Hubertian Penitential*, and the gift or service (*munus*) of the Council of Nantes, Regino of Prüm and Burchard of Worms.[103]

Some fifteen documents testify to the popularity of banquets at forbidden

[97] 22: 176. Placing models of feet at crossroads may also have been intended as magic to bring a traveller safely home, rather than as a votive offering.

[98] 19, 5.66, Schmitz 2: 424; 19, 5.94, *ibid.*, 430.

[99] 2, Caspari, 6.

[100] *Confessionale Ps.-Egberti* (c.950-c.1000) 2.22, *PL* 89: 419.

[101] *Syn. Autissiodorensis* (561-605) 3, *CCSL* 148A: 265; *Epítome hispánico* (c.598-610) 43.4: 190.

[102] *Admonitio Generalis* (789) 65, *MGH CapRegFr* 1: 59.

[103] *Indiculus superstitionum et paganiarum* (743?) 6, 7 and 11, *MGH CapRegFr* 1: 223; Ps. Boniface (8th century?) *S.* 6.1, *PL* 89: 855; *Poen. Hubertense* (8th century, 1st half) 24, *CCSL* 156: 11; *Conc. Namnetense* (7th or 9th century) 18, Aupest-Conduché, 54; Regino of Prüm, *De synodalibus causis* (c. 906) 2, 366: 353; Burchard of Worms, *Decretum* (1008-1012) 10.10, *PL* 140: 834.

places. We have seen that in the 6th century Caesarius of Arles and St. Columban were inclined to dismiss these as being motivated by greed, ignorance or resistance to ecclesiastical authority rather than idolatry.[104] However, from the 8th century on, penitentials (none from Ireland or Iberia) treated such repasts as particularly dangerous. At the same time, they create the impression that the perception of the nature of the rite had changed. Caesarius' word *convivia* put the emphasis on the social context in which the gourmandising occurred, while the later texts (more than a dozen penitentials and Charlemagne's decree for Saxony) tended to emphasise the actual eating and drinking with terms such as *edere, bibere*, *comedere*.[105] Even then, the communal aspect of such meals was sometimes recognised. It is suggested in the *Homilia de sacrilegiis* by the word *epulari* (to banquet) and in two nine-century penitentials by a peculiar expression, "If they eat festivities ...." It is made explicit in *Halitgar's Penitential* with its treatment of those "who go together to a feast, taking their food and eating it in pagans' loathsome places."[106] One early ninth-century penitential implies that toasts or pledges were drunk at such feasts.[107]

Burchard of Worms, unlike Caesarius of Arles, was not disposed to treat these feasts lightly. In a passage already cited several times, he placed them firmly in the context of religious rituals celebrated by springs, stones, trees and crossroads. Eating a meal there was not innocent conviviality but, like lighting candles and torches, making offerings and asking for cures, an expression of forbidden reverence and devotion.[108] For once, Burchard was more severe than his predecessors, calling for a penance of three years where they had thought a year or two enough–proof in itself of the seriousness of the offence in his eyes.

5.3 SACRED SPACE AND MAGICAL RITUALS

Trees, springs, crossroads, *etc.*, were sites to which people went to pray, make requests and promises, leave offerings and participate in communal feasts. There were in addition other places which were thought to have the potential for magic. Our records testify of the numinous quality of mills and basilicas and, most importantly, of the *domus*, a term which should be understood as the totality of the house with its dependent structures together with the family dwelling within it.

[104] Caesarius of Arles, *S.* 54.6, *CCSL* 103: 239- 40; *Poen. Columbani* 20, Bieler, 104.

[105] *E.g., Poen. Parisiense Simplex* (8th century, second quarter) 21, *CCSL* 156: 76; *Capitulatio de partibus Saxoniae* (775-790) 21, *MGH CapRegFr* 1: 69.

[106] *Homilia de sacrilegiis* (late 8th century) 2, Caspari, 6; *Poen. Valicellanum I* (beginning of the 9th century) 79, Schmitz 1: 303; *Poen. Vindobonense a* (late 9th century) 50, Schmitz 2: 354; *Poen. Halitgari* (817-830) 41, Schmitz 1: 480. See also *Anonymi liber poenitentialis*, *PL* 105: 722. The author of the *Homilia* may have chosen the word *epulari* to underline the religious nature of these banquets since in classical Latin, *epulae* was applied particularly to banquets on religious occasions.

[107] *Poen. Hubertense* 30, *CCSL* 156: 111).

[108] 19, 5.66, Schmitz 2: 424.

Mills and basilicas: Water-powered mills were introduced on large estates in northern Francia, especially those belonging to monasteries, during the 8th century.[109] An innovation that exploited still-venerated streams, rapidly flowing ones in particular in such peremptory fashion, may well have seemed extraordinarily powerful, as were the basilicas where the Christian mysteries took place. These structures were thought to be suitable places for magic to recover or, perhaps, protect fugitives (according Caspari, fugitive slaves): it was pagan, not Christian, claimed the *Homilia de Sacrilegiis*, to inscribe tablets or parchment for that purpose and place them near or affix them to mills or basilicas.[110]

The mill also appears to have had added to the curative powers of water–it has been noted above that, in Spain, washing below the dam (that is, in water that had passed through the mill over the wheel) was thought efficacious in case of illness.[111]

The domus: In Roman paganism, the house (*domus*) had been among the holiest of places, the centre of the religious life of the family.[112] The ancestors, the *lares* and *penates*, and the *genius* had all resided within the home; the door, the hearth, the table, the altars and the boundaries were all sacred. Early medieval pastoral texts do not leave the impression that this was so among the mostly Germanic or Celtic populations of Western Europe. In the minds of the authors, the cults of trees, springs and stones unquestionably took precedence. Nevertheless, the house with its surroundings retained a part of its holy character. To some of the faithful it seemed as appropriate a site for religious activities as the church. Rituals were

[109] B. H. Slicher van Bath noted that, because of the difficulty of transporting millstones, early mills "were often found in remote places by streams, at points where a sharp change of ground-level made the current run faster" (*The Agrarian History of Western Europe A. D. 500 -1850* [trans. Olive Ordish, London, 1963] 72). For a brief history of mills, see H. Leclercq, "Moulin" (*DACL* 12.1: 359-366). For the spread of the watermill during the Middle Ages, see Marc Bloch, "Avènement et conquêtes du moulin à eau," *Annales d'histoire économique et sociale* 7 (1935) 634-643; Jean Gautier-Dalché, "Moulin à eau, seigneurie, communauté rurale dans le nord de l'Espagne (IXe-XIIe siècles)," in *Mélanges E.R. Labande* (Poitiers, 1974), 337-349; John Muendel, "Mills and milling," in F. Mantello and A. Riggs, eds., *Medieval Latin. An Introduction and Bibliographical Guide* (Washington, 1996), 497-502. For beliefs associated with mills, see Bächtold-Stäubli, *s.v.* "Mühle."

[110] *Homilia de sacrilegiis* (late 8th century) 20, Caspari, 11 and 40. Caspari associated this with a practice described by Caesarius of Arles (*S.* 184.4: *CCSL* 104: 750-751), in which the victim of a theft went to a specific but unidentified place where an apparition would be raised to identify the thief. In Caesarius' text, the emphasis was on the expert's skill in raising the spirit. Here, by contrast, the power resided in the building itself, which responded to the magical properties of the symbols or written words of the document.

[111] *Poen. Albeldense* (c. 850) 100, *CCSL* 156A: 12; *Poen. Silense* (1060-1065) 193, *ibid.*, 36.

[112] For Roman family religion, see David G. Orr, "Roman domestic religion: The evidence of the household shrines," in *Aufstieg und Niedergang* 16.2: 1550-1591; Daniel P. Harmon, "Family festivals of Rome" *ibid.*, 1593-1603. See also Fowler, *The Religious Experience of the Roman People*, 68-91.

devised to protect it from sorcery. Leaving and entering it was portentous. Within it was safety from the malign spirits that roamed the world, but it had its own friendly spirits as well. Finally, even its appurtenances possessed the power to heal the members (particularly the children) of the household.[113]

On occasion, the home substituted for the parish church. The Synod of Auxerre (561-605) protested against the custom of making an offering (and, perhaps, keeping the vigil of saints' feasts) in one's own home.[114] This might have been meant to make up for absence from church[115] but, given the other actions described in the same clause (fulfilling vows in forbidden places, making models of the body), the offering in question might have been motivated by the desire to honour household numina rather than negligence and sloth.

Beginning in the second half of the 8th century, the Frankish hierarchy ruled repeatedly against the custom of celebrating mass unnecessarily in inappropriate places: private houses most of all, or little huts or shrines, and even gardens (which must have seemed particularly distasteful to clerics who remembered the old pagan preference for worshipping in groves or in the open). Even bishops were guilty of this practice, sometimes in response to pressure from lay magnates determined to maintain proprietary rights over the church built on their land.[116]

The Council of Laodicea had forbidden this practice centuries earlier but the first precisely datable record available from Western Europe is Chrodegang's *Rule for the Canons of Metz* (762).[117] Charlemagne's legislation of 789 required everyone to attend church on feast days and Sundays, and to refrain from "inviting" priests to their own houses to say mass.[118] A flurry of rulings followed in the 9th century, beginning with Ghärbald of Lüttich's ordinance in 801 and Charlemagne's in 802, warning priests against celebrating mass in houses or anywhere except in consecrated churches.[119] Shortly afterwards, Haito of Basel threatened priests who said mass in huts, unconsecrated churches or houses, unless they did so while visiting

[113] For a description of medieval houses, see Simone Roux, *La maison dans l'histoire* (Paris, 1976), 101-166, especially 120-125 for excavations of different kinds of peasant habitations. See also Jean Chapelot and Robert Fossier. *Le village et la maison au moyen âge* (Paris, 1980), 79-135.

[114] 3, *CCSL* 148A: 265. See also *Epítome hispánico* (c.598-610) 43.4: 190.

[115] This is Blaise's opinion, see *s. v.* "Compensus." *Cf.* Du Cange, *s.v.*

[116] For the growth of proprietary churches, see Friedrich Kempf, "The Church and the Western kingdoms from 900 to 1046," in *History of the Church* 3: 258-264; :Ulrich Stutz, "The proprietary church as an element of mediaeval Germanic ecclesiastical law," in *Mediaeval Germany 911-1250*, trans. and ed., Geoffrey Barraclough, (Oxford, 1967), 2: 35-70; K. Schäferdieck, "Eigenkirchen," *Reallexikon der germanischen Altertumskunde* 6: 558-561.

[117] 72, *Concilia Germaniae* 1: 118. See also *Benedicti capitularium collectio* (mid 9th century) Add. 3.13 and 3.70, *PL* 97: 880. See *Council of Laodicea* (380?) 58, Hefele-Leclercq, *Histoire des conciles* 1.2: 1025. This canon is to be found also in the *Collectio Hispana* 4.9 *PL* 84: 67, and in the *Hadriana* (57, Mansi 12: 868).

[118] *Duplex legationis edictum* (789) 25, *MGH CapRegFr* 1: 64.

[119] *Ghärbald of Lüttich, Capitularia* 1, 8, *MGH CapEp* 1: 18; *Capitula a sacerdotibus proposita* 9, *MGH CapRegFr* 1: 106.

the sick.[120] Theodulph of Orleans banned the saying of mass "in houses of any sort and in vile places."[121] The Council of Paris (829) dealt with the issue in detail, acknowledging that it was a practice difficult to correct. Priests were being forced to celebrate mass in gardens, houses or small private sanctuaries built nearby, and ornamented with hangings and fitted with altars. These shrines, it was explained, were, by a new custom, commonly called chapels (*capellae*).[122] The little chapels were popular: the Council urged "a good many laymen" to keep in mind the punishment that befell Uzzah when he trespassed on the domain of the priests.[123] At about the same time, Halitgar of Cambrai held a synod which also banned the celebration of mass in gardens and houses.[124] Two contemporary capitularies warned priests against offering the sacred mysteries "through contempt of the Church" anywhere except in consecrated places,[125] while a century later Ruotger of Trier allowed mass to be said elsewhere only during a journey or military campaign (*iter*), specifying that it had to be under a tent and on a consecrated altar.[126]

Interest declined after midcentury, but the Council of Mainz (852) forbade bishops or priests "to sing mass" in houses.[127] Pope Leo IV also preached against singing mass in houses or non-consecrated places, or doing so when alone.[128] Charles II's *Capitulary of Pavia* (876) insisted that both townsfolk and the inhabitants of hamlets and estates attend liturgical celebrations in the properly appointed places, and that no one should dare to celebrate "the secret offices of the mass" at home without due permission from the bishop.[129]

Popular sermons and penitentials ignored the abuse; only the ninth-century *Quadripartitus* mentioned it, reiterating the ruling of the Council of Mainz.[130] Evidently the hierarchy hoped to abolish the practice by appeals and threats to the clergy rather than exhortations to the faithful. A revealing canon in Benedict

[120] *Capitula ecclesiastica* (807-823) 14: *MGH CapRegFr* 1: 364.

[121] *Capitularia 1* (813) 11, *MGH EpCap* 1: 110. See also Radulph of Bourges, *Capitula* (between 853 and 866) 3, *PL* 119: 705.

[122] *Concilium Parisiense* (829) 47, *MGH Concilia* 2.2, 641; *Concilium Parisiense*, *epistola episcoporum* (829) 73, *ibid.*, 672. Benedict Levita, *Capitularium collectio* (mid 9th century) 1.334: *PL* 97: 746, and Hincmar of Reims, *Capitularia* 5 (c. 852) 7, *MGH CapEp* 2, 88, include a ban against the construction of *capellae* without the bishop's permission.

[123] *Concilium Parisiense* (829) 47, *MGH Concilia* 2.2, 641.

[124] *Diocesan Synod I* (829/831) 10, Hartmann, 384=*Capitula Neustrica prima* (829) 10, *MGH CapEp* 3: 55.

[125] *Statuta Bonifacii* (1st half, 9th century 2, *MGH CapEp* 3: 360). See also *Capitulare Vesulensia* (c. 800) 24: *ibid.*, 352.

[126] *Capitulary of Ruotger of Trier* (first half of the 10th century) 3, *MGH EpCap* 1: 63. For portable altars, see Victor Elbern, "Tragaltar," *Lexikon des Mittelalters* 8: 931-932.

[127] 24: *MGH CapRegFr* 2, 191.

[128] Leo IV (d. 855) *Homilia*, 12, *PL* 115: 677.

[129] 7, *MGH CapRegFr* 2, 102.

[130] 293, Richter, 33.

Levita's *Collection of Capitularies* put the issue into the context of a struggle between the episcopacy and the inferior clergy

> who are known to have secular protectors against their bishops. Those priests who, against their own bishop's will, have living lay protectors against the bishops and assemble the people separately, who perform their wayward ministry, not in churches but in wild places and little houses, and who disturb churches, are to be shunned, deprived of their own honours and sent under penance to prison or to a monastery for the rest of their lives to expiate the evil that they have done.[131]

The usual prohibitions were enshrined in the *Vetus Gallica,* Atto of Vercelli's *Capitulare*, Burchard of Worms' *Decretum*[132] and, most strikingly, in Benedict Levita's *Collectio* which displayed the importance of this problem to Carolingian authorities by its many repetitions of related ordinances. It contained sanctions against the unauthorised celebration of mass in homes or non-consecrated buildings;[133] it affirmed that *capellae* were not to be set up even in the palace without the local bishop's consent or command; it forbade the celebration of mass in oratories in houses without the bishop's permission;[134] again and again it insisted that mass could be said in unconsecrated places only on voyages or military campaigns, within tents and on properly consecrated stone tables.[135]

Some reasons for the disapproval of the Church are obvious: neglect of the parish church, disregard of the authority of the hierarchy, evasion of the sermons which were supposed to be preached every Sunday and feast day to edify the great as much as the humble. We may well believe that having Mass said on their private property expressed the arrogance of powerful men who used this method to proclaim their superiority over the local ecclesiastical establishment. Other grounds for opposition were spelled out by Regino of Prüm and Burchard of Worms: the use of unconsecrated tables and the presence of scrambling dogs and bevies of harlots, by which "the sacred mysteries are polluted rather than consecrated."[136] Finally, it is possible that proprietors wished to exploit sacramental magic, to use what they saw as the most potent of all rituals, to protect the house, its belongings and environs. But it is clear that these masses were also an expres-

[131] 3. 144: *PL* 97: 81.

[132] *Collectio Vetus Gallica,* 27.8 and 40.2k, Mordek, 453 and 496; Atto of Vercelli (d. 961) *Capitularia*, 8, *PL* 134: 30; Burchard of Worms, *Decretum* (1008-1012) 3.58 and 3.61, *PL* 140: 685 and 686.

[133] 1.55, *PL* 97: 710; 1. 178, *ibid.,* 723. See also Add. 3, 12, *ibid.,* 874.

[134] 1.334: *PL* 97: 746; 1.383 and 2.102, *ibid.,* 752 and 760.

[135] 3. 136, *PL* 97: 812. See also 3.396, *ibid.,* 848. The same text is repeated almost in its entirety in 3.431, *ibid.,* 852-854; also in Add. 1, 6, *ibid.,* 863-864.

[136] Regino of Prüm, *De synodalibus causis* I, 134: 83. See also Burchard of Worms, *Decretum* (1008-1012) 3.61, *PL* 140: 686. Regino attributed this to the Council of Laodicea, Burchard to a council of Orleans; its actual source has not been identified.

sion of a sincere piety, demonstrated by the care taken to provide chapels and decorate them appropriately.

Making vows and offering mass were wrong only because of the unsuitableness of the place, but other rituals performed in the house were considered bad in themselves. Pirmin of Reichenau's *Dicta*, Herard of Tours' *Capitula* and the *Ps.-Theodorian Penitential* condemned a variety of celebratory rites in and around the house, such as dancing and vaulting, lewd songs, jesting or pranks, races or games, make-believe and, in the case of the penitential, a rite which may have been circumambulation.[137] A single penitential, the ninth-century *Quadripartitus,* forbade *agapes* whether held in dwellings or in churches (*in domiciliis sive ecclesiis*).[138] These *agapes* may have been convivial gatherings including some form or imitation of Eucharistic celebrations, but it is more likely that a scribe made a mistake in copying the Council of Laodicea's *in domiciliis divinis, id est, in Ecclesiis* ("in the houses of God, that is, in churches"), the version found in Burchard of Worms' collection.[139]

As the church offered sanctuary, so did the house, if proper precautions had been taken. Choosing the right moment in putting it up was a matter of anxious deliberation. Martin of Braga's canon forbidding the pagan custom of observing the heavens to determine the time for building a house was repeated throughout our period in continental capitularies, collections and penitentials (although in neither Iberian nor insular sources).[140] The terms are formulaic, but there is no doubt that at least in tenth-century Piedmont astrologers were able to persuade architects to pay heed to horoscopes – Atto of Vercelli drove home the vanity of such measures with the example of buildings recently destroyed by fire in Pavia.[141]

Choosing the propitious time for building was not enough to guarantee safety. Evil insinuated into the house was particularly dreaded. Bede reported that King Ethelbert insisted on receiving St. Augustine and his fellows out of doors because of superstitious fear that indoors they would be able to overcome him by their black arts.[142] The same idea is suggested in Frankish texts by passages dealing implicitly with the exclusion of unfriendly forces, whether physical or ghostly, from the precincts of the house or the village. The *Homilia de sacrilegiis* altered the usual text concerning astrology and construction by connecting the observation of the lunar cycle (specifically of the "countermoon") with the "enclosure" of the house. This seems to have been the hedging in of the house with a purely physical barrier, since only the timing of the work is condemned, not the work itself. By

[137] *Dicta Pirmini* (724-753) 28: 188); Herard of Tours, *Capitula* (858) 114: *PL* 121: 772); *Poen. Ps.-Theodori* (mid 9th century) 38.9, Wasserschleben, 607.

[138] 312, Richter, 36.

[139] *Decretum* 3.82, *PL* 140: 690.

[140] *Canones ex orientalium patrum synodis* (572) 72, Barlow, 141.

[141] *S.* 3, *PL* 134: 837.

[142] *HE* 1.25: 74

contrast, the furrows which, according to the *Indiculus superstitionum,* were plowed around habitations (*villae*) were meant keep out more nebulous forms of evil.[143] The same is probably true of the *circus* which the *Ps.-Theodorian Penitential* described as being performed in houses, roads and other places.[144] Danger prowled beyond the furrows, beyond the magic circle, especially in the dark, so that Burchard of Worms' Hessians did not dare go out until the cock's crow drove away and laid the unclean spirits of the night.[145]

If, thanks to the malignity of enemies or sorcerers, evil made its way into the house, people turned to cunning men to track it down and get rid of it. Martin of Braga borrowed a canon from the Council of Ancyra (314) to impose five years of penance on anyone "who, following pagan custom, brought soothsayers and fortune-tellers into his house, as if to expel evil or uncover hexes (*maleficia*) or perform pagan lustrations."[146] This text is found in some form in about a score of documents from every part of the continent and England, in capitularies, synodal decisions, canonic collections and penitentials. Others limited the magicians' activities to finding the hexes and getting rid of the evil.[147] Some omitted mention of lustrations; the specialists were there to perform an act of atonement.[148] The same ruling seems to be the basis of a penitential canon concerning the introduction of soothsayers into the house.[149] Penitentials in the Theodorian tradition recognised that the clergy as well as laymen resorted to these means.[150] This practice is not

[143] 23, *MGH CapRegFr* 1: 223. *Villa* is an ambiguous term, meaning country house, rural grouping of houses or even village–the furrows probably ringed a number of houses rather than individual dwellings. The Romans and Etruscans had been accustomed to plow a furrow around a town at its foundation, but this is the only hint in our sources of a similar custom among the Germans or Celts. See Homann, *Der Indiculus,* 122-123. On the importance of boundaries around property, see Alfons Kirchgässner, *La puissance des signes. Origines, formes et lois de culte* (trad. Sr. Pierre-Marie, rev. M. A. Barth; Paris, 1962) 446-464 and Czarnowski, "Le morcellement de l'étendue."

[144] 38.9, Wasserschleben, 607. For the magically protective function of circumambulation, see Dieter Werkmüller, "Recinzioni, confini e segni terminali," in *Simboli e simbologia nell'alto medioevo,* SSAM 23 (Spoleto, 1976) 640-659.

[145] *Decretum* (1008-1012) 19, 5.150, Schmitz 2: 442.

[146] *Canones ex orientalium patrum synodis* (572) 71, Barlow, 140. See Mansi 2: 522. See also the simplified form in the late 6th/early 7th century *Epítome hispánico,* 1.71: 103.

[147] *Poen. Casinense* (9th/10th century) unnumbered, Schmitz 1: 431. See also Regino of Prüm, *De synodalibus causis* (c. 906) II, 355: 349.

[148] *Collectio Vetus Gallica* (7th century) 44.4a, Mordek, 523. See also Halitgar of Cambrai, *De Poenitentia* (9th century, 1st half) 4.25, *PL* 105: 685; *Poen. Quadripartitus* (9th century, 1st half) 139, Richter, 18); *Synod of Coetleu* (848-849) 3, *MGH Concilia* 3: 188; *Poen. Ps.-Theodori* (mid 9th century) 27.17, Wassershleben, 597; Burchard of Worms, *Decretum* 19, 5.60, Schmitz 2: 422.

[149] *Poen. XXXV Capitolorum* (late 8th century) 16.1: Wasserschleben, 516.

[150] *Poen. Theodori* (668-756) I, 15.4: Schmitz 1: 537-538. See also *Excarpsus Cummeani* (early 8th century 7.16, Schmitz 1, 633-634; *Poen Remense* (early 8th century), 9.17, Asbach, 58; *Poen. Vindobonense b* (late 8th century) 17, Wasserschleben 497, 482; *Poen. XXXV Capitolorum* 16.2, Wasserschleben, 516; *Poen. Martenianum* (9th century) 48, Wasserschleben, 523; Rabanus Maurus,

found in Irish documents; the *Collectio Hibernensis*, the only one to invoke this clause, leaves out any mention of the house.[151]

Some seasonal practices were also meant to protect the house, or to appease the powers dwelling within. The well-documented domestic rites of the Calends of January, such as ornamenting or walking around the house with boughs of laurel, pouring offerings on the hearth, preparing a table to ensure good fortune, the refusal to share the fire of the hearth with outsiders, were all meant to ensure the prosperity of the household. Similarly, the tenth-century Piedmontese practice of hanging baptised turves and boughs in the house surely had an aspect of protective magic.

The house was thought particularly vulnerable while a corpse lay within. Burning grain for the "health of the living and of the house" was first described in the *Penitential of Theodore* (668-756) and then in numerous other penitentials down to Burchard of Worms' time.[152] All but three specify protection of the house as the reason.[153] The *Penitential of Silo* (1060-1065) blamed this practice specifically on women, but it is the only text to do so.[154] In its origins, the practice was probably meant to be an offering to the dead, but here its significance seems purely magical, an attempt to protect the household from the angry ghost or the evil that lingered around the corpse.

To leave the shelter of the house for a prolonged period entailed risk, to be obviated in sixth-century Arles by observing propitious days for doing so. Caesarius of Arles' injunction against this practice was copied into the *Hibernensis* and into sermons by St. Eligius and Rabanus Maurus.[155] Some thought that the moment itself when one left or returned to the house was portentous, so that every accidental occurrence, an overheard word or a bird's song, was potentially an omen.[156]

Poenitentiale ad Heribaldum (c. 843) 20, *PL* 110: 491; *Capitulary of Radulph of Bourges* (853-866) 38, *MGH EpCap* 1: 262-263.

[151] 64.5, H. Wasserschleben, 231.

[152] *Poen. Theodori* I, 15.3, Schmitz 1: 537. See also *Canones Gregorii* (late 7th/8th century) 117, Schmitz 2: 535; *Canones Cottoniani* (late 7th/8th century) 149, Finsterwalder, 281; *Poen. Remense* (early 8th century) 9.15, Asbach, 57; *Excarpsus Cummeani* (early 8th century) 7.15, Schmitz 1: 633; Poen. *Vindobonense b* (late 8th century) 17, Wasserschleben, 497, 482; *Poen. XXXV Capitolorum* (late 8th century) 16.4: *ibid.,*, 517; Rabanus Maurus, *Poenitentiale ad Heribaldum* (c. 843) 20, *PL* 110: 491; *Poen. Ps.-Theodori* (mid 9th century) 27.15, Wasserschleben, 597; *Poen. Parisiense I* (late 9th century or later) 22, Schmitz 1: 684; Regino of *Prüm De synodalibus causis* (c. 906) 2.368: 353.

[153] The exceptions are: *Double Penitential of Bede-Egbert* (9th century?) 35, Schmitz 2: 683; *Confessionale Ps.-Egberti* (c.950-c.1000) 1.32, *PL* 89: 409; Burchard of Worms, *Decretum* (1008-1012) 19, 5.95, Schmitz 2: 430.

[154] 198, *CCSL* 156A: 36.

[155] Caesarius of Arles [502-542] *S.* 54.1, *CCSL* 103: 236. See also *idem*, *S.* 193.4: *CCSL* 104: 785); *Collectio Hibernensis* (late 7th / early 8th century) 64.8, H. Wasserschleben, 232; *Vita Eligii* (c.700-725) *MGH SRM* 4: 705; Rabanus Maurus (d. 856), *Hom.* 43, *PL* 110: 81.

[156] Ps.-Eligius, *De rectitudine catholicae conversationis* 7, *PL* 40: 1174.

Fortified by magical rituals, the house provided a safe haven from the malice of unfriendly spirits and sheltered good-natured ones. Although the hosts of the dead roamed the night sky on New Year's Eve, one could peer into the future in relative safety while perched on the rooftop. This is suggested in three eighth-century sermons and is made explicit in Burchard of Worms' question: "Did you sit on the roof of your house ... so that you could see and find out what would happen to you in the coming year?"[157] Potentially friendly sprites (*satyri, pilosi)* who managed to breach the ring of safety to get into the storehouses and granaries were given a little *douceur* in the form of shoes and toys, to bribe them into stealing the goods of others to enrich the master of the house. The mistress of the house tried to induce more ominous spirit-visitors, the *Parcae*, to show her favour by spreading a table for them with food and drink and three little knives.[158]

Finally, various healing techniques seem to be based in part on the belief that the house sheltered presences, ancestral spirits, willing and able to protect their descendants–the house stood for the kin, dead and alive. We have already seen that sick children were placed on the roof and in the oven, more rarely over the well and, in one case, by the wall.[159] If the oven (*fornax*) is in fact the hearth, the appeal is clearly made to the ancestors. By contrast, the roof, wall and door were transitional points between the domain of the family and the outside world to which the disease could be transferred.

[157] *Vita Eligii* (c.700-725), *MGH SRM* 4: 705; *Dicta Pirmini* (724-753) 22: 173-174; *Anonymous sermon* (late 8th/9th century), Levinson, 310-311); Burchard of Worms, *Decretum* 19, 5.62, Schmitz 2: 423.

[158] *Decretum* 19, 5.103: Schmitz 2: 432; 19, 5.153, *ibid.*, 443.

[159] *Poen. Casinense* (9th/10th century) 57, Schmitz 1: 412. See the discussion in the next chapter, 6.3.2.

6
Magic, Magicians and Beneficent Magic

6.1 CUNNING MEN, CUNNING WOMEN AND MAGICIANS

Almost fifty different technical terms to identify practitioners of magic or possessors of skills unauthorised by the Church are found in our texts: *ariolus, aruspex, augur, caragius, cauclearius, cocriocus, divinus, herbarius, incantator, magus, maleficus, necromanta, obligator, praecantator, praedicator, pithon, somnarius, sortilegus, suffitor, tempestarius, vaticinator* and *veneficus*, their potentially significant variants, and their feminine equivalents: *auguriatrix, divina, herbaria, incantatrix, malefica, pithonissa, sortiaria, tempistaria* and *venefica.* The *striga* (witch) and werewolf, together with those who engage in dream flights and dream battles, stand apart from these. Passages using such terms are found in documents produced in all Christianised parts of Western Europe but are very rare in Latin-language texts of Irish origin.[1]

The interpretation of these terms is difficult. Those used most often come from pre-Christian Latin, where they had a precise meaning. Isidore of Seville defined some of them for the Middle Ages[2] It would be a mistake, however, to assume that they bore the same meaning throughout the entire period. The multiplicity of terms, one piled on another, characteristic of pastoral literature from the time of Caesarius of Arles on, points to an uncertainty as to meaning and a general tendency to use them indiscriminately, which may disguise an even more profound uncertainty as to the nature of their practices.[3] Diviners, astrologers, interpreters of omens, healers, poisoners, herbalists, enchanters are all lumped together as practitioners of forbidden arts, or the carriers of the remnants of the pagan past.[4] In the case of Caesarius of Arles, this may well have been a deliberate, contemptuous jumbling. Writers following him copied his style but many must have been ignorant of the original meaning of the words.

To the extent that they described real magicians and real magical practices, the fine Latin words often disguised a quite different vernacular reality, specific to each ethnic group, for which a technical vocabulary was lacking. The case of the *mathematicus* has already been cited. To Caesarius of Arles in the 6th century, Isi-

[1] The principal exception is the *Collectio Canonum Hibernensis*, which reflects continental influences, and which gives six different terms (*aruspex, caragius, divinus, magus, maleficus, pithonicus* and *sortilegus*).

[2] *De magis*, *Etymologiae* VIII, 9: 712-719.

[3] Churchmen also found a precedent for such lists in *Jer* 27 9.

[4] *E.g.*, in Benedict Levita's *Capitularium collectio* 2.215, *PL* 97: 774.

dore of Seville in the 7th and Atto of Vercelli in the 10th, he was merely an astrologer but to the authors of penitentials in the 8th and 9th, he was a magician with malign powers over the human mind. Similarly, the *aruspex*, who in Etruscan and Roman paganism had performed divinations from the examination of entrails, became an augur to many medieval authors.

The words themselves pose other problems. There are some seven different variations of spelling for the term that ostensibly refers to a person who casts or interprets lots. The preferred form is *sortilegus*, and perhaps all the other forms mean the same thing. But it is possible that some spellings carry a nuance that escapes us, particularly since the practitioner in question occasionally seems to be less a fortune-teller than a sorcerer or sorceress, as is the *sortiaria*. A word may be new, with no classical background: *caragius* appears in the 6th century and dwindles away in the 8th (unless he reappears as the *cauculator* of Carolingian capitularies). From the context we know that the *caragius* was a cunning man of some sort since he is defined in one passage as a *divinator,* but it is only a guess that he was invariably a soothsayer. In still another case, a neutral word may have taken on another, more sinister meaning in the popular mind–the *mangones* of Carolingian legal texts stands for ordinary, secular rascals; but the word may have suggested the name "Magonia," the mysterious land of airborne harvest-thieves.

To a lesser degree, the same confusion exists as to the exact meaning of the terms used to identify magical acts. It is often hard to know whether words like *incantatio, maleficium, fascinatio, divinatio* and *ariolare* are used in a precise technical sense, to mean incantation, destructive magic, the evil eye and soothsaying, or in a more general sense, equivalent to *ars magica.*

Women's participation in magic is difficult to gauge.[5] Its importance in Germanic culture is evident in Tacitus and Burchard of Worms, but only a handful of terms such as *auguriatrix* or *malefica* testify to it. Less than fifteen per cent of the technical terms for cunning folk are feminine. When it comes to the participation of ordinary persons in magic, women tend to vanish behind the inclusive masculine words in all but approximately six per cent of the texts studied. The Councils of Narbonne (589) and Paris (829) took the trouble to explain that cunning folk may be women as well as men, and some two dozen texts raised the question of women who practiced divination or incantation,[6] but most authors evidently did not intend to single them out as the practitioners *par excellence* of magic. Burchard

[5] Heide Dienst correctly emphasises women's dominant role in magic concerning sexuality and human fertility, but underrates men's involvement in love magic; see "Zur Rolle von Frauen in magischen Vorstellungen und Pratiken nach ausgewählten mittelalterlichen Quellen," in *Frauen in Spätantike und Frühmittelalter*, ed. Werner Affeldt (Sigmaringen, 1990), 173-194. See also Daniela Gatti, "Curatrici e streghe nell'Europa dell'alto medioevo," in *Donne e lavoro nell'Italia medievale*, Maria Giuseppina Muzzarelli *et al.* (Turin, 1991), 127-140.

[6] *E.g.*, *Poen. Theodori* (668-756) I, 15.4, Schmitz 1: 537-538. These passages are usually to be found in penitentials and, more rarely, in conciliar legislation or capitularies; the only sermon to include them an the *Anonymous sermon* from Trier (c. 850-882) 5, Kyll, 10.

of Worms himself, who time and again identified women as the principal practitioners of magic, never gave the feminine form of any word for magician. With some authors this may represent a cultural bias: women's magic might have been perceived as less important in the Mediterranean world than in northern Europe; the authors, to a certain extent the products of an urban milieu and fairly segregated from feminine society, might have been more concerned about peasant rituals and techniques.[7] On the other hand, this reflects a certain detachment with respect to women: in general their misdeeds receive relatively little attention in pastoral texts. In most penitentials, for example, except for contraception and abortion, women's sexual sins are given small play in comparison to men's.

Identifying the magicians and their specialty is often difficult. Many magicians are named in the texts without any particular act of magic being ascribed to them, while such acts are often mentioned without being attributed to what may be called a professional magician with a title such as *sortilegus, ariolus, incantator*. It is significant that among the thirty titles of the *Indiculus superstitionum*, filled as they are with "superstitious," "pagan" and magical practices, only one mentions magicians.[8] In most cases, therefore, we cannot be sure whether the texts target individuals with specific training or all those who occasionally practiced magic. While our documents give a distinct impression that certain people were recognised by the community as having special skills and were regularly consulted in times of crisis, there undoubtedly also existed a fund of traditional techniques at hand for daily use by ordinary men and women who were neither blackened nor honoured with a professional title. In many cases, the terms "cunning folk," "cunning man" or "cunning woman" convey the ambiguous status of the practitioners of magic better than the more technical terms used by the authors.

Before turning to the individual types of cunning folk and their magic, it is worth considering a text that illustrates the difficulties presented by the documents. It is taken from the pastoral letter of the bishops at the Council of Paris (829) included in their report to Louis the Pious. Introduced by the words "concerning the perpetrators of various evils," it goes on to list magicians, soothsayers, casters of lots, poisoners, diviners, enchanters and interpreters of dreams as being doomed to punishment by divine law. It then launches into a catalogue of their crimes:

> There is no doubt that, as has been noted by many people, minds are being infected with certain kinds of trickery and diabolical illusions by means of love potions, foods and amulets, so that, unaware of their own shame, they are considered by many to have succumbed to madness. There are those who claim that they can disturb the air with their spells, bring on hail, foretell the future, take away produce and milk and give it to other people. They are reputed to do countless such things. Whoever, man or woman, who is known

[7] Boglioni, "La religion populaire dans les collections canoniques occidentales."

[8] 14, *MGH CapRegFr* 1: 223.

> to be of this sort is to be punished particularly severely ... It has been written of such people under title 23 of the Council of Ancyra: Whoever seeks divinations and follows the customs of pagans or introduces such men into his house to find something by witchcraft or to carry out a purification [or avert some omen?] shall fall under the rule of five years [of penance] ....[9]

Here only the masculine form is used in the terms for magicians, but we are told that they are both men and women. All categories of cunning folk are suspected of being capable of an undifferentiated mass of practices, illusions, weather magic, love magic, fortune-telling, black magic, and countermagic; all are to be punished harshly without regard to the type involved. Distinguishing between the rumor-mongers, the victims of the magic and the magicians themselves is difficult. What is clear, however, is that there existed an atmosphere haunted of formless dread of malign magic, a dread which haunted laity and clerics alike.

The different kinds of magicians/cunning folk and their magic are discussed in this and the following chapter. The nature of the texts makes it difficult to develop a logically coherent system of classification. A simple division into white magician/magic and black will not quite do, since the authors considered all these practitioners as evil-doers more or less knowingly in cahoots with the devil. At the same time, they drew a much more complex picture than they could have intended. In the eyes of ordinary people, some of these experts were beyond question beneficent; others, more uncanny, could use their powers for good or for ill; a few were, apparently, entirely malign. The dividing lines are vague: any practitioner could move easily from one to another and even the most baneful sorcerer had clients whom he benefited.

This nebulousness favours the more promising method of classification proposed by Raymond Firth, in which magic is divided into "productive" (*e.g.*, for fertility, love, rain-making), "protective" (*e.g.*, for curing and protection and as counterspells) and "destructive" (*e.g.*, storm-making, death-dealing) types. These are defined in terms of "practical ends, whether the promotion of human welfare, the protection of existing interests, or the destruction of individual well-being through malice or the desire for vengeance." The first two are "socially approved" and performed either by and for private individuals or by experts for the benefit of others or for the group. Firth doubted that the third, which he called witchcraft, is ever actually performed, although it may be attempted and it is thought to have occurred. Sorcery he places between the first two, since it may be either approved or disapproved by society.[10]

I have been obliged to combine these two methods of classification, with some fairly large-scale adjustments to allow for the difficulties presented by our

[9] *MGH Concilia* 2.2: 669-670. Also as addition to the capitularies of Louis the Pious in *Episcoporum ad Hludowicum imperatorem relatio,* 54 (XX), *MGH CapRegFr* 2: 44-45. For the circumstances surrounding this Council of Paris, see Hefele-Leclercq, *Histoire des conciles* 4.1: 54-60.

[10] "Reason and unreason in human beliefs," in *Witchcraft and Sorcery,* 38-40.

documents. Since one can seldom distinguish clearly between experts in productive and in protective magic, or even sometimes between productive and protective techniques, they will be examined together under the heading of "beneficent magic." This category includes soothsayers, healers and the associated techniques.

Even productive and protective magic has dangerous, deeply hostile aspects in societies such as the one described by Burchard of Worms, where the success of one person was seen to be obtained at the expense of another.[11] A large number of cunning folk employed their skills in ways that must have sometimes been seen by ordinary people as beneficial, at other times as destructive. These belong to a group which I term practitioners of "ambivalent magic." The group takes in different kinds of enchanters, sorcerers, experts in herbs and weather magicians, together with their arts. Love magic, indubitably productive from many points of view, is put into this category since a significant body of the documents associate and almost identify it with murder and abortion. Also included here are some individuals about whose activities no information is given.

Last to be considered are *strigae*, who are presented in our texts as being unmitigatedly evil and who function mainly on the level of dream or hallucination, and werewolves. *Strigae* and werewolves have some innate quality differentiating them from other people. They come closest to the anthropological definition of witch.

6.2 SOOTHSAYERS AND DIVINATION

Soothsayers form the largest category among cunning folk. They include *sortilegi, divini, arioli, haruspices, auguri, caragii, etc.*, whom we find crowded together in various combinations throughout the early Middle Ages, first in Caesarius of Arles' sermons,[12] then in penitentials[13] and other sermons,[14] more rarely in canon law[15] and pastoral letters.[16] Although in classical Latin some of these were engaged in specific forms of divination, in our texts these terms seem to be interchangeable, and any generalisation about one can be considered applicable to all at any given period.

It must be emphasised again that not everyone labelled *sortilegus, divinus, etc.*, was in fact a soothsayer. When Caesarius of Arles and others wrote that clients

[11] The experience of some African societies under the cultural and economic pressure of Europeans seems particularly relevant; see Audrey Richards, "A modern movement of witch-finders," in *Witchcraft and Sorcery*, 164-177.

[12] *E.g., S.* 12.4, *CCSL* 103: 61; *S.* 19.5, *ibid.*, 90.

[13] *E.g., Poen. Vindobonense b* (late 8th century) 7, Wasserschleben, 496-497; Halitgar of Cambrai, *De Poenitentia* (9th century, 1st half) 4.21, *PL* 105: 685.

[14] *E.g., Anonymous sermon* (late 8th/9th century), Levison, 308; Rabanus Maurus (d. 856), *Hom. 45: De fide, spe et charitate, PL* 110: 83.

[15] *E.g., Conc. Toletanum IV* (633) 29, Vives, 203.

[16] *E.g.*, Gregory I to the notary Adrian, *Ep.* 11.33, *MGH Ep* 2, 302.

sought out certain cunning folk in order to question, consult and interrogate them, it is obvious that they were dealing with specialists in information. Authors of the 6th and 7th century used the words correctly, as did some later writers, such as Rabanus Maurus, Atto of Vercelli and occasionally Burchard of Worms. But by the 8th century, these precise verbs often disappear from penitentials and legal texts, to be replaced by vaguer ones, such as to observe, honour and follow. Significantly, the canon of the Council of Paris (829) quoted above lists four soothsayers together with three other types of magicians without attributing a single act of divination to them; only at the end of the paragraph does it cite an almost five hundred year old ruling against divination. The technical meanings of the words, therefore, cannot be assumed to be still valid.[17]

Most soothsayers were probably not educated men and women; not one text found accuses members of the clergy of being diviners as such, even though in practice they were necessarily the ones to interpret the *sortes biblicae*, lots drawn from the Bible. The Council of Narbonne testified that any man or woman, slave or free, could possess the skills required[18]–"some art" according to the fourth Council of Toledo (633), "meddlesome [or inquisitive] art" according to Rabanus Maurus. It is known from Caesarius of Arles' sermons that they could be summoned or sent for, which might contain a hint of low social status. On the other hand, in the 8th century, the faithful were warned against going to cunning men –some may have been too exalted to summon, others too disreputable to admit into one's house.[19]

Whatever their social status, they seem to have enjoyed at times the trust of every segment of the population. The bishops of Narbonne had to take measures against men and women of every ethnic group, Goths, Romans, Syrians, Greeks and Jews, who invited such people into their houses to question them.[20] No other text matches this in precision, but repeated injunctions make it certain that the appeal of soothsayers continued unabated throughout the centuries. Worse, it was not merely the ignorant laity who consulted them but the clergy as well, including the highest ranks. If the words of the Spanish bishops at Toledo IV, ordering the removal of bishops, priests, deacon or other clerics who had recourse to soothsayers and condemning them to perpetual penitence, was repeated in the 9th, 10th and 11th centuries, it was probably not from a spirit of antiquarianism, but be-

[17] This may reflect ambivalence on the part of the Church toward divination; while anything seen to involve consultation of demons was wholly rejected, other forms were tolerated. Islam, by contrast, had no such problem since divination was considered to be a strictly secular science. See Pierre Boglioni, "L'Église et la divination au moyen âge, ou les avatars d'une pastorale ambiguë," *Théologiques* 8 (2000): 37-66, and Toufic Fahd, *La divination arabe. Études religieuses, sociologiques et folkloriques sur le milieu natif de l'Islam* (Leiden, 1966).

[18] See n. 20 below.

[19] *E.g., Anonymous sermon* (8th century), Morin, 518; *Homilia de sacrilegiis* 5, Caspari, 6-7.

[20] *Conc. Narbonense* (589) 14, *CCSL* 148A: 257. The *Epítome hispánico* (c. 598-610) mentions only the soothsayers' Catholic and Jewish clients in its version of this clause (39.14: 185).

cause they continued to be relevant.[21] The problem existed in England as well: Archbishop Egbert of York judged that consultation of soothsayers and enchanters was grounds to prevent ordination or to remove a bishop, priest or deacon from his charge.[22]

Despite soothsayers' popularity, their clients seem to have consulted them with some embarrassment. Caesarius of Arles drew a lively picture of his flock as, with mixed feelings, they turned to various cunning folk and their wares. Supposedly sensible Christian women, the mothers of ailing children, pretended to be above such lamentable practices ("I don't meddle in such things," they claimed), but at the same time encouraged their friends and servants to do so, promising to pay the costs–"as though thus indeed they can exonerate themselves of so heinous a crime." Not so, Caesarius said: they were all guilty of the sin of sacrilege.[23] Others he suspected of reveling in the thrill of defying Christian teaching, of "sacrilegious delight," when they dared to summon or to question such men.[24]

Some pastors were aware that, to their flocks' confusion and anxiety, the soothsayers were sometimes right. God permits this, Caesarius explained, to test their Christian faith; those who give way in this then fall more readily prey to the devil.[25] The *Hibernensis* followed Isidore of Seville in acknowledging that the demons were experts in many things and so were able to provide answers; their superior knowledge was owing partly to human weakness, partly to their own more acute senses and longer life, and partly to divinely ordained revelation.[26] Ps.-Eligius' *De rectitudine catholicae conversationis* affirmed the Augustinian concept that spirits flying through the air may easily see future events.[27] Soothsayers were to be avoided, said Rabanus Maurus, because they consorted with demons.[28] On the other hand, an English penitential dismissed divination as being altogether folly.[29] Most authors, however, chose to ignore the question of soothsayers' reliability.

Why did people turn to soothsayers?[30] According to the Carolingian *Ordo de catechizandis rudibus*, disorderly passions like vain curiosity fuelled a desire for

[21] Rabanus Maurus, *Poenitentium Liber ad Otgarium* (842-843) 123, *PL* 112: 1417; Atto of Vercelli (d. 961) *Capitularia*, 48, *PL* 134: 37-38; Burchard of Worms, *Decretum* (1008-12) 10.48, *PL* 140: 851.

[22] *Dialogues of Egbert* (732-766) 15, Haddan and Stubbs 3, 410.

[23] *S*. 52.6, *CCSL* 103: 232.

[24] *S*. 19.4, *CCSL* 103: 90.

[25] *S*. 54.3 *CCSL* 103: 237-238.

[26] *Collectio Hibernensis* (late 7th/early 8th century) 64.7, H. Wasserschleben, 232). *Cf.* Isidore of Seville, *Etymologiae* VIII, 11.15 and 16: 232.

[27] 9, *PL* 40: 1175.

[28] *Hom*. 45, *PL* 110: 83.

[29] *Confessionale Ps.-Egberti* (c. 950-c. 1000) 2.23, *PL* 89: 419.

[30] Carole Myscofski's analysis of magic in early modern colonial Brazil offers useful insights on various aspects of magic, especially on its social functions, which may throw light on early medieval magic as well ("The magic of Brazil. Practice and prohibition in the early colonial period, 1590-1620," *History of Religions* 40 [2000]: 153-176, esp. 167-176).

forbidden knowledge and remedies.[31] But more often it was sheer need that drove people to such experts. In various kinds of difficulties, in case of theft, fear of malign influences, and especially illness (as is natural when illness was supposed to be caused by hostile forces), they turned to soothsayers for advice, reassurance and help. "There are those who seek out lot-casters for whatever kind of illness, they question haruspices and diviners, they summon healers, they hang diabolical amulets and symbols on themselves .... I admonish and urge you again and again not to question *caragi*, diviners and lot-casters nor interrogate them about any matter or illness," wrote St. Caesarius, and his words are echoed in sermons up to the 9th century.[32] But for the actual circumstances, only Caesarius gave details:

> When the children of some women are tormented by various kinds of trials or illnesses, the weeping mothers run about in a frenzy ... They say to themselves: "Let us consult that soothsayer or diviner, that caster of lots, that herbwoman [*erbaria*]; let us sacrifice one of the patient's garments, a belt to be inspected or measured; let us offer some symbols [*caracteres*], let us hang some protective charms on his neck."[33]

The frantic mothers were ready to grasp at any hope, to turn to any quack with a reputation for expertise. Here the soothsayers' input appears to be primarily diagnosis rather than magical healing, but no doubt they also made suggestions for treatment and perhaps supplied amulets and *caracteres*. In another sermon, Caesarius described how the devil, acting through seemingly kind friends, urged a sick man or one in some other difficulty to go to forbidden experts:

> What usually happens, brothers, is that a persecutor sent by the devil comes to some sick man, and says, "Had you summoned that healer you would be better already; had you been willing to apply those symbols you could already have been cured." Perhaps someone comes and says, "Send to that diviner, give him your belt and headband/breastband to be measured, and he will inspect it."... And someone else says: "That fellow knows how to fumigate well, for when he did it for such-and-such, he promptly got better and all trouble vanished from his house." ... And hereabouts the devil is accustomed to deceive careless and lukewarm Christians, so that if a man has suffered a theft, that cruelest persecutor goads him through his friends, saying: "Come secretly to that place and I shall summon forth a person [raise an apparition? –*excitabo personam*] who will tell you who stole your silver or money; but if you want to find this out, do not cross yourself."[34]

Here the diviners not only find the cause of the trouble but also provide the

[31] 63, Bouhot, 223-4.

[32] *S.* 50.1, *CCSL* 103: 225); *ibid.*, *S.* 54.1, 235-236. See also *Vita Eligii* (c.700-725), *MGH SRM* 4: 705-706; *ibid.*, 707; *Homilia de sacrilegiis* (late 8th century) 27, Caspari, 16; *Anonymous sermon* (8th century), Morin, 518; Rabanus Maurus (d.856), *Hom.* 43, *PL* 110: 81.

[33] *S.* 52.5, *CCSL* 103: 232.

[34] *S.* 184.4, *CCSL* 104: 750-751.

cure. Caesarius adds an independent witness to the many repetitions of the ruling of the Council of Ancyra against bringing magicians into the house to purify it and expel evil. The last sentence, one of the few hints of necromancy to be found in our literature, will be considered presently.

The beginning of any enterprise also called for divination. It has been already seen that the single most important occasion was the New Year, and that heavenly signs were considered when building a house or getting married; they were doubly important when starting on a voyage, from the point of view of both the risks of travelling and the eventual success of the purpose of the trip. But they were used at other critical times as well. In Visigothic Spain, men ambitious of backing the likeliest candidate for royal power, or of gaining it for themselves, were suspected of reading the future during the king's lifetime.[35] During the 8th century, lots were sometimes drawn before the duel called the *wehadinc*. The Council of Neuching forbade them, not because they were pagan or magical in themselves, but because they gave an opportunity for the use of some kind of spell, trick or magical art against one or other of the combatants.[36] Finally, two Carolingian texts dealt with sorceresses who performed certain "divinations" to make their husbands more loving. It is likely, however, that divination here simply means enchantment, the usual potions and charms described so frequently in other contemporary documents.[37]

6.2.1 *Specialists*

Sortilegus,[38] *sortilogus, sortilecus, sortilicus, sortilocus, sorticularius, sortiaria*: Ostensibly these words refer to those who cast and interpreted lots.[39] The word *sortilegi* is defined more narrowly by Isidore of Seville to mean those who performed divination by examining writings of some sort, in particular the *sortes sanctorum*, the "lottery of the saints.".[40] Not one of the almost seventy passages in pastoral literature to mention the *sortilegus* follows Isidore in this respect.[41] No type of divination or of magic was assigned to him alone. Indeed an eighth-century penitential seems to use the word as a general term for diviner or magician ("*arioli,* that is, *sortilegi*").[42]

[35] *Conc. Toletanum* V (636) 4, Vives, 228.

[36] *Conc. Neuchingense* (772) 4, *MGH Concilia* 2.1: 100.

[37] *Capitulary 2 of Ghärbald of Lüttich* (between 802 and 809) 10, *MGH CapEp* 1: 29. See also *Capitula Silvanectensia prima* (9th century, 1st half) 11, *MGH CapEp* 3: 83.

[38] It has not always been possible to discriminate between *sortilegus* (magician) and *sortileg[i]um* (magic); when in doubt, I have taken it to refer to the person.

[39] *Dict. étymologique, s.v.* "Sors."

[40] *Etymologiae* VIII, 9.28: 716.

[41] However, a very faint hint of this is found in the *Poen. Silense* (1060-1065) 105 (*CCSL* 156A: 27) which at least includes the *sortilegi* and the examination of writings in the same list of magical practitioners and practices. It is the only text to do so and it may be significant that it is of Spanish origin .

[42] *Poen. Hubertense* (8th century, 1st half) 25, *CCSL* 156: 111.

Almost invariably, the *sortilegus* appears in combination with at least one other practitioner, most often the *divinus*. The Council of Narbonne identifies *sortilegi* and *caragii* as *divinatores*; in the *Epítome hispánico*, the *sortilegus* and the *divinus* were the magicians brought in to purify the house.[43] Gregory the Great in effect treats such persons as ministers of idols when he calls for vigorous measures against Sardinian "followers of idols or, if you prefer [*vel*], of *sortilegi* and *haruspices*."[44]

The only one to mention the *sortilegus* in isolation makes a direct link between him and malevolent magic: "If anyone is a *sortilegus* or at any rate [*aut*] if he disturbs men's minds by means of certain spells, [he is to do penance for] three years."[45] But this is probably a slip of the pen for *mathematicus*, who often appears alone in the penitentials of this period and was well known for his malicious attacks on mental stability. In the same way, the *sortiaria*'s name suggests a female version of the *sortilegus*, but it is evident from the actions ascribed to her (causing illness and death) that she is a sorceress rather than a fortune-teller.[46]

Insular authors paid scant attention to the *sortilegus*. The *Hibernensis* included him in a passage borrowed directly from Caesarius of Arles.[47] The *Confessional* of Ps.-Egbert referred to him indirectly in a clause concerning those who "offered sortilege or divination."[48] The *Double Penitential* of Bede-Egbert also mentioned *sortilegi*, but that is probably a mistake for *sortilegia*, which may have been offerings of some sort rather than divination.[49]

Divinus, vir et mulier divinator, adivinator, divina: *Divinus* is only slightly less common than *sortilegus*, appearing in approximately sixty passages, usually in a list of similar terms for practitioners of forbidden arts.[50] An unimpeachably classical term validated by use in the Vulgate (*e.g., Deut* 18 11), it was explained by Isidore of Seville with more psychological insight than any of the other terms for magicians: the *divini* get their name "because they are as it were full of the god; they make themselves out to be full of divinity and, by a kind of dishonest shrewdness, they guess the future, for there are two kinds of divination, one which comes from art, the other from prophetic frenzy."[51] Most authors implicitly agreed with his anal-

[43] *Epítome hispánico* (c. 598-610) 1.71: 103; *Poen. Casinense* (9th/10th century) unnumbered, Schmitz 1: 431. See also Regino of Prüm, *De synodalibus causis* (c. 906) II, 355: 349; Burchard of Worms, *Decretum* (1008-1012) 10.6, *PL* 140: 834A.

[44] Gregory I, *Ep.* 9. 204 (59)], *MGH Ep* 2:192. Included in abbreviated form in Burchard of Worms' *Decretum*, 10.3, *PL* 140: 833.

[45] *Poen. Sangallense Simplex* (8th century, 1st half) 11, *CCSL* 156: 120.

[46] *Capitulare Carisiacense* (873) 7, *MGH CapRegFr* 2: 345.

[47] *Collectio Hibernensis* (late 7th/early 8th century) 64.1, H. Wasserschleben, 230-231.

[48] *Confessionale Ps.-Egberti* (c.950-c.1000) 4. -, *PL* 89: 426A.

[49] 18, Schmitz 2: 682.

[50] The *Collectio Hibernensis* is the only Irish document to mention them, again in texts based on Caesarius of Arles (64.1 and 64.2, H. Wasserschleben, 230-231).

[51] *Etymologiae* VIII, 9.14: 714.

ysis. Insofar as diviners and their ilk did not work by some kind of cunning,[52] their knowledge derived from demons. Men believe that divination occurs through human skill, observed an eighth-century penitential, but it is a demon at work.[53] The *divini* are the instruments through which the demons give replies; this is why their clients must do penance for five years.[54] The *divini* were not said actually to invoke demons–that was left to the *mathematici*. Rather, the demons came of their own volition when the diviner practiced his art. Diviners were right on occasion, as we have seen, and thus enjoyed a prophet-like prestige.[55]

More frequently than any other cunning man, the *divinus* is defined by means of some other technical term that emphasises his function. A number of eighth- and ninth-century continental penitentials describe him as a kind of *ariolus* who performs divinations.[56] In one text, Burchard of Worms presents *divinus* as the more general term, the *ariolus* as a type.[57] A tenth- or eleventh-century penitential defines him as an *adivinator*.[58] Three penitentials refer to *divini praecantatores*, either *divini* who specialised in healing or two different kinds of cunning men. The term is in the middle of a passage borrowed from Caesarius of Arles ("Why would one slay his soul by means of soothsayers and diviners, healing quacks and diabolical amulets?"),[59] and then presented as a unit in garbled compendia of superstitious practices drawn mostly from his sermons.[60] An *incantator divinus* appear in the midst of a list of other diviners drawn up by Regino of Prüm. This too may refer to two different types of cunning men, or to a magician who combined the arts of an enchanter with those of a soothsayer.[61]

Women's membership in this profession is recognised in four documents. In addition to the male and female diviners of the Council of Narbonne and the general reference to women by the Council of Paris, the *Epítome hispánico* used the

[52] *E.g., Conc. Toletanum IV* (633) 29 (Vives, 203) and numerous other texts up to and including Burchard's *Decretum* (19, 5.60, Schmitz 2: 422).

[53] *Poen. Oxoniense I* (8th century, 1st half) 21, *CCSL* 156: 90.

[54] *Homilia de sacrilegiis* (late 8th century) 5, 6; *Poen. Remense* (early 8th century) 9.5, Asbach, 56; *Poen. Halitgari* (817-830) 35, Schmitz 1: 479.

[55] Burchard of Worms, *Decretum* (1008-1012) 19, 5.60, Schmitz 2: 422.

[56] *Poen. Remense* (early 8th century) 9.5, Asbach, 56; *Excarpsus Cummeani* (early 8th century) 7.5, Schmitz 1: 632; *Poen. Burgundense* (8th century, 1st half, *CCSL* 156: 64; *Poen Oxoniense I* (8th century, 1st half) 21, *ibid.,* 90; *Poen. Floriacense* (late 8th century) 23, *ibid.,* 99; *Poen. Merseburgense* (late 8th century) 23 *ibid.,* 132; *Poen. XXXV Capitolorum* (late 8th century) 16.1, Wasserschleben, 516; *Poen. Halitgari* (817-830) 35, Schmitz 1: 479; *Poen. Ps.-Theodori* (mid 9th century) 27.11, Wasserschleben, 597; *Anonymi liber poenitentialis* (n.d.), *PL* 105: 722.

[57] *Decretum* (1008-1012) 10.30, *PL* 140: 837.

[58] *Poen. Arundel* (10th/11th century) 90, Schmitz 1: 462.

[59] *S.* 19.5, *CCSL* 103: 90.

[60] *Liber de remediis peccatorum* (721-731) 3, Albers, 411. See also *Poen. Egberti* (before 766) 8.4, Schmitz 1: 581; Burchard of Worms, *Decretum* (1008-1012) 10.33, *PL* 140: 837-838.

[61] Regino of Prüm, *De synodalibus causis* (c. 906) II, 5.42: 212.

term *divina* in summarising Narbonne.[62] Finally, the *Homilia de sacrilegiis* defined *divina* as "pythoness."[63]

Ariolus (hariolus): In classical Latin, *(h)ariolus* was a general term for soothsayer. It had fallen out of use in the early Empire,[64] but was given new currency by St. Jerome's translation of the Old Testament.[65] It appeared about forty-five times in our documents, and was particularly popular with the authors of continental texts particularly from the 8th century onward, but it is not found in Irish or English sources. In the earliest medieval period, it was used but seldom: once by Caesarius of Arles, in a context that makes clear that the *ariolus*, like other soothsayers, was sought in times of illness,[66] and once by Toledo IV, in a warning to clerics of all ranks against recourse to magicians.[67] Otherwise *arioli* are absent from Iberian texts[68]–a surprising omission, since Isidore of Seville provided a definition and an etymology: "The *arioli* are given that name because they utter their vile prayers around the altars of the idols, and they offer wicked sacrifices, through the celebrations of which they receive answers from the demons."[69] A late ninth-century penitential unexpectedly connects *arioli* with the *sortes sanctorum*, a technique that Isidore of Seville had attributed to *sortilegi.*[70]

Generally the *ariolus* appears in a list of magicians, and although he is sometimes grouped with malign practitioners,[71] his usual character as soothsayer is made abundantly clear. We have noted above that a number of penitentials from the early 8th century on refer to "the *arioli* who are called diviners" and perform

[62] *Epítome hispánico* (c.598-610) 39.14, 185.

[63] 5, Caspari, 6-7.

[64] *Dict. étymologique*, *s.v.* "Haruspex."

[65] *E.g.*, "Let no one be found among you who ... accepts *arioli* and observes dreams and omens, nor one who is a sorcerer or enchanter nor one who consults pythons [necromancers?] or soothsayers" (*Deut* 18 10). These words are used in Charlemagne's *Admonitio Generalis* (789) 65, *MGH CapRegFr* 1: 58-59, and *Capitulare missorum generale item speciale* (802) 40, *ibid*, 104. Carolingian legislation concerning sorcery and magic is discussed in Hubert Mordek and Michael Glatthaar, "Von Wahrsagerinnen und Zauberern. Ein Beitrag zur Religionspolitik Karls des Grossen," *Archiv für Kulturgeschichte* 75 (1993) 33-64.

[66] *S*. 52.5, *CCSL* 103: 232.

[67] *Conc. Toletanum* IV (633) 29, Vives, 203. See also the *Poen. Quadripartitus* (9th century, 2nd quarter) 142, Richter, 18, and Atto of Vercelli (d. 961) *Capitularia* 48, *PL* 134: 38.

[68] Pirmin of Reichenau's inclusion of this term in his exhaustive lists of cunning folk (*Dicta Pirmini* [724-753] 22 and 28, 172 and 178) probably owes more to his experience in the Frankish empire than to his Visigothic Spanish background.

[69] *Etymologiae* VIII, 9.16: 714. The first part of this text is repeated by the *Poen. Ps.-Gregorii* [2nd quarter, 9th century) 16, Kerff, 177. Du Cange defines *ariolus* (*s.v.*) as "one who reveres pagan altars, or a soothsayer."

[70] *Poen. Vindobonense a* 97, Wasserschleben, 422. This clause is not in Schmitz's edition of this penitential.

[71] *E.g., Poen. Ps.-Theodori* (mid 9th century) 27, Wasserschleben, 595-596).

divinations.[72] Two penitentials show that they were brought into the house to do so.[73] A Frankish or Burgundian penitential identifies them with *sortilegi* and maybe *praedicatores* and *herbarii* as well; their arts won them a devoted following.[74] Another penitential imposed three years of penance on an *ariolus* who employed other soothsayers, performed sorcery or interpreted omens. Here perhaps the word denotes "magician" in general.[75] When Rabanus Maurus discussed the penance appropriate for various kinds of magical activity, he invoked as precedent the punishment due to practitioners of the magic arts, such as *magi* and *arioli.*[76]

Aruspex (auruspex, haruspex): About forty references to haruspices are found in passages drawn from both continental and insular sources; they retained thei popularity throughout our period. Isidore of Seville provided a fanciful etymology for the word, but described their original function accurately: "Haruspices are given that name as though they were the examiners of hours, for they keep watch over the days and hours in which business and work is to be done, and they pay attention to the observances that man should follow at any given moment. They also examine the entrails of animals, and foretell the future from them."[77]

Isidore's definition had little apparent influence on our texts.[78] None mention the examination of entrails. The *haruspex* in the mid fifth-century Synod of St. Patrick was not a soothsayer at all, but a great religious authority, a representative of the divinity, according to Bieler a druid, in front of whom people were accustomed to take oaths in "pagan fashion."[79] To Caesarius of Arles, however, he was simply another diviner, undifferentiated from the rest, to whom one turned in times of trouble.[80] This is generally how he figures in more than a dozen lists of cunning folk. Undoubtedly Gregory I used the word in its technical, classical sense when he urged the Bishop of Sardinia to take action against the clients of such

[72] *E. g.*, *Poen. Merseburgense a* (late 8th century) 23, *CCSL* 156: 132. Two other recensions of this penitential suggest that some penitents might have tried to make a distinction between amateur and professional soothsayers: "If anyone was an *ariolus* or, what is the same thing (*vel*), performed any divinations ..."

[73] *Poen. XXXV Capitolorum* (late 8th century) 16.1, Wasserschleben, 516. See also *Poen. Valicellanum C.6* (10th/11th century) 59, Schmitz 1: 379.

[74] *Poen. Hubertense* (8th century, 1st half) 25, *CCSL* 156: 111.

[75] *Poen. Sangallense tripartitum* (8th century, 2nd half) 20, Schmitz 2: 181.

[76] Rabanus Maurus, *Poenitentiale ad Heribaldum* (c. 853) 30, *PL* 110: 491.

[77] *Etymologiae* VIII, 9.17: 714.

[78] Even in the 5th century, this word had a more varied meaning, see Maximus of Turin's description of a drunken half-naked *haruspex* who was evidently a priest, not a diviner (*S.* 107.2, *CCSL* 23: 420-421).

[79] 14, Bieler, 56-57. Druids appear as *magi* in the *Collectio Hibernensis* (see Wasserschleben, 212 and below). Druidism is mentioned without any clarifying detail in the 8th century *Old-Irish Table of Commutations* (5, Binchy in Bieler, 278).

[80] *S.* 52.1, *CCSL* 103: 230. Caesarius mentioned the *haruspex* six other times.

people.[81] But about ten eighth- and ninth-century penitential texts confused *aruspex* with *augur*, and accused him of the sacrilegious act of divining from birds and "other things" or by "some other wicked device."[82] Only in the northwestern parts of Carolingian territory is there even a slight, probably haphazard, connection made between the *haruspex* and the observance of time, in Ghärbald of Lüttich's enquiry into lot-casters, *aruspices*, and " those who observe months and the times."[83]

A most unexpected use of the word is found in the ninth-century *Hubertian Penitential*: "If anyone commits sacrilege, that is, what they call *aruspices* ..." Either this telescopes the phrase "which is performed by those whom they call *aruspices*" or *aruspices* is a mistake for *aruspicia*, that is, omens. In either case, the word may have a wider meaning which includes the other practices listed in this text: frequenting forbidden places, offering sacrifices and questioning soothsayers as well as augury from birds or from "some other wicked device."[84]

Augur, augurator, auguriosus, agurius, auguriatrix: Augurs are mentioned but seldom in pastoral literature although they appear in the Vulgate as astrologers (*Is* 47 13; *Jer* 27 9). Caesarius of Arles spoke of the *auguria* interpreted by soothsayers, but never of *augures*.[85] Isidore of Seville followed classical tradition in describing them as diviners who observe the flight and cries of birds and other signs; these, especially the behaviour of birds, are called auspices, and observed particularly by travellers.[86]

The word crops up in a mere half dozen lists of practitioners.[87] Given the small impact made by Isidore's treatise *de magis* on pastoral literature in general, it is curious to find echoes of this definition in two penitentials and a Carolingian sermon which (in a very garbled discussion of objectionable practices) mentions an *agurius*, and continues with a brief list of omens: "little singing birds, ridiculous sneezes and signs."[88] The Council of Toledo of 693 linked *auguratores* with idolaters

[81] *Ep.* 9. 204 (599), *MGH Ep* 2:192.

[82] *E.g.*, *Poen. Remense* (early 8th century) 9.3, Asbach, 56.

[83] *Capitulary 2* (between 802 and 809) 10, *MGH CapEp* 1: 29. See also *Capitula Silvanectensia prima* (9th century, 1st half) 11, *MGH CapEp* 3: 82; *Anonymous sermon* (c. 850-882) 5, Kyll, 10.

[84] 24, *CCSL* 156: 110. But when the same expression "commit the sacrilege *called aruspices*" is used in an anonymous *Liber poenitentialis* (9th century?), the editor suggests plausibly that "sacrilege" is a mistake for "sortilege" (*PL* 105: 722).

[85] *S.* 189.2, *CCSL* 104: 772.

[86] *Etymologiae* VIII, 9. 18 and 19: 714-716. The same ideas are more concisely expressed in the definition of *augures* given in the *Poen. Ps.-Gregorii* (2nd quarter, 9th century) 16, Kerff, 177 and in the *Poen. Valicellanum C.6* (10th/11th century) 59, Schmitz 1: 379.

[87] *Conc. Toletanum* IV (633) 29, Vives, 203; *Legatine Synods–Report of the Legates George and Theophylact of their proceedings in England* (787) 3, Haddan & Stubbs, 449; *Poen. Quadripartitus* (9th century, 2nd quarter) 142, Richter, 18; Atto of Vercelli (d. 961) *Capitularia*, 48, *PL* 134: 37-38; *Poen. Valicellanum C.6* (10th/11th century) 59, Schmitz 1: 379; Burchard of Worms, *Decretum* (1008-1012) 10, *PL* 140: 831.

[88] *Anonymous sermon* (late 8th/9th century), Levison, 310-311.

and *praecantatores* as dupes of the devil. An *auguriosus* appears among Pirmin of Reichenau's list of cunning folk.[89] Finally, women who practiced the science of augury made two appearances as *auguratrices*, once in a list of malefactors in Charlemagne's directive to his *missi*,[90] and once in Arno of Salzburg's 806 sermon urging the faithful to refrain from turning to superstitious practices and magicians when in distress.[91]

Caragus/ caragius/ charagius/ caraus/ karagius/ ceraius: The existence of the *caragius* is first recorded by Caesarius of Arles, who mentioned him a dozen times in his catalogues of cunning men. The importance of the *caragius* is underscored by the fact that his name comes first eight times. But neither in these sermons nor in the approximately fifteen other passages in which he appears up to the middle of the 9th century is there any clue about his distinctive characteristics, if any.[92] Since the Council of Narbonne identified him as *divinator*, and since he always appears in company with soothsayers, it is safe so say that he was a fortune-teller of some sort.[93]

Caesarius' influence is unmistakable in all subsequent mentions of the *caragius* in continental and insular sources up to and including the 9th century.[94] It is perceptible not only in the terms used but also in the kinds of documents, chiefly legal texts (conciliar decrees and collections) and especially sermons, as well as in a letter addressed by St. Boniface to the pope. Most penitentials, a form of literature probably unknown to Caesarius, ignore the *caragius*. In this regard, Caesarius seems to have had a particularly powerful effect on his English readers. The three penitentials that use this word are Anglo-Saxon in origin, the anonymous sermons, although written on the continent, may have had English authors, and St. Boniface himself, of course, was an Englishman born and bred.

The form *ceraius* ("karajoc" in one manuscript) is found only in the eighth-

[89] *Conc. Toletanum* XVI (693) 2, Vives, 498-499; *Dicta Pirmini* (724-753) 28, 188-189.

[90] *Capitulare missorum generale* (802) 25, *MGH CapRegFr* I, 96. *Is* 57, 3 provides the biblical precedence for *auguriatrix*.

[91] Arno of Salzburg, *Synodal Sermon* (c. 806), Pokorny, 393-394.

[92] No evidence has been found in these texts to justify the interpretation of *caragius* to mean juggler (McNeill and Gamer, *Medieval Handbooks of Penance*, 69) or as a magician specialising in magical symbols (Blaise, *s.v.* "Caragius"). In his *Saint Césaire d'Arles* (Namur, 1962), 73 n., Blaise associates him with the Provençal *caraque* or gypsy. See also Dag Norberg, *Manuel pratique de latin médiéval* (Paris, 1968), 104.

[93] 14, *CCSL* 148A: 257.

[94] *Syn. Autissiodorensis* (561-605) 4, *CCSL* 148A: 265; *Collectio Hibernensis* (late 7th/early 8th century) 64.2, H. Wasserschleben, 231; *Vita Eligii* (c. 700-725), *MGH SRM* 4: 705 and 707; *Liber de remediis peccatorum* (721-731) 3, Albers, 411; *Dicta Pirmini* (724-753), 22 and 28: 172 and 188; Boniface to Pope Zachary, *Ep.* 50 (742) *MGH EpSel* 1: 84; *Anonymous sermon* (8th century), Morin, 518; *Anonymous sermon* (8th century) Nürnberger, 43; *Anonymous sermon* (late 8th/9th century), Levison, 308; *Double-Penitential of Bede-Egbert* (9th century?) 30.3, Schmitz 2: 694-695; Rabanus Maurus (d. 865) *Hom.* 43 and *Hom.* 45, *PL* 110: 81 and 83. The *Hibernensis* attributed the text to St. Augustine.

century *Poen. Egberti,* English as well.[95] This spelling is so peculiar that, although it appears in the same Caesarian context as the other forms of *caragius*, one may wonder if the author connected his practices in some way with wax (*cera*), perhaps divination using melted wax or wax-covered writing-tablets.

Mathematicus: Some twenty-five passages in pastoral texts, dating from the beginning of the 6th to the second half of the 10th century, speak of *mathematici*. This, reported Isidore of Seville, was the name given in common speech to those who studied the constellations, that is, the position of the stars at the time of one's birth.[96] In late antiquity, *mathematicus* meant astrologer more generally. The word was used in this sense during the 6th century but only rarely afterwards, when it came to mean a sorcerer who specialised in producing mental disturbances. *Mathematici*, said Caesarius of Arles, claimed to foretell the length of one's life; they and the Manichees, Satan's spokesmen, tried to absolve one of the responsibility for sin and the duty of confession: "Has a man sinned? The stars were so placed that they made him sin."[97] The Council of Braga (560) stated that *mathematici* observed the twelve signs of the zodiac, which the Priscillianists believed controlled the soul and the different parts of the body, and to which they ascribed the names of the patriarchs.[98] The concept of them as diviners of some sort still exists in in Ps.-Eligius' admonition that "*mathematici* are to be spurned, auguries abhorred and dreams despised." Finally, Atto of Vercelli used the word in the ancient sense; the *mathematici* taught people that the stars presided over births and ruled over marriages and the construction of houses.[99]

Pithon/phiton, fitonis, pitonissa, necromanta: Isidore of Seville's longest and most detailed description of magicians and their sinister craft is dedicated to necromancers:

> Necromancers are those by whose spells the dead appear to be raised in order to prophesy and to answer questions, for "corpse" is *nekros* in Greek, and divination *mantia*. Blood is thrown on a corpse to raise the dead because demons are said to love blood. Therefore, whenever necromancy is practiced, blood is mixed in water to obtain more easily the colour of blood.[100]

Necromancers are mentioned by name once only, if at all, in early medieval pastoral literature–a very late recension of an eighth-century Bavarian council calls

[95] 8.4, Schmitz 1: 581. For "karajoc," see the *Excarpsus Egberti* (Schmitz 2: 668) and McNeill's comments in *Medieval Handbooks of Penance*, 69.

[96] *Etymologiae* VIII, 9. 24. See Du Cange, *s.v.* "Mathematicus."

[97] S. 18.4, *CCSL* 103: 85; S. 18.4, *ibid.*, 259-260.

[98] 10, Vives, 68.

[99] Ps.-Eligius, *De rectitudine catholicae conversationis* 7, *PL* 40: 1174; Atto of Vercelli (d. 961) S. 3 (In festo octavae Domini), *PL* 134: 837.

[100] *Etymologiae* VIII, 9.11: 714.

for poisoners and women necromancers to undergo the ordeal by hot iron.[101] Nevertheless, there are other hints of such specialists. Caesarius of Arles described an expert supposedly able to conjure up an apparition (*persona*) to find stolen goods; clients must come in secret and avoid making the sign of the cross when approaching its haunts.[102] This is proof of the being's diabolical nature, since Martin of Braga affirmed that the sign of the cross was a sure protection against the assaults of malign demons and evil spirits.[103]

Isolated references to divination through "pythons" may also have some connection with necromancy. The Old Testament associates pythons with the realm of the dead: in *Is* 29, 4, the cry of humbled Jerusalem is like that of the python from the earth, the very home of the dead; in *I Sam* 28, 7, Saul visits a woman "having a python" that she may raise Samuel from the dead.[104] Two Carolingian capitularies make use of the mention of pythons in *Deut* 18, 10-11 to justify repression of very contemporary miscreants (Benedict Levita transcribed *python* as *fitonis*).[105] Two other passages are more promising. In marshalling the authorities on auguries, the *Collectio Hibernensis* advances *Lev* 20, 27 against men or women possessed by a python or the spirit of divination.[106] Since (as far as I know) this is a text quoted in no other comparable document, the compilers may have had particular reasons for including it in this collection. The impressive testimony of the *Homilia de sacrilegiis* evokes images of both Caesarius' victim of theft and Saul:

> And whoever consults seers or seeresses, that is, pythonesses, through whom devils make reply, who goes to question them and believes what they say, or goes to listen to them attentively in order to hear anything from demons –he is not Christian, but pagan.[107]

But necromancy may lie hidden in other texts as well. R.L.M. Derolez suggested that the enigmatic clause "concerning the sacrilege performed over the

[101] *Concilium Rispacense* (799/800) *Iordani Recensio* (1550) 9, *MGH Concilia* 2.1: 219.

[102] *S.* 184.5, *CCSL* 104: 750-752.

[103] *De correctione rusticorum* (572) 8, Barlow, 188.

[104] This connotation is not evident in *Deut* 18, 11 and is wholly missing in *Acts* 16, 16, the only New Testament reference to pythons. Nor is there any suggestion of necromancy in Gregory of Tours' use of the term (see *HF*, 5.14 and 7.44, *MGH SRM* 1: 210 and 364-365). In the latter passage, a woman "having the spirit of a python" practices her craft of finding stolen goods quite openly for her masters and, after she is freed, is richly rewarded by her admiring clientele. Her bishop established that her powers were owing to her being possessed by a demon but since he was unable to expel the demon, he let her go, presumably to continue her trade under the unspeakable Fredegund's protection.

[105] *Admonitio Generalis* (789) 65, *MGH CapRegFr* 1: 58-59. See also *Capitulare missorum generale item speciale* (802) 40, *ibid.,* 104 and *Benedicti capitularium collectio* (mid 9th century), 2.374, *PL* 97: 701.

[106] 64. 3, H. Wasserschleben, 231. The other authorities cited in this section are "Agustinus" (*i.e.*, Caesarius of Arles), the Council of Ancyra and Isidore of Seville.

[107] 5, Caspari, 6-7.

dead, that is, *dadsisas*" of the *Indiculus superstitionum* might refer to a Germanic necromantic rite corresponding to *valgaldr*, the spell with which Odin awoke the *völva* (seeress).[108] The divinations at tombs and graves or funeral pyres (*busta*) described in the *Penitential of Arundel* appear to be connected with necromancy. Clearly the rite required special skill or sacramental powers, since the diviner was assumed to be a bishop, priest or deacon. Equally clearly, it was a dangerous proceeding, for the author considered it likely that the diviner was constrained to do so by force or fear of an enemy.[109] The dead were invoked in Scotland too: a spurious law attributed to Kenneth MacAlpin (853-859) condemned "invokers of the shades [*manes*], familiars of wicked demons or those who seek their help" to death by burning.[110]

The *Hibernesis*' arguments disproving the notion that the souls of the dead (even of martyrs) are aware of goings-on in this world imply that attempts were being made to get in touch with the dead for counsel or help. St. Augustine is quoted that men are sometimes led into great errors by false appearances: "It is agreed that the dead do not know what happens here, but they hear about it afterwards from those who, by dying, go to them." The authority of St. Jerome is also cited: those who migrate from the flesh forget earthly matters. He is quoted further on the question whether, when necessary, the dead see events in this world: "Some think that, by the revelation of the Spirit, the souls of the dead can see other things that happen, not merely the present but the past and future as well, when it is needful either for them or for us." The answer is less than categorical: "Not all the dead can see what happens here, just as not everyone, except for prophets, foresees all things in this world."[111]

Finally, some element of the blood rituals at which Isidore of Seville hinted may lie behind the insistent reiteration of the biblical taboos on blood.

Somniarius, coniector somniatorum and the interpretation of dreams:[112] Interpreters of dreams appear late and seldom in the texts. Only in the documents of the Council of Paris (829) and the capitularies of Benedict Levita (mid 9th century) and Herard of Tours (858) do they figure among other cunning men and magicians as *somniatorum coniectores* or *somniarii*.[113] Dreams are altogether absent from our sources until the beginning of the 8th century. From then on, however, about two dozen texts include them among auguries and omens to be avoided–according to one author-

[108] 2, *MGH CapRegFr* 1: 223. Derolez, "La divination chez les Germains," 284-285.

[109] 88, Schmitz 1: 461. On necromancy as a clerical specialty, see Richard Kieckhefer, *Forbidden Rites. A Necromancer's Manual of the Fifteenth Century* (University Park, Pa., 1997), 10-13.

[110] 18, Haddan and Stubbs 2: 123.

[111] 49.12, 49.13, 49.14, H. Wasserschleben, 207.

[112] See Jacques Le Goff, "Christianity and dreams (second to seventh century)," in *The Medieval Imagination*, trans. Arthur Goldhammer (Chicago and London, 1988) , 193-231.

[113] *Conc. Parisiense*, 69 (II), *MGH Concilia* 2.2: 669; *Benedicti capitularium collectio* 2.214, *PL* 97: 774; Herard of Tours, *Capitula* 3, *PL* 121: 764.

ity, particularly by the clergy.[114] Most common are the repetitions, in continental texts, of the clause from the *Penitential of Theodore* imposing penances on clerics and laymen for the observance of dreams and other forms of divination.[115] Warnings against dreams in capitularies and other penitentials of the same period are equally vague.[116] A fifth-century Carthaginian canon which Burchard incuded in his *Decretum* may explain the prevalence of this theme. According to this canon, dreams and other revelations were invoked by private individuals as an authority higher than the bishop's in deciding where altars were to be set up. Dreams take their place here as a part of the laity's arsenal in the struggle to maintain some control over sacred places against the centralising efforts of the hierarchy.[117]

The categorical condemnation in pastoral literature of this form of divination stands in contrast to the more nuanced position of Holy Scripture and the Fathers and the positive attitude of hagiographers. On the one hand, the Bible condemns the interpretation of dreams (*e.g.*, *Deut* 13, 3 and *Jer* 29, 8). On the other, both the Old and New Testaments show that God's messages are conveyed in dreams. The very term used by the Council of Paris was flung scornfully at Joseph by his brothers: "Behold the dreamer (*somniator*) comes!" (*Gen* 37 19). The *Collectio hibernensis* quotes St. Augustine and Gregory I to the effect that the dead appear to the living in dreams, to give consolation and teach of the rewards of heaven and the pains of hell.[118] It continues with St. Gregory's analysis of the six reasons for dreams: a full stomach, hunger, illusion, thoughtfulness and illusion together, revelation, and, finally, thoughtfulness and revelation together.[119] Prophetic dreams are common in hagiography. They frequently herald the birth of saints: St. Columba's pregnant mother dreamed of a royal robe that covered the world; before she conceived, the mother of St. Ciaran of Saigir dreamed that a star fell into her mouth; the mothers of St. Bernard and St. Dominic had dreams of carrying dogs in their wombs.[120] St. Jerome himself dreamed a dream which proved a turning point in his life.[121]

[114] *Statuta Bonifacii* (1st half, 9th century) 20, *MGH CapEp* 3: 363. See also *Capitula Vesulensia* (1st half, 9th century) 33, *ibid.*, 351.

[115] *Poen. Theodori* (668-756) I, 15.4, Schmitz 1: 537-538.

[116] *E.g.*, *Capitulare missorum generale item speciale* (802) 40, *MGH CapRegFr* 1: 104.

[117] 3.54, *PL* 140: 683.

[118] 51.1-4, H. Wasserschleben, 209.

[119] 51.6, H. Wasserschleben, 210. See Gregory the Great, *Dialogues* IV, 50.2, *SC* 265: 172.

[120] *Vita Columbae,* in *Adomnan's Life of Columba*, trans. and ed. Alan Orr Anderson and Marjorie Ogilvie Anderson (Oxford, 1991), 183-185; *Vita Ciarani de Saigir* 1, *VSH* I, 217; Jacob of Voragine, *Legenda Aurea* (ed. Maggioni) 2: 812 and 719. For an analysis of dreams of this sort, see Francesco Lanzoni, "Il sogno presago della madre incinta nella letteratura medievale e antica," *Analecta Bollandiana* 45 (1927): 225-261. See also Pierre Saintyves, *En marge de la Légende Dorée. Songes, miracles et survivances* (Paris, 1930), especially 3-33.

[121] See P. Antin, "Autour du songe de S. Jérôme," in *Recueil sur saint Jérôme* (Brussels, 1968) 71-100.

Nevertheless, there were good grounds for opposing dreams. Not only were they subversive, a challenge to the monopoly of the clergy over religious matters, but also (as it was explained in another context) they could well be inspired by the devil who, disguised as an angel of light, used them to seduce the ignorant.[122] Holy men, explained Gregory the Great, know the difference between illusion and revelation by "a certain inmost flavour, " but "if one does not bear a cautious mind with respect to this, the spirit is plunged into many follies by the Deceiver who is accustomed now and then to predict true things, so that he will be able to entangle the soul at last by another falsehood."[123] No doubt dreams were doubly suspect because they were tainted with the traditions of the non-Christian past. Both Celts and Germans valued divinatory dreams, and had developed techniques to evoke them. An Irish rite for the selection of the king of Tara involved a dream-vision induced by a ritual meal and the chanting of druids;[124] in Scandinavian sagas, *seidhr* was a form of divination entailing quasi-shamanistic dream voyages, shape-shifting or the evocation of spirits.[125] Churchmen, then, had ample reason to distrust dreams.

Vaticinator: *Vaticinatores* appear in the Visigothic legal code, but among our authors, only Benedict Levita and Burchard of Worms mentioned them, describing them as false soothsayers who claimed to know the future.[126] This word derives from the very old *vates*, probably of Gaulish origin, but was well-known in classical Latin.

6.2.2 *Divinatory techniques*[127]

Adjurations against consulting soothsayers and practicing divination, such as "Whoever makes use of auguries or enchantments is to be separated from the

[122] Regino of Prüm, *De synodalibus causis* (c. 906) 2, 371: 355-356.

[123] 51.6, H. Wasserschleben, 210-211

[124] See Green, *Dictionary of Celtic Myth and Legend, s.v.* "Tarbhfhess." Le Roux suggests that this may have been the basis of the bull-sacrifice described by Pliny (*Historia Naturalis* 16: 95) in his account of the harvest of the mistletoe ("La divination chez les Celtes," 252-253).

[125] Derolez, "La divination chez les Germains," 276-278. For the interpretation of dreams among the Germans, see Grimm, *Teutonic Mythology*, 1145-1147 and 1647-1649.

[126] *Lex Visigothorum* 6.2, 1, *MGH Leg* 1.1: 257; Benedict Levita, *Capitularium collectio* 3.222, *PL* 97: 825; Burchard of Worms, *Decretum* 10.22, *PL* 140: 836.

[127] For divination in general, see H. J. Rose et al., "Divination," (*ERE* 4: 775-830) and Bächtold-Stäubli, *s.v.* "Wahrsagen." For divination among European peoples, see Raymond Bloch, *La divination dans l'Antiquité* (Paris, 1984) and the articles by Bloch on "La divination en Étrurie et Rome," by Le Roux on "La divination chez les Celtes," and by Derolez on "La divination chez les Germains" in *La divination*,ed. A Caquot and M. Leibovici (Paris, 1968), 1: 197-232, 233-256 and 257-302. See also George K. Park, "Divination and its social contexts," in *Magic, Witchcraft and Curing* , ed. John Middleton (New York, 1967), 233-254. For purposes of comparison with divination in another culture, see Fahd, *La divination arabe.*

community of the Church,"[128] and variations of the form "do not observe omens" or "do not look for divinations" abound. Generally these may be taken as blanket condemnations of all forms of divination, but at times the formulaic phrases cover specific techniques. This was the case with the lots (*sortes*) about which the Breton bishops consulted Pope Leo IV. The pope found them similar to, but evidently not identical with, practices condemned by the Council of Ancyra centuries earlier: divination and reliance on magicians to clear the house of hidden evil.[129]

Sermons and penitentials contain numerous explicit references to different techniques of divination. A few of these deal with preparations for divination and many more with the observation of signs without attributing them to what may be called professional soothsayers. Ordinary men and women, who did not consider themselves and whom their pastors obviously did not consider to be *sortilegi*, *arioli* or any other kind of diviner, mastered many of them. Some have a written tradition dating back to the classical period; others are described probably for the first time in these sources.

Preparations for divination: Hints of techniques to put oneself into the right frame of mind for divination appear in a handful of documents. The puzzling phraseology of two continental penitentials, "If anyone by soothsaying performs divinations [by magic?–*ariolando*]...," may refer to preparatory rites.[130] A third penitential, after condemning soothsaying by means of the *sortes sanctorum* or "any other wicked device," proposed a considerably lighter penance for eating and drinking "for the same reason" (that is, for divination).[131] Overeating and the use of hallucinogens, narcotics or alcohol to induce prophetic trances may have been envisioned. Finally, the Synod of Erfurt (932) issued a directive (repeated by Burchard of Worms) against unauthorised fasting:

> Let no one, without the permission of his own bishop or his representative, impose a fast on himself, choosing one day over the others under pretext of religion. This is not pleasing and we forbid it for the future, because it is perceived as being done more for the sake of divination [magic?–*ariolandi*] than as a supplement to Catholic law.[132]

Techniques of divination–from human beings: Sermons testify to a popular belief that spontaneous or involuntary actions on the part of human beings were portentous.

[128] *Statuta Ecclesiae Antiqua* (c. 475) 83 [LXXXIX], *CCSL* 148: 179. Repeated in some dozen documents down to the 11th century.

[129] *Ep.* 16, *MGH Ep* 5: 594.

[130] *Poen. Sangallense tripartitum* (8th century, 2nd half) 24, Schmitz 2: 181 and *Poen. Casinense* (9th/10th century) 70, Schmitz 1: 414. Both penitentials impose a penance of five years for this practice.

[131] *Poen. Vindobonense a* (late 9th century) 29, Schmitz 2: 353.

[132] *Gesta Synodalia*, 5, *MGH Leg* 6.1: 109. See also Burchard of Worms, *Decretum* (1008-1012) 13.27, *PL* 140: 889-890. But note that in 19.33 (*ibid.*, 986), once the officially prescribed penance has been completed, extra voluntary fasting is considered meritorious.

Most frequently cited are sneezes. Caesarius of Arles had complained that sneezes were observed as an omen in his diocese, and warned that they were "not only sacrilegious but also ridiculous." Some eight preachers from Martin of Braga to Rabanus Maurus faithfully recorded the same custom, as did the compiler of the *Indiculus superstitionum*.[133] The eighth-century *Homilia de sacrilegiis* added hisses expressing disapprobation and cries of joy; another sermon, random occurrences or disapproving words, heard while entering or leaving the house.[134]

Other conclusions also were to be drawn from observing the human body and its actions. The dimensions of a person's head, waist or chest gave signs of future trouble and provided clues to explain illness. In sixth-century Arles, certain soothsayers had the reputation of being able to advise preventive measures by inspecting and measuring the vulnerable person's headband or bandage, and belt or girdle.[135] Stumbles were an evil omen, at least if one was on the way to commit some wicked deed such as adultery or theft–Arlesians tried to undo the damage by tracing the sign of the cross over their face or lips as a counterspell.[136] Another, apparently unrelated custom described by Martin of Braga and then Pirmin of Reichenau was the "observation of the foot;" this probably meant taking care to start out on the right foot.[137]

Signs were sought in more nebulous matters as well. The *Epítome hispánico* threatened a beating for those who, "either alone or with others," practiced recondite forms of divination, including one from the "names of brothers" (implying perhaps that the practitioners were monks).[138] Seeing or hearing a cleric or monk was a bad sign. The *Homilia de sacrilegiis* warned those who "abominated a soldier of Christ" and considered the sight or sound of him to be abominable (literally, a bad omen) were "not only pagan but possessed by demons as well." The same sermon described a form of chiromancy: watching to see whether a man's hand was "heavy or light" at the moment when he took a chalice and "looked into it." It also cited the belief that a man's shadow (*umbra*) could be good or evil, that is, portend good or bad luck or exert a benign or malign influence.[139]

[133] *S.* 54.1, *CCSL* 103: 236. See also *Indiculus superstitionum et paganiarum* (743?) 13, *MGH CapRegFr* 1: 223. Plutarch recorded a belief that the direction from which the sneeze came foreshadowed the success or failure of a venture (Lynn Thorndike, *The First Thirteen Centuries,* 1, *The First Thirteen Centuries* [1923; New York, 1953], 208).

[134] *Homilia de sacrilegiis* (late 8th century) 9, Caspari, 7; Ps.-Eligius, *De rectitudine catholicae conversationis* (8th century?) 7, *PL* 40: 1174. See also *Anonymous 8th-century homily*, Nürnberger, 43. Tacitus stated that the Germans predicted the outcome of a battle from the sound of the warcry known as *baritus* (*Germania*, 3).

[135] *S.* 52.5, *CCSL* 103: 232; S. 184.4, *CCSL* 104: 750-1.

[136] *S.* 13.1, *CCSL* 103: 65-5; S. 134.1, *CCSL* 104: 550.

[137] *De correctione rusticorum* (572) 16, Barlow, 198. See also Dicta Pirmini (724-753) 22: 172.

[138] 4.13, 114.

[139] 11, Caspari, 8; *ibid.*, 6: 7; *ibid.*, 7: 7. According to Caspari, having another man's shadow fall over one is a sign of coming misfortune (*ibid.*, 20-21), but it can also work good: in *Acts* 5, 15, those on whom St. Peter's shadow (*umbra*) falls are cured of their illness.

Techniques of divination–from animals: The song and flight of birds were filled with omens.[140] Caesarius of Arles urged the faithful to pay no attention to "little [*i.e.*, contemptible] singing birds" while travelling, nor to find "diabolical divinations" in their song.[141] An eighth- or ninth-century tract lists the idle chatter of birds among the signs observed while leaving or entering the house.[142] In ninth-century Trier, omens for "months and seasons and the beginning of the year" were sought in the behaviour of birds and wild beasts.[143] According to a confusingly phrased clause in Burchard of Worms' penitential, the location and direction of a crow's song was thought to announce the success of journeys: "While [travelers] are on the road, they hope to enjoy a prosperous voyage unless they hear a crow on their left hand singing to them toward the right." This either deduces a favourable outcome from the absence of a presage, which is odd, or it implies that the crow's song coming from or directed toward another direction is a good sign; this is odd, too, since the crow is normally considered to be an ill-omened bird.[144] In the same clause, the appearance of a "mousecatcher" (owl? hawk?–*muriceps*) flying across the road was a reassuring sign for the prospects of a night's lodging.[145]

The reference to little birds, or little singing birds, of three eighth-century sermons are evidently owing to Caesarius' influence, although none mention voyages.[146] But the sparrows and "other birds" specified by the eighth-century *Homilia de sacrilegiis* as a source of omens have no antecedents in our texts.[147] Martin of Braga warned of the dangers of a misplaced trust in "mere birds" (*avicelli*). It was demons who spoke through their voices; men forfeited their Christian faith by such "silly and futile things" and at the same time "incurred the catastrophe of unforeseen death" (presumably through recklessness prompted by over-conf-

[140] Since they move freely through the air, birds are considered to be important for divination in many cultures; see Edward A. Armstrong, *The Folklore of Birds* (2nd ed., New York, 1970). See also R. I. Best, "Prognostication from the raven and the wren," *Ériu* 8 (1915): 120-126.

[141] *S.* 54.1, *CCSL* 103: 236. See also *Vita Eligii* (700-725) *MGH SRM* 4: 705, and Rabanus Maurus (d. 856) *Hom.* 43, *PL* 110: 81.

[142] Ps.-Eligius, *De rectitudine catholicae conversationis* 7, *PL* 40: 1174. The verb *garrire* used here for the sound of the birds links this text to Isidore of Seville's *Etymologiae*, which explains *auguria* as deriving from the chatter (*garria*) of birds.

[143] *Capitula Treverensia* (before 818) 5, *MGH CapEp* 1: 55. See also *Anonymous sermon* (c. 850-882) 5, Kyll, 10.

[144] An omen known as *emponemb,* mentioned by Du Cange as taken from Michael Scotus' *De physionoma* 56, functioned in a similar way: it was a propitious omen for business if a man or bird crossed in front of one going from the left side to the right, and moved out of sight without stopping (*s.v.* "Emponemb").

[145] 19, 5.149, Schmitz 2: 441-442.

[146] *Dicta Pirmini* (724-753) 22: 172; *Homilia de sacrilegiis* (late 8th century) 27, Caspari, 16; *Anonymous homily*, Nürnberger, 43.

[147] 9, Caspari, 7.

idence).[148]

Other animals have a lesser place in the texts. Instead of Tacitus' awe-inspiring technique of divining from the whinnies and snorts of sacred horses, the *Indiculus superstitionum* records the practice of looking for auguries in the droppings of birds (from the context, surely barnyard fowl), horses and cattle.[149] The *Homilia de sacrilegiis* mentions the barking of dogs as an omen. The "days of moths and mice" appear only in Martin of Braga's and St. Eligius' sermons; the destructiveness of these pests, manifested on rags and crumbs placed in a box, were *praefigurationes* for the New Year.[150] In Burchard of Worms' diocese, it was considered that if, on approaching a sick man's house, one found a stone nearby and turned it over, one could read the prognosis for the patient's recovery from insect activity beneath: wriggling earthworms, flies or ants meant that the patient would get better, but lack of movement signified death.[151] Of the vast repertoire of divinatory techniques based on animals available in a peasant society, these are the only clues.

Techniques of divination–from bread: References to divination from bread are found in sources from the 6th century to the 11th.[152] The custom of placing bread into a spring or well described by Martin of Braga and Pirmin of Reichenau might have been, as Meslin suggested, a form of ordeal, but perhaps this should be considered as divination from water rather than from bread. We have also seen that Burchard of Worms was aware of a local custom of baking breads in one's own name during the Calends of January. How well they rose and how dense and high they became heralded one's fortunes in the coming year.[153] The "figure [or effigy] made of sprinkled flour [dough?]" that appears in the *Indiculus superstitionum* may also have been a tool for divination.[154]

The sixth-century Synod of Auxerre specifically condemned lotteries using bread.[155] Continental sources do not explain this form of lottery, but it is known that the Anglo-Saxons practiced a type of ordeal in which an accused person was given a piece of bread that he would be unable to swallow if he was guilty.[156]

[148] *De correctione rusticorum* (572) 16, Barlow, 200; *Anonymous sermon* (late 8th/9th century), Levison, 310-311; *De correctione rusticorum*, 12, Barlow, 191.

[149] 13, *MGH CapRegFr* 1: 223; Tacitus, *Germania*, 10.

[150] *De correctione rusticorum* (572) 11, Barlow, 190-191; *Vita Eligii*, *MGH SRM* 4: 705. Oddly enough Pirmin of Reichenau, who relied so heavily on Martin's text, failed to mention this practice.

[151] *Decretum* (1008-1012) 19, 5.102, Schmitz 2: 432.

[152] See Bächtold-Stäubli, *s.v.* "Brot."

[153] Martin of Braga, *De correctione rusticorum* (572) 16, Barlow, 198; *Dicta Pirmini* (724-753) 22: 172; Burchard of Worms, *Decretum* (1008-1012) 19, 5.62, Schmitz 2: 423.

[154] C. 26, *MGH CapRegFr* 1: 223.

[155] *Syn. Autissiodorensis* (561-605) 4, *CCSL* 148A: 265.

[156] See Du Cange, *s.v.* "Corsned" and Sarah Larratt Keefer, "*Ut in omnibus honorificetur Deus*: The *Corsnæd* Ordeal in Anglo-Saxon England," in *The Community, the Family and the Saint*, 237-264.

Possibly, therefore, the Synod of Auxerre, though far removed in time and place from the circumstances observed by the papal legates George and Theophylact some two hundred years later, throws light on the "lots in the pagan style" which the English used to settle litigation.[157]

Techniques of divination–from liquids: Isidore of Seville had described hydromancy as a form of necromancy, the calling up of demons by the contemplation of water, to watch their likenesses or their mockeries, and to obtain information from them; hydromancers added blood to the water to summon the dead for questioning.[158] Although the word does not appear in our texts, the technique of gazing into liquids for divination was in use in sixth-century Visigothic Spain and in eighth-century Francia. In the *Epítome hispánico*, one looked into altar vessels to "detect" information.[159] The *Homilia de sacrilegiis* denounced as sacrilegious those who looked into the chalice when taking it. The author used the technical term *calix* rather than a more mundane word–an indication that this was practiced during mass at consecration or communion.[160] When he spoke of those who "pretended" to divine from another type of container, he used the word *orcii, i.e., urcei* (water-pots).[161]

Techniques of divination–from wood, wool and metal: A prohibition of "lotteries made of wood" appears among the canons of the Synod of Auxerre. This German practice is well established from other sources. Tacitus had described a technique of divination which involved randomly picked up slips of wood cut from the branch of a fruit tree, marked with runes and tossed on a white cloth or garment. The Frisian Code enshrined a comparable method in Christian form, which established guilt by drawing lots of marked sticks which had been placed, wrapped in clean woollen cloth, on an altar or on relics.[162]

Ostensibly, the practice of divination *cum lanas et acias* in the *Homilia de sacrilegiis* describes a method of drawing tufts of wool and needlefuls of thread for the purpose of divination: presumably thread was drawn from woven fabrics. A similar practice may be intended in an eleventh-century Spanish penitential which imposed five years of penance on any Christian "who practiced or scrutinized ...

[157] *Legatine Synods* (787) 19, Haddan & Stubbs, 458-459.

[158] *Etymologiae* VIII, 9.12: 714.

[159] 4. 13: 114.

[160] 6, Caspari, 7. The choice of the term "sacrilegious" to condemn the practitioner of this form of divination, rather than the usual "not Christian but pagan," may be another indication that one is dealing with the misuse of a sacred object. This word is employed only two other times in the twenty-seven clauses of the *Homilia*, in c. 7 (fortune-telling from scriptures) and c. 22 (practicing magic to cure a demoniac); the usage is inconsistent but suggestive.

[161] 9, Caspari, 7.

[162] *Syn. Autissiodorensis* (561-605) 4, CCSL 148A: 265; Tacitus, *Germania*, 10; *Lex Frisionum*, Tit. 14, *Lex Frisionum, MGH Fontes Iuris Germanici Antiqui* 12 (Hanover, 1982) 56. See Du Cange, *s.v.* "Tenus."

dreams, woolwork or sorcery."[163] But if the *Homilia*'s *acias* can be taken to be a slip of the pen for *acies* (sharp or pointed objects, swords), then we have some variation of the Frisian practice, this time using blades either tossed on or wrapped in woollens.

Techniques of divination–from the hearth: The hearth also provided omens. Burchard of Worms described as fairly common a divinatory rite ("which many perform"), in which grains of barley were placed "where fire is usually made in the house" after it had been swept clean; if the grain skipped about on the still hot surface, danger was on the way, but all was well if the grain did not move.[164] Conceivably the clause "concerning pagan customs with respect to the hearth or at the beginning of any activity" in the *Indiculus superstitionum* dealt with "observation" (divination) also, rather than "observance" (a protective or other rite).[165]

Techniques of divination–from thunder and the stars: Reading" thunder and astrology are cited in the *Homilia de sacrilegiis*.[166]

Techniques of divination–from sacred texts (sortes biblicae, sortes sanctorum): In numerical importance, all other techniques pale in comparison to divination from written texts.[167] Some reference to this method is found in approximately one third of the total number to deal with divination, that is, in about fifty passages, mainly in

[163] 9, Caspari, 7; *Poen. Silense*, 105, *CCSL* 156A: 27. Caspari was certain that this is divination using wool and thread; he related it to texts from Martin of Braga and Burchard of Worms involving women's woolworking magic (*ibid.*, 24). If this is the correct reading, then the *Homilia* is describing a type of lottery practiced by women. The use of the masculine terms *iste, christianus* and *paganus* is not necessarily significant, since the sentence as a whole deals with a variety of practices, but (although a feminine pronoun appears nowhere in this work) one might have expected to find *quae* rather than *qui* in a phrase dealing specifically with women. But Caspari was surely mistaken. The practices mentioned by Martin and Burchard are concerned with the process of spinning and weaving, not with divination.

[164] 19, 5.101, Schmitz 2: 431.

[165] 17, *MGH CapRegFr* 1: 223.

[166] 8, Caspari, 7. The Lombard king Agilulf, consulted a boy with divinatory gifts for the meaning of a thunderbolt which struck a tree in the royal courtyard (Paul the Deacon, *Historia Longobardorum* 3.30, *MGH Hist. Rer. Long.*, 110). For divination from thunder, see Ps-Bede, *De tonitruis* in *The Complete Works of the Venerable Bede*, ed. J.A. Giles (London, 1843-1944) 4: 343-348 (= *PL* 90: 610-612); see also Flint, *Rise of Magic*, 191-192. For divination from heavenly bodies, see chapter 3.

[167] See the lengthy entry in Du Cange, *s. v. Sortes sanctorum*. For divinatory as well as other uses of written documents in the Middle Ages, see J.-C. Poulin, "Entre magie et religion: Recherches sur les utilisations marginales de l'écrit dans la culture populaire du haut moyen âge," in *La culture populaire au moyen âge*, ed. Pierre Boglioni (Montreal, 1979), 123-143; M. Mostert, "La magie de l'écrit dans le haut moyen âge. Quelques réflexions générales," in *Haut moyen âge. Culture, Éducation et Société. Études offertes à Pierre Riché*, ed. M. Sot (Nanterre, La Garonne-Colombes, 1990), 273-281; Wolfgang Hartnung, "Die Magie des Geschriebenen," in *Schriftlichkeit im frühen Mittelalter*, ed. Ursula Schaefer (Tübingen, 1993), 109-126. See also Flint, *The Rise of Magic*, 221-224, 273-287 and *passim*.

conciliar decrees and penitentials. Only two Frankish sermons mention it. Even Caesarius of Arles, who presided over the Councils of Agde and Orleans which condemned it vigorously, and who had so much to say against divination in his sermons, failed to preach about it.

Two or three different types of documents used for divination are found in our literature: the *sortes biblicae*, the *sortes sanctorum*, and unspecified prophetic texts which may in fact be the same as the *sortes sanctorum*. Both lay people and clerics were accused of this sin. The ability to read was not essential for their use, but indubitably the clergy were the principal practitioners of the art, both as producers and interpreters. Although this seems a serious abuse of their special training and status, only one document recognised this by imposing a weightier penance on them than on lay people.[168]

The *sortes biblicae* were a Christianised form of the classical method of seeking guidance in the first words seen on a page, opened at random, in the works of Virgil or Homer. Christian practitioners replaced the pagan authors with Sacred Scripture. St. Augustine had disapproved mildly of this practice, although he himself owed the final impetus to conversion to exactly this kind of revelation of the Divine Will.[169] In his letter to Januarius, he wrote that it went against the grain for him to transfer the words meant for the next life to the vanity of earthly affairs, but conceded that drawing lots from the gospels was preferable to consulting demonic works.[170]

Augustine's words were largely ignored. Churchmen (especially when electing bishops or deciding on the burial place of saints), princes and, no doubt, ordinary folk, had frequent recourse to this form of divination to settle doubts and conflicts. Even notes written in the margins of Scripture could give necessary information. Gregory of Tours described a fairly typical method used by Chilperic's son Merovech: he placed the Psalter, the *Book of Kings* and the Gospels on the altar of St. Martin of Tours and after three nights of watching, fasting and praying, opened the books only to learn that his chances for inheriting his father's kingship were doomed.[171]

Perhaps because of the general tolerance for this practice, the *sortes biblicae* are

[168] As an alternative to excommunication, the early eighth-century *Liber de remediis peccatorum* called for three years of penance for clerics, but only one for laymen (8, Albers, 410).

[169] Augustine's conversion is related in *Confessions* VIII, 12.29. For an analysis of this episode, see Pierre Labriolle, "Source chrétienne et allusions païennes de l'épisode du 'Tolle, lege'," *Revue d'Histoire et de Philosophie Religieuse* 32 (1952): 170-200.

[170] *Ad inquisitiones Januarii, Ep.* LV, 18.37, *PL* 33: 222.

[171] *HF* 5, 14, *MGH SRM* 1: 212. Gregory himself had previously received intimation of the ruin coming to Merovech through opening the *Book of Solomon* (*ibid.*, 210). See also *ibid.*, 4.16: 149-150; Sulpicius Severus, *Vita Martini,* 9.5, *SC* 133: 273, and Fontaine's commentary on this episode, *SC* 134: 654-656. These and other examples of divination through Scriptures are discussed in Labriolle, "L'enfant et les 'sorts bibliques'," *Vigiliae Christianae* 7 (1953): 194-220. For further references, see Poulin, "Entre magie et religion," 130-136.

seldom mentioned in our texts, never in precisely that form and always associated with other types of divination. The earliest condemnation is found in a Carolingian edict. It forbade the use of the *Psalter* and the *Gospel* (along with tables and codices and "other things") for the purpose of casting lots or of divination.[172] The roughly contemporary *Homilia de sacrilegiis* denounced those who expected to learn their future from the indications of Sacred Writ. The tenth- or eleventh-century *Penitential of Arundel* imposed a penance of forty days for divination or lot-casting from the Gospel, other books or holy objects (*sanctuaria*).[173] By contrast, the *sortes sanctorum* were condemned in repeated denunciations, first by church councils and then by penitentials. These *sortes* were drawn from collections of prophecies, advice and admonitions, one of which was known in the 5th or 6th century as the "the book called the lottery of the apostles;"[174] passages were selected randomly, for example, by rolling dice or flipping open the booklet by means of strings attached to the pages.[175]

The Council of Vannes (461-491) was the first to condemn the *sortes sanctorum* as a grave nuisance to the Catholic faith: certain clerics studied omens and, under the guise of a made-up religion, practiced divination by "the lottery of the saints;" they pretended to foretell the future from inspecting "whatever kind" of written documents. The writings in question were not biblical, for a church council would hardly speak so dismissively of Holy Writ.[176] The Council of Agde (506) extended the ban to include laymen as well as clerics in terms repeated in English, Irish and Frankish texts well into the 9th century.[177] The Council of Orleans (511) used a different formula for the same practice, but condemned both those who practiced the art and those who merely believed in it; this proved less popular with compilers.[178] In the second half of the century, the churchmen meeting at Auxerre

[172] *Duplex legationis edictum* (789) 20, *MGH CapRegFr* 1: 64. See also Regino of Prüm, *De synodalibus causis* (c. 906) Appendix III, 55: 487; Burchard of Worms, *Decretum* (1008-1012) 10. 26, *PL* 140: 836-837, and 19, 5.67, Schmitz 2: 425.

[173] *Homilia de sacrilegiis* (late 8th century) 8, Caspari, 7; *Poen. Arundel* (10th/11th century) 96, Schmitz 1: 463.

[174] The earliest reference in juridical texts is contained in the so-called *Decretals* of Gelasius, which date from the late 5th or early 6th century (see *Oxford Dictionary of the Christian Church, s.v.* "Decretum Gelasianum").

[175] More direct approaches to the saints for advice seem to have been less successful. Chilperic had tried to get St. Martin of Tours to write him a letter about expelling Guntram, who had taken sanctuary at Tours, but the saint remained obstinately silent (*HF* 5.14, *MGH SRM* 1: 212).

[176] *Conc. Veneticum* (461-491) 16, *CCSL* 148: 156)\.

[177] *Conc. Agathense* (506) 42, *CCSL* 148: 210-1. See also *Collectio Hibernensis* (late 7th/early 8th century) 64.4, H. Wasserschleben, 231; *Liber de remediis peccatorum* (721-731) 1, Albers, 410; *Collectio Vetus Gallica* (8th/9th century) 44.3, Mordek, 522; *Poen. Egberti* (before 766) 8.1, Schmitz 1: 581; *Poen. Quadripartitus* (9th century, 2nd quarter) 282, Richter, 32; *Capitulary of Radulph of Bourges* (between 853 and 866) 38, *MGH CapEp 1:* 262-263; Rabanus Maurus, *Poenitentium Liber ad Otgarium* (842-843) 123, *PL* 112: 1417; *idem*, *Poenitentiale ad Heribaldum* (c. 853) 31, *PL* 110: 491-492.

[178] *Conc. Aurelianense* (511) 30, *CCSL* 148A: 12. See *also Collectio Frisingensis Secunda* (late 8th cen-

once again condemned it together with a variety of other divinatory practices.[179]

Church councils lost interest in the *sortes sanctorum* after the 6th century although the old canons continued to be repeated in collections, and Isidore of Seville included them in the *Etymologies* as the specialty of the *sortilegi*. But the authors of penitentials, ever conservative, continued faithfully to record the practice. Over twenty mentioned it, usually in some variation of a form that first appeared in the 8th century: "If anyone invokes the lots which are irrationally called the *sortes sanctorum*, or who makes use of or venerates other forms of lots or who soothsays using any other device, he shall do penance for three years, one of them on bread and water."[180]

It is impossible to determine whether variations were deliberate or merely the fruit of scribal error. Some are insignificant–for example, "have," "divine" or "augur" instead of "venerate." But occasionally the changes raise questions. When the eighth-century *Poen. Oxoniense I* replaces "venerate" with "generate," it is possible that the author was protesting against the production of these booklets (although in that case one would expect a harsher penalty than the usual three years).[181] The copyists of two penitentials may have had in mind a particular work known as "the soothsayer's lottery of the saints" (*sortes sanctorum arioli*), since *ariolus* does not appear in this context anywhere else.[182] The same may be true of the "the lottery of the saints whom they call the patriarchs or apostles."[183] Toward the end of our period, one penitential seems to suggest that hunters and fishermen used the *sortes sanctorum* and other magical devices (to determine time, place, implements?): "Si quis sortes sanctorum ... fecerit, aut aliquid mala sorte venatus aut piscatus fuerit. ..."[184] Quite possibly *venatus* (hunted) is a mistake for *veneratus*, with *piscatus* (fished) added for good measure. But it may also be that the *sortes sanctorum* was no longer a form of divination at all, but a charm or phylactery, with a blessing, curse or prophecy written on it. They could be tucked into the bosom or pinned over one's heart "for either good or evil" as, in fact, they are explained in a gloss to the *Poen. Valicellanum I* and in an early eleventh-century collection.[185]

tury) 74, Mordek, 628; *Collectio Vetus Gallica* (8th/9th century) 44.4, Mordek, 523; *Poen. Quadripartitus* (9th century, 2nd quarter) 369, Richter, 40.

[179] *Syn. Autissiodorensis* (561-605) 4, *CCSL* 148A: 265.

[180] *Poen. Remense* (early 8th century) 9.4, Asbach, 56. Also in six other 8th-century penitentials, in the *Poen. Valicellanum II* (10th/11th century) 58, Schmitz 1: 379 and in an anonymous 9th-century penitential (*PL* 105: 722).

[181] *Poen. Oxoniense I* (8th century, 1st half) 23, *CCSL* 156: 90.

[182] *Poen. Merseburgense a,* 3rd recension (late 8th century) 107, *CCCM* 156: 169; *Poen. Vindobonense a,* 97, Wasserschleben, 422 (the Schmitz ed. of the latter does not include this clause).

[183] *Poen. Arundel* (10th/11th century) 91, Schmitz 1: 462.

[184] *Poen. Parisiense I* (late 9th century or later) 20, Schmitz 1: 684.

[185] *Poen. Valicellanum I* (9th century, beginning) 111, Schmitz 1: 327. See also *Collectio Canonum in V Libris* (1014-1023) 3.328, *CCCM* 6: 488.

Techniques of divination–from other writings: Similar divinatory techniques were practiced with miscellaneous documents. The *Epítome hispánico* included writings (*scripturae*) along with altar vessels and names as means for "detecting" information.[186] Charlemagne's edict against divination banned the use of unspecified works (*tabula*, *codices*) as well as the psalter and gospels.[187] There is also a brief, confusing reference in the so-called *Statutes of St. Boniface*: "If any priest or cleric practices auguries ... or lots or phylacteries, that is, scriptures ...." Here, "scriptures" appears to be in apposition with both lots and phylacteries.[188] The *Homilia de sacrilegiis* condemned as sacrilegious persons who believed "those useless writings usually called the lots of the saints."[189] A little later, Arno of Salzburg urged the diocesan clergy to admonish the faithful not to look to false writings (*falsae scripturae*) for help in times of crisis.[190]

6.3 Healers, Healing and Protection[191]

6.3.1 *Specialists*

Pastors were concerned about cunning folk who used magical means, especially amulets and verbal charms (spells), not about legitimate doctors and their techniques, although these were in fact hardly less dependent on magical means.[192] The former and their arts are a constant theme in pastoral literature throughout our period. With the usual exception of the *Collectio Hibernensis*, which borrowed from Caesarius of Arles for its single reference to phylacteries,[193] no mention is made of healers, amulets or any kind of spell, either beneficent or malevolent, in our Irish texts.

The principal terms for cunning folk engaged in healing and protection are

[186] 4. 13: 114.

[187] *Duplex legationis edictum* (789) 20, *MGH CapRegFr* 1, 64. See also Regino of Prüm *De synodalibus causis* [906] Appendix III, 55: 487; Burchard of Worms, *Decretum* (1008-1012) 10.26, *PL* 140: 836-837 and 19.5.67, Schmitz 2: 425.

[188] 20, *MGH CapEp* 3: 363. See also *Capitula Vesulensia* (1st half, 9th century) 33, *ibid.*, 351.

[189] 7, Caspari, 7.

[190] *Synodal Sermon* (c. 806), Pokorny, 393-394.

[191] See Hand, *Magical Medicine*; Hugues Breton, *Médecine traditionnelle et magique en milieu rural* (Châteauguay, 1987); C. S. Myers *et al.,* "Disease and medicine," in *ERE* 4: 723-772. For the state of medicine in Merovingian Gaul, see Caterina Lavarra, "Il sacro cristiano nella Gallia merovingia tra folklore e medicina professionale," *Annali della Facoltà di Lettere e Filosofia*, Bari (1989): 149-204; here, 183-204. For Anglo-Saxon folk medicine, see Bonser, *The Medical Background of Anglo-Saxon England;* Jolly, *Popular Religion in Anglo-Saxon England* and "Anglo-Saxon charms in the context of a Christian world view," *Journal of Medieval History* 11 (1985): 279-293; M. L. Cameron, "Anglo-Saxon medicine and magic," *Anglo-Saxon England* 17 (1988): 191-215. For Old German healing and protective charms, see K. A. Wipf, "Die Zaubersprüche im Althochdeutschen," *Numen* 22 (1974): 42-69, and Edwards, "German vernacular literature," 163-169.

[192] See Thorndike, *History of Magic and Experimental Science* 1: 566-615 and *passim.*

[193] 64.2, H. Wasserschleben, 231.

praecantator and *incantator*; *praedicator* (probably a variation of *praecantator*) and *suffitor* are also used.[194] *Praecantator* and *incantator* seem to be used interchangeably and, as they never appear together in the same passage, they were probably often considered to be synonymous.[195] Nonetheless, a difference in function between the two practitioners and their arts may be detected: the *praecantator* and *incantator* can perhaps be said to represent two aspects of the same enchanter. Invariably, when any information is given, the *praecantator* is connected with healing and protection against disease and other ills. *Praecantare* and *praecantatio* carry equally positive meanings. The *incantator*, on the other hand, was ambiguous–his charms and *incantationes* could be used either to heal and protect or to harm. The uncanny aspects of the *incantator* and his art will be examined in the next chapter; here he is considered only in his character of specialist in healing and protective magic.

Praecantator/praedicator: The *praecantator* appears to have been much sought after in sixth-century Provence. Caesarius of Arles mentions him some fifteen times (only *divini* outnumber them in his sermons), usually in a list of cunning men, but often standing slightly apart from the soothsayers who dominate the lists: "No one should summon *praecantatores*, no one should question [or seek out]" various soothsayers is typical.[196] His role was not to interpret omens, foretell the future and diagnose diseases, but actually to cure sicknesses–he seems to have been particularly efficacious in cases of snakebite.[197] Since Caesarius frequently mentioned him in the same breath as *phylacteria*, *ligaturae*, *caracteres* and *herbae*[198] (on one occasion summarised as *praecantationes*),[199] we can assume that amulets were his stock in trade.

Caesarius' successors paid less attention to the *praecantator*. References to him are found in about fifteen passages from XVI Toledo (693) to Burchard of Worms, from both the continent and England. He appears alone in one.[200] Otherwise he is invariably grouped with soothsayers. As noted above, three documents

[194] Hand found that in folk medicine the "gift of healing" was acquired through a specially conferred gift (*e.g.*, from God), by an innate quality (*e.g.*, birth order) or because of "some unique condition" (*e..g.*, nudity) (*Magical Medicine*, 43-56).

[195] Du Cange (*s. v.*) defines *praecantator* as "enchanter;" he gives no definition specifically for *incantator* but *incantare* is defined as "to trick magically." For Blaise, *praecantator* is enchanter, magician and healer, while *incantator* is enchanter, magician, sorcerer or snake-charmer (*s.v.v.*). Neither was current in classical Latin, but St. Augustine used the former (*e.g., In evangelium Iohannis tractatum*), and the Vulgate, the latter (*Deut* 18 11 and *Is* 47 12). Isidore of Seville defined *incantatores* as those who used words to achieve their ends (*Etymologiae* VIII, 9.15: 714), but none of our texts associates the magic of either *incantatores* or *praecantatores* exclusively with verbal charms.

[196] *E.g.*, S. 1.12, *CCSL* 103: 8.

[197] *S*. 54.3, *CCSL* 103: 237.

[198] *E.g.*, *S*. 14.4, *CCSL* 103: 71-2; *S*. 50.1, *ibid.*, 225; *S*. 204.3, *CCSL* 104: 821.

[199] *S*. 52.5, *CCSL* 103: 232.

[200] Rabanus Maurus (d. 856) *Hom. 16: In Sabbato Sancto Paschae*, *PL* 110: 34.

mention *divini praecantatores*, indicating perhaps that some practitioners combined the arts of diagnosis and healing. St. Eligius' sermon connects *praecantatores,* like soothsayers, with help in time of illness or other problems, but elsewhere their specialty is no longer made explicit.[201] Nevertheless, it seems probable that their reputation continued, for the name is associated with amulets in seven of these lists.[202]

In two texts, the usual *praecantatores* is replaced by a new term. An anonymous preacher of the 8th century warned his flock against frequenting soothsayers or *praedicatores* in the case of illness in their homes "because he who does so goes to the devil."[203] In another contemporary penitential, *praedicatores* are found in the midst of soothsayers and herbal practitioners.[204]

Incantator, incantatrix: Caesarius used the term *incantatores* only twice, in lists where normally he put *praecantatores*; it presumably had the same meaning for him.[205] Arno of Salzburg inserted *incantatores* among the wicked men and women whose advice was sought in case of sickness in humans or animals, and Rabanus Maurus simply took one of Caesarius' texts and substituted *incantator* for *praecantator.*[206]

Caesarius' use of the word "enchantresses" (*incantatrices*) is more unexpected and may be more significant:

> Now and then the devil ... injures various members of the body so as to cause a headache or bring on some other ailment. Then indeed the Christian proves himself if he does not forsake God during that persecution, if he does not send for a soothsayer, if he does not make bindings, he does not admit any enchantresses. The woman enchants, the serpent enchants.[207]

The context plainly is sickness, but the last sentence suggests that in this case *incantatrices*' services were not limited strictly to medical practice. Indeed, when they appear in two Carolingian documents, they are not associated with healing.

Suffitor: *Suffitores,* as we have seen, appear in an anonymous Carolingian letter (c.814), in a difficult passage concerning those who mime the doe and the yearlings, "that is, *suffitores*" and who put spells on horns.[208] The equation of *suffitores*

[201] *Vita Eligii* (c.700-725), *MGH SRM* 4: 705.

[202] That is, "amulets" are mentioned either directly before or after *praecantator; e.g., Dicta Pirmini* (724-753) 28: 188; *Poen. Egberti* (before 766) 8.4, Schmitz 1: 581.

[203] *Anonymous sermon* (8th century), Morin, 518.

[204] *Poen. Hubertense* (8th century, 1st half) 25, *CCSL* 156: 111.

[205] *S.* 189.2 and *S.* 197.2, *CCSL* 104: 772 and 796.

[206] *Synodal Sermon* (c. 806), Pokorny, 393-394; Rabanus Maurus (d.856), *Hom.* 43, *PL* 110: 81).

[207] *Les épreuves du juste*, Étaix, 275. This and a single mention of a herbwoman are Caesarius' only references to women who practiced the forbidden arts professionally.

[208] 5, Mansi 13: 1095.

with maskers is highly improbable. Conceivably the word is a mistake for *sifilatores*, but if so, this is the only reference to whistling in our texts. If, on the contrary, *suffitores* is correct, the copyist must have dropped a line, and the meaning is fumigators.[209] Caesarius of Arles spoke of an expert with a reputation for fumigation, and in the late 8th century, smoke was a treatment for possession by the devil.[210] The *suffitor*, therefore, may have been a specialist in healing physical and spiritual ills.

6.3.2 *Techniques*[211]

Christians could look to two permissible sources of help in time of sickness: God, Whose help was invoked by the Creed and the Lord's Prayer and came through the application of holy oils or the relics of the saints, and medical science. For Caesarius of Arles, the former was far preferable, but he was prepared to tolerate the latter if it kept his flock from forbidden methods, such as those used by the anxious mothers of sick children:

> They do not request medicine from the Church nor Christ's Eucharist from the Author of salvation and, when, as it is written, they should anoint themselves with oil blessed by the priest and place all their hope in God, they do otherwise; while they are seeking health for the body, they find death for the soul. Oh, would that they would seek that health, even from the straightforward skill of the doctors![212]

Far from resigning themselves to God's will or the doctors' skill, people often took matters into their own hands for themselves and their children, and used techniques that were magical, both in their own eyes and in those of churchmen. Amulets and spells were seemingly the most popular, but the afflicted also resorted to various folk remedies and non-medical techniques. Our texts undoubtedly record only a few of the many available; the rest our authors either did not know or lacked the inclination to record. Near the end of our period, the author of the Frankish *Penitential of Arundel*, confronting the same situation as had Caesarius at the beginning, virtually acknowledged as much:

> Let anyone who has put his child on the roof or in the oven to bring him back to health, or dragged him through a hole dug in the ground, or has done anything of the sort, or has committed himself to charms (spells?–*carmina*),

[209] Quicherat and Daveluy, *s.v.*; Blaise, *s.v.*, defines it on the basis of this text as those who disguise themselves as beasts. In Pliny (*Historia Naturalis* 34.19), the sculptor Lycius is said to have made a statue of a *puer suffitor*, which Rackham translates as "a boy burning perfume" (186-187).

[210] Caesarius of Arles (502-542) *S.* 184.4, *CCSL* 104: 750; *Homilia de sacrilegiis* (late 8th century 22, Caspari, 12.

[211] For forbidden magical means and Church-sanctioned methods for healing sickness and diabolical possession, see Adolph Franz, *Die kirchlichen Benediktionen im Mittelalter* (1901; Graz, 1960), 399-615. For folk medicine, see Hand, *Magical Medicine* and *Medicina antiqua*, which may be taken as representative of early medieval learned texts.

[212] *S.* 52.5, *CCSL* 103: 232.

symbols, soothsaying invention or any kind of devilish art, and not to Divine power or the assured liberal art of medicine, do penance for forty days.[213]

Amulets: Among the tools used for healing or for protection against disease, amulets, objects worn or kept to repel misfortune or to ensure good fortune, are mentioned most often, usually as *phylacteria* (*philacteria, fylacteria, filacteria*).[214] In addition to St. Caesarius' sermons, about forty texts mention them explicitly, mostly without detail beyond the occasional addition of Caesarius' epithet "diabolical." Sometimes *ligaturae* and *ligamina* (knots, bindings, threads or amulets attached with ligatures) seems to have the same meaning–Caesarius used the terms interchangeably. In one Carolingian text, ligatures and "lying writing" appear to be types of phylacteries, but the *Indiculus superstitionum* treats them as distinct from each other.[215] *Carmen* (normally a song or a spell) is used at least once as a synonym for amulet. For Benedict Levita, these objects all constituted "the signs of magic."[216]

Amulets were readily available. Archaeological evidence suggests that women used them more than men: many more are found in female burial sites than in male, according to studies by Meaney for Anglo-Saxon England and by Salin for the continent.[217] But men used them as well–English and continental sources testify that even clerics relied on them.[218]

Not only did monks and clerics use them, they also produced and dispensed them, thus validating them in the eyes of the faithful. Caesarius warned his flock that such clergy were the devil's henchmen; their amulets contained not Christ's

[213] 97, Schmitz 1: 464.

[214] For amulets during the early Middle Ages, see Audrey L. Meaney, *Anglo-Saxon Amulets and Curing Stones* (Oxford, 1981), esp. 3-15 and 239-273, and Salin, *La civilisation mérovingienne* 4: 49-118. For photographs of amulets, see Ludwig Pauli, "Heidnische und christliche Bräuche," in *Die Bajuwaren von Severin bis Tassilo 488-788*, ed. Hermann Dannheimer and Heinz Dopsch, (Salzburg, 1988), 274-279. For the concepts underlying magic stones, see Maria Grazia Lancellotti, "Médecine et religion dans les gemmes magiques," *Revue de l'histoire des religions* 218 (2001): 427-456. See also B. Frire *et al.*, "Charms and amulets," *ERE* 3: 392-472; W.J. Dilling, "Knots," *ERE* 7, 747-752; H. Leclercq, "Phylactère" (*DACL* 14.1: 806-810); K. Ranke *et al.*, "Amulett," *Reallexikon der Germanischen Altertumskunde* 1: 269-274; E.A.W. Budge, *Amulets and Talismans* (reprint, New York, 1968). See also Mostert, "La magie de l'écrit," Poulin, "Entre magie et religion" and Hartnung, "Die Magie des Geschrieben."

[215] *Benedicti capitularium collectio* (mid 9th century) 72, *PL* 97: 758-759; *Indiculus superstitionum et paganiarum* (743?) 10, *MGH CapRegFr* 1: 223.

[216] *Benedicti capitularium collectio* 72, *PL* 97: 758-759.

[217] Meaney suggests that Anglo-Saxon men may have considered amulets as "women's stuff," and preferred to protect themselves by ornamenting their weapons and armour (*Anglo-Saxon Amulets and Healing Stones*, 240).

[218] 3, Albers, 411. See also the *Poen. Ps.-Theodori* (mid 9th century) 27.24, Wasserschleben, 598; *Double Penitential of Bede-Egbert* (9th century?) 30.3, Schmitz 2: 694-695; Burchard of Worms, *Decretum* (1008-1012) 10.33, *PL* 140: 837-838); *Statuta Bonifacii* (1st half, 9th century) 33, *MGH CapEp* 3: 365.

remedies but the devil's poison–efficacious no doubt in healing the body, but only in order to kill the soul with the "sword of infidelity." One became pagan by making, ordering or even agreeing to them.[219] Under his influence, the Council of Agde forbade priests and clerics to be magicians (*magi*) or enchanters or to make "the things called phylacteries which greatly ensnare souls."[220] The sixth-century *Epítome hispánico* laid down the principle that "no cleric should make incantations (*incantaturae*) or amulets." .[221] Martin of Braga likewise forbade clerics to be magicians or to make *ligaturae* because these are "a binding together of souls."[222] Such prohibitions fell on deaf ears. Many clerics declined to believe that charms (at least those containing holy objects) were evil, and continued to recommend, hand out and manufacture them. Even Gregory the Great sent an amulet to the Lombard Queen Theodolinda, and St. Oyend of the Jura and his monks were well known for their production of efficacious phylacteries.[223]

In the earliest period, these bans targeted only clerics, perhaps because the ready availability of amulets and the Church's incomplete hold over the general population made it futile to try to stop the production of them altogether. Caesarius had urged people not to use the *praecantatores*' wares, but he had never tried to prevent the *praecantatores* from making them. The gradual appearance of injunctions applying to lay people as well as priests seems to signal the increasing confidence of the Church and its determination to bring the laity into line.[224] Nevertheless, even in the mid 8th century, women continued to sell and flaunt amulets publicly in the very heart of Christendom.[225]

In rare cases, the term *phylacteria* was used in its technical sense, to mean recipients containing script or other objects, such as relics and scraps from Holy Writ.[226] Such were the *ligamina* and *cartellae* which an anonymous eighth-century homily described as being bound about the necks of men or animals.[227] St. Boniface's statutes and the Capitulary of Vesoul banned phylacteries, "that is, little writings (*scriptulae*),"[228] while Benedict Levita' cited a canon condemning the "phylacteries, deceptive writings or ligatures" used by foolish people against leprosy and other diseases. Ghärbald of Liège ordered his priests to seek out those who

[219] S. 50.1, *CCSL* 103: 225. See also *Anonymous homily* (8th century), Nürnberger, 44.

[220] *Conc. Agathense*–Additiones (506) 21 (68), *CCSL* 148: 228.

[221] 7.35: 122. This is the only example found of the word *incantatura*.

[222] *Canones ex orientalium patrum synodis* (572) 59, Barlow, 138. See also Rabanus Maurus, *Poenitentium Liber ad Otgarium* (842-843) 123, *PL* 112: 1417.

[223] Poulin, "Entre magie et religion," 136.

[224] *Collectio Vetus Gallica* (8th/9th century) 44.4d, Mordek, 524; *Benedicti capitularium collectio* (mid 9th century) 2.72, *PL* 97: 758. See also *Poen. Ps.-Theodori,* 27.8, Wasserschleben, 596.

[225] Boniface to Pope Zachary, *Ep.*50 (742), *MGH EpSel* 1: 84-85.

[226] Caesarius of Arles (502-542) *S.* 50.1, *CCSL* 103: 225.

[227] Nürnberger, 44.

[228] *Statuta Bonifacii* (1st half, 9th century) 33, *MGH CapEp* 3: 365); *Capitula Vesulensia* (1st half, 9th century) 33, *ibid.*, 351.

wore such objects "inscribed with [or containing] heaven knows what words." Phylacteries bearing writing circled the necks of soothsayers, according to another roughly contemporary capitulary.[229] It has been seen that *sortes sanctorum* were worn in phylacteries in one's bosom or over one's heart to bring about good or avert evil.[230] "Solomonic writings" and *characteres* (symbols, perhaps runes), according to a difficult passage of the *Homilia de sacrilegiis*, were inscribed on a document or parchment or thin pieces of copper or bronze, iron or lead, or on "something Christian," and tied around the necks of people or of dumb brutes.[231] Written charms together with incantations were particularly effective against ague. Amulets against this complaint were inscribed with angelic or Solomonic texts (or names?) or symbols and hung from the neck.[232] The "letters for ague" in a mid ninth-century list of forbidden practices may also have been worn as protective charms.[233]

But other types of amulets were probably much more common. Caesarius himself had spoken of the custom of hanging *characteres*, herbs and amber on oneself or one's family and possessions; his words echoed into the 11th century.[234] St. Eligius admonished his hearers against hanging bunches of herbs on their own necks or on those of animals, and warned women in particular against wearing amber.[235] The clause in the *Indiculus superstitionum* "concerning *petendo* (perhaps a mistake for *pendente*, 'pendant') which good people call St. Mary's" refers, suggest McNeill and Gamer, to amulets made from a plant called St. Mary's bedstraw.[236] According to the *Homilia de sacrilegiis*, people tied the antlers and hide of stags on themselves to drive off serpents.[237] In ninth-century Tours, bindings of bones and herbs were thought to cure sick people or sick, lame or dying beasts; the clergy were urged to warn the faithful that these and other types of magic were of no

[229] *Benedicti capitularium collectio* (mid 9th century) 72, *PL* 97: 758-759; *Capitulary 2 of Ghärbald of Liège* (802-809) 10, *MGH CapEp* 1: 29; *Capitula Silvanectensia prima* (9th century, 1st half) 11, *MGH CapEp* 3: 82.

[230] *Poen. Valicellanum I* (9th century, beginning) 111, Schmitz 1: 327. See also *Collectio Canonum in V Libris* (1014-1023) 3.328, *CCCM* 6: 488.

[231] 19, Caspari, 11. See *ibid.*, n. 2.

[232] *Homilia de sacrilegiis* (late 8th century) 15, Caspari, 9-10.

[233] Herard of Tours, *Capitula* (858) 3, *PL* 121: 764.

[234] Caesarius of Arles (502-542) *S.* 13.5 and *S.* 14.4, *CCSL* 103: 68 and 71-72; *Liber de remediis peccatorum* (721-731) 3, Albers, 411; *Dicta Pirmini* (724-753) 22: 173; *Poen. Ps.-Gregorii* (9th century, 2nd quarter) 27, Kerff, 183-184; Burchard of Worms, *Decretum* (1008-1012) 19, 5.92, Schmitz 2: 429, and 10.33, *PL* 140: 837-838. "Characteres" is missing from the *Double Penitential of Bede-Egbert* (9th century?) 30.3, Schmitz 2: 694-695. The 8th-century *Poen. Egberti* (8.4, Schmitz 1: 581) substituted "misdeed" (*facinus*) for "amber" (*sucinum*) in this passage, in Schmitz's opinion, by mistake.

[235] *Vita Eligii* (700-725) *MGH SRM* 4: 706. See also the anonymous 8th-century homily, Nürnberger, 44.

[236] 19, *MGH CapRegFr* 1: 223. See McNeill and Gamer, *Medieval Handbooks of Penance*, 420, and n., but Homann considers this interpretation to be unlikely (*Der Indiculus*, 108-110).

[237] 21, Caspari, 11.

avail.[238] But some amulets were acceptable: the *Penitential of Theodore* allowed "stones and vegetables" (gems, such as jet or topaz, and roots, dried leaves or seeds?) to those suffering from diabolical possession, so long as no incantations were used. About ten eighth-century texts and Burchard's *Decretum* repeated the permission.[239]

Other writers are less informative. In the early 8th century, Pirmin of Reichenau merely urged his Alemannian flock neither to trust in a diabolical charm (*carmen*) nor to put it over themselves.[240] More than a hundred years later, Rabanus Maurus asked Rhenish godparents to warn their charges against amulets and symbols and the cunning folk who, presumably, made them.[241]

Amulets were generally worn on the person. Where *phylacteria* or a similar word is accompanied by a verb, the most common throughout our period imply that they were hung, usually around the neck, or tied on: *adpendere* (the word preferred by Caesarius of Arles) *impendere, suspendere, ligare, portare ad/circa collum* appear in about fifteen passages. Among these were written charms against fevers and serpent's tongues worn on the neck.[242] Roman women, according to St. Boniface, wore them openly on bracelets and anklets. By contrast, an amulet worn to foil the Judgment of God was hidden carefully. Regino of Prüm was the first to take note of this custom, but it was Burchard who described it in detail: they were made "of herbs, [written] words, wood, stone or any [other object of] stupid confidence;" they were kept in the mouth, sown into one's clothing, tied around one, or applied by "some other trick."[243]

Amulets may have been put in trees, or on roofs, road signs or other structures. In Reichenau, people placed them "over" themselves.[244] St. Eligius's sermon spoke of a healing practice involving the use of phylacteries by wells, trees and crossroads.[245] Three continental penitentials warn against hanging ligatures made from herbs or some other object enchanted by "wicked means."[246]

It is not always clear how the amulets were used. Verbs like "use" (in five texts) and "have," "observe" or "practice" (in two each) are not informative. Cer-

[238] *Conc. Turonense* (813) 42, *MGH Concilia* 2.1: 292. See also *Benedicti capitularium collectio* (mid 9th century) Add. 3.93, *PL* 97: 882-883. In Burchard's *Decretum*, the reference to the ligatures of bones and herbs was left out (10.39, *PL* 140: 839).

[239] II, 10.5, Schmitz 1: 544.

[240] 22: 176.

[241] *Hom. 16: In Sabbato Sancto Paschae*, *PL* 110: 34.

[242] *Homilia de sacrilegiis* (late 8th century) 15, Caspari, 9-10. Meaney suggests that the "serpent's tongue" may have been a fossilised shark's tooth (*Anglo-Saxon Amulets*, 12).

[243] Regino of Prüm, *De synodalibus causis* (c. 906) II, 5.50, 213; Burchard of Worms, *Decretum* (1008-1012) 19, 5.167 Schmitz 2: 445).

[244] *Dicta Pirmini* (724-753) 22: 176.

[245] *Vita Eligii* (c.700-725), *MGH SRM* 4: 707.

[246] *Poen. Merseburgense a* (late 8th century) 36, *CCSL* 156: 136. See also the *Poen. Valicellanum I* (early 9th century) 89, Schmitz 1: 312; *Poen. Vindobonense a* (late 9th century) 39, Schmitz 2: 353.

tainly they were held in great confidence, since four texts require people to put no trust in them. The strongest testimony comes from Pirmin. In a passage listing all manner of pagan customs and in which he had already twice mentioned different kinds of amulets, he gave them the place of honour in a final charge: "Do not believe in every diabolical phylactery nor in the other things that I have mentioned, nor adore them, nor fulfill vows to them, nor pay them any honour."[247] But here phylacteries may be small idols or fetishes, rather than talismans meant to be worn.

Spells: Spells, the verbal charms that Wilfrid Bonser has called the "liturgy of magic,"[248] were used alone or together with other devices. Three groups of terms are used: *praecantatio* and *praecantare*, *carmen* (poem) and associated words (*carminare*, *carmen cantare*), and *incantatio* and *incantare*. The first is not found in combination with the others, but *carmen* and *incantatio* sometimes appear together; they are, therefore, not necessarily synonymous.

Praecantatio and *praecantare* have no connotations of harmful magic. A text attributed to a mid seventh-century Council of Châlons and repeated by Burchard condemns those, practitioners and their clients, who used protective spells against the evil eye or some other form of bewitchment (*praecantare ad fascinum*), or those who chanted or had others chant any *praecantationes* except for authorised prayers.[249] Here the magic is clearly verbal, but in other cases, *praecantationes* may have stood for other types of protective techniques as well.[250] A sixth-century synod at Eauze in southwestern Aquitania condemned *incantatores* and those who were said to put spells (*praecantare*) on horns (gestures or charms against the evil eye? amulets made of deer horn against snakebite? instruments used in war or in the hunt or to frighten away the monster attacking the moon? drinking horns?).[251] A north Italian penitential imposed a penance on men and women who either practiced divination or *praecantationes* themselves or ordered them to be performed–indication that *praecantationes* sometimes required professional expertise.[252]

Carmen and *incantatio* are more ambiguous. *Carmen* is sometimes an object: Pirmin used it to mean charm, *i.e.*, amulet, and this may be its meaning in the *Penitential of Arundel* also.[253] Likewise, the *incantatura* of the *Epítome hispánico* is probably

[247] 22: 176.

[248] *The Medical Background of Anglo-Saxon England*, 117.

[249] 10, Mansi 10: 1197; see also Burchard of Worms, *Decretum* (1008-1012) 10.49, *PL* 140: 851. For the use of amulets against the evil eye, see Daremberg and Saglio, *s.v.* "Fascinum." On Christian teaching as to the malign power of envy, see Matthew W. Dickie, "The Fathers of the Church and the evil eye," in *Byzantine Magic*, ed. Henry Maguire (Dumbarton Oaks, 1995), 9-34.

[250] *Poen. Floriacense* (late 8th century) 42, *CCCM* 156: 100-101; *Poen. Ps.-Gregorii* (2nd quarter, 9th century) 27, Kerff, 183-184.

[251] *Synodus Aspasii Episcopi* (551) 3, *CCSL* 148A: 164.

[252] *Poen. Oxoniense II* (8th century) 24, *CCSL* 156: 195.

[253] 97, Schmitz 1: 464.

an amulet fortified by a cleric's spell.[254] In some cases, as with the clause "concerning incantations" of the *Indiculus superstitionum*, they may be spells, enchanted objects or enchantment in general.[255] Elsewhere, they are verbal charms or formulas combined with some magical act. Of themselves, they are neutral, and only from the context can it sometimes be determined if they are benevolent or malign. *Incantationes* are undoubtedly benevolent spells when churchmen propose to replace them by "incantations" like the Creed or the Lord's Prayer or the Sign of the Cross, as did Martin of Braĝa in *De correctione rusticorum*.[256] Elsewhere Martin specified that these were the only permissible *incantationes* to be used while collecting medicinal herbs. The same ordinance was repeated more than half a dozen times in continental penitentials and canonic collections from the 9th century onward.[257] An anonymous eighth-century sermon is exceptional in forbidding any kind of incantation over herbs and (uniquely) trees. Perhaps what is intended here is hunters' and herdsmen's well-known practice of putting enchanted herbs or bread in trees (see below).[258]

Carmina and *incantationes* are benevolent in the *Homilia* when they were added to the authorised prayers chanted over animals and men, or when they were recited or sung (*carminari*) against snakebite and worms in vegetables or other crops.[259] Such beneficent *carmina* or *incantationes* were used against a host of ills: *ad fascinum* (evil eye), *ad spalmum* (apparently a mistake for *spasmum*, cramps), *ad furunculum* (boils), *ad dracunculum* (canker sores), *ad aluus* (diarrhea), *ad apium* (bee sting), *ad uermes, id est lumbricos, que [in] intrania hominis fiunt* (worms, "that is, the intestinal worms which get into men's guts"), *ad febres* (fevers), *ad frigurae* (ague), *ad capitis dolorem* (headache), *ad oculum pullinum* ("foals' eye"–walleye?), *ad inpediginem* (impetigo?), *ad ignem sacrum* (rash[260]), *ad morsum scvrpionis* (scorpion's sting), and *ad pullicinos* ("chicks?"–possibly a mistake for *ad pediculos*, lice).[261] The same sermon puts *incantationes* among the therapeutic means in cases of diabolic possession.[262] Their popularity is underscored by the insistence of the *Parisian Penitential* that demoniacs

[254] 7.35: 122.

[255] 12, *MGH CapRegFr* 1: 223.

[256] *De correctione rusticorum* 16, Barlow, 199-200. two *Anonymous sermons* (10th/11th and 12th centuries), ed. Caspari , *Kirchenhistorische Anecdota* 1: 200-201 and 207, contain an abbreviated version of this.

[257] *Canones ex orientalium patrum synodis* (572) 74, Barlow, 141. An indication of how permissible incantations were performed can be found in Ratherius of Verona's *Praeloquia* I, 24 (*CCCM* 46A), trans. and comm. Giovanni Polara, "Il racconto del servo che guarisce il giovane epilettico nel I libro dei 'Praeloquia' di Raterio," in *Raterio da Verona* (Todi, 1973), 187-195.

[258] Nürnberger, 44.

[259] 14, Caspari, 9.

[260] *Ignis sacer*: skin diseases, such as erysipelas or herpes in humans, anthrax in animals. Since this list deals with relatively minor ailments, anthrax seems ruled out.

[261] 15, Caspari, 9-10.

[262] 22, Caspari, 12.

must manage without recourse to incantations.[263]

The priests of Tours were required to teach their flocks in the early 9th century that incantations were useless against the miseries afflicting humans and animals.[264] But Ps.-Egbert evidently thought that they were effective on occasion, since he condemned English wives for using them, apparently successfully, in order to conceive.[265]

Dirges and lamentations over the dead can be viewed as protective *carmina* as well. The Council of Toledo of 589 forbade them absolutely at the funerals of clerics although it recognised the difficulty of preventing them altogether.[266] They reappear in Regino of Prüm and Burchard of Worms' collections, usually as being performed during a nocturnal wake.[267]

Carmina present a more ambiguous aspect in cowherds, swineherds and huntsmen's practice of reciting "diabolical spells" over bread, herbs and "certain wicked bands" which were then hidden in trees or flung in the crossroads. Here the spells had a dual role: they protected the spellcasters'own beasts from pestilence and disease and at the same time brought ruin on the herds of others.[268] This practice, first described as "beyond doubt idolatry" by a putative seventh-century Council of Rouen, was recorded by Regino of Prüm and several times by Burchard., who introduced additional elements in his penitential. He extended the practice to other folk, implied that hunters were less given to such practices than cowherds and swineherds, and at the same time made clear that ligatures and spells were only a part of the rite. "Did you," he asks, "make knots and incantations and perform *those varied magical acts* [*fascinationes*] which wicked men, swineherds or cowherds and *now and then* hunters perform while they are saying diabolical spells...."[269] Such acts could involve amulets, poisoning the bread, laying a snare or baiting a trap, the success of which depended on the efficacy of the spells used. In an abbreviated form found in the Arundel Penitential, these proceedings have only protection from disease as their goal; the attack on the property of others is missing.[270]

[263] *Poen. Parisiense I* (late 9th century or later) 14, Schmitz 1: 683.

[264] *Conc. Turonense* (813) 42, *MGH Concilia* 2.1: 292. See also Burchard of Worms, *Decretum* (1008-1012) 10.39, *PL* 140: 839.

[265] *Confessionale Ps.-Egberti* (c.950-c.1000) 4, *PL* 89: 426.

[266] *Conc. Toletanum III* (589) 22, Vives, 132-133.

[267] *E.g.*, Regino of Prüm, *De synodalibus causis* (c. 906) II, 5.55: 213.

[268] *Conc. Rothomagense* (650) 4, Mansi 10: 1200. See also Regino of Prüm, *De synodalibus causis* (c. 906) II, 5.44: 212 and Burchard of Worms, Decretum (1008-1012) 1.94, 43 and 10.18, *PL* 140: 576 and 836. Czarnowski reported having frequently seen rags containing "la misère" suspended from branches of trees overhanging boundaries in his native Poland ("Le morcellement de l'étendue," 348-349).

[269] *Decretum* (1008-1012) 19, 5.63, Schmitz 2: 423-424.

[270] 94, Schmitz 1: 463. Mention of specific occupations is rare in pastoral literature. The activities of swineherds and other herdsmen were worrisome since their work was not performed

An eighth-century Frankish council provides another example of a *carmen* which can be viewed as either benevolent or malevolent. It forbade drawing lots before the *wehadinc* (duel) "lest they lie in wait with spells, diabolical tricks or magical arts."[271] This wording suggests that some interested third party took the opportunity to do mischief rather than that the combatants used such means. The *Arundel Penitential* also testifies to the use of magic to determine the outcome of a duel or legitimate ordeal.[272]

Human and animal by-products: In addition to approved medicines and enchanted herbal potions, folk medicine[273] provided an arsenal of other types of remedies. Important among these were preparations made from human and animal by-products.[274] Prohibitions against blood and polluted food and drink are common, especially in penitentials. Often they are made on the authority of the Bible, but other times they appear to be based on the idea that people considered these substances to have magical powers as remedies. Theodulph of Orleans' *Capitulary 2* of 813 distinguished between the bewitched objects and the unclean things that were drunk for medical purposes, but elsewhere it is difficult to know if the authors differentiated between the hygienic and the magical aspects.[275]

The first notices of this type come from Irish and English texts of the 7th century in prohibitions echoed and amplified in insular and continental penitentials down to the end of the millennium. The *Irish Canons* called for lengthy fasting and the imposition of the bishop's hands as penance for drinking blood or urine; later penitentials show that these could be of either human or animal origin.[276] In a text repeated some dozen times, the *Penitential of Cummean* imposed a lighter penance for "eating one's own skin, that is, scurf, or the little worms known as lice, not to mention drinking one's own urine or eating excrement."[277] In the Rhineland of Burchard's day, scurf, urine and excrement were used for a variety of ailments,

under the eyes of the parish priest and provided them with an excuse for neglecting their normal religious duties. A Rhenish or Lotharingian *Admonitio synodalis* of c. 813, repeated by Burchard, urged priests to insist that such men assist at Sunday Mass. Burchard also adds other men whose work takes them away from the village and who "live like beasts" in the fields and forests (78, Amiet, 64; *Decretum* 1.94, Interrogatio 63, *PL* 140: 577; *idem* 2.71, *PL* 140: 638.

[271] *Conc. Neuchingense* (772) 4, *MGH Concilia* 2.1: 100.

[272] *Poen. Arundel* (10th/11th century) 78, Schmitz 1: 457.

[273] See Hand, *Magical Medicine.*

[274] For internal and external application of such substances as remedies, see Pliny, *Natural History*, Bks 28-32.

[275] 19, *MGH CapEp* 1: 178. All "diabolical remedies" were denounced in the *Ordo de catechizandis rudibus* (796) 63, Bouhot, 223-4.

[276] *Canones Hibernenses* (mid 7th century) 1.12, Wasserschleben, 137; *Poen. Bigotianum* (late 8th-late 9th century) I,5.1, Bieler, 216; *ibid.*, I, 5.3: 216. See also the *Old Irish Penitential* (c. 800), which imposes a penance for eating horse-flesh and for drinking the blood of a cat (1.2 and 1.21, Binchy in Bieler, 259 and 261).

[277] 18, Bieler, 128.

lice for diarrhea.[278] "Some little thing" (*aliquid paruum*–perhaps a mistranscription for *aliquid prauum,* that is, "something wicked") drunk "with incantation" in the *Homilia de sacrilegiis* may be this sort of substance.[279] This practice is often found in midst of dietary taboos (although not in Cummean's Penitential) and might be taken for a hygienic regulation. It is only at the end of our period that the use of these remedies is lumped together with pagan practices and identified explicitly as sacrilegious.[280]

The *Penitential of Theodore* and seven penitentials of the 8th and 9th century imposed three years of penance for drinking semen and blood.[281] Some twenty imposed the relatively light penalty of forty days or less on women who tasted their husband's blood as a remedy. In a continental penitential, the woman "mixed" her husband's blood before using it. Another differentiates between "tasting" one's husband's blood for medical purposes and "mixing" his semen into food to increase his love –possibly, therefore, the blood was sometimes drunk or eaten as it was, and other times combined with food or drink.[282] Women did not necessarily keep this remedy for themselves. A late English penitential recommended forty days of penance for one who gave (betrayed?–*tradit*) her husband's blood as some kind of medication. The maybe inadvertent choice of the word *tradere* evokes the idea of the wife's disloyalty inherent in this act of generosity.[283] Drinking one's own blood was far worse, meriting an unusually harsh penalty (five years) in the same document. The use of the masculine pronoun indicates that here at least the culprit was not invariably a woman.[284] Eating or drinking animal blood was treated as equivalent to the consumption of human blood in another, somewhat earlier English penitential; both required three years of penance. For Burchard, however, the offence was trivial and called for only five days.[285]

Blood could be applied externally; nose-bleed was stopped by putting the blood on the forehead.[286]

[278] Burchard of Worms, *Decretum* (1008-1012) 19, 5.127, Schmitz 2: 437.

[279] 15, Caspari, 10.

[280] *Poen. Arundel* (10th/11th century) 95, Schmitz 1: 463.

[281] *Poen. Theodori* (668-756) I, 7.3, Schmitz 1: 530.

[282] *Canones Gregorii* (late 7th/8th century) 190, Schmitz 2: 541; *Poen. Parisiense I* (late 9th century or later) 91, Schmitz 1: 691; *Poen. Casinense* (9th/10th century) 61, Schmitz 1: 413.

[283] *Confessionale Ps.-Egberti* (c.950-c.1000) 1.31, *PL* 89: 409.

[284] *Confessionale Ps.-Egberti* (c.950-c.1000) 1.16, *PL* 89: 405. The harshness of the penalty is evidently determined by the equation of drinking blood to drinking semen, a practice limited here to women.

[285] *Double Penitential of Bede-Egbert* (9th century?) 26, Schmitz 2: 682; Burchard of Worms, *Decretum* (1008-1012) 19, 5.178, Schmitz 2: 448. This is probably not a dietary taboo, since it is placed between two texts dealing with unquestionably magical healing practices.

[286286] *Homilia de sacrilegiis* (late 8th century) 15, Caspari, 9. An Anglo-Saxon charm against nose-bleed called for marking the forehead with the Sign of the Cross with Greek words taken from the mass of St. John Chrysostom; we are not told that blood was to be used (Cameron, "Anglo-Saxon medicine and magic," 213).

Although semen, like menstrual blood, was generally used for love magic, it may have been used for its curative powers as well. The *Parisian Penitential* asks whether anyone has made a man or woman drink his or her blood or his semen "for love or for another purpose." Burchard of Worms implies that semen, like blood, was drunk for a variety of reasons, but he adds a significant word: "If anyone has drunk blood or semen *knowingly* for any reason ..." It might, therefore, have been slipped into one's food or drink without one's knowledge or consent.[287]

A tonic of a quite different type appears in the 9th century. In a paragraph devoted to feminine magic, Rabanus Maurus described a preparation of burnt human skulls made by women to administer to their husbands as protection against illness. Regino of Prüm copied this as it stood, but Burchard expanded it in his *Corrector*: women "took" the skull, burnt it, then mixed the ashes in a liquid and gave it to their husbands to drink for health.[288] Possibly there is some connection between this practice and the puzzling clause *de cerebro animalium* of the *Indiculus superstitionum,* which may then refer to a medicinal preparation compounded of animals' brains.[289]

Locations with healing powers:The exploitation of certain locations for healing purposes has already been described. As far as our literature goes, the most important was the technique of putting children (son, daughter, baby) on the roof or in the oven to cure a fever or some other ailment, first described in the *Penitential of Theodore* and then in some form in over twenty-five English and continental documents.[290]

[287] *Poen. Parisiense I* (late 9th century or later) 18, Schmitz 1: 683; Burchard of Worms, *Decretum* (1008-1012) 19.91, *PL* 140: 1003.

[288] Rabanus Maurus, *Poenitentiale ad Heribaldum* (c. 853) 20, *PL* 110: 491; Burchard of Worms, *Decretum* (1008-1012) 19, 5.177, Schmitz 2: 448. See also Regino of Prüm, *De synodalibus causis* 2. 369: 354.

[289] 16, *MGH CapRegFr* 1: 223. A medieval remedy proposed for genital boils is an external application of boar's brains mixed with honey; the same preparation was also a protection against snakebite (*Medicina antiqua* I, 123, lines 23-25; II, 121). But Haderlein suggested plausibly that *de cerebro animalium* is a mistranscription of *de cervulo et annicula,* that is, the well-known mumming in animal masks ("Celtic roots," 5-8).

[290] *Poen. Theodori* (668-756) I, 15.2, Schmitz 1: 537. See also *Canones Gregorii* (late 7th/8th century) 116, Schmitz 2: 535; *Canones Cottoniani* (late 7th/8th century) 148, Finsterwalder, 280; *Liber de remediis peccatorum* (721-731) 2, Albers, 410; *Poen. Remense* (early 8th century) 9.15, Asbach, 57; *Excarpsus Cummeani* (early 8th century), 7.14, Schmitz 1: 633; *Poen. Egberti* (before 766) 8.2, *ibid.,* 581; *Poen. Sangallense tripartitum* (8th century, 2nd half) 32, Schmitz 2: 184; *Poen. Vindobonense b* (late 8th century) 17, Wasserschleben, 497, 482; *Poen. XXXV Capitolorum* (late 8th century) 16.4, *ibid.*, 517; *Poen Merseburgense a* (late 8th century) recension 1, 99, *CCSL* 156: 155; *Double Penitential of Bede-Egbert* (9th century?) 34, Schmitz 2: 682; *Poen. Valicellanum I* (beginning of the 9th century) 92, Schmitz 1: 316; *Poen. Martenianum* (9th century) 49.7, Wasserschleben 292, 482; *Poen. Ps.-Gregorii* (2nd quarter, 9th century) 23, Kerff, 180; Rabanus Maurus, *Poenitentiale ad Heribaldum* (c. 853) 20, *PL* 110: 491; *Poen. Ps.-Theodori* (mid 9th century) 27.14, Wasserschleben, 597; *Poen. Parisiense I* (late 9th century or later) 21, Schmitz 1: 684; Regino of Prüm, *De synodalibus*

In general, Theodore was quoted verbatim, but a number of texts introduced variations. Some undoubtedly were due to mistakes on the part of the copyist; others may have been inserted deliberately, to describe local custom. The tenth-century English penitential merely set the child on top of the house or oven.[291] Other variations are more imaginative. In one recension of the seventh- or eighth-century *Canones Gregorii,* roof (*tectum*) is replaced by bed (*lectus*), an easy mistake in the accusative.[292] The second recension of the late eighth-century *Poen. Merseburgense a* gives *pectum*, probably a double error for the accusative of *pectus* (bosom)–although laying a sick child on his mother or nurse's bosom could hardly have been deemed superstitious. It also mentions the unquestionably magical practice of putting the child *in funem super puteum* (in a net over the well?). Still another recension condemns women who place the child on their breasts (*super pectos suos*–here the meaning is unmistakable) or who suspend the child over a well–it is no longer even a question of a sick child. The same practices reappear in a text of the following century.[293] These may be a matter of a scribe's carelessness and the attempts of his successors to make sense of them. It is more difficult, however, to dismiss the evidence of the somewhat later *Poen. Casinense*, which adds "over the well" and "by the wall or anywhere else," to the usual roof and fireplace.[294] Here the evidence for a belief in the therapeutic efficacy of walls gives unexpected weight to the dubious information about wells: the former cannot be a copyist's error, especially since the usual verb *ponere* (put) has been replaced by *transponere* (moved, carried over). Such practices were based on confidence in the magical beneficent power of the family home and its belongings.

Other man-made structures with healing power were "monuments, that is, the ancient *Sarandas* [sacred places?] which they likewise call the Greater Ones" in the case of Frankish demoniacs in the late 8th century and, for sick people in ninth- and eleventh-century Spain, the mill or, rather, the water below it.[295]

More popular than these, however, were trees, earth and roads, where a variety of rituals were practiced to to evoke the power of nature, enact symbolic death and rebirth or transfer their sicknesses to passersby. Testimony is sporadic but impressively varied both in detail and geographic distribution. From the

causis (c.906) 2.368: 353; *Confessionale Ps.-Egberti* (c. 950-1000) 1.33, *PL* 89: 409; Poen. *Arundel* (10th/11th century) 97, Schmitz 1: 464); Burchard of Worms, *Decretum* (1008-1012) 10.14 (*PL* 140: 835) and 19, 5.95, Schmitz 2: 430. According to 16th and 17th century inquisitorial records, parents in 16th and 17th century Modena continued to place their sick children on the roof (personal communication from Mary O'Neil).

[291] *Confessional of Ps.-Egberti* 1.33, *PL* 89: 409.

[292] Document G, 117, Finsterwalder 265. The version in Schmitz 2 has the usual *tectum* (535).

[293] *Poen. Merseburgense a* (late 8th century) 99 (all recensions), *CCSL* 156: 155; Poen. *Vindobonense a* (late 9th century) 79, Schmitz 2: 355.

[294] *Poen. Casinense* (9th/10th century) 57, Schmitz 1: 412.

[295] *Homilia de sacrilegiis* (late 8th century) 22, Caspari, 12; *Poen. Albeldense* (c. 850) 99 and 100, *CCSL* 156A: 12; *Poen. Silense* (1060-1065) 192 and 193, *ibid.,* 36.

eighth-century *Vita Eligii*, we know that in northwestern Frankish territory phylacteries were employed near or passed through springs, trees or crossroads in the case of illness, and that cattle were driven through hollow trees or openings pierced in the ground. In the following century, a Frankish penitential described a ritual of passing through an opening in the ground and afterwards blocking it with thorns, for the benefit of a sick child. A tenth-century English penitential denounced the "greatly pagan" custom of taking a child to crossroads and dragging him along the ground (more probably, through an opening in the ground). According to Burchard, the women in his diocese dug a hole in the ground and pulled their whining children through it to stop their crying. Sick men in eleventh-century Spain were placed in doorways and roads–since this appears in a list under the heading "Concerning different types of homicide," the intention was clearly to transfer the sickness to a passerby.[296]

Other techniques: The *Homilia de sacrilegiis* is a rich source of information about forbidden magical treatments for ailments, in late eighth-century Francia. Gold rings were placed around wounds, and that any kind of ring tied over the afflicted part was thought good for sore eyes.[297] It gave particular attention to curing diabolical possession: in addition to the pilgrimage to ancient sites and the incantations mentioned above, methods included fumigation, roots (worn as an amulet or eaten?), herbal potions, iron rings and bracelets worn or borne on the body, the placing of some iron object in the house to terrify demons, driving coloured rods or twigs into the ground or iron nails into the ground under the demoniac's bed.[298] The concept that metal was efficacious in driving off demons and illness appears to derive from the late Neolithic, but the meaning of coloured rods (*virgae coloriae*) escapes me.

The *Needfire* appears in eighth-century Frankish legislation. It is first mentioned in the *Indiculus superstitionum* in a title "concerning the fire rubbed from wood, that is, *nodfyr*,"[299] and then is condemned in a list of pagan practices as "those sacrilegious fires which they call *niedfyr*," in Carloman's legislation (*nedfratres* in Benedict Levita's version).[300] In all likelihood this was a rite intended to protect cattle from disease.[301] The fact that royal law treated this issue suggests that it was

[296] *Vita Eligii* (700-725) *MGH SRM* 4: 707; *ibid.*, 706; *Poen. Ps.-Theodori* (mid 9th century) 27.16, Wasserschleben, 597; *Confessionale Ps.-Egberti* (c.950-c.1000) 4, *PL* 89: 426; Burchard of Worms, *Decretum* (1008-1012) 19, 5.179, Schmitz 2: 448; *Poen. Silense* (1060-1065) 89, *CCSL* 156A: 25. For doors and healing magic, see Bächtold-Stäubli 8: 1195-1197, *s.v.* "Tür"."

[297] *Homilia de sacrilegiis* (late 8th century) 21, Caspari, 11.

[298] 22, Caspari, 12. For the popular response to an outbreak of diabolical possession in the 9th century, see Agobard of Lyons, *De quorum inlusione signorum* 1, *CCCM* 52: 237.

[299] 15, *MGH CapRegFr* 1: 223.

[300] *Conc. in Austrasia habitum q. d. Germanicum* (742) 5, *MGH Concilia* 2.1, 1: 3-4=*Karlomanni principis capitulare* (742) 5, *MGH CapRegFr* 1: 25; *Benedicti capitularium collectio* 1.2, *PL* 97). *Cf. Conc. Quinisextum* (692) 75, Mansi 11: 974

[301] The old fires of the community were extinguished, and new fire was made by rubbing

accompanied by a good deal of rowdiness that made it necessary for the authorities to intervene. It is noteworthy that as yet no record of it has been found in sermons or penitentials despite its markedly pagan elements.

A passage in a letter from Pope Pelagius I to one of Caesarius' successors to the See of Arles may deal with a healing technique also. The pope had heard of an "idol" patiently shaped of dough, of which the different parts (ears, eyes, hands and other members) were distributed to the faithful, "to each as though according to merit [need?–*pro merito*]"–perhaps the pieces corresponded to diseased parts of the body.[302] Since the artisans in question appear to be clerics, the pieces may have been distributed as part of some religious service. The pope's use of the singular *idolum* might imply that this happened once only, and "patiently" suggests an unusual technique or unusually hard-to-come-by ingredients. Odette Pontal suggested that the Council of Tours (567) might have targeted a similar practice in forbidding the representation of Christ's body in "fanciful form" on the altar.[303] There may also be a relationship–at a remove of two centuries and many hundreds of kilometres–between this rite and the clause "concerning the figure/effigy made of sprinkled flour" in the *Indiculus superstitionum*.[304]

Finally, a single penitential, continental in origin, offers testimony that fasting "in honour of the moon" was considered beneficial for health.[305]

Weaving-women's magic: Perhaps the most obscure of the almost two hundred clauses in Burchard's penitential deals with the magic used by women at difficult junctures in their wool work. The grammar is so complex that any translation is largely a matter of guesswork; even the general meaning is controversial. However, by studying this passage in the context of other texts on weaving magic, one may risk a translation which (it is hoped) is faithful to the sense:

> Did you participate in or consent to the vanities which women practice in their wool work [spinning?–*in lanificiis*], in their weaving? When they are setting up the web, they hope to succeed in both with a spell at the beginning, as the warp and woof may become so tangled together that if they did not remedy these with other counterspells of the devil, the whole thing would be ruined. If you participated in this or consented to it, you shall do penance for

sticks together. It was used to light bonfires through which cattle were driven, and over which the men leaped. The ashes and charcoal were used to mark trees and scattered over the fields and in mangers, the embers, to relight the household fires (Homann, *Der Indiculus*, 96-98). Peter Chrysologus had condemned a comparable custom of the Calends of January in fifth-century Ravenna, where livestock was driven around the fire (*Homilia de pythonibus*, *PG* 65: 27-28).

302 *Ep.* to Sapaudus, Bishop of Arles (558-560), *MGH Ep* 3: 445.

303 3, *CCSL* 148A: 178. See Odette Pontal, "Survivances païennes, superstitions et sorcellerie au moyen âge d'après les décrets des conciles et synodes," *Annuarium historiae conciliorum* 27/28 (1995/96): 129-136; here 133.

304 26, *MGH CapRegFr* 1: 223.

305 *Poen. Ps.-Theodori* (mid 9th century) 27.26, Wasserschleben, 598.

thirty days on bread and water.[306]

That is, the women use spells at the beginning of the work to keep the work straight to prevent difficulties that would require other spells to repair the damage. A passage elsewhere in the *Decretum* asks women whether they have said or done anything, except in the name of God, while they were at their weaving or laying the web.[307] In fact, laying the web correctly is critical; if the tension is wrong, the threads do indeed get tangled so badly that the whole work may be ruined.

It has been suggested that Burchard refers here to malevolent magic meant to bring about the complete destruction of an unnamed victim.[308] This is implausible. Were it correct, the text would describe magic for murder. Now murder calls for much more than the almost trivial thirty days Burchard assigns here–his typical penance for homicide, magical or otherwise, is five to seven years or more; even belief in the evil eye calls for two years of penance. We are obviously in the realm of some minor beneficent magic, not murder.

A review of earlier texts dealing with the habits of weaving women reinforces this view.[309] The first references come from late sixth-century Galicia. In a canon attributed to the oriental fathers (repeated some half a dozen times later), Martin of Braga instructed women not to practice foolish customs in their wool work, but to invoke the aid of God, "who gave them the knowledge to weave."[310] The *Epítome hispánico* paraphrased this as "Women may not observe rituals at their wool

[306] I give the text in full: Interfuisti, aut consensisti vanitatibus quas mulieres exercent in suis lanificiis, in suis telis, quae, cum ordiuntur telas suas, sperant se utrumque posse facere, cum incantationibus et cum agressu illarum, ut et fila staminis, et subtegminis in invicem ita commisceantur [ut], nisi his, iterum aliis diaboli incantationibus econtra subveniant, totum pereat? Si interfuisti, aut consensisti, triginta dies poeniteas in pane et aqua (19, 5.64, Schmitz 2: 424). This translation depends on replacing the editor's *ut* ("in order to") after *commisceantur* with *ne*, which has the opposite meaning ("lest"). McNeill and Gamer's translation follows the text faithfully, but it posits that the women deliberately set out to ruin the work that they were starting, which seems unlikely (*Medieval Handbooks of Penance*, 330).

[307] 1.94, Interrogatio 52, *PL* 140: 577. See also Regino of Prüm, *De synodalibus causis* (c. 906) II, 5.53: 213.

[308] *Cf.* Meaney's translation: "when they begin their weaving [they] hope to be able to bring it about with incantations and with their actions that the threads of the warp and of the woof become so intertwined that unless [someone] makes use of these other diabolical counter-incantations, he will perish totally" (*Anglo-Saxon Amulets and Curing Stones*, 185). See also Flint, *The Rise of Magic*, 227.

[309] There is no obviously magical purpose to the practice of idling on Thursday in honour of Jupiter, of which Caesarius of Arles accused the weaving women of his diocese (*S*. 13.5 and *S*. 52.5, *CCSL* 103: 68 and 230-231).

[310] *Canones ex orientalium patrum synodis* (572) 75, Barlow, 141. See also *Collectio Vetus Gallica* (8th/9th century) 44.4f, Mordek, 524; Halitgar of Cambrai, *De Poenitentia* (9th century, 1st half) 4.26, *PL* 105: 686; *Poen. Quadripartitus* (9th century, 2nd quarter) 146, Richter, 19; *Capitulary of Radulph of Bourges* (between 853 and 866) 38, *MGH CapEp*, 262-263; Regino of Prüm, *De synodalibus causis* (c. 906) II, 5.53: 213; Burchard of Worms, *Decretum* (1008-1012) 10.19, *PL* 140: 836.

work."[311] The ritual specified, we find, was "naming" Minerva while they were at the loom.[312] The sermon attributed to St. Eligius of Noyon goes further:

> Let no woman dare to name Minerva or other ill-omened personages [*infaustas personas*] while weaving or dyeing or doing any other work, but let them wish to be in the grace of Christ in all their work, and trust wholeheartedly in the power of His name.[313]

This cannot be interpreted as a curse against "unfortunate persons" woven into the textiles.[314] *Minervam* and the *infaustas personas* are both accusatives governed by the same verb, "to name." They are being invoked, not cursed: the *infaustae*, like Minerva, have the power to help the weavers at their work. This is made clear in a very similar sermon which replaced *infaustae personae* with "diabolical names."[315] Clearly, when the fathers urged women to invoke God's help or to call on the name of Christ, they were not telling them to refrain from a murderous act, but offering them acceptable alternatives to the pagan deities on whom they had traditionally relied.[316]

Christian rituals and blessed objects: The power inherent in the sacred symbols of Christianity was a temptation in times of trouble to use the sign of the cross, holy chrism, the sacred species and altar vessels in illicit, magical ways.

Caesarius of Arles protested against what he saw as a blasphemous use of the sign of cross as a protective charm, by the many who made the sign of the cross when they stumbled while on their way to commit a theft or adultery or by those who blessed themselves before eating sacrilegious food, or who signed themselves repeatedly when they saw something frightening at the circus.[317] Other authors ignored this practice, probably not because the faithful had abandoned it but because using the sign of the cross and various prayers at critical times was an "incantation" recommended by such authorities as Martin of Braga.[318]

From the beginning of the 9th century onward, Frankish canon law and capitularies manifested great interest in possible abuses of holy chrism. In addition to repeated insistence that only bishops could consecrate it and that priests should keep it safely locked away, about a dozen texts deal with its unlawful use. A Carolingian capitulary condemned clerics, nuns and lay persons to various punishments

[311] 1.75: 104.

[312] Martin of Braga, *De correctione rusticorum* (572) 18: 203.

[313] *Vita Eligii* (c.700-725), *MGH SRM* 4: 706-707.

[314] Meaney, *Anglo-Saxon Amulets and Curing Stones*, 185.

[315] *Anonymous homily* (8th century), Nürnberger, 44.

[316] For a sidelight on the symbolical and magical aspects of textiles, see Henry Maguire, "Garments pleasing to God: The significance of domestic textile designs in the early Byzantine period," *Dumbarton Oaks Papers* 44 (1990): 215-224.

[317] *S.* 13.1, *CCSL* 103: 64-65; *S.* 54.6, *ibid.*, 240; *S.* 134.1, *CCSL* 104: 550.

[318] *De correctione rusticorum* 8 and 16, 188 and 199.

for giving or receiving too much chrism.[319] The Council of Arles (813) and the somewhat later Statutes attributed to St. Boniface required priests to guard it under seal and to refrain from handing it out "under pretext of medicine or for some other excuse" for "it is a kind of sacrament, to be touched only by priests."[320] Here the concern seems to be not with the magical use of the chrism, but with the indignity that it would suffer from being handled by laymen; its efficacy as medicine is not questioned. Other documents of 813 though, replaced "some other excuse" with "because of certain unbelievers" or, more commonly, with "sorcery" (*maleficium*). The fear that chrism would be used this way persisted throughout our period.[321]

This *maleficium* was intended principally to cheat the Judgment of God. Carolingian capitularies threatened priests with dismissal from their office and the loss of their hand (the punishment for perjury) for giving the chrism in order to subvert the ordeal.[322] "It is thought by many," explained the Council of Tours, "that guilty people who were anointed with chrism or had drunk it can in no way be detected in any test." Accordingly, Regino of Prüm enquired about the use or provision of chrism for subverting the Judgment of God.[323]

Chrism may have been used for other reasons as well. Early ninth-century diocesan capitularies forbid priests, deacons and clerics to hand out chrism "for any necessity whatsoever."[324] In the *Double Penitential of Bede-Egbert*, a question about eating or drinking chrism follows immediately on a question about love potions; it is possible that it also was used for that purpose. The penitential unexpectedly offers a choice of eight, five or three years of penance for these sins.[325]

The magical use of the Eucharist is more difficult to establish for the early Middle Ages. Peter Browe collected and published ample evidence of it for the

[319] *Capitula post a. 805 addita* (806-813) *MGH CapRegFr* 1: 142. This is one of the two texts that I have found for the involvement of nuns in a practice which may be magical (on the supposition that the clause "concerning the handmaids of Christ: those who arrived at 'deformity' with them ..." in the Poen. Cordubense (128, *CCSL* 156A: 64) deals with fornication, and not some other practice.

[320] *Conc. Arelatense* (813) 18, *MGH Concilia* 2.1: 252. See also *Statuta Bonifacii* (1st half, 9th century) 5, *MGH CapEp* 3: 360; *Benedicti capitularium collectio* (mid 9th century) Add. 3.99, *PL* 97: 885.

[321] *Admonitio synodalis* (c. 813) 78, Amiet, 64; *Concilium Moguntinense* (813) 27, *MGH Concilia* 2.1: 268. See also *Appendices ad concilia anni 813*, A.17, *ibid.*, 2.1: 296; *Capitula a canonibus excerpta* (813) 1, *ibid.*, 1: 174; Leo IV (d. 855) *Homilia*, *PL* 115: 683; Hincmar of Reims, *Capitularia II* (c. 850) 8, *MGH CapEp* 2: 48; Isaac of Langres, *Capitula* (860-869) 4, *ibid.*, 232; *Conc. Treverense* (927/928) *MGH Concilia* 6.1: 8.

[322] *Capitulare Aquisgranense* (809)10, *MGH CapRegFr* 1: 149. See also *Capitulare Missorum* (809) 21, *ibid.*, 150; Isaac of Langres, *Capitula* (860-869) 5, *MGH CapEp* 2: 232.

[323] *Conc. Turonense* (813); 20, *MGH Concilia* 2.1: 289; Regino of Prüm, *De synodalibus causis* (c. 906) I, 304: 145. See also Burchard of Worms, *Decretum* (1008-1012) 19, 5.167, Schmitz 2: 445.

[324] *Capitula in diocesana quadam synodo tracta* (c. 803, 804) 11, *MGH CapRegFr* 1: 237. See *also Capitula Corbeiensia* (c. 805/806) 13, *MGH CapEp* 3: 14.

[325] 30, Schmitz 2: 682.

period beginning with the 12th century, but there is no such convincing evidence for our period.[326] Nevertheless, certain passages raise the suspicion that the consecrated species may have been used in this way earlier than thought. The most suggestive deal with unconsumed consecrated hosts. An early Spanish collection and Benedict Levita's collection included a ruling of the first Council of Toledo calling for the expulsion of persons who did not consume the Eucharist after receiving it.[327] No more is heard of this sin until the 11th century, when it was treated with great rigour by a continental penitential.[328] Nevertheless, Burchard of Worms had noted that certain priests, unwilling to consume the consecrated species themselves, handed the chalice over to women (characterised dismissively as *mulierculae*) and to laymen who failed to esteem the Lord's body properly, "that is, who did not know how to distinguish between spiritual and fleshly food." He insisted that the priest and other ministers of the altar consume the excess, and that the priest place the Eucharist with his own hand into the mouth, not the hands, of the faithful.[329] For Burchard, the issue is clearly ignorance and irreverence, not magic, but it cannot be ruled out that some of the laity used the host in unsuspected ways.

The possibility of magic is also raised in the issue made in many penitentials of the disposal of unused and unusable consecrated hosts and wine. An eighth-century Frankish penitential stated that only as much as would be used immediately should be consecrated; the superfluity was to be consumed by the ministers, not kept over for another day; anything left over should be burnt and the ashes hidden, and the priest should do a year's penance.[330] Disposing of consecrated hosts that had been dropped to the ground, made grubby by age or gnawed by mice, or of consecrated wine that had been spilled, was another problem. A few of the penitentials in the Theodorian tradition demanded merely that whatever had been spoiled by "foul age" should be burnt.[331] But almost twenty penitentials from all parts of Western Europe, and Burchard's Decretum, followed the seventh-century Penitential of Cummean in displaying great concern that no trace of the sacred species should survive. The polluted hosts, the sweepings of the ground where the host had fallen and, if possible, even the flooring where it had fallen should be burnt, and the ashes hidden beneath the altar. The negligent priest should lick up any spill and drink the water used in washing the cloths touched by the wine.[332] Although reverence for the Eucharist is unquestionably the dominant

[326] "Die Eucharistie als Zaubermittel im Mittelalter," *Archiv für Kulturgeschichte* 20 (1930): 134-154.

[327] *Epítome hispánico* (c.598-610) 31. 14: 173. See also *Conc. Toletanum I* (397-400) 13, Vives, 23; *Benedicti capitularium collectio* (mid 9th century), 3.473, *PL* 97: 866.

[328] *Poen. Valicellanum F. 92* (11th century) 10, Wasserschleben, 686.

[329] Decretum (1008-1012) 3.76, *PL* 140: 689-690.

[330] *Poen. Merseburgense a* (late 8th century) 3rd recension, 48, *CCSL* 156: 136.

[331] XII, 6, Schmitz 1: 534. See also *Canones Cottoniani* (late 7th/8th century) 46, Finsterwalder, 273; *Canones Basilienses* (8th century) 87, Asbach, 87.

[332] 20, 23, 24, 26 and 27, Bieler, 130-132.

reason for these measures, here too an underlying fear may have existed lest the ashes and the water be used for some inappropriate purpose, such as curing ailments or increasing the fertility of fields and livestock.[333]

The possible use of altar vessels in hydromancy has already been mentioned. Concerns about this and other magical uses of altar vessels and paraphernalia may form a subsidiary motive for the repeated injunctions against the appropriation of such objects for non-liturgical purposes. The *Epítome hispánico* demanded that worn-out altar vessels and cloths be burnt and the ashes submerged in the holy water font. A late seventh-century Iberian council was outraged to learn that some bishops or priests (*sacerdotes*) used altar vessels for their own feasts and placed their own food in them; it pronounced a sentence of perpetual excommunication for laymen and deposition for priests who knowingly used these and other belongings of the altar for themselves, or "who believed that they could be sold or given away to others."[334] Twenty years later, the Council of Toledo was sure that this was the case, for it attacked "the indecent desire and ill-omened recklessness" of bishops who not only handed over altar furnishings and other church ornaments to others "to be abused in their foulest activities" but, even worse, did not fear to appropriate them for their own use.[335] Similar worries may lie behind the condemnation in Theodulph of Orleans' *Capitulary* of 813 of the priest or layman who risked suffering Balthasar's fate by diverting the chalice and paten to uses other than the divine cult.[336] The purchaser or recipient of objects infused with such sanctity and power might easily be tempted to employ them for other than ordinary household purposes, as did the Breton count and Lombard duke who, according to Gregory of Tours, used them to bathe their sore feet.[337]

Finally, Regino of Prüm's insistence that the pyx or other container be kept locked up carefully "because of mice and wicked men" may have been motivated by fear lest consecrated hosts be stolen for magical purposes.[338]

[333] In the Ardennes, the ashes of several idols and *simulacra* burned by St. Hubert were preserved and venerated sacrilegiously by "fanatical men" (*i.e.*, frequenters of *fana*) (*Vita Hugberti* 3, *MGH SRM* 6: 484-485). Stéphane Boulc'h has suggested that the regulations concerning the disposing of the Sacred Species may possibly carry implications of elements of pagan mysticism ("Le repas quotidien des moines occidentaux du haut moyen âge," *Revue Belge de philologie et d'histoire* 75 [1997]: 287-328).

[334] *Epítome hispánico* (c.598-610) 3: 193; *Conc. Bracanense III* (675) 2, Vives, 374.

[335] *Conc. Toletanum XVII* (694) 4, Vives, 530).

[336] 18, *MGH CapEp* 1: 115. See also Atto of Vercelli (d. 961) *Capitularia* 13 and 14, *PL* 134: 31: and Burchard of Worms, *Decretum* (1008-1012) 3.105, *PL* 140: 694.

[337] *Gloria Martyrum* 84, 95, *MGH SRM* 1: 545.

[338] Regino of Prüm, *De synodalibus causis* [c. 906] I, 71: 56. See Browe, "Die Eucharistie als Zaubermittel," 135 n. 1.

7
Ambivalent and Destructive Magic

7.1 Ambivalent Magic

In anthropological terms, cunning folk whose magic is sometimes perceived as productive and beneficial, sometimes as destructive or harmful, are sorcerers and their magic is sorcery. These can be divided into three groups. Makers of weather (*emissores* and *immisores tempestatum, tempestarii, tempistariae*) and herbal experts (*herbarii, herbariae*) were specialists who engaged in one kind of magic only. Others may be termed generalists. Even finding an English equivalent for their names is difficult. *Incantator/incantatrix* may be translated as enchanter/enchantress, *maleficus/malefica, veneficus/venefica* and, sometimes, *mathematicus* as sorcerer/sorceress, and *magus* as wizard, but this is an arbitrary assignment of words. It is safer to remain with the Latin terms. The final group includes practitioners about whom nothing is known for certain but their names: *cauculator/cauclearius/coclearius, cocriocus* and *obligator*. The section concludes with a survey of ambivalent magic, variously labelled *incantatio, ars magica, maleficium, veneficium* and, rarely and late, *fascinatio*.

7.1.1 *Specialists*

Tempestarius/tempestaria, emissor/immissor tempestatum—weather magicians and weather magic: Sometime between 815 and 817, Agobard, the bishop of Lyons, sat down to compose *De grandine et tonitruis*,[1] a polemic against a belief held in his diocese by "nobles and commoners, townsmen and country folk, old and young," that hail and thunder were caused at will by certain persons. At the first sign of thunder or lightning, they announced that "the wind [air?–*aura*] was being raised" and, when pressed to explain themselves, they answered with varying degrees of embarrassment ("as is the custom of ignoramuses") that it was being raised by the spells of *tempestarii*.[2] Similar beliefs had been common in antiquity,[3] and there is evidence

[1] *CCCM* 52: 3-15. For an evaluation of Agobard's testimony, see Henri Platelle, "Agobard, évêque de Lyon (†840), les soucoupes volantes, les convulsionnaires," in *Problèmes d'histoire des religions: Apparition et miracles*, ed. Alain Dierkens (Brussels, 1991), 85-93, esp. 85-89. Paul Dutton sets the beliefs described in this tract in their political context in "Thunder and hail over the Carolingian countryside," in *Agriculture in the Middle Ages. Technology, Practice and Representation*, ed. Del Sweeney (Philadelphia, n.d.), 111-137. See also Schmitt, "Les 'superstitions'," 464-466, Cabaniss, *Agobard of Lyons*, 17-32, and Boshof, *Erzbischof Agobard von Lyon*, 171-185.

[2] *De grandine et tonitruis* 1, *CCCM* 52: 3.

[3] See Du Cange, *s.v.* "Tempestarius."

enough for such notions in hagiography,[4] but pastoral sources are silent about makers of weather until the early 8th century, when a veritable flood of material testifies to a suddenly awakened anxiety about such people.

Although texts dealing with weather magicians come from both English and Frankish territory, particularly the latter, it is possible that fear of them was not home-grown but brought into northwestern Europe by refugees from Visigothic Spain after its fall to the Moors. The only pastoral texts dealing with weather magic to predate the early 8th century are from Iberia. There, concern about weather magic went back at least to the days of Priscillian. Priscillianist beliefs in the devil as the author of unpleasant weather phenomena (thunder, lightning, storms and droughts) survived into the mid 6th century in Galicia and were still cited in the same region in the 11th, while the Visigothic Code dealt harshly with the "summoners of storms" whose spells brought in hailstorms to destroy the grape harvest and the crops.[5] The Spanish origin of Agobard himself and Pirmin of Reichenau, who also wrote of weather magicians, may be relevant here.

Except for Agobard's, only seven eighth- or ninth-century texts, principally capitularies, have been found to contain reference to *tempestarii.* A comprehensive list of cunning folk in Caspari's edition of the *Dicta Pirmini*, which contains the only mention of mistresses of the weather (*tempistariae*), is probably the earliest.[6] The authors of penitentials preferred other terms: the *emissor tempestatum*, who drove out storms, is mentioned in about fifteen documents,[7] and the *immissor tempestatum*, who brought them on, in another half dozen.[8] Some identified them as sorcerers (*malefici* and/or *maledici*).[9] Other documents avoided these technical

[4] St. Caesarius' crosier warded off hailstones and promoted fertility, although without his consent (*Vita Caesarii* 2, 27, *MGH SRM* 3: 494); St. Scholastica brought on a thunderstorm to keep her brother St. Benedict with her (Gregory the Great, *Dialogues* II, 33.3, *SC* 260: 232); another saint stirred up a storm to immobilise Attila's Huns (*Vita Aniani* 9, *MGH SRM* 3: 115). Numerous Irish saints had the power to cause rain and snow to fall selectively; for references, see *VSH*, cxxxviii, n. 12.

[5] *Conc. Bracanense I* (561) 8, Vives, 68. See also *Poen. Silense* (1060-1065) 207, *CCSL* 156A: 37; *Lex Visigothorum* 6.2, 3 (4), *MGH Leg* 1.1: 259.

[6] *Dicta Pirmini* (724-753) 22: 173; *ibid.*, 28: 188-189. References to *tempestarii* are found in Charlemagne's *Admonitio Generalis* (789) 65, *MGH CapRegFr* 1: 58-59, and *Capitulare missorum generale* (802) 40, *ibid.*, 104; see also *Ansegesis capitularium collectio* (9th century, 1st half) I, 62, *ibid.*, 402 and *Benedicti capitularium collectio* 2.374 and Add. 21, *PL* 97: 791 and 867. An anonymous eighth- or ninth-century sermon also uses this term (Levison, 308-309). *Tempestuarii* appear in Herard of Tours' list of cunning men (*Capitula* [858] 3, *PL* 121: 764).

[7] One of the earliest is found in the penitential ascribed to the Venerable Bede which called for seven years of penance for *emissores tempestatum* (*Liber de remediis peccatorum* [721-731] 14, Albers, 407). This term is used throughout our period.

[8] *E.g.*, *Poen. Hubertense* (8th century, 1st half) 20, *CCSL* 156: 110). This term appears also in Burchard's *Decretum* 10, 8 and 28 (*PL* 140: 834 and 837) as well as in 19, 5.68 (Schmitz 2: 425).

[9] *Poen. Burgundense* (8th century, 1st half) 20, *CCSL* 156: 64. See also *Poen. Parisiense simplex* 12, *ibid.*, 75; *Poen. Oxoniense I*, 15, *ibid.*, 90; *Poen. Floriacense* 19, *ibid.*, 99; *Poen. Hubertense* 20, *ibid.*, 110; *Poen. Sangallense tripartitum* 19, Schmitz 2: 181.

terms. Carolingian Councils referred to "those who make storms and [perform] other forms of sorcery"[10] and to men or women "who claim that they are able to stir up the air by their spells and bring on hail, foretell the future, transfer fruit and milk from one person to another," *etc.*[11] Ratherius of Verona (d. 974) denounced those who asserted that wicked men or the devil made storms, scattered hailstones, destroyed vineyards and fields, sent lightning and slaughtered beasts of burden, flocks and even men; he also denounced those who professed to avert storms by means of spells.[12]

The texts treat *emissor*, *immissor* and *tempestarius* as interchangeable. The last is obviously a general term, but in none of our documents do we find both of the other words together, to indicate that the authors were aware that *emissor* and *immissor* were not necessarily identical. Yet Agobard, like Ratherius, made clear that there was a difference. It was believed, on the one hand, that there were men and women who summoned storms and hail with the specific purpose of knocking crops and fruit down in order to sell them to aerial sailors, who would bear the harvest away in ships sailing the clouds to "Magonia. " The belief was so firmly entrenched that some people actually captured three men and a woman whom they believed had tumbled from the airships; Agobard barely managed to save these unfortunates from being publicly stoned to death.[13] Frustratingly for him, he could find no one who had actually seen magicians producing storms; rumour always placed the action in some other area. When at last he heard of someone who claimed to have witnessed such magic, intensive cross-examination proved that the so-called witness, who could name the guilty party, the time and the place, had not in fact seen it performed.[14] On the other hand, opposed to these malicious weather makers were other experts who claimed to be able, not to bring on storms, but to avert them. They evidently enjoyed the trust of the population, for Agobard complained that those who neither paid their tithe to the church nor gave alms to the poor readily set aside a portion of their crops, called the *canonicum,* to pay for their services.[15]

The *canonicum* may explain the preponderance of references to the *emissores tempestatum* (those who drive storms out) over the *immissores* (those who bring them in). The latter existed in rumour alone, and sensible clerics refused to believe in such bogey men of popular imagination; but the former existed in hard fact, and as practical men, pastoral writers were quite aware that they were rivals both morally and economically.

De grandine et tonitruis is particularly interesting for what it reveals about the

[10] *Statuta Rhispacensia Frisingensia Salisburgensia* (799, 800) 15, *MGH CapRegFr* 1: 228.

[11] *Conc. Parisiense, Epistola episcoporum* (829) 69 (II), *MGH Concilia* 2.2: 669-670.

[12] *S.* 1.4, *CCCM* 46: 52.

[13] *De grandine et tonitruis* 2, *CCCM* 52: 4.

[14] *De grandine et tonitruis* 7, *CCCM* 52: 8.

[15] *De grandine et tonitruis* 15, *CCCM* 52: 14.

limits of popular beliefs. If men can make hail, says Agobard, they can make rain too, "for no one has ever seen hail without rain." But when drought makes it impossible to sow the winter crop, no one expects the *tempestarii* to bring rain to irrigate the soil.[16] In this respect, magicians were feared for their ability to cause harm actively, but their power to do good was limited to defensive magic at best–they were unable to bring about a positive benefit. Agobard seems to suggest that even the malignant power of the *tempestarii* was thought to have bounds. The people of Lyons were prepared to believe that such persons could steal the crops and cause hail to fall over a wide-spread region, or concentrate in one area, either on a stream or a barren wood, or over the tub under which they (the sorcerers) were hiding. Unlike Ratherius' Veronese flock, however, they apparently doubted that *tempestarii* could batter their enemies to death with hailstones even when the latter were exposed to the force of the storm in the roads and fields.[17]

By the end of the 10th century, popular imagination had attributed new baleful powers to weather-magicians. According to Burchard of Worms, enchanters who claimed to be *immissores tempestatum* were thought to control not only storms but men's minds as well.[18] The *Arundel Penitential* treats of persons who "attempt to transform the serenity of the atmosphere by means of some incantation or who summon demons to trouble minds."[19]

As for the origins of Magonia and the Magonians, we have seen that hints may be found in other eighth- and ninth-century documents. Jecker's edition of the *Dicta Pirmini* follows the words "Do not believe in weather-magicians" with "nor give anything for that [storm-prevention?] to those who say that spirits [*manus*, *maones*] can take produce away." Jecker speculated that these were the *manes*. An anonymous sermon cited the belief that *mavones* were able to make off with the harvest and the vintage. Among the Picards, crops were supposedly carried away by *maones* (elsewhere, *dusi hemaones* or *dusi manes*).[20] The Magonians, then, may have been spirits or preternatural beings allied to or identical with the *manes* or the Gallic *dusii*.

But other texts offer a more mundane possibility. Carolingian authorities were concerned about the activities of *mangones* and *cotiones*, lawless peddlers (more accurately perhaps, swindlers) and tramps, who wandered about deceiving the population (these rascals seem remarkably like the *ergach* and *praeco* condemned in the *Canones Hibernenses*).[21] They were evidently more than ordinary thieves and frauds,

[16] *De grandine et tonitruis* 7 and 13, *CCCM* 52: 7 and 12.

[17] *De grandine et tonitruis* 7, *CCCM* 52: 7-8.

[18] *Decretum* 19, 5.68, Schmitz 2: 425. See also *idem*, 10.28, *PL* 140: 837, and the *Poen. Arundel* (10th/11th century) 82, Schmitz 1: 460.

[19] 82, Schmitz 1: 460.

[20] *Die Heimat des hl. Pirmin des Apostels*, 55; Anonymous sermon (late 8th/9th century), Levison, 311; *Vita Richarii* I, 2, *MGH SRM* 7: 445. See also chapter 2, under the heading *Mavones/Maones/Manes*.

[21] *Admonitio Generalis* (789) 79, *MGH CapRegFr* 1: 60-1. See also *Capitulare missorum generale item*

since legal texts associated them with self-proclaimed penitents who wandered about nude, weighed down with chains of iron. Their pilfering, extortions and cock-and-bull stories may well have given rise to tales of a mythological land where stolen goods were hidden away. The fact that Agobard's flock were satisfied that four flesh-and-blood humans were the culprits makes this the more likely explanation.

At times ordinary people took weather-management into their own hands. The terse "concerning storms and horns and snail shells" of the *Indiculus superstitionum* is amplified by other sources.[22] An anonymous eighth-century sermon urged the faithful to run to church instead of resorting to diabolical observances when overtaken by a storm - obviously a hailstorm, since they were to pray to God to turn it to rain.[23] The observances consisted of raising a clamour (as when the moon was darkened) or hanging magic symbols to drive off the hail. A Carolingian edict forbade baptising bells and hanging inscriptions on poles "because of hail."[24] The *Homilia de sacrilegiis* likewise banned the use of inscribed lead tablets as well as enchanted horns for this purpose, while another anonymous sermon of roughly the same period condemned making an uproar with clanging, trumpet blasts and yells.[25] Two penitentials of the eighth and ninth centuries threatened to deprive of communion those who beat drums or snail shells (*coclea*) because of thunder, or who made any sound "except the psalms or 'God have mercy on me' ... because drums are made to call people together, not to mitigate the wrath of God."[26] These texts are all from Frankish territory.[27]

Finally, although Agobard had been sure that professional magicians were not expected to be able to put an end to drought, Hessian women and girls in the 11th century believed that they could perform an effective rain-making ritual. It was painstakingly described by Burchard of Worms. "Did you do," he asked, " as some

speciale (802) 45, *ibid.*, 104; *Ansegesis collectio canonum* 34, *ibid.*, 447; *Benedicti capitularium collectio* 2.378, *PL* 97: 792; Regino of Prüm, *De synodalibus causis* (c. 906) II, 79: 245; *Canones Hibernenses* 1.4, Bieler, 160.

[22] 22, *MGH CapRegFr* 1: 223.

[23] Morin, 518.

[24] *Duplex legationis edictum* (789) 34 and gloss, *MGH CapRegFr* 1: 64; Regino of Prüm, *De synodalibus causis* (c. 906) Appendix III, 56: 487).

[25] *Homilia de sacrilegiis* (late 8th century) 16, Caspari, 10; *Anonymous sermon* (late 8th/9th century) Levison, 308-309.

[26] *Poen. Merseburgense a*, 3rd recension (late 8th century) 109, *CCSL* 156: 169. See also *Poen. Vindobonense a* (late 9th century) 99, Wasserschleben, 422. *Coclea* may be a mistake for *cochlea* (drinking vessel, see Du Cange *s.v.* "Cochlea"); a significant noise is produced more easily by striking a metal container than a snail shell.

[27] Elsewhere other techniques might have been used, such as those described by St. Bernardine of Siena (d. 1444) for Italy. They included making "conjurations," drawing swords, flinging hearth-chains in front of doors, putting the ashes left over from the Christmas fire outside houses, or, in the case of seamen, driving swords into the mast (see Montesano, "'Supra acqua et supra ad vento'," 9-10). *Cf.* techniques used against lunar eclipses, chapter 3.

women do?

> When there is no rain and rain is needed, they get together several girls, and they put one tiny little maiden forward as if their leader. They strip her naked, and lead her, naked as she is, out of the village to a place where they find henbane, the plant which is called *belisa* in the Teutonic language. They make the maiden, naked as she is, pull up the plant with the little finger of her right hand, and they have the pulled-up root tied with some kind of binding to the small toe of her right foot. And all the girls, each holding a twig in her hand, lead the said maiden, dragging the plant behind her, with them into the nearest river. They then sprinkle her with the twigs, which have been dipped in the river. They hope to obtain rain thus by their enchantments. And afterwards, with steps turned and transposed crabwise, they carry the maiden, naked as she is, in their hands [lead her by the hand?] back from the river to the village.[28]

The elements of this ritual are extraordinarily rich in symbolism and magical potency: procession from village to meadow to river by women at all stages of life, wives, unmarried girls and a child whose undoubted virginity makes her the most apt medium for magic; nudity; a toxic plant (among the Greeks, sacred to Apollo) with narcotic and hallucinogenic properties; the magical efficacy of the right side; the use of the smallest finger and smallest toe of the smallest participant; a reenactment of baptism by immersion and sprinkling, and the submersion of the plant; the return with twisted and reversed steps. One is struck, also, by the remarkably apt (if no doubt unconscious) wordplay on *virgo* and the other possible meanings (magic wand, penis) of *virga*, the words chosen to translate the vernacular terms for maiden and twig.[29]

Herbaria and herbarius–herbalists and magic potions: John Noonan remarked that "herbs were the special resources of the magician" and that the "homeopathic basis of a number of the potions ... is one of the oldest forms of magic."[30] Nevertheless the texts show little interest in specialists in this field. A herbwoman (*erbaria*) was among the cunning folk consulted by Caesarius of Arles' flock.[31] He

[28] *Decretum* (1008-1012) 19, 5.194, Schmitz 2: 452.

[29] For rainmaking magic, see Walter Puchner, "Zur Typologie des balkanischen Regenmädchen," *Schweizerisches Archiv für Volkskunde* 78 (1982): 98-125; Puchner notes a swing between the symbolic content of heathen and Christian vegetation cults in comparable rainmaking rites of southeastern Europe (124). See also M. Esther Harding, *Women's Mysteries,* 127-129. For ritual nudity, see Heckenbach, *De nuditate sacra.* For the magico-religious symbolism of right-sidedness, see Hertz, "The pre-eminence of the right hand." Raoul Manselli ("Simbolismo e magia nell'alto medioevo," 314) missed a point of some significance when he attributed a position of leadership to a specialist ("la donna *tempestaria*"). As far as Burchard's description goes, this is a rite organised collectively by the village women on the basis of shared knowledge.

[30] *Contraception: A History of its Treatment by the Catholic Theologians and Canonists* (enlarged ed., Cambridge, Mass., and London, 1986), 44.

[31] S. 52.5, *CCSL* 103: 232. See also Burchard of Würzburg, *Hom.* 25, Eckhart, 844.

referred to her but once, perhaps because he thought her to be less of a menace to faith than soothsayers and *praecantatores*. Two continental penitentials include *(h)erbarius* in a list of practitioners of obviously benevolent magic,[32] but two others consider the *herbarius* of either sex to be a murderer of infants. In fact, he or she was almost certainly an abortionist.[33] Professional herbalists may play such a small role in our texts because knowledge and use of herbs were widespread among the population at large. Some people may have had access to experts, but no doubt many more relied on traditional lore to make up their own concoctions.[34] Regino of Prüm and Burchard of Worms give hints of the transmission of this kind of forbidden knowledge with their questions: "Did you teach someone else to do this?" "Did you do, or consent to, or teach ...?" "Did you give or show ...?"[35]

Pastoral authors concentrated on suspect practices associated with herbs rather than on the herbs themselves.[36] Martin of Braga condemned only the observances and spells with which medicinal herbs were gathered, insisting that only the Creed and the Lord's Prayer were acceptable. Regino of Prüm and the authors of some half dozen penitentials including Burchard of Worms repeated his words, imploring their flocks to call upon God, not the devil.[37] St. Eligius also warned against enchanting (or chanting formulae over) herbs.[38] A frequently repeated clause from the *Penitential of Theodore* authorised persons possessed by the devil to use vegetables and stones, but without spells.[39] The *Medicina Antiqua* amply justifies these concerns, with its copious recommendations for appropriate spells and rituals to be used while collecting and employing herbs.[40]

[32] *Poen. Hubertense* (8th century, 1st half) 25, *CCSL* 156: 111; *Poen. Merseburgense b* (c.774-c.850) 24, *ibid.,* 176.

[33] *Poen. Halitgari* (817-830) unnumbered , Schmitz 1: 487; *Poen. Oxoniense II* (8th/9th century) 43, *CCSL* 156: 198. The penalty of a fast lasting thirty or fifty weeks for the repentant herbalist is comparable to the penance advocated by Burchard of Worms for abortion before the child is quickened (*Decretum* [1008-1012] 19, 5. 162, Schmitz 2: 445), but light in comparison to the seven to ten years penance more typically imposed for abortion, or to the twelve years of fasting and a lifetime of penance recommended by Burchard for deliberate infanticide (19, 5.163, *ibid.*).

[34] See Véronique Charon, "The knowledge of herbs," in *The Pagan Middle Ages*, 109-128; here 119-123.

[35] Regino of Prüm, *De synodalibus causis* (c. 906) II, 5. 8: 209; Burchard of Worms, *Decretum* (1008-1012) 19, 5.159 and 19, 5.161, Schmitz 2: 444.

[36] For herbs, see F. Cardini, "Le piante magiche" in *L'ambiente vegetale nell'alto medioevo*, SSAM 37 (Spoleto, 1990), 623-658; Bonser, *The Medical Background of Anglo-Saxon England,* 306-340. Also useful is Christian Rätsch, *The Dictionary of Sacred and Magical Plants*, trans. John Baker (Bridport, Dorset, 1992).

[37] Martin of Braga, *Canones ex orientalium patrum synodis* (572) 74, Barlow, 141; Burchard of Worms, *Decretum* (1008-1012) 1.94, Interrogatio 51, *PL* 140: 577.

[38] *Vita Eligii* (c.700-725), *MGH SRM* 4: 706.

[39] *Poen. Theodori* (668-756) II, 10.5, Schmitz 1: 544.

[40] *E.g.*, while gathering bitter cucumber: calling on "exalted Hygia, the king of dragons," and relying on the spells of Asclepius, "doctor of herbs" (*Medicina antiqua* 1, 107v, lines 10-13; 2:

Our text provide ample evidence for the use of herbs and plant products, usually in liquid form, for forbidden purposes or in forbidden ways, both for malign or beneficent purposes. The most common term is some version of potion (*potiones*, *potationes*, *potus*); or it is stated that they were given in cups (*in pocula*) or were meant to be drunk. Potions could contain ingredients of animal or mineral origin as well as herbs, but there is no precise indication of the substances used.

Ps-Theodore included herbs in his catalogue of pagan practices, trusting apparently to the discretion of the parish priest to judge between legitimate and illegitimate uses.[41] Other documents are more precise. Potions were used above all to control fertility, usually for contraception and abortion, but also, very rarely, as an aid to conception.[42] The women of Caesarius of Arles' flock used herbs and potions along with mystic symbols and amulets as contraceptives and abortifacients, or husbands and wives took "heaven knows what sacrilegious medications and sap of trees" when they were desperate to have children.[43] The use of sap (amber?) is probably based on the hope that the would-be parents would absorb the fertility and life-force of the tree.

A considerable amount of uncertainty must have existed as to the toxicity and proper dosage of the preparations, since Caesarius noted that the women taking them risked self-murder.[44] He was aware that rich women who did not wish to divide their inheritance among too many children had recourse, as much as their own slaves, to "death-dealing poisons" to prevent conception or rid themselves of the infants that they were carrying.[45] Caesarius was unusual in his understanding of the psychological pressures on his charges, including, even more unusually,

112). A passage describes the proper method for picking basil - the gatherer must observe ritual purity (that is, be in good health, wear clean intact clothes and avoid the sight of menstruous women and unclean men); using an oak twig held in his right hand, he must asperge himself at the break of dawn with well water and then recite a prayer to *Sancta Tellus*. Beside this, added in another hand, is the remark: "You are lying, for it is against the Christian faith" (*Medicina antiqua* 1: 116v; 2: 114 - vol. 1 contains the reproduction of the text, vol. 2 the transcription).

[41] *Poen. Ps.-Theodori* (mid 9th century) 27.24, Wasserschleben, 598.

[42] See Noonan, *Contraception*, 8-170; J.-C. Bologne's, *La naissance interdite. Stérilité, avortement, contraception au Moyen-Âge* (Paris, 1988), deals mainly with the latter Middle Ages but has valuable insights for the earlier period as well. Caesarius of Arles' position is summarised in A.-M. Dubarle, "La contraception chez saint Césaire d'Arles," *La Vie Spirituelle, Suppléments* 17 (1963): 515-519.

[43] *S.* 19.5, *CCSL* 103: 91; *S.* 51.4, *ibid.,* 229; *S.* 51.1, *ibid,* 227; *S.* 52.4, *ibid.,* 231. The obverse of recourse to trees for aid in getting a child is apparent in a much later practice, by which a woman who "despised" her husband's seed placed the semen into a rotten tree to prevent him from begetting (or herself from conceiving?) offspring, a sin meriting seven years of penance (*Poen. Ps.-Theodori* [mid 9th century] 16.30, Wasserschleben, 576).

[44] *S.* 51.4, *CCSL* 103: 229. For the ingredients used in such preparations, see Noonan's index in *Contraception* under the heading, "contraceptive methods," 562.

[45] *S.* 51.4, *CCSL* 103: 229; *S.* 44.2, *ibid.*, 196.

women.[46] But he directed himself to the well-born ladies, mistresses of households, who attended his masses, and paid no attention to the motivation of their less fortunate sisters. The authors of a handful of penitentials, however, implicitly recognized the special burden of pregnancy on a poor woman by suggesting a lighter penance for the "poor little thing" (*paupercula*) who killed "someone" (*aliquem*) by means of her sorcery or a potion. Two replaced *aliquem* with *aliquos*, acknowledging that some women habitually resorted to abortion or infanticide.[47] Burchard of Worms explicitly differentiated between a poor woman who used contraception or aborted her child because she lacked the means to raise it and one who did so to hide her shame.[48]

These texts take for granted that the woman in question took the potion deliberately, but that was not necessarily the case. Regino of Prüm and Burchard of Worms implied that either a man or a woman might have drunk poisons of this sort unknowingly. He or she was merely the victim of the lust or malice of someone who had administered the poison or performed some other deed designed to cause infertility.[49]

In his attacks on attempts to control fertility, Caesarius had been concerned about ends, and he did not put the emphasis on the magical aspect of the means. Certainly other, non-magical means could have been used also. When the tenth-century *Confessional of Ps.-Egbert* envisions the uses of drinks and "other things" to destroy the child in the womb, there is no reason to assume that these other things, even the potions themselves, were magical.[50] But, although a dozen texts from the 8th century onward failed to spell out the connection between magic and the potions used for contraception or abortion, magic must often haven been suspected if not taken for granted. Three ninth-century texts intended for parish clergy put *potiones* in the midst of unquestionably magical practices,[51] while Bur-

[46] *Cf.* The lack of nuance in the *Collectio Hibernensis* (late 7th/early 8th century) 45.3 and 45.4, H. Wasserschleben, 180 and 181. *Poen. Martenianum* (9th century) 47, Wasserschleben, 291; *Poen. Valicellanum* C.6 (10th/11th century) 57, Schmitz 1: 379.

[47] *Liber de remediis peccatorum* (721-731) 7 and 8, Albers, 410; *Poen. Egberti* (before 766) 7.7 and 7.8, Schmitz 1: 580. See also *Poen. Ps. Bedae* (9th century?) 15.3, Wasserschleben, 266; *Poen. Ps.-Theodori* (mid 9th century) 21.6, *ibid.*, 587; *Poen. Martenianum* (9th century) 77.4, Wasserschleben, 300, 239.

[48] *Decretum* (1008-1012) 19, 5.159, Schmitz 2: 444.

[49] Regino of Prüm, *De synodalibus causis* (c. 906) 2, 88: 248; Burchard of Worms, *Decretum* (1008-1012) 17.57, *PL* 140: 933. Since Regino puts this under the heading of castration, the "other" method mentioned may have been literally castration. It is possible, however, that he was thinking of the usual potion or some form of magical constraint, or even onanism, a practice condemned in the *Poen. Hubertense* (8th century, 1st half) 56, and *Poen. Merseburgense b* (c. 774-c. 850) 12, *CCSL* 156: 114 and 174.

[50] 4, *PL* 89: 426.

[51] Ghärbald of Lüttich, *Capitulary 2* (between 802 and 809) 10, *MGH CapEp* 1: 29; *Capitula Silvanectensia prima* (9th century, 1st half) 11, *MGH CapEp* 3: 82-83; Anonymous sermon (c. 850-882) 5, Kyll, 10.

chard of Worms and the *Penitential of Arundel* explicitly connected contraception, abortion and infanticide with sorcery and herbal potions.[52] For Regino of Prüm, too, herbs were the first magical means to come to mind both for contraception and for murder:

> Did you drink any product of sorcery, that is, herbs or something else, so that you would be unable to have children, or did you give it to others, or did you want to kill a man with a death-dealing potion, or did you teach others how to do so? ... Has any woman murdered her husband with poisonous herbs or death-dealing potions, or taught someone else to do so?[53]

Often neither pastor nor practitioner could have differentiated between the magical and the natural result of the use of herbal preparations. J. C. Bologne pointed out the efficacy of herbs was as inexplicable by natural means as was the efficacy of amulets. Unless sanctified by orthodox prayers or the blessing of the priest, or vouched for by a physician, all such techniques were tainted with magic in the eyes of many, lay or clerical. From the point of view of the practitioner, this may not have been wholly true. Cameron insisted on the "observational and experiential pragmatism" of Anglo-Saxon leeches, and there is no suggestion of magic in Gregory of Tours' account of the poisonous beverage, composed of absinthe mixed with wine and honey, which Fredegund administered to a Rouenese dignitary.[54]

The preparations were not always meant to be taken orally. The mid sixth-century Council of Lerida imposed seven years of penance on adulterers of either sex who killed the fetus or the child or placed "any potions whatsoever" into the mother's womb. Although the Leridan assembly evidently believed that it was an abortifacient, a potion applied like this is more likely to be a pessary or spermicide. This canon was ignored for three centuries, to become important again in the 9th century, when it was inserted in the decrees of the Council of Mainz, in both of Rabanus Maurus' penitentials, and in two canonic collections.[55]

Herbs were used in the preparation of other magical *potiones* (although, seemingly, not of love potions, for which other substances were favoured). The *Homilia de sacrilegiis* lists roots and herbal preparations among the forbidden cures attempted for demoniacs.[56] An anaphrodisiac potion was brewed in ninth-century Spain.

[52] Burchard of Worms, *Decretum* (1008-1012) 19, 5.159, Schmitz 2: 444; *Poen. Arundel* (10th/11th century) 17, Schmitz 1: 443.

[53] *De synodalibus causis* (c. 906) I, 304: 145; *ibid.*, II, 5. 8: 209.

[54] Bologne, *La naissance interdite*, 91; Cameron, "Anglo-Saxon medicine and magic"; Gregory of Tours, *HF* 8.31, *MGH SRM* 1: 399.

[55] *Conc. Ilerdense* (546) 2, Vives, 55. See also *Concilium Moguntinum* (847) 21, *MGH CapRegFr* 2: 181; Rabanus Maurus, *Poenitentium Liber ad Otgarium* (842-843) 11, *PL* 112: 1411A; *idem*, *Poenitentiale ad Heribaldum* (c. 853) 9, *PL* 110: 474C; *Collectio Vetus Gallica* (8th/9th century) 50.2g, Mordek, 568; Burchard of Worms, *Decretum* (1008-1012) 19, 5.160, Schmitz 2: 444.

[56] 22, Caspari, 12.

To drink it as an aid to chastity was a relatively minor sin, calling for only a year of penance, but drinking it as a contraceptive merited twelve years.[57]

The use of herbs in other types of magic has already been noted. In sermons and penitentials, pastors from Caesarius of Arles to Burchard of Worms deplored the custom of hanging them as amulets on one's person or belongings.[58] St. Eligius' sermon refers to *ligamina* (perhaps bundles of herbs) hung on the necks of men and beasts.[59] Three penitentials of the eighth and ninth centuries denounced tying ligatures of herbs above Christians, and the Council of Tours warned that amulets of bones and herbs tied on sick, lame or dying animals were snares of the devil.[60] Herbs could also serve as charms that, if swallowed, held in the mouth, sewn into one's clothing, tied around one, or otherwise manipulated, would enable one fraudulently to escape the judicial ordeal.[61] Finally, it has been suggested that the clause *De petendo quod boni vocant sanctae Mariae* of the *Indiculus superstitionum* refers to a herb known to "good people" as "St. Mary's bedstraw."[62]

Herbs were associated with seasonal rites and weather. On St. John's Day in tenth-century Piedmont, women baptised *herbae* (clumps of grass?) and hung them in their houses. They were probably an ingredient in the potions or food consumed in the 8th century on the Calends of May or on its eve.[63] The women of the Rhineland used henbane to make rain.[64]

Herbs and potions had malicious uses as well. Caesarius compared the devil's wiles to the death-dealing sap of herbs mixed by evil-doers into savoury or sweet drinks.[65] A late ninth-century Hessian council denounced the "poison, herbs and sorcery of various kinds" used to bring about someone's ruin. Regino of Prüm worried that women might use poisonous herbs or "death-dealing potions" to kill their own husbands or other men.[66] In this respect, too, the effects of the poisons

[57] *Poen. Albeldense* (c. 850) 97 and 98, *CCSL* 156A: 12; *Poen. Silense* (1060-1065) 243, *ibid.*, 40. The Iberian clergy may have been particularly inclined to look for radical aids to chastity. Martin of Braga had made a distinction between those who were castrated by doctors for medical reasons or by enemies on the one hand, and, on the other, those who castrated themselves so as to be rid of the temptations of the flesh (*Canones ex orientalium patrum synodis* [572] 21, Barlow, 130). The *Vetus Gallica* calls for the removal of a cleric who castrated himself, or had himself castrated, without giving extenuating circumstances (16.2, Mordek, 409).

[58] Caesarius of Arles (502-542) *S.* 13.5, *CCSL* 103: 68.

[59] *Vita Eligii* (c.700-725), *MGH SRM* 4: 706. See Niermeyer, *s.v.* "Ligamen."

[60] *Poen. Merseburgense a* (late 8th century) 36, *CCSL* 156: 136. See also *Poen. Valicellanum I* (beginning of the 9th century) 89, Schmitz 1: 312; *Poen. Vindobonense a* (late 9th century) 39, Schmitz 2: 353; *Conc. Turonense* (813) 42, *MGH Concilia* 2.1: 292.

[61] Burchard of Worms, *Decretum* (1008-1012) 19, 5.167 Schmitz 2: 445.

[62] 19, *MGH CapRegFr* 1: 223. See McNeill and Gamer, *Medieval Handbooks of Penance,* 420.

[63] *Poen. Merseburgense a*, recension W10 (late 8th century) 32, *CCSL* 156: 135; *Poen. Vindobonense a* (late 9th century) 35, Schmitz 2: 353.

[64] Burchard of Worms, *Decretum* (1008-1012) 19, 5.194, Schmitz 2: 452.

[65] S. 167.2, *CCSL* 104: 684.

[66] *Concilium Triburiense* (895) 50a, *MGH CapRegFr* 2,.2: 241; Regino of Prüm, *De synodalibus causis*

were unpredictable, for Burchard distinguished between those who actually did harm by means of these preparations, and those who intended to do so but did not succeed (or, perhaps, who changed their minds).[67] It will be recalled that bewitched herbs were also used by herdsmen and huntsmen to inflict damage on the livestock belonging to rivals.[68]

7.1.2 *Generalists*

Incantator and incantatrix–enchanter and enchantress: The *incantator* as expert in healing and protective magic was discussed in the previous chapter. But in the majority of cases where *incantatores* are mentioned, their role is ambiguous. The texts give no clear indication of their function, whether beneficial or maleficent. In approximately fifteen lists beginning with the documents of the Council of Agde (506) and especially from the 8th century onwards, they appear sometimes among the usually beneficent soothsayers, some times between them and the usually malign *malefici* and *venefici*, or in midst of the latter set, or associated with other magicians whose nature is not quite clear.[69] The groupings may be random but they may also indicate the enchanters' ambivalent or malevolent character.

In other texts, enchanters (*incantatores*) are unambiguously presented as destructive–a Carolingian compilation, which treated them (and *haruspices*) merely as rather shady entertainers on a footing with prostitutes and actors, is the exception.[70] Gregory III seemingly distinguished between soothsayers on the one hand and, on the other, enchanters whom he pairs with *venefici*, "that is, sorcerers [*malefici*]." Rabanus Maurus identified enchanters and sorcerers as purveyors of (literally or figuratively) death-dealing poisons. A gloss to a ninth-century penitential explained *incantatores* as "the enchanters who invoke demons and make men unstable or disturb their minds."[71] They bore the same character in Burchard of Worms' question to penitents: "Did you ever believe or participate in the falsehood that there are enchanters ... who can move storms about or alter men's minds by the incantations of demons?"[72]

When the name of the enchanter is found in apparent connection with a magical practice or device, the relationship is ambiguous. The terms *incantator* and *phylacteria* appear together some nine times, but the enchanter is not necessarily the

(c.906) II, 5.8: 209. See also Burchard of Worms, *Decretum* (1008-1012) 1.94, Interrogatio 8, *PL* 140: 573.

[67] *Decretum* (1008-1012) 19, 5.165, Schmitz 2: 445.

[68] *E.g., Conc. Rothomagense* (650) 4, Mansi 10: 1200.

[69] *E.g., Conc. Agathense*–Additiones (506) 21 (68), *CCSL* 148: 228; Herald of Tours, *Capitula* (858) 3, *PL* 121: 764; *Admonitio Generalis* (789) 18, *MGH CapRegFr* 1: 55; Burchard of Worms, *Decretum* (1008-1012) 10.28, *PL* 140: 837.

[70] *Capitula Rotomagensia* (8th-10th centuries) 6, *MGH CapEp* 3: 370.

[71] Gregory III, *Ep 43 to the nobility and people of Hesse and Thuringia* (c. 738), *MGH EpSel* 1: 69; Rabanus Maurus (d. 856), *Hom*. 45, *PL* 110: 83; gloss to *Poen. Valicellanum* I (beginning of the 9th century) 80, Schmitz 1: 303.

[72] *Decretum* (1008-1012) 19, 5.68, Schmitz 2: 425.

supplier of the amulet: the texts leave open the possibility that clients either frequent *incantatores* or (*aut/vel*) have recourse to amulets. The pairing may be accidental, but it may also mean that phylacteries were used as a form of protection against the enchanters' magic.[73] Similarly, a clause of the Synod of Aspasius "concerning enchanters or those who, moved by the devil, are said to put spells [protective spells?–*praecantationes*] on horns" may be interpreted either as simply explaining the enchanters' practices, or as opposing malevolent to benevolent magicians.[74] The *ligaturae* associated with *incantatores* in late sixth-century Iberia and the Rhineland of the 9th are equally ambivalent.[75] That they could be used either to work harm or to do good is evident from the accusations that herdsmen and hunters recited spells over *ligamenta* both to protect their own animals and to bring pestilence on the herds of others.

The enchanter is almost invariably designated by the masculine noun *incantator* which appears about forty times, compared to three for *incantatrix*. We have already seen Caesarius of Arles' reference to the enchantresses of doubtful character who were summoned in the case of illness. The others are found in Charlemagne's *General Admonition* and Ansegesis' *Collectio* in which a canon, attributed to the Council of Laodicea, urged bishops not to tolerate the presence of various sorcerers and enchantresses.[76]

This does not mean that the authorities elsewhere ignored enchantresses. The Council of Paris (829) made this clear in its canon "concerning the perpetrators of divers evils," which used only masculine terms in its list of enchanters, but went on to demand punishment for all such experts, men or women. In fact, women were more likely to be suspected of practicing the arts of enchantment than men. Some dozen penitentials followed Theodore's in imposing penances of a year or three *quadragesimae* on women who performed incantations and "diabolical divinations."[77] The *Dacheriana* proposed the identical penance for women who "chanted"

[73] *E. g.*, *Conc. Romanum* (721) 12, Mansi 12: 264. Exceptionally, Gregory III put phylacteries in direct juxtaposition with enchanters and sorcerers (*filacteria et incantatores et veneficos*) in the address to the nobility and people of Hesse and Thuringia, see n. 71 above.

[74] *Synodus Aspasii Episcopi* (551) 3, *CCSL* 148A: 163.

[75] Martin of Braga, *Canones ex orientalium patrum synodis* (572) 59, Barlow, 138. See also *Epítome hispánico* (c. 598-610) 1.59: 103 and Rabanus Maurus, *Poenitentium Liber ad Otgarium* (842-843) 123, *PL* 112: 1417. This canon, which forbade clerics to be enchanters or to make ligatures "because they were a binding of souls," is based on the Council of Laodicea (380?) 36 (Hefele-Leclercq, *Histoire des conciles*, 1.2: 1018).

[76] *Admonitio Generalis* (789) 18, *MGH CapRegFr* 1: 55. See also *Ansegesis capitularium collectio* (1st half of 9th century) 1.21, *ibid.*, 399.

[77] *Poen. Theodori* (668-756) I, 15.4, Schmitz 1: 537-538. See also *Canones Cottoniani* (late 7th/8th century) 145, Finsterwalder, 280; *Liber de remediis peccatorum* (721-731) 6, Albers, 410; *Poen. Remense* (early 8th century) 9.13, Asbach, 57; *Excarpsus Cummeani* (early 8th century) 7.12, Schmitz 1: 633; *Poen. Egberti* (before 766) 7.6, Schmitz 1: 580; *Double Penitential of Bede-Egbert* (9th century?) 30.2, Schmitz 2: 694; *Poen. Martenianum* (9th century) 49.1, Wasserschleben, 292; Rabanus Maurus, *Poenitentiale ad Heribaldum* (c. 853) 20, *PL* 110: 491; *Poen. Ps.-Theodori* (mid 9th century)

enchantments, that is, spells.[78] A tenth-century English penitential also identified women, specifically married women, as performers of magic, enchantment and sorcery.[79] By contrast, except for the few texts that included the reference to herdsmen and hunters first appearing in the 7th century, only one document attributed the use of enchantments to men who were not explicitly identified as magicians. A capitulary drawn up by the Italian bishops together with Charlemagne and his son Pippin urged the duty of investigating "those men who perform any kind of divination or enchantment or who do things of the sort."[80] Experts, at least in the eyes of churchmen, might have been men but, among ordinary people, enchantment was typically the province of women.

Incantatio–incantation and enchantment: As has been seen, *incantationes* may be interpreted as incantations (spells), enchantments (magical acts or rituals) or both. It is seldom clear which meaning is intended. One ninth-century penitential condemned divinatory practices made in the course of *incantationes*; here *incantatio* plainly meant magic in general, with no connotation of verbal formulae.[81] In a capitulary attributed to Gregory II, *incantationes* together with *fastidiationes* may stand for a whole range of magical acts and seasonal rituals. It is not clear if these two are simply the first in a list of forbidden practices or if they are general terms for practices detailed in the rest of the clause: the varied observances of the Calends, sorcery, the illusions caused by magicians and the use of lots and divination.[82] *Incantationes* could also mean authorised prayers, and at it must have been difficult at times for simple parish priests to tell the difference between those and the other kind – a late ninth-century ordinance reminded them neither to practice "magical incantations" and sorcery themselves nor to permit others to do so.[83]

The purpose of this type of magic is equally vague. Unspecified enchantments are condemned either by themselves or as one of a variety of undesirable customs in over thirty pastoral texts drawn from both the continent and England. A recommendation to expel those who made use of auguries or *incantationes* was taken from a fifth-century collection by the Council of Agde (506), and is found in Spanish, Gallic and Rhenish collections, and in two ninth-century penitentials.[84] Look-

27.13, Wasserschleben, 597; Burchard of Worms, Decretum (1008-1012) 10.8 and 10.24, *PL* 140: 834 and 836.

[78] *Capitula Dacheriana* (late 7th/8th century) 147, Wasserschleben, 158.

[79] *Confessionale Ps.-Egberti* (c.950-c.1000) 1.29, *PL* 89: 408.

[80] *Capitula cum Italiae episcopis deliberata* (799-800?) 2, *MGH CapRegFr* 1: 202.

[81] *Poen. Ps.-Gregorii* (2nd quarter, 9th century) 16, Kerff, 177.

[82] *Capitulare Gregorii Papae II* (731) 9, *Concilia Germaniae* 1: 37. Du Cange defines *fastidiacio* (*s.v.*) merely as "a kind of superstition", and posits a derivation from *fasti dies*.

[83] *Capitula Ottoboniana* (889 or later) 21, *MGH CapEp* 3: 129.

[84] *Statuta Ecclesiae Antiqua* (c. 475) 83 (LXXXIX), *CCSL* 148: 17). See also *Epítome hispánico* (c. 598-610) 19.83; *Collectio Vetus Gallica* (8th/9th century) 44.1, 44.4g and 55.2, Mordek, 521, 526 and 578; Halitgar of Cambrai, *De Poenitentia* (9th century, 1st half) 4.27, *PL* 105: 686: *Poen. Quadripartitus* (9th century, 2nd quarter) 136, Richter, 17; *ibid.*, 143:18; Burchard of Worms, *De-*

ing into *incantationes* and lots appears together with idolatry and an assortment of condemned beliefs and practices in a sermon ascribed to St. Boniface.[85] The faithful are told in another sermon neither to make use of nor to believe in omens, *incantationes* and ligatures because they are the "tokens of the devil, not the commands of the Lord."[86] A mid eighth-century Germanic Council urged the spiritual and secular authorities to cooperate in suppressing *incantationes* together with other examples of "all the filth of paganism."[87] A title *De incantationibus* appears in the *Indiculus superstitionum*. Typical of all of these is the blanket condemnation of incantations or enchantments, irrespective of content or function.

It cannot be assumed that authors who included the same canon without giving the context necessarily had the same practices in mind, or ascribed the same function to them. Burchard's *Decretum* and ten penitentials treated women's "divinations and diabolical *incantationes*" in isolation. But the clause is modelled on the *Penitential of Theodore* where such women were clearly thought to be engaged in protective magic. Theodore equated them with the practitioners described in canon 24 of the Council of Ancyra (314), who investigated hexes in the house: "Concerning this, it is said in the canon: Those who observe auguries or omens or dreams or any kind of divination according to the customs of pagans, or bring such kinds of men into their houses to seek out [something done by] the art of sorcerers. ..." That later authors separated the two parts of the Theodorian clause means presumably that they took a more general view of the nature of women's magical activities.

In some cases, it is obvious that the authors had some specific if undefined practice in mind. A Carolingian letter worried about priests willing to receive penitents who had been repeatedly guilty of a number of idolatrous practices, including that of putting enchantments on horns.[88] A tenth-century English penitential stated that no meeting should be held with any *incantatio* except the proper prayers, the Lord's Prayer or Creed "or some other prayers which refer to God." Backsliders to "this foolishness" (toasting perhaps) must perform a triple penance.[89] The *Arundel Penitential* dealt with the observance of the Calends of January with *incantationes* and *maleficia*.[90] Burchard prescribed three years of penance for those who celebrated nightly sacrifices to demons or summoned demons to do their will "artfully by means of *incantationes*."[91] These texts probably refer to both spells and other non-verbal rituals.

cretum (1008-1012) 10.7, *PL* 140: 834.

[85] Ps. Boniface (8th century?) *S*. 15.1, *PL* 89: 870.

[86] Ps. Boniface (8th century?) *S*. 8, *PL* 89: 859.

[87] *Conc. in Austrasia habitum q. d. Germanicum* (742) 5, *MGH Concilia* 2.1: 1, 3-4. See also Carloman's *Capitulare* (742) 5, *MGH CapRegFr* 1: 25 and Charlemagne's *Capitulare primum* (769 vel paullo post), *ibid*., 45.

[88] *Epistola Canonica sub Carolo Magno* 5, Mansi 13: 1095.

[89] *Confessionale Ps.-Egberti* (c.950-c.1000) 2.23, *PL* 89: 419.

[90] 93, Schmitz 1: 462.

[91] 10.31, *PL* 140: 837.

Magi–magicians: For Isidore of Seville, *magi* was the general term that embraced all kinds of magicians, but in early medieval pastoral literature, it usually denotes only one practitioner among many others. Despite their positive role in the New Testament,[92] *magi* are treated as reprehensible figures in approximately twenty-five passages between the 6th and 11th centuries. Beginning with the fourth Council of Toledo (633), they appear in over a dozen continental lists of the cunning folk (most often seers) who were consulted even by clerics of the highest rank.[93] Ministers of the altar were warned not to be *magi* or *incantatores* in four texts.[94] After the middle of the 9th century, fears of the clergy being magicians apparently abated for this ordinance was no longer repeated, but Atto of Vercelli in the 10th century and Burchard of Worms in the 11th found it necessary once again to warn all ranks of the clergy against consulting such folk.[95] Burchard also cited a letter from St. Cyprian, concerning a *magus* who, however, seems to have been an entertainer or prestidigitator specialising in indecent performances rather than a magician in Isidore of Seville's sense.[96] In none of these texts are they placed in a context that identifies them conclusively with either benevolent or malevolent magic.

Of particular interest are two texts of Irish origin, where *magi* are associated not with the cunning folk of continental lists (there are no such Irish lists) but with a more puzzling group of evil-doers. The mid seventh-century *Canones Hibernenses* imposed seven years of penance on a *magus* or a person who had dedicated himself to evil or who was cruel, "and the same for an *ergach*, crier, cohabiter, heretic or adulterer."[97] A generation or so later, the *Collectio Hibernensis* reduced the penance due for manslaughter committed in a sacred place which had already been polluted by the presence of a *laicus,* that is, a murderer, thief, perjurer, crier or *magus.*[98] We learn here that Irish *magi* (*i.e.*, druids, as Bieler translated the word in another context) wore their hair in a tonsure like that of Simon Magus.[99]

Of the *magi*'s techniques we know only that they consisted at least partly of protective spells and amulets. According to Maximus of Turin, their spells were used to free the moon from its enemies.[100] Martin of Braga upbraided those who

[92] *Matt* 2, 7 and 16. In *Lev* 19,31, they are linked with *arioli* as suspect figures.

[93] *E.g., Conc. Toletanum* IV (633) 29, Vives, 203.

[94] *Conc. Agathense - Additiones* (506) 21 (68), *CCSL* 148: 228. See also the *Vetus Gallica* (8th/9th century) 44.4d, Mordek, 524; Rabanus Maurus (842-843) *Poenitentium Liber ad Otgarium* 123, *PL* 112: 1417; *Poen. Ps.-Theodori* (mid 9th century) 27.8, Wasserschleben, 596.

[95] Atto of Vercelli (d. 961) *Capitularia* 48, *PL* 134: 37-38; Burchard of Worms, *Decretum* (1008-1012) 10.48, *PL* 140: 851.

[96] *Decretum* (1008-1012) 5.21, *PL* 140: 756-757.

[97] 1.4, Bieler, 160. *Cf.* Wasserschleben, 136, which, less plausibly, gives "credulous" (*credulus*) instead of Bieler's "cruel" (*crudelis*). Bieler translates *ergach* as "hawker" (161).

[98] *Hibernensis* (late 7th/early 8th century) 44.8, H. Wasserschleben, 176-177. For the nuances of *laicus* in early medieval Ireland, see chapter 9.3.

[99] *Collectio Hibernensis* (early 7th century) 52.3, H. Wasserschleben, 212.

[100] *S.* 31.1, *CCSL* 23: 121; *S.* 31.3, *ibid.,* 122.

despised the sign of the cross and clung instead to the incantations "invented by magicians and sorcerers" rather than the *incantatio* of the Creed or the Lord's Prayer.[101] For Rabanus Maurus, those who used spells or amulets were *magi* and enchanters. Women who used various love and healing charms, he declared, should do penance as did the *magi* and *arioli* "who are known to practice the magic arts," but it is not clear that he believed that the professionals dabbled in exactly these charms.[102] The "magicians' sorcery and exploits" referred to in the so-called *Capitulary of Gregory* II seem to be no more than the usual spells and seasonal rituals.[103]

Ars magica–magic: In about a dozen passages, the term "art of magic" (*ars magica*) takes in a variety of forbidden practices used for both productive and destructive purposes: healing, protection, tampering with minds, soothsaying, infanticide.[104] It appears late in our documents, most commonly in legal texts.

The earliest precisely datable reference is in a Carolingian edict forbidding the drawing of lots before a duel (*wehadinc*) was fought, lest perhaps they (the combatants? interested by-standers?) lie in wait "with spells, diabolical tricks and the magical arts."[105] The Council of Tours (813) wanted the faithful to learn that neither the arts of magic nor incantations could work a cure for sick men or beasts.[106] A gloss to a Carolingian edict against hanging bits of writing on trees or poles (as protection against hail?) explains that they are inscribed with "symbols or magic."[107] In Regino of Prüm and Burchard of Worms, those starting on a piece of work were prone to use magic arts.[108]

These were fairly beneficent uses of magic. Others were more sinister. A mid ninth-century Pavian synod had ample testimony proving that the "pestiferous roots and remnants of the magic arts" still survived and were exercised by evil women: "Certain sorceresses (*maleficae*) are said to put illicit love into the minds of some people, but hatred into the minds of others. Others are such poisoners (*venenariae*) that rumours are rife that people have died."[109]

For Rabanus Maurus, magic took in a host of women's practices. Replying to

[101] *De correctione rusticorum* 16, Barlow, 199.

[102] *Poenitentium Liber ad Otgarium* (842-843) 123, *PL* 112: 1417; *Poenitentiale ad Heribaldum* (c. 853) 20, *PL* 110: 491).

[103] *Capitulare Gregorii Papae II*, 9, *Concilia Germaniae* 1: 37.

[104] This is in addition to lengthy excerpts taken from St. Augustine and Isidore of Seville on the nature of demons and magic which Burchard included in the *Decretum* 10.41-10.47, *PL* 140: 839-851.

[105] *Conc. Neuchingense* (772) 4, *MGH Concilia* 2.1: 100.

[106] 42, *MGH Concilia* 2.1: 292; see also Burchard of Worms, *Decretum* (1008-1012) 10.39, *PL* 140: 839.

[107] Gloss to *Duplex legationis edictum* (789) 34, *MGH CapRegFr* 1: 64.

[108] Regino of Prüm, *De synodalibus causis* (c. 906) II, 5.52: 213; Burchard of Worms, *Decretum* (1008-1012) 1.94, Interrogatio 51, *PL* 140: 577.

[109] *Synodus Papiensis* (850) 23, *MGH Concilia* 3, 229.

Heribald's question concerning those who fed their husbands food and drink laced with menstrual blood, or dosed them with tonics made of burnt human skull, or who drank their semen, Rabanus recommended the same penalty as for *magi* and *arioli*.[110] He justified this by citing Theodore of Canterbury "concerning those who exercise *magica ars* [a term absent from Theodore's *Penitential*] and listen for auguries and practice divination. ... He who sacrifices to demons in trivial matters is to do penance for forty days, but who does so in important ones, for ten years." The examples are: putting sick children on the roof or in the fireplace, burning grain in the vicinity of a dead body, women's enchantments and soothsaying, and the consumption of idolothytes. This marks a shift in meaning of "sacrificing to demons." In Theodore's *Penitential*, each of the practices mentioned, including the sacrifice to demons, is a separate item under the heading "Concerning the cult of idols."[111] To Theodore, therefore, idolatry consisted both of formal acts of cult and of magical practices. In Rabanus Maurus' eyes, however, magical acts themselves were the acts of cult. By contrast, in a roughly contemporary capitulary of Rouen, "magical arts" appear to be merely tricks of entertainers, practiced by actors, diviners and enchanters.[112]

The Latin translation of a tenth-century English penitential appears to offer a definition in a clause concerning women's magic: "Concerning a wife who uses the art of magic. If a woman practices the art of magic, or enchantment, or sorcery, or something of the sort...." The penances proposed were relatively light, ranging from forty days to a year of fasting. Love magic is the exception: in such cases the author recommended much longer periods of penance (three years, seven years).[113] In other clauses too, heavier penances were imposed on women who used different kinds of love philters. The term *magica ars* was also applied to what is announced as infanticide ("Concerning a woman who kills her newborn baby by magic art") but is in fact abortion, induced or spontaneous.[114]

Sorcerers and sorcery: *Maleficus*, *veneficus*/ *beneficus*, *venenarius* and their feminine counterparts, *malefica*, *venefica* and *venenaria* together with the etymologically unrelated *sortiaria*, appear to be interchangeable terms for sorcerers and sorceresses, that is, practitioners who might have been involved in a variety of magical practices, but who specialised principally in the destructive magic (*maleficium*, *veneficium*).

The *maleficus* is found in some fifty passages while another half dozen or so contain a reference to *malefica*. As a masculine noun, it is found throughout the period and in documents from all the parts of Europe under consideration, in its feminine form from the sixth-century *Penitential of Finnian* to canons of the ninth-century Synod of Pavia. This is a Christian term–it did not exist as a noun in

[110] *Poenitentiale ad Heribaldum* (c. 853) 20, *PL* 110: 491.

[111] I, 15.1-15.5, Schmitz 1: 537-538.

[112] *Capitula Rotomagensia* (8th-10th centuries) 6, *MGH CapEp* 3: 370.

[113] *Confessionale Ps.-Egberti* (c. 950-c. 1000) 1.29, PL 89: 408.

[114] *Confessionale Ps.-Egberti* (c. 950-c. 1000) 1.31, PL 89: 408-409.

classical Latin.[115] St. Jerome used it in the Vulgate in lists of magicians, and St. Augustine in *The City of God* as a pejorative term for magician.[116] It was thanks to them, rather than to Caesarius of Arles, that *maleficus* entered the medieval vocabulary, for Caesarius used the word but twice, clearly to indicate a beneficent cunning man. It appeared in only one of his lists, where a *maleficus* is one of those who are sought for their ability to help people escape the troubles of this world.[117] Moreover, Caesarius' single use of the word *maleficium* identifies it as protective magic used against a lunar eclipse: "when the moon is hidden, they rely with sacrilegious daring on their outcries and *maleficia* to protect themselves."[118]

In later texts, the *maleficus* presents a different face. Much more than cunning folk like *sortilegi, praecantatores, incantatores, magi* and *herbarii, malefici* were involved with what may be called "sorcery," that is, magic which results in harm to others.[119] To Isidore of Seville, they were the most uncompromisingly malevolent of all the practitioners of the forbidden arts. They got their name from the serious nature of their crimes: they agitated the elements, disturbed men's minds and killed by the force of their spells alone, without recourse to poison.[120] The authors of our texts, however, ignored this definition. Although they sometimes used *malefici* as a general term for magicians engaged in harmful magic, they preferred to allocate weather magic to *tempestarii* and attacks on mental stability to *mathematici.* Moreover, they were convinced that *malefici* achieved their ends by means of poison as well as spells.[121] To Remedius of Chur they were equivalent to those guilty of sacrilege. Their heads were to be shaved and covered with pitch, and they were to be led, while being beaten, around the countryside on the back of an ass for a first offence; for a second, their tongues and noses were to be cut off; if they persisted yet again, they were to be handed over to secular authority. This harsh and

[115] *Maleficus* (wicked, criminal, harmful) is an adjective in classical Latin; it has no magical connotation. For the origins of the Christian usage of this word, see Blaise, *s.v.* "Maleficus" II.2. *Maleficus* as a noun sometimes had a meaning in Carolingian usage closer to common criminal or malefactor than sorcerer. The *Capitulare de villis* (c. 800) couples *malefici* with brigands and charges the civil authorities with the responsibility of repressing both, without the reference to bishops which one would expect if sorcerers were intended (53, *MGH CapRegFr* 1: 88). In the *Capitulare missorum generale* (802), *malefici* are confusingly placed between thieves, murderers and adulterers on the one hand and cunning men on the other–again without reference to clergy (5, *ibid.*, 96). See also *Benedicti Capitularium Collectio* (mid 9th century) 3.181, *PL* 97: 821.

[116] *Deut* 18 10; Jer 27 9; Augustine, *Civitas dei* 10, 9.

[117] *S.* 70.1, *CCSL* 103: 296. *Maleficus* also appears in lists in about a dozen other texts, *e.g.,* in Charlemagne's *Admonitio generalis* (789) 18, *MGH CapRegFr* 1: 55.

[118] *S.* 13.5, *CCSL* 103: 67-68.

[119] For sorcery as malevolent magic in general, see *Witchcraft and Sorcery, passim.*

[120] *Etymologiae* VIII, 9.9: 714.

[121] Gregory III, *Ep. to the nobility and people of Hesse and Thuringia* (c. 733) *MGH EpSel* 1: 69; *Poen. Burgundense* (8th century, 1st half) 10, *CCSL* 156: 63; *Homilia de sacrilegiis* (late 8th century) 18, Caspari, 11; *Synodus Papiensis* (850) 23, *MGH Concilia* 3, 229.

degrading punishment was meant undoubtedly to underline their baseness.[122]

The term sometimes stood for those guilty of general malfeasance. The compilers of the *Tripartite Penitential of St. Gall* attributed to *malefici* an impressive range of pagan and superstitious practices and some practices with no obvious association with sorcery, magic or even paganism. Eleven items appear under a heading *De maleficis* (concerning sorcerers): being a summoner of storms or soothsayer; looking for omens, practicing sorcery or telling fortunes; walking during the Calends of January in the pagan manner "with" a stag or some old woman; harming people by calling on demons; making and fulfilling vows at trees, wells or in any place other than church; making divinations by soothsaying; reading lots with the aid of the *sortes sanctorum*; arson; robbing a grave; miscarrying on purpose (or causing a miscarriage); and, finally, mutilation (cutting off a "member").[123] The same scribes followed an equally elusive system of classification in a section taken loosely from Theodore, where seven more items are presented under a second heading "Concerning *malefici*": putting one's daughters on the roof or in the stove to cure them from fever; making love within forty days of childbirth; making love on Sunday; eating unclean flesh which had already been "eaten" by a she-wolf; tasting one's husband's blood by way of a remedy; eating before receiving Holy Communion (but penitents were assured that blood from the teeth was not sinful).[124] Such a seemingly indiscriminate bundling together of practices is a warning against accepting the classifications presented in the texts at face value.

Maleficium, the sorcerer's art, required considerable skill. When Caesarius of Arles wished to describe how subtly demons could entrap souls "with the most delicate seductions," he compared them to *malefici*, to those who administered "death-dealing juices of herbs" with their flavour disguised in spicy or sweet dishes.[125] Although laymen and women could be sorcerers and, more frequently, practice *maleficium*, some eight texts between the 6th and 9th centuries pinpoint clerics, the mostly highly educated members of society, as the principal experts in the production of amulets and poisons in particular. Clerics' mastery of words of power presumably gave them the advantage over lay practitioners since the effectiveness of charms and poisons was enhanced by the appropriate spells. Popular opinion suspected them at least occasionally of using this skill magically to further illicit aims.[126] It must be remembered, however, that these penitentials were intended primarily for use in monastic communities and naturally concentrated on clerics' sins.

In our texts, the two principal areas in which *malefici* and their alter egos the

[122] *Capitula* (800-806) *MGH Leg* 5: 164. See also *Lex Romana Raetica Curiensis Additamenta Codicis S. Galli* 2, *ibid.*, 164.

[123] *Poen. Sangallense tripartitum* (8th century, 2nd half) 19-29, Schmitz 2: 181.

[124] 34-38, Schmitz 2: 184.

[125] S 167.2, *CCSL* 104: 684. The influence of this sermon is unmistakable in Rabanus Maurus' reference to "the deadly poison of sorcerers and enchanters" (*Hom.* 45, *PL* 110: 83).

[126] *E.g., Concilium Wormatiense* (868) 9, *Concilia Germaniae* 2: 312.

venefici exercised their skill is destruction and love, but even their love magic was thought likely to result in death or serious harm. The earliest Irish penitentials served as models. The *Penitential of Finnian* was the first to call attention to the "monstrous sin" of clerics and (apparently lay) sorceresses or sorcerers of using magic to ruin or inflict serious mischief (*decipere*[127]) on another person, and the evidently less serious offence of using magic for the purpose of seduction if the victim was not injured:

> If any cleric or sorceress or sorcerer ruins anyone by his/her sorcery (*maleficium*), it is a monstrous sin, but it can be redeemed through penance. Let him do penance for six years, three on a measured amount of bread and water, and for the others with abstinence from wine and meat. If however, he did not ruin that person, but gave it for the sake of wanton love, let him do penance for a whole year on a measured amount of bread and water.[128]

The monstrousness of the sin must derive from the use of magic. Finnian used the term for no other sin, even though some called for equivalent or heavier penance.[129] The *maleficium* in question appears to be a potion or a charm; it is not clear whether the sorcerer himself administers this poison for his own ends, or if he provides it for another person's use, nor whether the ruin that it causes is primarily physical or spiritual. In the next clause, however, Finnian uses the same term *maleficium* to describe abortionists' techniques.[130]

Only one penitential copied Finnian on love magic. Some half dozen others preferred to follow the modifications made in the *Penitential of Columban.* Here *perdere* sometimes replaces Finnian's *decipere*,[131] and the first part does not attempt to identify the user of magic in any way: "If anyone ruins another person with his sorcery...." The term *maleficus* is not used. But when it comes to love magic, Columban explicitly calls the practitioner a sorcerer, and assumes that he is a man, probably in holy orders:

[127] In classical Latin, *decipere* means to "entrap," "deceive," *etc.*, but Blaise, *s.v.*, gives to "kill" or "suppress" as primary meanings in Church Latin, along with "deceive," "seduce," *etc.*

[128] *Poenitentiale Vinniani* (1st half, 6th century) 18 and 19, Bieler, 78. See also *Poen. Vindobonense b* (late 8th century) 18, Wasserschleben, 494, 112. Bieler considers that "sorcerer" is in apposition with "cleric", and consequently that this clause refers to clerics and women only. He translates *decipere* as "to lead astray," but that interpretation makes it difficult to understand why Finnian would have made an exception for using magic "for the sake of wanton love"–surely he considered seduction to be "leading astray" as well.

[129] *E.g.*, If a cleric begets a child, whether he murders it or not, he must perform three years of penance on bread and water, three more years of abstinence from wine and meat, and be exiled for seven years; perjurers must undergo seven years of penance, in addition to other mortifications; clerics who are homicides must be exiled for ten years during which they are to fast and abstain (*Poen. Vinniani* 12-13, 22-23, Bieler, 76-78, 80-82).

[130] 20, Bieler, 78-80.

[131] In Bieler's opinion, Finnian's *decipere* and Columban's *perdere* mean the same thing ("Aspetti sociali, 123).

> But if he is a sorcerer for the sake of love and does not destroy anyone, let that cleric do penance on bread and water for an entire year, a layman half a year, a deacon two, a priest three. Particularly, if he destroys the woman's child, let each man add six more periods of forty days, lest he be guilty of homicide.[132]

The sorcerer evidently acted on his own behalf; he used an aphrodisiac or narcotic potion that resulted in a miscarriage or in other serious harm to the woman herself.[133] Here again, the penance differentiates among intentionally destructive magic, love magic and love magic which has resulted in miscarriage. But even when the *maleficus* is envisioned as using his arts only for love magic, death or miscarriage is not ruled out.[134]

In such penitentials, miscarriage may have been either deliberate or inadvertent, but in the *Homilia de sacrilegiis,* which grouped together the sorcerer, the poisoner and those who caused sterility and miscarriages, sterility and abortion were deliberately induced.[135]

The majority of these texts insist on the role of men in sorcery. It was women, however, who caught the attention of the Synod of Pavia (850). As we have seen, the bishops took seriously the rumours about the survival of the magic arts. *Maleficae* were said to put illicit love or hatred into the mind of their victim, and *venenariae* caused the death of several persons. Here again, it is open to question whether the deaths were deliberate or the result of the poisoners' incompetence.[136] A generation later, the *Capitulary of Quierzy* expressed no doubt about the malign intentions of the sorcerers and sorceresses (*sortiariae*) who had sprung up in many places. Their *maleficia* had already sickened many people and killed several. Such people and their accomplices were to be destroyed as the scriptures ordained,

[132] *Paenitentiale Columbani* (c. 573) 6, Bieler, 100. When he talks of abortion, Columbanus uses the same verb, *decipere*, that Finnian used; in this context it clearly means "to kill" and may do so in Finnian's text as well. Similar passages are found in *Poen. Burgundense* (8th century, 1st half) 9 and 10, *CCSL* 156: 63-64; *Poen. Sletstatense* (8th century, 1st half) 9, *ibid.*, 83; *Poen. Oxoniense I* (8th century, 1st half) 7 and 8, *ibid.*, 89; *Poen. Hubertense* (8th century, 1st half) 10 and 11, *ibid.*, 109; *Poen. Halitgari* (817-830) 32, Schmitz 1: 479; *Poen. Parisiense* (late 9th century or later) 17, Schmitz 2: 683.

[133] It was possible, however, that a cleric might use sanctified if unorthodox means "for the sake of love" to cause a woman to abort a pregnancy, as did an Irish abbot after rescuing his foster sister who had been abducted and raped: "Seeing that that woman's womb was swelling with child, the man of God blessed her womb. Her belly thereupon shrank and the embryo vanished from her womb" (*Vita Sancti Ciarani de Saigir* 10, *VSH* 1: 221).

[134] *Poen. Oxoniense I* (8th century, 1st half) 8, *CCSL* 156: 89.

[135] 18, Caspari, 11.

[136] 23, *MGH Concilia* 3, 229. The accused were to be examined carefully and punished with the utmost severity, although they were to be given the opportunity to repent and die in the state of grace. There was nothing new in the severity with which *malefici* were treated: *cf.* Gregory I's congratulatory letter to Paul, the Scholastic of Sicily, that he and the bishop of Catania were taking harsh measures against sorcerers (*Ep.* 14.1 [603], *MGH Ep* 2: 419).

unless they could exculpate themselves by the oath of reliable witnesses or by successfully undergoing the Judgment of God.[137]

An even more uncanny power was attributed to sorceresses, according to the tenth- or eleventh-century *Penitential of Arundel.* They were suspected of lifting people up into the air in the dead of night, a terse reference to the night ride of Diana's hordes described in such detail by Regino of Prüm and Burchard of Worms. Like them, the author was skeptical, imposing penance not for witchcraft, but for the belief in witchcraft.[138]

Continental authors sometimes labelled the practitioners of beneficent magic *malefici.* During an outbreak of varied misfortunes (animal disease, the plague, other illnesses), the people of the diocese of Salzburg, despairing of the efficacy of the means offered by the Church and orthodox physicians, resorted to forbidden agencies including "wicked men and women, seeresses, sorceresses and enchanters." They were warned that the only legitimate human help came from devout doctors who could be trusted to work "without enchantment."[139] Ghärbald of Lüttich used *malefici* as the general term to describe all who used forbidden techniques:

> Let lot-casters and fortune-tellers be sought out, as well as those who observe the months and favourable moments and who observe dreams, and the people who carry around their necks those amulets which are inscribed with who knows what kind of words, and *veneficae,* that is, women who administer different potions in order to abort a pregnancy and who perform certain divinations so that they will be loved the better by their husbands as a result. Have all *malefici,* of whatever they be accused, be brought in front of us so that their cases may be heard before us.[140]

Similarly, in the *Penitential of Ps-Gregory,* all those who put their feverish children on the roof or in the oven, protected themselves during the darkening of the moon with an uproar and *maleficia* "according to sacrilegious usage," or maintained pagan customs in their observances of Thursday and the Calends of January, were *malefici* and *maleficae.*[141]

Sorcerers and sorcery: veneficus/beneficus, venefica, venenarius, venenaria: Over a score of texts between the 7th and 11th century used the term *veneficus* for sorcerer. The *Capitulary of Ghärbald of Lüttich* alone used the word *venefica,* defining her as a maker of abortifacient potions and love potions for women to use on their husbands.[142]

[137] *Capitulare Carisiacense* (873) 7, *MGH CapRegFr* 2: 345.

[138] 84, Schmitz 1: 460.

[139] Arno of Salzburg, *Synodal Sermon* (c. 806), Pokorny, 393-394.

[140] *Capitulary 2* (between 802 and 809) 10, *MGH CapEp* 1: 29. See also *Capitula Silvanectensia prima* (9th century, 1st half) 11, *MGH CapEp* 3: 82-83.

[141] *Poen. Ps.-Gregorii* (2nd quarter, 9th century) 23, Kerff, 180.

[142] See n. 140 above.

Despite the fact that *maleficus* and *veneficus* are found together in about half of the dozen lists which include the latter, the only obvious difference between them is that one is more specialised than the other. A gloss to a ninth-century penitential text concerning harm done *per veneficium* explains that *veneficium* is *maleficium* using poisons or herbs.[143] The case was otherwise in classical Latin, where *veneficus* could be either poisoner or sorcerer, or in the Vulgate, where he is simply a sorcerer and *veneficium* is the general term for sorcery.[144]

In the familiar question "If anyone became a sorcerer for love ...," some dozen texts substituted *veneficus* for Finnian and Columban's *maleficus*,[145] thereby establishing the identity of the two. In addition, two texts replaced *veneficus* with *beneficus*.[146] These dabblers in love potions were not explicitly accused of causing miscarriages on purpose. However, in two versions of the eighth-century *Poen. Merseburgense a*, the usual "If he destroys [*decipere, perdere*] the woman's child by this [sorcery]" is replaced with "If a woman kills the child by this ...." The woman here appears to use the potion as an abortifacient on herself or on another woman.[147] In one probably corrupt text, the *veneficus* may be the woman's paramour: "If [a wife who practices the art of magic] has intercourse with a sorcerer ...." The guilt is evidently hers, since she is the one who has to do penance for "seven winters" for having had dealings with him. But surely *veneficus* is a mistake for *veneficium*, sorcery, that is, love magic.[148]

Venenarius and *venenaria*, each of which is used only once, are treated as synonymous with *maleficus* and *malefica*.[149]

Mathematicus and invocator daemonum: Some fifteen Frankish penitentials of the 8th and 9th centuries applied the term *mathematicus* to sorcerers who invoked demons in order to cause psychic damage: he gained control of minds, drove them into a frenzy, emptied or changed them, or took them away.[150] Twice he is asso-

[143] *Poen. Valicellanum I* (beginning of the 9th century) 84, Schmitz 1: 307. The gloss explains that the sorcery is called *veneficium* because poison or herbs were used.

[144] See *Dict. étymologique*, *s.v.* "Uenenum." See also Franck Collard, "*Veneficiis vel maleficiis*. Réflexion sur les relations entre le crime de poison et la sorcellerie dans l'Occident médiéval," *Le moyen âge* 109 (2003): 9-58. For Biblical usage, see *Apoc* 21, 8 and *Gal* 5, 20. The *Lex Visigothorum* appears to treat *venefici* simply as poisoners rather than as sorcerers (*MGH Legum* 1.1: 259).

[145] *E.g.*, *Poen. Remense* (early 8th century) 9.2, Asbach, 56.

[146] *Poen. Valicellanum I* (beginning of the 9th century) 83, Schmitz 1: 306; *Poen. Casinense* (9th/10th century) *ibid.*, 429-430. *Veneficus* is added by way of explanation in the margin of the former.

[147] 10, recensions V23 and W10, *CCSL* 156: 128-129.

[148] *Confessionale Ps.-Egberti* (c. 950-c. 1000) 1.29, *PL* 89: 408. But the use of *cum* in the phrase *cum venefico coeat* rather than the ablative of means is suprising if poison is intended.

[149] *Homilia de sacrilegiis* (late 8th century) 18, Caspari, 11 and *Syndous Papiensis* (850) 23, *MGH CapRegFr* 2: 122, respectively.

[150] *Poen. Burgundense* (8th century, 1st half) 36, *CCSL* 156: 65; *Poen. Floriacense* (late 8th century) 33, *CCSL* 156: 100; *Poen. Ps.-Theodori* (mid 9th century) 27.20, Wasserschleben, 597-598; *Poen. Vindobonense a* (late 9th century) 27, Schmitz 2: 353. *Mathematici* are most frequently accused of

ciated with storm-makers,[151] but generally he is alone. Regino of Prüm appears to have had *mathematici* in mind when he wrote of those "who disturbed men's minds by invoking demons." Burchard of Worms' penitential, however, attributed the attack on mental stability to *incantatores* and the summoners of storms.[152] *Mathematici* presumably used the same methods as other practitioners who addled minds–spells according to the authors of *Poen. Sangallense Simplex*, love potions, foods, amulets, tricks and illusions according to the Council of Paris–and, no doubt, hallucinogens and other poisons as well.[153]

7.1.3 *Obligator, cauculator/ cauclearius/ coclearius, cocriocus*

Obligatores and *cauculatores* appear in lists of malign magicians in three Carolingian legal texts.[154] The former evidently dealt in ligatures or binding spells. For Du Cange, the term applied to makers of magical ligaments to cure illness, but they might have dealt in other kinds of enchantment, too: one who is *obligatus* is bound, held, conquered or stimulated to desire or lust by magical means.[155]

Du Cange believed that *cauculatores* were dealers in love magic who addled their victims' brains. Thus they were related to both *malefici* and *mathematici.* He based this definition, however, on passages in the *Theodosian Code* and the *Visigothic Laws*, but neither this nor any comparable term appears in either.[156] Grimm more plausibly traced *cauculator* to *caucus, i.e.*, drinking bowl, which implies that the *cauculatores* were hydromancers. But Grimm also suggested that *cauculator* is the origin of *gaukler* (juggler); this would make *cauculatores* sleight-of-hand artists.[157] The medieval penchant for reading mystical significance into numbers suggests still another possibility, that they were numerologists.[158]

Both Du Cange and Grimm treat *cauclearius* and *coclearius* as variations of

seizing men's minds and of driving them into a frenzy. A gloss to the *Poen. Valicellanum I* (beginning of the 9th century) defines *mathematici* as "enchanters who call upon demons and make men inconsistent and overthrow minds;" a marginal note adds "frenzied" (Schmitz 1: 303).

[151] *Poen. Vindobonense b* (late 8th century) 7, Wasserschleben, 496-497. See also *Poen. Ps.-Theodori* (mid 9th century) 27, Wasserschleben, 595-596.

[152] Regino of Prüm, *De synodalibus causis* (c. 906) II, 360: 351; Burchard of Worms, *Decretum* 10.28, *PL* 140: 837 and 19, 5.68, Schmitz 2: 425.

[153] *Poen. Sangallense Simplex* (8th century, 1st half) 11, *CCSL* 156: 120.

[154] *Capitulare missorum generale item speciale* (802) 40, *MGH CapRegFr* 1: 104; *Admonitio Generalis* (789) 65, *ibid.*, 58-59; *Ansegesis capitularium collectio* 1.62, *ibid.*, 402.

[155] Du Cange, *s.v.* "Obligator."

[156] *Codex Theodosiani* 16.3, ed. Th. Mommsen and P.M. Meyer (Berlin, rep. 1954), 460; *Leges Visigothorum* 3, 5.13, *MGH Leg* 1.1: 257.

[157] Du Cange, *s.v.v.* "Cauculatores, cauclearii, coclearii, caucularii, circulatores"–this last does not seem to me to belong with the others. See also Grimm, *Teutonic Mythology,* 1037-1038.

[158] See Vincent Foster Hopper, *Medieval Number Symbolism. Its Sources, Meaning and Influence on Thought and Experience* (New York, 1938), esp. 89-135.

spelling for *cauculator*, since they appear in similar lists.[159] Here again one is tempted to look elsewhere, to a connection between *coclearius* and *coclear, coclearium*, which Du Cange (*s.v.*) defines as bell tower. Since we know from other sources[160] that snail shells (*cochleae, cocleae*) were used in magic rites against storms, the *coclearius* might well have been an *emissor tempestatum*, a benevolent weather-magician,.

Coming from a different time and cultural context is the *cocriocus* (not defined by Du Cange) of a supposedly English penitential. He is placed, puzzlingly, in the midst of practices and personages that are otherwise drawn wholly from Caesarius of Arles. He is not a *caragius*, because *caragius* appears in the same list.[161] Of his nature and activities nothing is known.

7.2 THE USES OF ENCHANTMENT, MAGIC AND SORCERY

___While magic, enchantment and sorcery are frequently condemned in general terms,[162] it is evident that in many cases the authorities had in mind specific practices that they found needless or unwise to spell out in detail. Below, we shall consider the texts dealing with love magic, magic to kill or inflict serious bodily injury, and magical theft and property damage, but without claiming that these categories exhaust the possibilities. Many noncommittal references to *incantatio* and *maleficium* (and, in the 10th or 11th century, *fascinatio*) must conceal some highly specific practice unknown to us. Such a one is the *maleficium* known in early seventh-century Sicilian vernacular as *canterma*. It is clear that it was deeply distressing to Gregory the Great and his fellow bishop Maximian, that it was practiced by "depraved men" and that it was heinous enough to call for imprisonment but did not justify the death penalty; we may guess that it was sexual in nature (it "defiled" the practitioners); other than that, we are in the dark.[163] But here, at least, there is a trace. Others left none.

7.2.1 *Love magic*

Love magic[164] (meaning all the techniques, spells and potions exploited to win,

[159] *Admonitio Generalis* (789) 18, *MGH CapRegFr* 1: 55; *Ansegesis capitularium collectio* (1st half of 9th century) 1.21, *ibid.*, 399.

[160] *Indiculus superstitionum et paganiarum* (743) 22, *MGH CapRegFr* 1: 223; *Poen. Merseburgense a*, 3rd recension (late 8th century) 109, *CCSL* 156: 169; *Poen. Vindobonense a* (late 9th century) 99, Wasserschleben, 22.

[161] *Liber de remediis peccatorum* (721-731) 3, Albers, 411.

[162] *E.g., Capitula de examinandis ecclesiasticis* (802) 15, *MGH CapRegFr* 1: 110.

[163] Gregory I, *Ep. Cypriano Diacono*, *MGH Ep* 1: 313. The editor, P. Ewald, disavows all previous attempts to explain *canterma* (*ibid.*, n 2).

[164] Raoul Manselli's examination of the use of symbolism in early medieval magic is particularly illuminating in reference to love magic ("Simbolismo e magia nell'alto medioevo"). For love magic in hagiography, see Thomas Ezzy, *Daemon Amoris. Amatory Magic vs. Thaumaturgy in Late Antique Hagiography* (M.A. thesis, Université de Montréal, Montreal, 1996). See also Richard Kieckhefer, "Erotic magic in medieval Europe," in *Sex in the Middle Ages. A Book of Essays*, ed.

hold or increase love in another, presumably without his or her knowledge and against his or her will) was practiced by enchanters and enchantresses, sorcerers and sorceresses, sometimes for their own benefit, sometimes for the sake of clients. In addition to these experts, ordinary people also used *incantatio, magica ars, maleficium* and *veneficium* for amorous purposes. Approximately eighty texts deal with love magic, proof either of its popularity or of the anxiety it roused in the clergy. Predictably, penitentials provide most of the information, although some love magic found its way into canon law as well.

In these texts, it is usually women, ostensibly wives (*uxores, mulieres*), who practice these techniques, almost invariably on their own husbands (*viri sui, mariti*). The exact legal and social status of the women in question is open to doubt however. The terms *mulier* and *uxor* could apply either to a legitimate wife or to a woman with no legal claim to wifehood. Germanic customary law provided for three different types of marriage. Formal, socially sanctioned marriage was by purchase or acquisition, from the woman's kin, of the *mundiburdium* over the woman. Other, irregular marriages were formed by elopement (that is, with the woman's consent but without that of her family) or by capture; the last two could be regularised by composition with the woman's kindred. The situation of a wife by irregular marriage was precarious: she was in effect no more than a concubine and, although her children could inherit, she could be discarded and replaced by a legitimate wife. The temptation must have been very strong for a woman in this position to use all means to strengthen her hold on her partner's affections.[165]

Philters are the only form of aphrodisiac magic to be described explicitly. The most common ingredient in English and continental penitentials from the 7th century onward is semen. Many penitentials, beginning with Theodore's, mentioned that some (the masculine or common gender pronoun *qui* is used) drank it or blood for unspecified purposes. But in other passages, the same documents made clear that it was the wife who was the most likely to use the semen, that she mixed it in food, and that her purpose was to secure her husband's love.[166] A score of texts in penitentials and canonic collections echo this verbatim or with more or less significant variations.

Joyce E. Salisbury (New York and London, 1991), 30-55.

[165] See Philip Lyndon Reynolds, *Marriage in the Western Church. The Christianization of Marriage during the Patristic and Early Medieval Periods* (Leiden, 1994), esp. 66-117, and Michel Rouche, "Des mariages païens au mariage chrétien. Sacré et sacrament," in *Segni e riti nella Chiesa altomedievale occidentale*, SSAM 33 (Spoleto, 1987), 835-80.

[166] *Poen. Theodori* (668-756) I, 14.15, Schmitz 1: 536. See also *Capitula Dacheriana* (late 7th/8th century) 87, Wasserschleben, 153; *Canones Gregorii* (late 7th/8th century) 191, Schmitz 2: 541; *Poen. Remense* (early 8th century) 5.100, Asbach, 41; *Excarpsus Cummeani* (early 8th century) 1.36, Schmitz 1: 618; *Poen. Sangallense tripartitum* (8th century, 2nd half) *Judicia Theodori Episcopi* 26, Schmitz 2: 184; *Poen. Vindobonense b* (late 8th century) 3, Wasserschleben, 494, 468; *Poen. XXXV Capitolorum* (late 8th century) 23.2, Wasserschleben, 519; *Poen. Ps.-Theodori* (mid 9th century) 16.30, Wasserschleben, 576; *Poen. Vindobonense a* (late 9th century) 83, Schmitz 2: 356; Burchard of Worms, *Decretum* (1008-1012) 17.29, *PL* 140: 891, 924.

Some merely alter the phraseology slightly or increase the penance, but others provide revealing details. The *Merseburgense a* and *Valicellanum I* penitentials add a vague "or who does forbidden things" to Theodore's text.[167] Rabanus Maurus ascribes the clause to the Council of Ancyra rather than to Theodore; oddly, he relates it to Ancyra's ruling on lesbian activity: "Concerning women who practice fornication among themselves."[168] According to an English penitential, some women hoped that this charm would work on other men as well as on husbands: "If a woman mixes her husband's semen with food and does so that she will be more attractive to men, she is to fast for three years."[169] The author of a fairly late penitential thought that some women might be naively unaware that such practices were sinful; if so the penalty was reduced from three years to forty days.[170] We learn, too, that it was the woman herself who consumed the potion: "Did you drink, did you taste of your husband's semen?" or "whoever drinks or tastes of her husband's semen ...." Such formulations are to be found in half a dozen texts.[171]

Women consumed other substances as well in their bid for their husband's love. According to the *Valicellanum E. 62*, a woman might mix his urine and excrement into food, but it is not clear which of them was expected to eat the preparation or who was supposed to be the more loved as a result; presumably it was she.[172] In the *Penitential of Paris*, potions containing one's own blood and semen could serve purposes other than love and be given to someone else (man or woman) to drink. Conceivably the man himself prepared and administered the potion, but it is more likely that the author had in mind two separate cases–blood probably for healing, semen for love.[173] This is the only hint offered by our documents that men might use their own semen as a love potion, although we have somewhat later evidence that they used it for other magical purposes.[174]

In the cases above, the woman absorbed elements of her husband's body to bind him to her. Other love philters, which he consumed, not she, drew their attractive power from her own body. Burchard of Worms is the authority on

[167] *Poen. Merseburgense a* (late 8th century] 103, *CCSL* 156: 156) and *Poen. Valicellanum I* (beginning of the 9th century) 90, Schmitz 1: 314.

[168] *Poenitentiale ad Heribaldum* (c. 853) 25, *PL* 110: 490.

[169] *Confessionale Ps.-Egberti* (c.950-c.1000) 1.29, *PL* 89: 408.

[170] *Poen. Arundel* (10th/11th century) 54, Schmitz 1: 453.

[171] *E. g.*, Burchard of Worms, *Decretum* (1008-1012) 19, 5.166, Schmitz 2: 445; *Confessionale Ps.-Egberti* (c.950-c.1000) 1.16, *PL* 89: 405. The somewhat later *Penitential of Silo* is even more explicit: "If a wife puts her husband's semen in her mouth or mixes it in food, she is to do penance for three years" (200, *CCSL* 156A: 36).

[172] *Poen. Valicellanum E. 62* (9th/10th century) 29, Wasserschleben, 560.

[173] *Poen. Parisiense I* (late 9th century or later) 18, Schmitz 1: 683.

[174] Gilbert of Nogent (c.1053-c.1124) relates that a young monk who wanted to learn the black arts was required by the devil first to offer him a libation of this "most precious" substance and then to drink some himself (*Monodiae* 1.26, in *Autobiographie*, ed. E.-R. Labande [Paris, 1981], 200-204).

philters of this type which, he was convinced, some women used as a matter of course. "Have you done as some women are accustomed to do? They take a live fish and put it into their vagina and keep it there until it dies, then boil or roast the fish and hand it to their husbands to eat; and this they do so that they will burn the more with love of them." "Have you done as some women are accustomed to do? They prostrate themselves on their face, bare their buttocks and order that bread be kneaded on their bare buttocks and give the baked bread to their husbands; this they do so that they will burn the more with love of them." (It should be noted that this required an ally.) "Have you done as some women are accustomed to do? They take their menstrual blood, and mix it into food or drink, and give it to their husbands to eat or drink, so as to be loved the more by them."[175]

These practices were known to the author of the *Penitential of Arundel* as well, although he preferred to express himself in less detail: "If anyone should give her husband to eat or drink a fish that died in her vagina, or bread kneaded on her *vases* [probably an error for *nates*, buttocks], or her menstrual blood, she is to do heavy penance for five years."[176] Rabanus Maurus was aware only of the use of semen and menstrual blood in mixtures of food or drink.[177] A continental penitential discreetly refrained from giving the recipe and contents itself with general terms: "If a woman mixed her husband's food with any kind of sorcery so that she be more loved as a result ...."[178] Potions consisting of menstrual blood were malign magic of the most dangerous sort as, thanks to Isidore of Seville, the early Middle Ages were well aware. At any contact with it, he explained, "plants fail to germinate, must turns sour, grass dies, trees drop their fruit, rust eats iron, bronze turns black. If dogs eat of it, they get rabies. Asphalt, which resists both iron and water, crumbles immediately when polluted by that gore."[179]

But philters were not the only available form of love magic. The bishops reporting to the emperor from the Council of Paris (829) implied that amulets were used as well as love potions and foods. Ghärbald of Lüttich called for an inquiry into *veneficae* who used "divinations" in order to increase their husbands' love.[180] The *Confessional of Ps.-Egbert*, always the source of unusual formulations, suggests three other possibilities. As we have seen, one clause prescribed seven winters of fasting to a woman ostensibly for frequenting a poisoner or sorcerer or making love to him; more probably her misdeed was using a love potion.[181] Further in the penitential, the woman was chastised for using "some art" as an aid to intercourse. This again may refer to magic for seduction or to a charm against impotence, or

[175] *Decretum* 19, 5.172 and 5.173, Schmitz 2: 447; 5.176, *ibid.*, 448.

[176] 81, Schmitz 1: 459.

[177] *Poenitentiale ad Heribaldum* (c. 853) 20, *PL* 110: 491.

[178] *Poen. Casinense* (9th/10th century), Schmitz 1: 429.

[179] *Etymologiae* 2, XI, 1.140: 36-38.

[180] *Capitulary 2* (between 802 and 809) 10, *MGH CapEp* 1: 29.

[181] *Confessionale Ps.-Egberti* (c.950-c.1000) 1.29, *PL* 89: 408.

merely to unorthodox erotic techniques; the penalty is meant to teach her how polluting the offence is.[182] A third passage concerns the use by men of unspecified enchantments to procure love. It is not clear whether the author intends to refer only to additives to food or drink, or whether he has in mind other forms of magic as well. The clause follows the model established by the Irish penitentials, punishing a cleric more severely than a layman, and a priest more severely still.[183]

Possibly certain spells or formulas were sung or recited in order to induce a man or woman to fall in love against his or her will. From Caesarius of Arles on, prohibitions of love songs and obscene words abound. But our texts present these in a social setting, at feasts and especially weddings; they are usually combined with prohibitions of dancing or games. The clergy objected to the bawdiness of the language, but do not appear to have considered such songs to be magical incantations.

The earliest texts about love magic concerned only the magic itself, not its results, although one penitential displayed skepticism about its efficacy: "If anyone made use of sorcery or enchantment for the sake of love, and achieved the purpose of his crime, let him repent for seven years; if not, for three years."[184] In the first half of the 9th century, however, two councils took notice of its psychological effects. The Synod of Pavia (850) dealt with the direct, intended results of love magic: forbidden love and hatred implanted by *maleficae* into the minds of their victims.[185] The Council of Paris, on the other hand, was concerned with a side effect of the magical methods. It reported a common belief, shared by the bishops themselves, that some people were driven mad with "deception and diabolical illusions" by means of love potions, foods and amulets to the point of being unaware of their degradation.[186] The Council may have meant either that the individuals in question were induced by trickery and fraud to use the magical substances, or that, having used them, they were afflicted by hallucinations. The hallucinations may have included the illusion of flight: a tenuous connection between love magic and illusions of flight can be found in the collections of Regino of Prüm and Burchard of Worms, where the same canon combines references to women's supposed power to influence minds to love or hatred, and other magic, including their claim to ride through the night with a horde of demons disguised as women.[187]

[182] *Confessionale Ps.-Egberti* (c.950-c.1000) 1.31, *PL* 89: 409.

[183] *Confessionale Ps.-Egberti* (c.950-c.1000) 4 -, *PL* 89: 425-426.

[184] *Poen. Arundel* (10th/11th century) 89, Schmitz 1: 462. Note that here the envisioned user of the magic may be a man: the masculine (or common) pronoun is used.

[185] *Synodus Papiensis* (850) 23, *MGH Concilia* 3, 229.

[186] *Conc. Parisiense* (829) Epistola episcoporum 69 (II), *MGH Concilia* 2.2: 669. See also *Benedicti capitularium collectio* (mid 9th century), Add. 21, *PL* 97: 867. According to Plutarch, Lucullus' mental disintegration was attributed to a potion administered by his freedman to increase his affection for him (*The Lives of the Noble Grecians and Romans*, 421).

[187] Regino of Prüm, *De synodalibus causis* (c. 906) II, 5.45: 212; Burchard of Worms, *Decretum* (1008-1012) 1.94, Interrogatio 44 and 10.29, *PL* 140: 576 and 837, and 19, 5.69, Schmitz 2: 425.

7.2.2 *Magical harm*

The efficacy of sorcery to do harm appears to have been questioned seldom during the early Middle Ages even by the best educated members of society. There is no trace of skepticism in Gregory of Tours' report of the public rumour that bishop Droctigisilus of Soissons' fainting fits were brought about by *maleficia* launched against him by his archdeacon.[188] The Council of Seville (619) accepted without hesitation that the bishop of Cabra's ungrateful freedman Eliseus made his patron sick by means of sorcery (he also destroyed his church, possibly by the same means).[189] When the Council of Merida (666) took to task priests who tortured and otherwise afflicted the members of their households on suspicion of having caused sickness magically, it was not for believing in magic, but for inappropriate behaviour–they should have referred the accused to the bishop instead of taking matters into their own hands.[190]

Penitentials in particular devote great attention to the intentional use of magic by ordinary people to inflict damage on each other. Murder and bodily harm were the principal concerns, but impotence was also thought to be caused at times by magic. Crimes against property were less likely to be magical, but it is clear that women in particular were suspected of using magic to steal or damage other people's goods. There is some evidence that clerics made use of the liturgy to do harm to others, and it is possible that fasting, too, might have been used for malign purposes.

Murder and bodily harm: In addition to the texts about *malefici* and *venefici*'s nefarious activities, more than fifty passages refer to the use of potions, poisons and various forms of witchcraft to bring about ruin or death. When the verb used is the ambiguous "destroy" (*perdere*), it is possible that the victim was merely made ill or injured in some other way. But verbs like *interficere* or *occidere* leave no doubt that at least sometimes the magic was not only intended to kill, but actually brought about the death of some unfortunate. A murder carried out like this was, in some clerics' eyes, the unpardonable sin, since it implied the worship of demons. The equation between idolatry and magical murder, originally made in Canon 6 of the Council of Elvira, was repeated in the 9th and 11th centuries.[191]

The most frequently repeated text concerning magical injury is drawn from the *Penitential of Theodore*. In its chapter on the varied circumstances of manslaughter (to avenge one's kin or brother, homicide, the slaying of a churchman, killing under orders, in battle, from hatred or rage, by accident or in a brawl), it called for

[188] *HF* 9, 37, *MGH SRM* 1: 457-458.

[189] 8, Vives, 168.

[190] 15, Vives, 336.

[191] *Poen. Quadripartitus* (9th century, 2nd quarter) 95, Richter, 14. See also Burchard of Worms, *Decretum* (1008-1012) 6.26, *PL* 140: 771. The *Penitential of Silo* (1060-1065) also refused deathbed communion to those who committed murder with sorcery, but did not identify this crime with idolatry (96, *CCSL* 156A: 26).

a penance of seven years or more for one who has killed a man by a poisonous draught or "by art of some sort."[192] Minor differences appear in about a dozen penitentials, English, Spanish and continental, such as the *Parisian Penitential* which redundantly specifies "art of wickedness."[193] An Old Irish version assigns the same penance to "anyone who gives drugs or makes a bogey or gives a poisonous drink so that someone dies."[194]

Another half dozen penitentials gave a significant twist to the original. The *Penitential of Theodore* had in mind primarily the killing of men by men–killing for revenge, or under a superior's orders, or in the midst of a brawl, was a man's province, not a woman's. Moreover, it is not certain that this clause referred to magical murder at all: the poisoned drink may not have required malign spells in its concoction, "art of some sort" could have been simply cunning.[195] But these penitentials shifted the focus to women and removed any doubt about the magical nature of the art. One of the earliest, the *Liber de remediis peccatorum*, proposed the usual seven years of penance for women who used "malefic art, that is, a poisoned draught or any kind of art" to kill anyone. This refers to abortion or infanticide, because the next clause goes on to reduce the penance if the woman happens to be poor.[196] The *Ps.-Theodorian Penitential* substitutes "malice" for "malefic art," thereby eliminating the magical element.[197] Two other penitentials referred to plural victims of this crime, in recognition that some women resorted regularly to such means to cope with unwanted births.[198]

Murderous women were thought to be particularly prone to resorting to poison. Among the questions to be directed to a married woman about various magical brews, a ninth-century English penitential asks whether she intended to

[192] *Poen. Theodori* (668-756) I, 4.7, Schmitz 1: 528. This is comparable with the penances due to the other forms of manslaughter: seven to ten years to avenge kin, three to ten years to avenge a brother, seven to ten years for homicide, seven years or the king's sentence for a churchman, forty days for killing in obedience to a lord or in war, three year for killing in a rage, one if by accident, and ten years if in course of a brawl.

[193] *Poen. Parisiense I* (late 9th century or later) 50, Schmitz 1: 687.

[194] 5. 7, Binchy in Bieler, 272.

[195] The poison mentioned in two clauses by the *Poen. Valicellanum B. 58* (11th century) does not seem to be magically enhanced (Schmitz 1: 783 and 785). The emphasis is on the motive, not the means.

[196] Liber de remediis peccatorum (721-731) 7 and 8, Albers, 410. See also *Poen. Ps. Bedae* (9th century?) 15. 3, Schmitz 1: 266; *Poen. Ps.-Theodori* (mid 9th century) 21.6, Wasserschleben, 58. *Poen. Egberti* (before 766) 7.7, Schmitz 1: 580. The 11th century *Penitential of Silo* made the connection with infanticide (newborn or older) unambiguous, but dropped the explicit reference to sorcery (83, *CCSL* 156A: 24).

[197] *Poen. Ps.-Theodori* (mid 9th century) 21.6, Wasserschleben, 587.

[198] *Poen. Egberti* (before 766) 7.7, Schmitz 1: 580 and *Poen. Martenianum* (9th century) 77.4, Wasserschleben, 300, 239. The *Martenianum* also included *Theodore*'s clause (51.7, Wasserschleben 293, 188).

use one to kill a man.[199] Regino of Prüm was worried that some wives might kill their husbands and other men with poisonous herbs or death-dealing potions, or teach others to do so.[200] "Did you make deadly poison and kill any woman (*aliquam*) with it?" is found among the questions addressed to women in Burchard's penitential. A woman who had done so was to fast for forty days a year for seven years and do perpetual penitence, but if she had not succeeded in the murder, the period of penance was reduced to a year.[201]

The formula "If anyone used his sorcery to destroy anyone" appears in over twenty texts between the 6th and 11th centuries. As noted above, this is more ambiguous than Theodore's clause since the verbs translated as "destroy" may mean leading astray, abortion, murder, or some other form of serious injury. It may even refer to spiritual harm: the Council of Trebur (895) used the usual word for destroy to describe the effects of magic on faith.[202]

Beginning in the 9th century, the belief that murder could be accomplished by sorcery found its way into a handful of capitularies and ecclesiastical decrees and even into a sermon to parish clergy. A capitulary from Trier called for an investigation into soothsayers, persons who observed forbidden customs and women who "hand out potions to abort fetuses, and who perform divinations, or furnish *veneficia* in order to kill a man."[203] This text apparently differentiates between potions for abortion and unspecified forms of sorcery for murder. Its ambiguous wording also raises doubts. The instructions as a whole apply explicitly to both men and women but the pronouns in the nominative are feminine–were both sexes suspected of using magic to kill, or only women? Are divinations distinct from the *veneficia*, or a part of the process for murder?

The Council of Trebur (895) saw witchcraft as a possible outcome of marital infidelity. It denounced adulterous husbands' practice of promising to marry their mistresses after their wives' death, since this led to various sins culminating in sorcery and murder, "for some have been slain by poison, some by the sword, some by diverse forms of witchcraft for the sake of such illicit love."[204] Regino of Prüm and Burchard of Worms also feared that adultery would lead to murder, by poison or magic; but it was the wife that they suspected of killing her husband. They were evidently more concerned about the social implications of the murder than about the means and motive - she was to withdraw permanently from the world because "she has killed her master and her lord." Killing a husband "by any kind of

[199] *Double Penitential of Bede-Egbert* (9th century?) 30, Schmitz 2: 682. Here, as usual, it is not clear whether "man" means person in general, or specifically a male.

[200] *De synodalibus causis* (c.906) II, 5. 8: 209.

[201] *Decretum* (1008-1012) 19, 5.165, Schmitz 2: 445. See also *idem*, 1.94, Interrogatio 8, *PL* 140: 573.

[202] 50, *MGH CapRegFr* 2: 241.

[203] *Capitula Treverensia* (before 818) 5, *MGH EpCap* 1: 55. See also *Anonymous sermon* (c. 850-882) 5, Kyll, 10.

[204] 40, *MGH CapRegFr* 2: 236-237.

diabolical art or trick" or consenting to it, was equally harshly punished in another, contemporary penitential.[205]

The *Homilia de sacrilegiis* furnishes an example of magic akin to cursing tablets. It berated those who either wrote letters or fixed them on a mill or on basilicas "on account of fugitives."[206] There is nothing to identify the runaways (slaves, prisoners, outlaws, an eloping couple?) nor the purpose of the magic (to harm, to bring back, maybe even to help?). We may, however, assume that the writer was a cleric working in his own interest or for another, since the practice involves a written text, however primitive, not merely crude markings inscribed on a tablet or a piece of parchment.

Burchard described precisely magical methods used to incapacitate or get rid of an enemy. Some women "filled with the devil's teachings" used the imprints left by their victims to destroy them: "they observe the footmarks and traces of Christians and take turf from their footmark and observe it, hoping thereby to take away their health or their life." This called for five years of penance.[207] Other women caused their husbands to pine away and weaken (become impotent?) by feeding them bread made of magically prepared flour:

> They remove their clothes, anoint their nude body all over with honey, and roll their behoneyed body this way and that in grains of wheat sprinkled on a sheet on the ground; with extreme care, they collect the grain stuck to their damp body, place it in a mill and, turning the millstone against the direction of the sun, grind it into flour from which they make bread to give to their husbands to eat.[208]

A simpler way to inflict harm on another person was to tie knots in a dead man's belt, apparently while the body was still laid out in the house. This technique was evidently used by either men or women.[209] The harm done by the two latter methods fell considerably short of murder, since the penances were quite light (forty and twenty days respectively).

Although sorcery is not mentioned in our text, almost certainly it was suspected during an outburst of mass panic about diabolical possession in the diocese of Uzès, especially in the vicinity of the church of a certain St. Firmin. Afflicted individuals collapsed to the ground in the manner of epileptics and demoniacs. Presently they (or others) developed wounds or, rather, brand marks that looked and felt like the result of sulphur burns - appropriate sign of the devil's work. Men and women were seized by such irrational terror that they rushed spontaneously

[205] Regino of Prüm, *De synodalibus causis* (c. 906) II, 84: 246-247; Burchard of Worms, *Decretum* (1008-1012) 6.39, *PL* 140: 774; *Poen. Valicellanum C.6* (10th/11th century) 7, Schmitz 1: 352.

[206] 20, Caspari, 11.

[207] *Decretum* (1008-1012) 19, 5.175, Schmitz 2: 447.

[208] *Decretum* 19, 5.193, Schmitz 2: 451.

[209] *Decretum* 19, 5.95, Schmitz 2: 430.

to unnamed places to make offerings of gold or silver, livestock or other valuables. The rich were no more immune to the panic than the poor. The episode is described in a letter written between 828 and 834 by Agobard of Lyons in response to a request for advice from Bartholomew of Uzès.[210] Bishop Bartholomew may have shared Agobard's aversion to magical explanations since this time Agobard failed to present his case against the folly of blaming sorcerers for misfortunes.

The belief that malice and magic underlay sickness and misfortune can be seen in other, less spectacular acts. The familiar canon concerning the introduction of cunning men into homes so that they can rid the house of evil is quite explicit that this is a measure against sorcery.[211] *Praecantationes* were used against the evil eye (*fascinum / fascinus*) in a late eighth-century penitential and in Burchard of Worms' Decretum, and *carmina* and *incantationes* for the same purpose in the *Homilia de sacrilegiis*.[212] These passages show signs of Caesarius of Arles' influence, but they reversed the meaning of *fascinum*, which Caesarius had used in the sense of protective charm.[213]

Magic to cause impotence: Burchard of Worms and the *Penitential of Arundel* blame the malice of discarded concubines for the impotence of newly wed husbands: "Have you done as certain adulteresses are accustomed to do?" asked Burchard. "When first they learn that their lovers want to take legitimate wives, they extinguish the men's passions by some sort of sorcery, so that they are not potent with their wives nor can they have intercourse with them." Burchard worried that this was a skill that women passed on to each other: "If you have done this or taught others how to do so, you should do forty days of penance on bread and water." The *Arundel* took a much harsher view, imposing a combined penance of seven years on any woman "who deprives men of the possibility of having intercourse, so that they cannot exercise [the rights of] legitimate marriage."[214]

These discreet statements can be amplified from the work of Hincmar of Reims (d. 882). In a letter to the ecclesiastical authorities of Aquitaine, he merely mentioned the fact that a man's failure to consummate a marriage may result from physical causes or from the devil's work through the agency of sorceresses–"as

[210] *De quorum inlusione signorum* 1, *CCCM* 52: 237.

[211] *E.g., Poen. Casinense* (9th/10th century), Schmitz 1: 431.

[212] *Poen. Floriacense* (late 8th century) 42, *CCSL* 156: 100-101; Burchard of Worms, *Decretum* (1008-1012) 10.49, *PL* 140: 851 (Mansi attributed this to a Neustrian council: *Conc. Cabilonense* [c. 650] 10, Mansi 10: 1197); *Homilia de sacrilegiis* (late 8th century) 15, Caspari, 9.

[213] *S.* 184.4, *CCSL* 104: 750-1. The 8th century *Penitential of Egbert* also used *facinus* in the protective sense: (8.4, Schmitz 1: 581.

[214] Burchard of Worms, *Decretum* (1008-1012) 19, 5.186, Schmitz 2: 450; *Poen. Arundel* (10th/11th century) 85, Schmitz 1: 460. Gunhild in *Njal's Saga* practices this kind of magic when she puts a spell on her faithless lover to make him impotent with his new wife (trans. Magnus Magnusson and Hermann Pálsson [Baltimore, Md., 1960], 48-53).

usually happens."[215] But, in a treatise written during the controversial divorce proceedings of Lothair, King of Lorraine, against his wife Teutberga, he considered the validity of the common opinion that certain women were able with their *maleficium* to put implacable hatred between husband and wife and inexpressible love between man and woman.[216] He went into painstaking detail on the techniques of creating aversion magically.

With modest reluctance, he describes objects and practices that he and his colleagues found were used to prevent the consummation of married love: "objects bewitched by spells, compounded from the bones of the dead, ashes and dead embers, hair taken from the head and pubic area of men and women, multicoloured little threads, various herbs, snails' shell and snake bits." The magic was sometimes conveyed in clothes and coverings, sometimes in food or drink prepared by sorceresses (*sorciariae*) to drive their prey mad, sometimes in the spells of witches (*strigae*). Men were debilitated by vampires (*lamiae*) or whores, and women embraced by male demons (*dusii*) who took the form of their lovers. There were other practices as well, which Hincmar refuses to describe partly from distaste, partly from fear of spreading unlawful knowledge.[217]

A jealous mistress was not the only one who could be suspected of resorting to magic to unman a would-be husband. Hincmar tells the story of a young nobleman who found himself unable to consummate his marriage although he had been perfectly potent before with a concubine. Without explicitly accusing the bride's mother of bewitching her son-in-law, Hincmar implies that she had done so because the marriage had been settled by her husband against her wishes.[218] We have it from Gilbert of Nogent that malign men could also impose impotence: sorcerers kept his parents from consummating their marriage for more than seven years. The magic was eventually lifted by "some old woman." [219]

____*Theft and property damage*: Magical theft appears in only three of our sources. Representations made to the Council of Paris accused cunning folk of using sorcery to transfer produce and milk from their owners to others. These evildoers seemed to have been working for others, not directly for themselves.[220] For Burchard of Worms, this was women's belief and practice; he assumed that they would steal this way for their own profit as readily as for the benefit of another. Confessors were to ask women whether they had done "what certain women are accustomed to do and firmly believe, that is, that if their neighbour has plenty of

[215] *Ep.* 22, *De nuptiis Stephani et filiae Regimundi comitis*, *PL* 126: 151.

[216] For the question of the marriage of Lothair and Teutberga, see Hefele-Leclercq, 4.1, 313-326 and 365-391.

[217] *De divortio*, XV Interrogatio, *MGH Concilia 4 Supplementum*, 206.

[218] *De divortio*, XV Interrogatio, *MGH Concilia 4 Supplementum,* 205.

[219] *Monodiae* 1.12: 84.

[220] *Concilium Parisiense* (829) Epistola episcoporum 69 (II), *MGH Concilia* 2.2: 669. See also *Benedicti capitularium collectio* (mid 9th century), Add. 21, *PL* 97: 867.

milk and honey, they can transfer all his plenty of milk and honey to themselves and their own livestock, or they believe that they can transfer it for life to whomever they wish, by means of their witchcraft [*fascinationes* and *incantationes*]."[221] Sorcery (*incantatio* and *maleficium*) to remove or to obtain milk, honey, and other belongings also appears in the *Penitential of Arundel*.[222]

Thefts were done for economic gain, but sheer malice added another motive. Regino of Prüm and Burchard of Worms recorded the belief that some women, not content merely to snatch away or pilfer goods magically, used their powers to destroy the property of others.[223] The malicious magic of herdsmen and hunters will be remembered. They were thought to hide bewitched bread, herbs and knots in trees and at crossroads, in part to protect their own herds from plague and disease, in part to bring these misfortunes down on the livestock of others. These *fascinationes* appeared in Burchard's penitential as well as the *Arundel*.[224] Some Rhenish women boasted that they, too, could do harm to their neighbours' livestock, although on a smaller scale. More confined in their movements than men, they had to restrict their activities mainly to barnyard fowl. The sheer baleful emanations of their personality were enough to harm young animals without their being obliged to resort to the more usual magical techniques. "Did you believe as certain women are accustomed to?" asks Burchard.

> They claim that when they step into a house, they can make away with and destroy goslings and the chicks of peafowl and of hens, even piglets and the young of other animals, by word or look or hearing? If you did so or so believed, you should do penance for a year on the appointed days.[225]

It is not clear whether Burchard meant that some women claimed this power or that they actually had it :"Did you believe or did you make away with and destroy ... ?" But the author of the *Penitential of Arundel* did not doubt that this was merely a wrong-headed belief, one, moreover, which was not limited to women: "He who believes that he bewitches (*obfascinare*) or destroys anything in any way by approaching it or with a look is to do a year's penance."[226] The ability to destroy in

[221] *Decretum* (1008-1012) 19, 5.168, Schmitz 2: 446. For a discussion of the importance of bees in Germanic culture and the possibility of stealing swarming bees by magic, see James B. Spamer, "The Old English bee charm: An explication," *Journal of Indo-European Studies* 6 (1978): 279-294. By contrast, the Lorsch bee charm was meant to control a swarm occurring naturally (Edwards, "German vernacular literature," 165).

[222] 79, Schmitz 1: 459.

[223] Regino of Prüm, *De synodalibus causis* (c.906) II, 5.45: 212. See also Burchard of Worms, *Decretum* (1008-1012) 1.94, Interrogatio 44, *PL* 140: 57; *idem*, 19, 5.69, Schmitz 2: 425.

[224] *Concilium Rothomagense* (650) 4, Mansi 10: 1200. See also Regino of Prüm, *De synodalibus causis* (c. 906) II, 5.44: 212; Burchard of Worms, *Decretum* (1008-1012) 1.94, Interrogatio 43, *PL* 140: 576, and 10.18, *PL* 140: 836; *idem*, 19, 5.63, Schmitz 2: 423-424. The *Penitential of Arundel* (94, Schmitz 1: 463) however, mentioned only the protective aspects of this magic.

[225] *Decretum* (1008-1012) 19, 5.169, Schmitz 2: 446.

[226] 80, Schmitz 1: 459.

this way comes close to the anthropologists' definition of witchcraft, that is, a power inherent in the witch, rather than a learned technique.

Various agricultural disasters during the latter part of Charlemagne's reign may have been blamed on magic. Around the turn of the century, he charged his magistrates to prevent wicked men from hiding his seed grain underground or elsewhere, thereby reducing the harvest, and from performing other *maleficia* as well (*maleficia* here may mean crimes only, or sorcery, or both).[227] Some ten years later, mobs in Aquitaine evidently lynched people accused of having spread a deadly dust (a fungus?), for Charlemagne ordered his *missi* into action "against the murders that have occurred among common people during the present year because of the powder."[228] Similar hysteria swept the Lyonnais at about the same time. An epidemic among cattle was blamed on the Emperor's enemy, Grimaldus Duke of Beneventum. His agents were accused of having sprinkled fields, mountains, meadows and wells with a dust fatal to cattle. The outraged Agobard of Lyons was witness that, despite the patent absurdity of the charges, the supposed agents when captured readily admitted their guilt and continued to do so despite beatings, torture and death. Some were killed outright, others had tablets tied to them and were thrown into the river to drown.[229] There can be little doubt that popular opinion ascribed the effectiveness of such malice to magic.

7.2.3 *Harmful Christian magic, liturgy and fasting*

The most convincing evidence in our texts for the use of the liturgy to do harm comes from Visigothic Spain.[230] The thirteenth Council of Toledo (683) threatened to remove and disgrace any cleric who vindictively subtracted anything from the altar furnishings or the divine service, or who replaced the usual altar cloths with funereal cloths and brought in some funereal object:

> Concerning those who in the event of quarrels dare to strip the altars and to remove the lights of the church .... Henceforth any bishop or member of the clergy moved by any sort of resentment or bitterness, who dares to strip the divine altar of its sacred cloths and who covers it with any kind of other, mourning cloth, or who takes the customary tribute of sacred lamps from God's temple or who orders them to be extinguished, or who introduces something appropriate to mourning into God's temple and who, what is worse, brings it about that the customary services are not said in God's temples or that the offering of the individual sacrifice seems to be defrauded in any respect, should know that he is deprived of the dignity and honour of his station.[231]

[227] *Capitulare de villis* (800 vel ante) 51, *MGH CapRegFr* 1: 88.

[228] *Capitulare missorum Aquisgranense alterum* [810] 4, *MGH CapRegFr* 1: 153.

[229] *De grandine et tonitruis* (815-817) 16, *CCCM* 52: 14-15.

[230] But see Lester K. Little, *Benedictine Maledictions. Liturgical Cursing in Romanesque France* (Ithaca and London, 1993) for the prevalence of the custom elsewhere in Western Europe.

[231] 7, Vives, 423-424.

This is not essentially similar to the rituals described by Patrick J. Geary, by which Frankish monastic communities of the high Middle Ages put pressure on their opponents, and humiliated and coerced saints who had not fulfilled their duties of patronage satisfactorily.[232] The symbolism of the extinguished lights and the mournful altar cloth suggests rather that this was an attempt to will an enemy to die–according to his hagiographer, St. Fructuosus of Braga succeeded in doing precisely this to his brother-in-law by using similar tactics assisted by prayer and fasting.[233]

If this was case, the prohibition was disregarded. A little more than a decade later, the seventeenth Council was forced to deal explicitly with those who offered the Mass for the Dead, apparently publicly, in order to bring about the death and damnation of someone who was still alive:

> Concerning those who dare malevolently to celebrate the mass of the dead for the living ... They are nowise afraid to perpetrate this in the holy basilicas, on the altar of the Lord, in the presence of God, for by a fallacious vow they seek to celebrate, for living men, the mass instituted for the repose of the dead, for this reason alone, that the person for whom that same sacrifice was offered should incur the danger of death and perdition by means of that most holy offering; and they request that that which was given to obtain the remedy of health for all, should be performed with an evil motive, to obtain an untimely death for certain individuals.[234]

It is evident that the magical use of the mass is the target here, not the murderous intention. Invoking God to do harm to enemies among one's fellow Christians did not shock the Spanish clergy in principle. The saintly bishop Masona of Merida prayed to have his ruthless archdeacon predecease him; when the hapless man's pious mother pleaded for him, Masona refused her, saying, "What I have prayed for, I have prayed for."[235]

The tendency to use the mass for such purposes may have lingered on in Spain well after the Moorish conquest. In the early 9th century, Halitgar of Cambrai and the author of an anonymous penitential apparently based on his declared that "We ought to offer [mass] for good things [or causes – *bonis rebus*] and not at all for evil ones"[236] – a statement with no evident connection to the Spanish ritual, but rather to a clause in the *Penitential of Theodore* quoting Dionysius the Areopagite

[232] "Coercion of saints in medieval religious practice," in *Living with the Dead*, 95-124.

[233] *The Vita Sancti Fructuosi* 3, ed. Clare Frances Nock (Washington, 1946), 93.

[234] *Conc. Toletanum* XVII (694) 5, Vives, 531-532.

[235] *The Vitae sanctorum patrum Emeretensium* 5.13, trans. and ed. J. N. Garvin (Washington, 1946), 248-253. *Cf.* the quotation attributed to St. Maedoc of Ferns ("Woe to that person whose neighbor is an angry saint") with which Little concludes his book (*Benedictine Maledictions,* 239).

[236] *Poen. Halitgari* (817-830) 61, Schmitz 1: 483. See also *Anonymi Liber poenitentialis*, *PL* 105: 723; *Poen. Theodori* (668-756) II, 5.7, Schmitz 1: 531.

to the effect that it was blasphemous to offer masses for an evil man. Schmitz, in fact, relates it to one of Gildas' canons, which had urged the duty of offering mass for good kings (*pro bonis regibus*) but under no circumstances for wicked ones (*pro malis*).[237] But the only other penitentials to echo Halitgar in this respect are Spanish, both from the 11th century. One insists that the mass should be offered for good servants (faithful servants of God?) and not for bad ones, the other for good things only, not for evil.[238]

Magic or fear of magic might also have been the basis of the "utterly horrible" custom of some Visigothic priests to refuse to receive Holy Communion – perhaps in an attempt to invalidate the Mass by the omission of an essential element.[239]

Fasting, one of the privileged forms of Christian discipline, could also be used in unorthodox ways. We are told of beneficial, although forbidden, uses: in honour of the moon, for the sake of health and for divination.[240] In addition to these, however, an impressive corpus of texts, covering our entire period from every region except Ireland,[241] treats nefarious motives for fasting, and it is just possible that, in some cases, these and other magical uses are behind the prohibition: we have already seen that St. Fructuosus of Braga brought about his brother-in-law's death by *(inter alia)* fasting against him.

Prohibitions against fasting on Sunday go back to the Council of Gangra (c. 345), which opposed excesses of asceticism, and, in Europe, to Martin of Braga, for whom it was associated with Manicheism.[242] The *Penitential of Theodore*, however, introduced a new concept. After dealing with those who fasted on Sunday "from negligence" it cited fasting "to damage (or condemn) the day"; the culprit was to be "detested as a Jew" by all the Catholic churches. Eight other penitentials included this condemnation.[243] Another formulation, "Let it not be believed that

[237] *Praefatio Gildae de Poenitentia* (early 6th century) 23, Bieler, 62.

[238] *Poen. Cordubense* (early 11th century) 11, *CCSL* 156A: 53; *Poen. Silense* (1060-1065) 17, *ibid.*, 19.

[239] *Conc. Toletanum* XIII (683) 9, Vives, 426.

[240] *Conventus et Synodus Erfordiensis* (932) Gesta Synodalia 5, *MGH Leg* 6.1: 109; *Poen. Ps.-Theodori* (mid 9th century) 27.26, Wasserschleben, 598.

[241] The *Bigotian Penitential* (late 8th-late 9th century) I, 9.1 (Bieler, 218) deals with Sunday fasting "by negligence" only, and the *Old Irish Penitential* (c. 800) I, 16 (Binchy in Bieler, 261) with fasting "through carelessness or austerity."

[242] Martin of Braga, *Canones ex orientalium patrum synodis* (572) 57, Barlow, 138. See also Burchard of Worms, *Decretum* (1008-1012) 13.20, *PL* 140: 888. The Council of Gangra banned fasting on Sunday, a practice of the Eustathians (18, Hefele-Leclercq, 1.2: 18; 1040-1041, see also 1031). Priscillianists probably fasted on Sundays during the twenty-one days before Epiphany and during Lent (Chadwick, *Priscillian of Avila*, 14).

[243] *Poen. Theodori* (668-756) I, 11.3, Schmitz 1: 534. See also *Canones Gregorii* (late 7th/8th century) 57, 58, Schmitz 2: 528; *Excarpsus Cummeani* (early 8th century) 12.7, 12.8, Schmitz 1: 640; *Poen. Remense* (early 8th century) 14.10, Asbach, 70; *Poen. XXXV Capitolorum* (late 8th century) 33, Wasserschleben, 524; *Poen. Merseburgense a* (late 8th century) 92, *CCSL* 156: 153-154; *Poen.*

he who fasts assiduously on Sunday is a Catholic," appears in both Frankish and Spanish documents.[244] The Council of Aix-la-Chapelle (816), in a clause copied by Burchard of Worms, suggested that Sunday fasting was done for the sake of continence or out of obstinacy – an anathema was pronounced in either case;[245] Burchard imposed a much lighter penalty on those who fasted for the sake of abstinence or religion.[246]

7.3 WITCHES AND WEREWOLVES

Cunning folk and magicians of different sorts were ordinary people who had acquired and exercised special knowledge which enabled them to help themselves and their neighbours or to do them harm. But there were other human beings as well who, while they usually appeared to be quite ordinary, were believed and believed themselves to live a secret life in which they killed and ate people, harmed babies and livestock, assumed the shape of animals, rode through the air in the train of pagan goddesses and fought battles in the clouds. These people were called *strigae*, *striae* or *astriae* (witches) and shape-shifting, primarily into werewolves.[247]

7.3.1 *Witches and illusions*

Striga, stria, astria: The word *striga* was unknown to classical Latin. It derived from *strix, strigis* (screech-owl) which had been used figuratively to mean sorceress. In our texts, *striga* appears principally in laws and sermons. It made its first appearance in pastoral texts in the mid fifth-century Synod of St. Patrick, where it was identified with the *lamia*, *i.e.*, witch or vampire. Christians, declared the Synod, were forbidden to believe in them.[248] This is the only reference found in

Valicellanum I (beginning of the 9th century) 108, Schmitz 1: 325; *Poen. Martenianum* (9th century) 59.6, Wasserschleben, 296; *Poen. Ps.-Theodori* (mid 9th century) 38.14, Wasserschleben, 608; *Poen. Casinense* (9th/10th century), Schmitz 1: 430-431.

[244] *Epítome hispánico* (c.598-610] 19: 80. See also *Collectio Frisingensis Secunda* (late 8th century) 42, Mordek, 624; *Collectio Vetus Gallica* (8th/9th century) 23.8, *ibid.*, 442; *Poen. Quadripartitus* (9th century, 2nd quarter) 339, Richter, 38

[245] *Conc. Aquisgranense* (816) 68, *MGH Concilia* 2.1: 365; Burchard of Worms, *Decretum* (1008-1012) 13.19, *PL* 140: 888.

[246] *Decretum* (1008-1012) 19, 5.78, Schmitz 2: 427.

[247] See the stimulating studies of witches and werewolves in European history are Carlo Ginzburg's wide-ranging *Ecstasies* and *The Night Battles.Witchcraft and Agrarian Cults in the Sixteenth and Seventeenth Centuries*, trans. John and Anne Tedeschi (Baltimore, 1992). Ginzburg has proved conclusively that the fighters of night battles and werewolves were in many cases attempting to protect the crops against adversaries. Éva Pócs adds supporting evidence from Hungarian witchcraft trials in *Between the Living and the Dead. A Perspective on Witches and Seers in the Early Modern Age* (trans. Szilvia Redey and Michael Webb; Budapest, 1999). See also Piero Camporesi, *Bread of Dreams; Food and Fantasy in Early Modern Europe,* trans. David Gentilcore (Chicago and Cambridge, 1980), for the relationship between hunger, the shifts to which the starving resorted for food, and collective fantasies in societies at subsistence levels.

[248] *Synodus I S. Patricii* (c.457) 16, Bieler, 56. The only other appearance of *lamiae* in our litera-

our documents drawn from the earliest period or from areas inhabited primarily by Celtic populations.

In the Germanic world, the word was used for a specific kind of witch, one who cooked and devoured men. Belief in such beings was firmly entrenched in folk culture and law. The Salic and the Alamannic legal codes penalised those who wantonly accused others of being *strigae* or their assistants, but did not question the existence of *strigae*.[249] The *Salic Law* fined *striae* heavily if it was proven that they had eaten human flesh.[250] The *Edict of Rotharius*, however, having forbidden the Lombards to kill "a freedwoman [*aldia*] ... as though she were a *striga*, who is called a *masca*," went on to declare that it was impossible for a Christian mind to believe that a woman could eat the viscera of a living man.[251] The belief continued in Saxony to the very end of the 8th century. Charlemagne thought that the Saxons murdered witches – significantly, by burning them to death – and ate them. He threatened his newly conquered subjects with the death penalty: "If any dupe of the devil believes in the pagan fashion that any man or woman is a *striga* and eats men, and sets fire to her because of this or gives her flesh to be eaten, or if he eats her, he is to be punished with the sentence of death."[252] This passage, with the witches being devoured in cannibalistic rituals, stands in contrast with the more common belief described by Burchard, in which it was the witches who devoured their victims. (It is noteworthy that while this clause assumes that either a man or a woman may be accused of being *strigae*, it considers only the case of women who are actually burned to death and eaten.)

The list of the devil's works given in a sermon attributed to St. Boniface includes belief in witches and "imaginary wolves" (*strigae, ficti lupi*).[253] An anonymous sermon of the 8th century testifies to the prevalence of the belief in such witches, and the struggle to suppress it: many people asserted that there were *astriae* who ate babies, cattle and horses and were guilty of other types of evil-doing. These ideas are not worthy of belief, maintains the sermon, because they were refuted by the wise; there never were *astriae*; and such talk is a wile of the devil, who speaks through stupid, witless, unhallowed people in order to trick mankind.[254] Another roughly contemporary sermon lumps *striae* with various preternatural denizens of the conceptual world of the uncouth: "There are certain oafish men who believe that some women must be *striae,* as it is called by the

ture is in Martin of Braga's *De correctione rusticorum*, where they are demons who, having been expelled from heaven, took up residence in rivers as naiads (8, Barlow, 188).

[249] *Pactus Legis Salicae* 64.1, *MGH Leg* 1.4.1 and 1.4.2: 230-232; *Leges Alamannorum Pactus* 31, *MGH Leg* 1.5.1: 23.

[250] *Pactus Legis Salicae* 64.3, *MGH Leg* 1.4.1, 232.

[251] *Edictum Rotharii* 376, in *Le leggi dei Longobardi,* ed. Claudio Azzara and Stefano Gasparri, (Milan, 1992), 100.

[252] *Capitulatio de partibus Saxoniae* (775-790) 6, *MGH CapRegFr* 1: 68-9.

[253] S. 15.1, *PL* 89: 870.

[254] Anonymous sermon (8th century), Morin, 518.

people, and that they can do harm to babies and livestock, or that goblins or nixes or protective spirits must exist."[255] Still another sermon reports that some people believed that the moon was pulled down by *striae*, and that that was what caused the eclipse.[256]

But the popular image of the witch was not limited to her cannibalistic and animal-harming activities. In the northern parts of the Frankish realm, it also took in a complex of ideas that, by the end of the Middle Ages, had developed into the witches' sabbath – flight through the air, secret battles and the walking dead. These ideas, which, it seems, were vigorously defended and stubbornly adhered to by significant numbers of people, especially women, were based on experiences which the clergy of the early middle ages dismissed as illusions held by foolish people unable to tell the difference between dream and waking reality.

Illusions: The *Canon episcopi* of Regino of Prüm's *De synodalibus causis* presents full-blown the first available account in medieval literature of the witches' ride and shape-shifting. Regino (seemingly) ascribes the canon to the Council of Ancyra, but this is an error, perhaps deliberately made to give it the weight of ancient tradition.[257] It is evidently taken from a Frankish capitulary or the rulings of a Frankish Council of which the records are now lost. This document must have originated in the second half of the 9th century since it is all but inconceivable that, had it been held much earlier, some hint of it would not have emerged before that time.

As it is, there is evidence that the church fathers of the 9th century were worried about undefined mental phenomena. We have already see that the Council of Paris (829) had connected "tricks and diabolical illusions" with the use of love potions, foods and amulets which drove people mad.[258] Belief in "the extremely filthy visions" induced by demons called for seven to ten years of penance in a contemporary penitential.[259] A mid century collection contains a possible mention of sorceresses who made or feigned to be monsters (of this more later).[260] Most significant is a canon promulgated toward the end of the century by the Council of Trebur in Hesse which considered certain "diabolical illusions" to be a serious danger to faith:

> Concerning the case of a person who causes another to turn away from the true faith or ruins him, and [whether] he should be doubly punished: If indeed anyone, man or woman, turns someone from the right belief and the Catholic faith by a diabolical illusion, "it is right," as the evangelist proclaims,

[255] Anonymous sermon (late 8th/9th century), Levison, 310.

[256] *Homilia de sacrilegiis* (late 8th century) 16, Caspari, 10.

[257] The preceding canon is marked *Ex Concilio Anquirensi* and under the title of the *Canon episcopi* is written *Unde supra*. Burchard also attributed it to the Council of Ancyra.

[258] *Epistola episcoporum*, 69 [II], *MGH Concilia* 2.2: 669-670.

[259] *Poen. Ps.-Gregorii* (2nd quarter, 9th century) 27, Kerff, 184.

[260] Herard of Tours, *Capitula* (858) 3, *PL* 121: 764.

> "that a millstone be hung around his neck and he be plunged into the depths of the sea." For it would be far better for him who entices others into sin so that they offend God and turn their back on the true faith and salvation, that he, being defiled by filth, be punished for the vices of his own life, than that he be plunged into hell for his own sins and those of others. And if he has ruined anyone by poison or herbs or sorcery of various kinds, he should be punished with a double penance as though for the worst homicide.[261]

This canon, enigmatic as it is, is highly suggestive. The illusion in question affected both the beliefs and the actions of those suffering from it, its consequences led to eternal damnation, and it was brought on by specific techniques. The last point in particular is interesting, because it is our only hint that the illusions were deliberately induced. Moreover, they were not inflicted on unwilling or unaware victims, but actively sought by persons who had been recruited to beliefs and practices involving the use of hallucinogens.[262]

Regino's text is far more circumstantial. Unlike the canon above, it defines the active participants as women exclusively, although both sexes take the illusion for reality, and it makes clear that the belief was widely held. It identifies the nature of the illusion as the ride of "Diana" and, almost incidentally, as shape-shifting. The experience described fits Gilbert Rouget's definition of ecstasy: immobility, silence, solitude, absence of crisis, sensory deprivation, remembrance and hallucination.[263] While Regino takes the doctrinal aspects of the belief quite as seriously as Trebur, he suggests that argument, not punishment, should be used to counter it. Because of its importance, it is worth quoting the text in its entirety:

> Concerning women who claim to ride with demons during the hours of night .... Let bishops and the bishops' ministers strive to work with all their might to uproot wholly a dangerous type of sorcery and witchcraft [*sortilega et malefica ars*], the invention of the devil, and if any man or woman is found to be an adherent of this kind of crime in a manner shameful to reputation,[264] he is to be driven out of their parishes. For the Apostle said: Shun a heretic after the first and second warning, knowing that he who is of this kind is damned. Those who, having abandoned their Creator, seek help from the devil are damned and held prisoner by the devil. And, therefore, Holy Church must be cleansed of this plague.
>
> It must not be omitted indeed that certain criminal women, who have turned back to Satan and are seduced by illusions of demons and by phan-

[261] *Conc. Triburiense* (895) 50, MGH *CapRegFr* 2: 241.

[262] See Michael J. Harner, "The role of hallucinogenic plants in European witchcraft," *Hallucinogens and Shamanism* (New York, 1973), 125-150; R. Gordon Wasson "What was the Soma of the Aryans?" in *Flesh of the God*, 201-213; Emboden, "Ritual use of cannabis sativa;" Francis Huxley, "Drugs," in *Man, Myth and Magic* 5: 711-716.

[263] *La musique et la transe*, 52.

[264] This passage may also be read as: "if any man or woman of standing is found to be an adherent shamefully of this kind of crime ..."

> tasms, believe and avow openly that during the night hours they ride on certain beasts together with the goddess of the pagans Diana and an uncounted host of women; that they pass over many lands in the silent dead of night; that they obey her orders as those of a mistress, and that on certain nights they are summoned to her service. But would that they only would perish in their treason and not drag many with them down into the ruin of infidelity! For a countless number, deceived by this false supposition, believe that this is indeed true and, thus believing, turn away from the right faith and return to the error of the pagans in thinking that there is any divinity or deity other than the one God.
>
> Accordingly, bishops must preach on every occasion to the people throughout the churches committed to their care, that they may know that this is utterly false and that such phantasms are inflicted on the minds of unbelievers not by the divine but by the malign spirit, to wit, by Satan himself. When he has taken hold of some trifling woman's mind, he transforms himself into an angel of light and subjects her to him through her lack of faith and false belief. Thereupon he transforms himself into the forms and likenesses of many persons and, tricking the mind that he holds captive in sleep [dreams?], leads it astray, showing it things merry and sad, persons known and unknown. While the spirit only experiences these things, the mind, lacking faith, thinks that they happened not in imagination but in the flesh. For who indeed has not been taken out of himself in sleep and in night visions, and seen many things while sleeping which, waking, he had never seen? For who indeed is so stupid and dull-witted that he thinks that things that exist only in spirit happen also in the flesh? ...
>
> Therefore, it is to be announced publicly to all that whoever believes in such and such-like things has lost his faith, and that whoever does not have the right faith in God does not belong to Him, but to the one in whom he does believe – that is, to the devil. For it is written of Our Lord: "All things are made by Him." Therefore, whoever believes that anything can be made or that any creature can be changed into something better or worse, or be transformed into another form or likeness except by the Creator Himself, Who made all things and by Whom all things are made, is beyond doubt an infidel.[265]

Burchard of Worms repeated this text twice. In Book 10, he quoted Regino almost verbatim, with an important addition: the women ride in the company of Diana or of Herodias. In the penitential of Book 19 (where Herodias is not mentioned), the penitent is questioned whether he has believed this or participated in this false belief.[266] But he made significant variations in other, briefer versions of these beliefs.[267] In the midst of proposed investigations of women's misdeeds

[265] *De synodalibus causis* (c. 906) II, 371: 354-356.

[266] *Decretum* (1008-1012) 10.1, *PL* 140: 831-833; 19, 5.90, Schmitz 2: 429. For versions of this clause in subsequent pastoral literature, see Vogel, "Pratiques superstitieuses," 754, n. 5.

[267] For an analysis of the differences among Burchard's versions, see Roberto Bellini, "Il volo

in Books 1 and 10, he refers briefly to a ride composed not of a host of women but of demons: "Query ... whether any woman says that she rides in company with a horde of demons transformed into the likeness of women and that she is numbered as one of their company." Book 10 recommends that such a woman be beaten with broomsticks and chased out of the diocese.[268] In his penitential, he identifies the popular local name of the leader of the Ride as the *striga* Holda (or Frigaholda).[269] It may be more than coincidence that the *Edict of Rotharius* dealt with accusations against a freedwoman (*aldia*) that she was a *striga*.

What happened during these nocturnal cavalcades? Burchard offered two accounts of related beliefs. In one, the women insisted that they did in fact participate in preternatural cannibal feasts like the *strigae*'s. They "maintain and firmly believe" that, while their bodies lie in bed in their husbands' arms ("with your husband lying on your bosom") behind locked doors, they take night journeys over many lands with like-minded women. During these voyages they slay Christians ("men baptised and redeemed by the blood of Christ") without visible weapons, and cook and eat their flesh. Afterwards they stuff "straw and wood or something of the sort in the place of the heart," so that, having eaten them, they can bring them back to life and give them a "respite for living."[270] (We have seen that the *Indiculus superstitionum* may have already hinted at this under the title concerning the belief that women "can take the hearts of men."[271]) For this fantasy-cannibalism, they were to do penance for seven years, as compared to a mere two years for the belief or participation in the belief in the ride of Diana.

There are several suggestive elements in this account. Burchard is careful to state that the victims of these operations are Christian men. The emphasis on baptism may have been intended to underline the absurdity of belief in evil spirits' power over Christian souls, or the specifically anti-Christian character of the rite. "Men" can stand for both sexes, but here the victims may have been exclusively male. It is worth speculating on the actual moment when the wife indulges in this fantasy – in her dream, while the spouses are asleep in each other's arms, or while her husband takes his pleasure "lying on her bosom" and she endures in silence? There appears to be a strong element of vengeful wish-fulfillment directed against men. And what was thought to become of these "dead men on furlough," the murdered men temporarily brought back to life? Did they return to their daily lives, or join the ride of the witches, like the dead heroes who followed the Valkyrie?

A passage in an anonymous eighth-century sermon dealing with the trans-

notturno nei testi penitenziali e nelle Collezioni canoniche," in *Cieli e terre nei secoli XI-XII. Orizzonti, percezioni, rapporti* (Milan, 1998), 293-310.

[268] *Decretum* 10.29, *PL* 140: 837. See also 1.94, Interrogatio 44, *ibid.*, 576.

[269] *Decretum* 19, 5. 70, Schmitz 2: 425. For a discussion of this text, see 2.1.1.4.

[270] *Decretum* 19, 5.170, Schmitz 2: 446.

[271] 30, *MGH CapRegFr* 1: 223. See 3.1.1.2.

migration of souls and diabolical possession may have some bearing. A "heresy" entertained by "stupid folk" consisted of the belief that when the soul left one man it could enter into another man's body. The preacher assured his audience that this was impossible, but immediately hedged: "unless it is done by demons who talk through the man himself/men themselves. And we know that nothing happens other than what we see with the eyes, except for the demons themselves who fly through the very air and have a thousand ways of doing mischief."[272] It is not clear through whom the demons are supposed to talk: the dead or sleeping man, or the one into whom the soul has wandered, or those who believed in such things. The participants in this belief were male, but the reference to hordes of demons of the air who influence the whereabouts of souls suggests some connection with the beliefs that Burchard describes.

In another clause, Burchard evokes the night battles of Ginzburg's sixteenth- and seventeenth-century *benandanti,* in which foreordained individuals went out to fight for the harvest:

> Did you believe what certain women are accustomed to believe, that in the silence of the dead of night, you, together with other limbs of the devil, were raised past locked doors into the air up to the clouds, and there fought with others so that you both wounded other women and were wounded by them? If you have believed this, you are to do penance for three years on the appointed days.[273]

In the roughly contemporary *Penitential of Arundel*, the witches did not ride themselves but moved others (of whom some at least were men) through the air: "If anyone believes that he/she is raised into air in the still of the night by sorceresses, he/she is to do penance for two years."[274]

Effects of witchcraft beliefs: If people genuinely believed in the existence of such sinister beings, we should expect to find evidence that the belief had some influence on their behaviour. In this, as in so many other cases, Burchard is helpful. Two popular beliefs already considered are relevant. First, unclean spirits made the night too dangerous for people to leave the protection of the homestead until the cock's crow dispelled the malign presences. Second, the bold souls who could not resist the temptation to look into the future on New Year's Eve felt it necessary to shelter within a magic circle or on a bull's hide, presumably while they questioned the spirits.[275] It is easy to believe that among these impure spirits and

[272] *Anonymous sermon* (late 8th/9th century), Levison, 312). The notion of demons flying in the air is borrowed from St. Augustine's *De divinatione daemonum*.

[273] *Decretum* (1008-1012) 19, 5.171, Schmitz 2: 447. I am obliged to Pierre Boglioni for the suggestion that there might be a connection between these night battles and the beliefs in "Magonians," the aerial harvest thieves described in Agobard of Lyons' *De grandine et tonitruis*.

[274] 84, Schmitz 1: 460.

[275] *Decretum* (1008-1012) 19, 5.150 and 5.62, Schmitz 2: 442 and 423.

purveyors of the future were the dream cavalcades of the *Canon Episcopi* and Burchard's Penitential.

7.3.2 *Werewolves and shape-shifting*[276]

Ps.-Boniface's reference to "imaginary wolves" and Regino and Burchard's attack on the belief in shape-shifting are amplified in a single text in Burchard's penitential, which connects this with belief in the Fates presiding over birth:

> Have you believed what certain people are accustomed to believe, either that those women whom the people call *Parcae* exist, or that they have the power to do what they are thought to have done: that is, when any man is born, they are able to choose whatever they wish for him, so that whenever he wants, that man can transform himself into the wolf which is called werewolf in German, or into some other shape? If you believed that this ever happened or could be done, that the image of God could be transformed into another form or likeness by anyone except by almighty God, you must do penance for ten days on bread and water.[277]

The trifling penance, in contrast to the years merited by the beliefs in the witches' ride, shows that Burchard considered this to be relatively innocuous.

A faint trace of this belief can be discerned in a ninth-century collection from a diocese well to the southwest of Burchard's. Herard of Tours finished a list of cunning folk and magical practices with the words *de muliebribus veneficis et qua diversa fingunt portenta*, of which a translation may be "sorceresses and those who form various monsters."[278] If this is the correct interpretation, as opposed to "manufacture various/conflicting portents," the clause allocates the responsibility for making werewolves to mischief-making women, not to the preternatural Fates.

[276] For werewolf beliefs, see Richard A. Ridley, "Wolf and werewolf in Baltic and Slavic tradition," *Journal of Indo-European Studies* 4 (1976): 321-331; H. R. E. Davidson, "Shape-changing in the Old Norse Sagas," in *Animals in Folklore*, ed. J. R. Porter and W. M. S. Russell (Ipswich, Cambridge and Totowa, N. J., 1978), 126-142; W. M. S. and Claire Russell, "The social biology of werewolves," *ibid.,* 143-182. J.A. MacCulloch's article on "Lycanthropy" (*Encyc. of Religion and Ethics* 8: 206-220, esp. 206-209) is still useful as a survey of werewolf and shape-shifting beliefs.

[277] *Decretum* 19, 5.151, *PL* 140: 442.

[278] *Capitula* (858) 3, *PL* 121: 764.

8
Death

Vestiges of pre-Christian attitudes about death and the dead survived well into the Middle Ages, leaving traces in archaeology and written documents demons.[1] Although a scattering of data on this subject is available in our texts from Ireland, Merovingian Gaul and Iberia before the Moorish conquest, most come from the Carolingian period and beyond, principally from Frankish and Rhenish territory. The flood of condemnations of reprehensible practices beginning in the mid 8th century can probably be ascribed to the effects of St. Boniface's reforming zeal.

Some are general in tone, simply calling for the eradication of "pagan customs." Such is Pope Gregory III's exhortation (c.738) to the Rhenish gentry and people to abstain from and ban the "sacrifices of the dead" that used to be practiced in Hesse and Thuringia.[2] It appears from a papal letter of 748 that Christian priests of the previous generation had conducted such sacrifices, including bloody offerings of goats and bulls, and participated in communal banquets.[3] Carloman and Charlemagne ordered bishops and counts to persuade Austrasians to turn their backs on these rituals and "all the filth of paganism" in 742 and, in case of the latter, once again a generation later.[4] Louis the Pious instructed his bishops to

[1] See Geary, "The uses of archaeological sources for religious and cultural history," in *Living with the Dead,* 30-45; Bailey Young, "Paganisme, christianisme et rites funéraires mérovingiens," *Archéologie médiévale* 7 (1977): 5-81; Merrifield, *The Archaeology of Ritual and Magic*, especially 58-82. See also Xavier Barral I Altet, "Le cimetière en fête. Rites et pratiques funéraires dans la peninsule ibérique pendant l'antiquité tardive," in *Fiestas y liturgia. Fêtes et liturgie* (Madrid, 1988), 299-308; Donald Bullough, "Burial, community and belief in the early medieval West," in *Idea and Reality in Frankish and Anglo-Saxon Society,* ed. Patrick Wormwald (Oxford, 1981), 177-201; Alain Dierkens, "Cimetières mérovingiens et histoire du haut moyen âge," in *Histoire et méthode* (Brussels, 1981), 15-70; *idem,* "The evidence of archaeology,"; Peter Metcalf and Richard Huntingdon, *Celebrations of Death; the Anthropology of Mortuary Ritual* (2nd edition. Cambridge, New York, Port Chester, Melbourne, 1991). For the development of Christian rituals surrounding sickness, death, burial and commemorations, see Frederick S. Paxton, *Christianizing Death. The Creation of a Ritual Process in Early Medieval Europe* (Ithaca and London, 1990) and Cécile Treffort, *L'Église carolingienne et la mort. Christianisme, rites funéraires et pratiques commémoratives* (Lyons, 1996). See also Guy Halsall, "Burial, ritual and Merovingian society," in *The Community, the Family and the Saint,*" 325-338.

[2] *Ep.* 43, *MGH EpSel* 1: 69.

[3] Zacharias to Boniface, *Ep.* 80, *MGH EpSel* 1: 174-175.

[4] *Conc. in Austrasia habitum q. d. Germanicum* (742) 5, *MGH Concilia* 2.1.1: 3-4. See also *Karlomanni principis capitulare* (742) 5, *MGH CapRegFr* 1: 25; *Capitulare primum* (769 vel paullo post), *MGH CapRegFr* 1: 45; *Benedicti capitularium collectio* (mid 8th century) 1. 2, *PL* 97: 704.

eradicate the "superstitions which some people practice in certain places during the rites for the dead."[5] Sacrifices of the dead "around dead bodies at their tombs" or "over their tombs" are found in two anonymous exemplary sermons of the late 8th or 9th century.[6] Before going on to mention specific rites, a contemporary capitulary from Vesoul urged priests to instruct the faithful to pray for their dead "with devout mind and penitent heart" instead of performing actions left over from pagan tradition.[7]

These texts merely testify to the persistence in Christian territory of pagan mortuary customs. But others are so detailed and specific that it is possible to draw a vivid though incomplete picture of what was done at different periods and in different areas before, during and after the funeral – mourning rituals, preparations of the body, funeral processions, the actual burial, commemorations of the dead, together with appropriate magical techniques. In addition, they give glimpses of beliefs of survival after death.

8.1 Before the funeral

The texts allow us to distinguish two different types of pre-funeral rite to which the clergy objected: mourning rituals and magical techniques designed to protect the survivors from the evil that surrounded death.

Mourning:[8]Most frequently mentioned is the practice of raising clamour of some sort: keening, singing, wild laughter. The mid seventh-century *Irish Canons* condemn the custom of keening or singing dirges (*bardigus, bardicatio*) by a woman (servant girl? the priest's former wife? – *glantella*[9]) over the dead, who are listed according to their status: a *laicus* or *laica*, a pregnant serving woman or her housemate (her lover?), a parish clergyman, an anchorite, a bishop, a scribe,[10] a great prince or a just king.[11] A century or so later, the *Bigotian Penitential* repeated these provisions and made clear that the serving woman had died in childbed. It added a clause singling out for special severity any nun who uttered such cries when moved (to grief). In these texts, *laicus* and *laica* should probably be translated as

[5] *Hludowici Pii capitulare ecclesiasticum* (818, 819) 28, *MGH CapRegFr* 1: 279. See also *Benedicti capitularium collectio* (mid 9th century) 1.229, *PL* 89: 855).

[6] *Anonymous sermon*, Scherer, 439; Ps. Boniface, S. 6.1, *PL* 89: 855.

[7] *Capitula Vesulensia* (c. 800 century) 23, *MGH CapEp* 3: 351.

[8] For mourning rituals, see Ernesto de Martino, *Morte e pianto rituale* (Turin, 1975), esp. 164-235.

[9] Bieler (*Irish Penitentials*, 161-162) and McNeill and Gamer (*Medieval Handbooks of Penance,* 121) translate this word as "serving woman," but, according to Kathleen Hughes, the *clentella* was the wife whom the priest was obliged to put aside when clerical celibacy was imposed in Ireland (*The Church in Early Irish Society* [London, 1966], 51).

[10] McNeill and Gamer identify this with "the principal teacher of the monastery;" abbots were frequently given the title "scribe" (*Medieval Handbooks of Penance*, 122, n. 33).

[11] *Canones Hibernenses* 1.26-1.29, Wasserschleben, 138. See also Beiler, 162-163 and McNeill and Gamer's *Medieval Handbooks of Penance*, 121-122.

"pagans" rather than as "layman" and "laywoman" – in the clause dealing with the pregnant dead woman and her housemate or mates, the Bigotian seems to discriminates between those who "have faith" (Christians, presumably) and those who have not.[12] The same prohibitions appear in the *Old Irish Penitential* under the heading "Anger" (*Ira*).[13]

It is not clear whether Irish clerics objected more to the pagan keening and dirges, or to the unsuitability of mourning for certain people. Keening was always sinful, but the degree of the sin was not always the same, since the penance varied with the standing of the person being mourned (from forty or fifty days if for laymen and laywomen to ten or fifteen days if for anchorites and dignitaries). However, the *Bigotian* confuses the issue by introducing approved, biblical examples of mourning (for Jacob and for Christ), and remarks finally that it is considered to be a bad mark against someone if no lament is made for him.[14]

In Frankish territory, dirges formed only part of the reprobated practices of the wake. A Carolingian text admonished bishops to prevent the "devilish" songs and frenzied laughter with which the masses were accustomed to observe night-watches over the dead.[15] In contemporary Trier, the clergy were concerned about dancers of both sexes who danced around the bodies of the dead, sang, and engaged in pranks or games. They evidently failed to put an end to these practices, since the diocesan clergy were asked about them again during the second half of the century.[16]

The cryptic second clause of the *Indiculus superstitionum* "concerning the sacrilege over the dead (*super defunctos*), that is, *dadsisas*" may refer to a wake ritual also. *Dadsisas* has given rise to various interpretations. Du Cange suggested hesitantly that it refers to funeral feasts (*dapes*) over the tomb. But since the *Indiculus* had already condemned practices at the tomb, and since *defunctus* suggests the dead body rather than the grave, *dadsisas* probably has a different meaning. Other interpretations of this word are "adjuring the dead," dirges and spells, wailing, circumambulation *daesil* (that is, following the orbit of the sun) of the corpse by a chorus of dancers, of the sort forbidden in Trier.[17]

The collections of Regino of Prüm and Burchard of Worms contain repeated condemnations of behaviour at wakes. Regino mentioned them in varying detail five times, Burchard four. The elements identified are: time (night), place (away

[12] *Poen. Bigotianum* (late 8th- late 9th century) IV, 6.6 and 6.3, Bieler, 230. Bieler considered *cohabitator/ trix* to be a gloss on *glandella* (*ibid.*, 257, n. 27).

[13] *Old-Irish Penitential* (c. 800) V, 17 Binchy in Bieler, 273.

[14] IV. 7, Bieler, 230.

[15] *Admonitio synodalis* (c. 813) 71, Amiet, 62. See also Leo IV (d. 855) *Homilia*, 40, *PL* 115: 681.

[16] *Capitula Treverensia* (before 818) 11, *MGH EpCap* 1: 56; *Anonymous sermon* (c. 850-882) 11, Kyll, 11.

[17] 2, *MGH CapRegFr* 1: 223. See Du Cange, *s.v.*; Grimm, *Teutonic Mythology*, 1873; Homann, *Der Indiculus*, 30-34; Derolez, "La divination chez les Germains," 284-285; Haderlein, "Celtic roots," 15-20.

from the church), rites ("devilish" songs, eating and drinking, jesting, dancing and wild laughter which distorted the mouths and faces of the participants) and, especially, emotional content (lack of reverence and brotherly love).[18] Unlike the Irish monks, Regino and Burchard had no objection to lamentations and wailing but only to what Burchard called the "unsuitable gaiety" and "noxious singing" which gave the appearance of rejoicing rather than grieving at the death of one's fellow.[19] Regino ascribed such customs to the *ignobile vulgus*, the rabble or common herd.[20]

Some half dozen continental penitentials from the late 8th to the late 9th century testify to self-mutilation by mourning kinfolk. Under the heading of "Lamentation for the dead," two prescribed penance for a mourner who "lacerates himself with his nails or sword over his dead or pulls his hair out or rends his garments."[21] In the others, the mourner cuts his hair and tears at his face with his nails or sword because of, or after, the death of parents or sons.[22] These records are not from England but, during their visit there in 787, Pope Hadrian's legates noted the "frightful scars" and dyes (tattoos?) sported by some of the Northumbrian and West Saxon gentry; these may have been the result of ritual self-laceration of the same sort.[23]

Protective magic: Mourning rituals no doubt propitiated the spirit of the dead man and protected the survivors, for the dead were feared as well as lamented. Other rituals were unmistakably protective. The practice, of burning grain in the house where a man had died or where he lay dead in about a dozen English and continental penitentials and collections from the 7th to the 11th century, was intended explicitly to safeguard the health of the living and (in most texts) of the house(hold).[24] This may have been either a propitiatory offering to the dead or,

[18] For the purposes of such rituals, see de Martino, *Morte e pianto rituale,* 220-231.

[19] Regino of Prüm, *De synodalibus causis* (c. 906) II, 5.55, 213; *ibid.,* I, 398: 180. See also Burchard of Worms, *Decretum* (1008-1012) 1.94, Interrogatio 54 and 10.34, *PL* 140: 577 and 838. Briefer condemnations are found in *De synodalibus causis* I, 304: 145 and II, 390: 363, and the *Decretum*, 1.94, Interrogatio 54, *PL* 140: 577 and 19, 5.91, Schmitz 2: 429.

[20] *De synodalibus causis* I, Notitium 73: 24.

[21] *Poen. Hubertense* (8th century, 1st half) 53, *CCSL* 156: 114. See also *Poen. Floriacense* (late 8th century) 48, *ibid.*, 101.

[22] *Poen. Merseburgense a* (late 8th century) 131, *CCSL* 156: 162; *Poen. Oxoniense II* (late 8th century) 40, *ibid.*, 197; *Poen. Valicellanum I* (beginning of the 9th century) 132, Schmitz 1: 338; *Poen. Halitgari* (817-830), Schmitz 1: 487; *Poen. Vindobonense a* (late 9th century) 85, Schmitz 2: 356.

[23] *Legatine Synods - Report of the Legates George and Theophylact of their proceedings in England* 19, Haddan and Stubbs 3, 458-459.

[24] *Poen. Theodori* (668-756) I, 15.3, Schmitz 1: 537. See also *Canones Gregorii* (late 7th/8th century) 117, Schmitz 2: 535; *Canones Cottoniani* (late 7th/8th century) 149, Finsterwalder, 281; *Poen. Remense* (early 8th century) 9.16, Asbach, 58; *Excarpsus Cummeani* (early 8th century) 7.15, Schmitz 1: 633; *Poen. Vindobonense b* (late 8th century) 17, Wasserschleben 497, 482; *Poen. XXXV Capitolorum* (late 8th century) 16.4, Wasserschleben, 517; *Double Penitential of Bede-Egbert* (9th century?) 35, Schmitz 2: 683; Rabanus Maurus, *Poenitentiale ad Heribaldum* (c. 853) 20: *PL* 110:

since demons were known to assemble at the approach of death, a fumigation to drive off malign spirits.[25]

Burchard added other customs, maybe peculiar to his diocese, which were probably intended for the same purpose. People (gender is not specified) struck together the combs "which trifling women use to card wool" over the body; as it was being carried out of the house, the dead cart was divided in two and the body placed on the dividing line. Carding combs evokes the idea of the *Parcae* (Norns) cutting the thread of life, but the connection, if any, is obscure. Dividing the dead cart at the last minute may have been meant to leave no time for evil to happen, while placing the body at its exact centre perhaps minimised the chance of danger spilling over on either side. Grimm's interpretation, splitting the cart in two and carrying the corpse out between the halves, seems improbably extravagant. [26]

Still another set of magical rituals accompanied the final removal of the body from the house. "Did you perform or consent to those inanities which silly women are accustomed to perform?" Burchard asks,

> While the body is still lying in the house, they run to the well and in silence bring back a pitcherful. As the body of the dead man is being picked up, they pour the water under the bier, and they watch that the bier is raised no higher than the knees at the moment that it is being borne out of the house - and this they do for the sake of a kind of health.[27]

Here again the elements of the ritual are difficult to interpret, but there is no mistaking the participants' anxiety. The women's silence as they fetch the water contrasts with the usually riotous wake.[28] The timing is crucial: the rites are performed in the last moments during which the dead man may be considered to belong to the house; the water is poured at the critical moment when the corpse is lifted; it is carried knee-high: precisely as it is borne out of the house, that is, over the threshold.[29] The water is poured out as a lustration, but it is not obvious why it should be poured beneath the bier; had it been across the threshold it might have served to keep the spirit from returning, but this is not the case here. The bier may have been carried so awkwardly low to prevent the dead man from seeing

491; *Poen. Ps.-Theodori* (mid 9th century) 27.15, Wasserschleben, 597; *Poen. Parisiense I* (late 9th century or later) 22, Schmitz 1: 684; Regino of Prüm, *De synodalibus causis* (c. 906) II, 368: 353; Burchard of Worms, *Decretum* (1008-1012) 19, 5.95, Schmitz 2: 430; *Poen. Silense* (1060-1065) 198, *CCSL* 156A: 36.

[25] Gilbert of Nogent, *Monodiae*, 1.25, 198-200.

[26] *Decretum* (1008-1012) 19, 5.95, Schmitz 2: 430. See Grimm, *Teutonic Mythology*, 1144.

[27] *Decretum* 19, 5.96, Schmitz 2: 430.

[28] Coincidentally, water could be removed from the spring on the sacred island of Fositeland only if it was done in silence (*Vita Willibrordi auctore Alcuino*, *PL* 101: 700).

[29] For the concept of liminality, see R. Hertz, "A contribution to the study of the collective representation of Death," in *Death and the Right Hand*, trans. R. Needham and C. Needham (New York, 1960); Czarnowski, "Le morcellement de l'étendue," 348. See also *Celebrations of Death*.

something, or to prevent something from seeing the dead man.

Burchard did not take any of these protective rituals very seriously, recommending a mere ten to twenty days penance for them.

8.2 THE FUNERAL

The preparation of the body for burial, funeral processions, the choice of burial site and the method of burial allowed further opportunities for dubious practices which were recorded in continental documents. Some testify equally to the loving care of the living for the dead and to their profound if unorthodox faith in the sacramentals of the Church as the means of salvation.

8.2.1 *Preparation of the body*

Sacraments, altar-furnishings, and kissing the dead: The faithful wanted the dead accompanied to the grave with sacraments and sacramentals, the most powerful protection imaginable. From the fourth-century council of Carthage onward, churchmen issued injunctions to their fellow clergy to refrain from giving Holy Communion to the dead.[30] The Synod of Auxerre (561-605) went into more detail, declaring that it was "not permitted to give the dead either the Eucharist or a kiss, nor to wrap their bodies in the altar cloth or corporal [vestments?]."[31] Iberian, Frankish and Rhenish collections between the 7th and 11th century also repeated rulings against communion for the dead,[32] but no evidence has been found for similar practices in insular sources. The so-called second synod of St. Patrick forbids the "oblation" for the dead, but it is evident that this meant a mass offered for an excommunicate or "layman," not the administration of communion to a corpse.[33]

Placing the consecrated host into the corpse's mouth required the participation of a priest, but lay people made use of their own representations of the sacred species also. While preparing a baptised baby's corpse for burial, eleventh-

[30] *Conc. Carthaginense* (397) 5, *CCSL* 149: 330. Alfred C. Rush showed that among both pagans and Christians, the concept of death as a journey led to the emphasis on the *viaticum* which, for pagans, was a coin for Charon, for Christians, the consecrated Host in one's mouth at the moment of death. The Church condemned the resultant practice of giving Communion to the dead not on the ground that it was superstitious, but because the corpse could not fulfill Christ's command to take and eat (*Death and Burial in Christian Antiquity* [Washington, D.C., 1941], 92-101). See also Elzbieta Dabrowska, "'Communio mortuorum'. Un usage liturgique ou une superstition," *Archiv für Liturgiewissenschaft* 31 (1989): 342-346; Peter Browe,"Die Sterbekommunion in Altertum und Mittelalter," *Zeitschrift für katholische Theologie* 60 (1936): 1-54, 211-240; Pierre Boglioni, "La scène de la mort dans les premières hagiographies latines," in *Le sentiment de la mort au moyen âge*, ed. Claude Sutto (Montreal, 1979), 183-210; here, 192-193.

[31] 12, *CCSL* 148A: 267. See also *Epítome hispánico* (c.598-610) 43.10 (in 2 recensions), 190.

[32] *Epítome hispánico* (43.10: 190; *Collectio Hispana, Excerpta Canonum*, 4.11, *PL* 84: 67; *Statuta Bonifacii* (1st half, 9th century) 18, *MGH CapEp* 3: 362; *Capitulary 2 of Theodulph of Orleans* (813) 30, *MGH CapEp* 1: 182; Burchard of Worms, *Decretum* (1008-1012) 4.37, *PL* 140: 734).

[33] *Synodus II S. Patricii* (7th century) 12, Bieler, 188.

century Rhenish women placed a wax paten with host in the right hand of the little body and a wax chalice with wine in the left, to be buried with it as guarantors of salvation.[34]

Burchard's *Decretum* contains two texts drawn from African Councils concerning the baptism of the dead, a practice attributable to "the weakness of the brethren" or the cowardice of priests.[35] The idea that this happened in Europe as well is quite plausible in principle. Christians were anxious to save their pagan ancestors and to foil the malevolence of the unbaptised dead. They might have seen baptism after the fact as an effective way to achieve these desirable ends. But, since no other document except the *Hispana* mentions this practice and since Burchard himself failed to mention it in his penitential, he must have inserted these passages into the *Decretum* from the desire to be thorough, not because he though them relevant to actual practice in his diocese.

Attempts were made, first in Merovingian Gaul and Visigothic Spain, to appropriate the grace immanent in liturgical vestments and altar furnishings for the benefit of the dead. Wrapping the body in these required the participation, or at least the consent, of the clergy officiating at the funeral mass, for it must have been done in the church itself. The Council of Clermont (535) took note of two separate but related practices. Dead bodies were wrapped in the altar cloths and vestments used in the divine service, and the corporal was placed (evidently by priests) on the corpse while it was being borne out to the grave. The Council opposed the latter not for fear of magical beliefs, but because a cloth that had been used to honour a corpse polluted the altar when it was restored to its liturgical functions.[36] We have seen that a generation or so later, the Synod of Auxerre condemned the wrapping of corpses in church hangings or altar cloths. This practice appears twice in the *Epítome hispánico*. One clause reiterates the rulings of both Clermont and Auxerre: "It is not permitted ... to put [any part of] the veils and cloths which are placed on the altar during Mass on the bier when the bodies go to be buried." The other, a brief and emphatic statement affirming that no dead person was to be wrapped in the altar cloth, implies that claims for special privilege were being made.[37] Such rulings were repeated in the *Vetus Gallica*, the *Statutes of Boniface* and Burchard's *Decretum*.[38]

We may infer from two penitentials of the 8th and 9th centuries that the clergy found it profitable to allow the practice. It appears among various depre-

[34] *Decretum* (1008-1012) 19, 5.185, Schmitz 2: 450.

[35] *Decretum* 4.37, *PL* 140: 734; *ibid.*, 5.31, *PL* 140: 758; *Conc. Carthaginense* (397) 5, *CCSL* 149: 330; *Canones in causa Apiarii* (419) 18, *ibid.*, 106. See also *Collectio Hispana, Excerpta Canonum*, 4.11, *PL* 84: 67.

[36] *Conc. Claremontanum seu Arvernense* (535) 3 and 7, *CCSL* 148A: 106 and 107.

[37] *Epítome hispánico* (c.598-610) 43. 10: 190; *ibid.*, Ex praeceptione beati Clementis 4: 193.

[38] *Collectio Vetus Gallica* (8th/9th century) 59.1, Mordek, 586; *Poen. Silense* (1060-1065) 190, *CCSL* 156A: 35; *Statuta Bonifacii* (1st half, 9th century) 18, *MGH CapEp* 3: 362; Burchard of Worms, *Decretum* (1008-1012) 3.107, *PL* 140: 694.

dations by priests, monks or other clerics on the property of the church (theft and sale of altar vessels, cloths, real estate and slaves).[39] On the other hand, Theodulph of Orleans and Atto of Vercelli saw this not as as an attempt to make profit, but as an example of the clergy's lack of proper reverence for the objects used during divine service.[40]

The ancient custom of giving the dead a kiss, described by the Synod of Auxerre, is found in lists of forbidden funeral customs in the *Epítome hispánico*, the *Statutes of Boniface* and Burchard's *Decretum*.[41] The practice, or its canonical tradition, was most strongly rooted in Spain, for it appears in two quite late penitentials, from Cordoba and from Burgos.[42] In the latter, kissing the corpse acquires an erotic nuance, since the author chose to place it, together with necrophilia, among sexual sins.[43]

Processions: Spanish funeral processions featured the singing or chanting of objectionable verses. The first great Visigothic Council, III Toledo (589), protested against dirges sung by the people (or in the vernacular? – *vulgo*), especially if sung while a cleric's body was being borne to the grave – these were to be prevented at all costs, even if the bishops were unable to ban them from the funerals of the laity. Only psalms sung by a choir were acceptable. The popular dirge was accompanied by a confusingly described action that can be no other than the well-known practice of mourning friends and relatives, of beating their breasts.[44] Echoes of this canon still reverberate in the eleventh-century Cordoban penitential which insisted that a cantor alone sing over the corpse, and that he sing only psalms and prayers for forgiveness (this is not connected explicitly to the

[39] *Poen. Hubertense* (8th century, 1st half) 43, *CCSL* 156: 112; Poen. Merseburgense b (c.774-c.850) 1, *ibid.*, 173.

[40] Theodulph of Orleans, *Capitulary I* (813) 18, *MGH EpCap* 1: 115. See also Atto of Vercelli (d. 961) *Capitularia* 13, *PL* 134: 31.

[41] *Synodus Autissiodorensis* (561-605) 12, CCSL 148A: 267; *Epítome hispánico* 10: 190; *Statuta Bonifacii* (9th century, 1st half) 18, *MGH CapEp* 3: 362; Burchard of Worms, *Decretum* 3.236, *PL* 140: 724. Roman ritual at death included the attempt by one's friends or family to catch the last breath of the dying man, and to give him a kiss (as did St. Ambrose with his brother). This was originally acceptable to the Church. Rush suggests that the Council of Auxerre condemned this practice because of a connection with a Germanic or Gallic custom that connected the kiss with paganism (*Death and Burial*, 101-104).

[42] *Poen. Cordubense* (early 11th century) 158, *CCSL* 156A: 67; *Poen. Silense* (1060-1065) 137, *ibid.*, 30. Neither penitential mentions giving Holy Communion to the dead nor covering them in liturgical vestments.

[43] *Poen. Silense* 133, *CCSL* 156A: 177. The first editors of this penitential, Pérez de Urbel and Vázquez de Parga, considered this to be a form of sorcery ("Un nuevo penitencial español," 16). For ritual necrophilia, see Azoulai, *Les péchés du Nouveau Monde*, 130.

[44] 22, Vives, 132-133. See also *Decretales Pseudo-Isidoriana* 22, Hinschius, 361. *Cf.* Felix Rodriguez's version in *Concilio III de Toledo, XIV centenario* (Toledo, 1991), 31. On breast-beating as a sign of mourning, see Gross, *Menschenhand und Gotteshand*, 59-61.

funeral procession).[45] Elsewhere, only the *Capitulary of Vesoul*, copied by Benedict Levita, described a comparable practice, the custom of raising a high-pitched wail while the body was being borne to the tomb.[46]

8.2.2 *Burial*

The burial site: About forty passages drawn from every region of Western Europe testify to the determination of the faithful to bury their dead in consecrated precincts, as close to the altar and the relics of the saints as possible ("in holy places," "in the baptistery," "within the basilica," "in the church"' "under the church," "in God's temple," "in God's house," "in the sacred precincts of a monastery") and the unsuccessful attempts of the hierarchy to prevent them.[47]

Canon law tried at first to keep bodies out of churches altogether. In 561, the Council of Braga categorically banned burials "within the basilica of the saints," but permitted them outside its walls.[48] The Synod of Auxerre forbade burial within the baptistery; the *Epítome hispánico* did the same and forbade placing the body in the church.[49] A seventh-century Frankish council denied lay people the right to be buried within a monastery, indeed to receive baptism or have a funeral mass said for them there, except with the bishop's permission.[50] Nevertheless, in a clause repeated by Theodulph of Orleans, the Council of Nantes (7th or 9th century) allowed burials within the adjuncts of the church, (churchyard, even the portico and choir) but not beneath the church proper, in the vicinity of the altar.[51]

Attempts to prevent burials in the church or baptistery were repeated in the 9th century,[52] but it proved impossible to keep all bodies out. Exceptions had to be made, for which a measured, if back-handed, justification was found in the

[45] *Poen. Cordubense* (early 11th century) 157, *CCSL* 156A 67.

[46] *Capitula Vesulensia* (c. 800) 23, *MGH CapEp* 3: 351; *Benedicti capitularium collectio* (mid 9th century) 2.197, *PL* 97: 771-772.

[47] See Yvette Duval, *Auprès des saints corps et âme. L'inhumation 'ad sanctos' dans la chrétienté d'Orient et d'Occident du IIIe au VIIe siècle* (Paris, 1988); *idem*, "Sanctorum sepulcris sociari," in *Les fonctions des saints dans le monde occidental (IIIe-XIIIe siècle)*. Actes du colloque organisé par l'École française de Rome avec le concourse de l'Université de Rome "La Sapienza," Rome 27-29 octobre 1988 (Rome, 1991), 334-351. For the medieval institution of burials in the churchyard, see Éric Rebillard, "Le cimetière chrétien. Église et sepulture dans l'Antiquité tardive (Occident latin, 3e-6e siècles)," *Annales HSS* 54 (1999): 1027-1046. See also Michel Lauwers, "Le cimetière dans le moyen âge latin. Lieu sacré, saint et religieux," *ibid.*, 1047-1072.

[48] 18, Vives, 75. See also *Collectio Hispana, Excerpta Canonum*, 4.38, *PL* 84: 73.

[49] *Synodus Autissiodorensis* (561-605) 14, *CCSL* 148A: 267; *Epítome hispánico* 43. 10: 190; *ibid.*, 32.16: 175). Treffort sees burials in baptisteries as reflecting the importance of baptism for the salvation of the dead (*L'Eglise carolingienne et la mort*," 35-43).

[50] *Conc. incerti loci* (post 614) 6, *CCSL* 148A: 287.

[51] 12, Aupest-Conduché, 51 = 6, Sirmond 3: 603. See also Theodulph of Orleans, *Capitulary 2* (813) 11, *MGH EpCap* 1: 153.

[52] *Capitulare Ecclesiastica* (810-813) 14, *MGH CapRegFr* 1: 179. See also *Ansegesis capitularium collectio* (1st half, 9th century) 1.153, *ibid.*, 1: 412; *Statuta Bonifacii* (1st half, 9th century) 19, *MGH CapEp* 3: 363.

writings of Gregory the Great. An early Irish collection of laws cited him as authorising the burial in church of those who had died in the state of grace, for the convenience of relatives who came to pray for them, but warned that those who had died in sin would be punished more severely if their graves were in the church. Riculf of Soissons also referred to Gregory to show that burial in church would not save one from rightful punishment - resting in holy places was of no avail to those who had died unshriven. Burchard of Worms related four anecdotes from the *Dialogues*, illustrating the grim fate of those who were buried, against their deserts, in church.[53] An eleventh-century penitential suggests that churchmen in Spain tried to restrict church burials to martyrs only.[54] In Carolingian territory, by contrast, burial in or beneath the church was conceded to members of the clergy (bishops, abbots, worthy priests) and, sometimes, to the pious laity from the 9th century on.[55] Hincmar of Reims' *Capitulary* of 852 insisted that priests could permit even this only after consultation with the bishop or on express instructions from the synod.[56]

Such measures were urgently required. The parish clergy in some areas had become so lax that, by the 9th century, churches were filled with tombs. Church burials were such a longstanding custom in Orleans, Bourges and Vercelli, that churches had become, in effect, cemeteries or mass graves. Here, too, diocesan capitularies ordered that henceforth only the bodies of priests and just men be given burial there. Bodies which had been there for a long time should be left in the church, but re-buried deep beneath the floor, leaving no visible signs. If there were too many dead bodies in the church for this to be done, the church itself should be left as a cemetery, and the altar set up in a more suitable place.[57]

Competition for space in the church could be vehement. Some families claimed the hereditary right to bury their dead there while bitterly opposing similar pretensions on the part of their rivals. Others, less presumptuous, buried their dead in the graves of their bolder neighbours. The Council of Meaux (845-846) firmly rejected such claims and reprimanded those who threw other people's

[53] *Collectio Hibernensis* (late 7th / early 8th century) 18.8, H. Wasserschleben, 58. See also Riculf of Soissons (888) *Statuta* 19, *PL* 131: 21-22; Burchard of Worms, *Decretum* (1008-1012) 3.152-156, *PL* 140: 703. See Gregory the Great, *Dialogues* IV, 52-56, *SC* 265: 176-184.

[54] *Poen. Cordubense* (early 11th century) 157, *CCSL* 156A: 67.

[55] *Conc. Moguntinum* (813) 52, *MGH Concilia* 2.1: 272; *Capitula a canonibus excerpta* (813) 20, *MGH CapRegFr* 1: 174; *Capitula Frisingensia tertia* (c. 840) 30, *MGH CapEp* 3: 229; Ruotger of Trier, *Capitulary* (10th century, 1st half) 4, *MGH EpCap* 1: 63; Burchard of Worms, *Decretum* 3.323, *PL* 140: 723. General references to the traditional rulings of the Church concerning such burials are found in *Conc. Arelatense* (813) 21, *MGH Concilia* 2.1: 252; *Notitia de conciliorum canonibus in Villa Sparnaco* (846) 19, *MGH CapRegFr* 2: 262; *Ansegesis capitularium collectio* (9th century, 1st half) *MGH CapRegFr* 1: 423.

[56] 12, *MGH CapEp* 2: 40.

[57] Theodulph of Orleans, *Capitularia 1* (813) 9, *MGH EpCap* 1: 109. See also Radulph of Bourges, *Capitularia* (between 853 and 866) 4, *ibid.*, 236-237; Atto of Vercelli (d. 961) *Capitularia* 23, *PL* 134: 33; Burchard of Worms, *Decretum* (1008-1012) 3.151, *PL* 140: 702-703.

bones out of "their" tomb or who "dared to violate another person's tomb in any way whatsoever." [58]

It was by no means only the bodies of Christians that were buried in the church. Demands to rid churches of the corpses of pagans appear in the late 7th, 8th and 9th centuries and in Burchard's *Decretum*. The *Capitula Dacheriana* stated categorically that dead pagans were to be thrown out of the places belonging to the saints, while a number of penitentials in the Theodorian school advised purifying the church and casting out the bodies of pagans.[59] Burchard, too, in permitting bodies to remain buried deep beneath the church, made pagans an exception: "unless they are the bodies of pagans."

In certain cases, the burials had taken place before the consecration of the premises, since some churches were erected on the site of mortuary monuments or of ruined and abandoned churches.[60] In the beginning, an attempt was made to prevent Mass from being said in a structure which housed the dead - a church was not to be consecrated if bodies had been buried in it previously.[61] Nevertheless, the authors of some penitentials allowed mass in buildings consecrated before the burials took place.[62] A problem arose if mortuary monuments in a desirable situation contained the remains of the pagan forefathers of the owners. The *Penitential of Theodore* declared that an altar was not to be consecrated in a church where pagans were buried, "but if it seemed suitable for consecration, the bodies should be removed, the wooden walls knocked down and washed, and the church rebuilt."[63] A later continental version added "heretics" and "perfidious Jews" to the unbelievers whose bodies were to be thus treated.[64] The choice of word for "remove," the forceful *evellere,* suggests how contemptuously the pagan remains were to be treated.

[58] *Conc. Meldense* (845-846) 72, *MGH CapRegFr* 2: 415). See also Hincmar of Reims, *Capitularia* 3 (852) 2, *MGH CapEp* 2: 74; Regino of Prüm, *De synodalibus causis* (c. 906) I, 124: 79-80; Burchard of Worms, Decretum 3.157, *PL* 140: 705.

[59] *Capitula Dacheriana* (late 7th/8th century) 59, Wasserschleben, 150; *Canones Gregorii* (late 7th/8th century) 150, Schmitz 2: 539. See also *Poen. Theodori* (668-756) II, 1.5, Schmitz 1: 538; *Canones Cottoniani* (late 7th/8th century) 58, Finsterwalder, 274; *Canones Basilienses* (8th century) 89, Asbach, 87.

[60] In the *Vita Frodoberti*, a new bishop orders that bodies be thrown out of the premises of a church built over the ruins of an earlier church; he is unaware that one of the bodies is that of his saintly predecessor (26, *MGH SRM* 5: 81-82). For the reburial of pagan ancestors in Christian sites, see Geary, "The uses of archaeological sources," 36-39.

[61] *Conc. incerti loci post A. 614*, 2, *CCSL* 148A: 287; *Capitula Dacheriana* (late 7th/8th century) 99, Wasserschleben, 154.

[62] *Canones Gregorii* (late 7th/8th century) 149, Schmitz 2: 539. See also *Canones Cottoniani* (late 7th/8th century) 56, Finsterwalder, 274; *Canones Basilienses* (8th century) 71a, Asbach, 85; Burchard of Worms, *Decretum* 3,14, *PL* 140: 676.

[63] II, 1.4, Schmitz 1: 538. See also *Benedicti capitularium collectio* (mid 9th century) 1.111, *PL* 97: 715; Burchard of Worms, *Decretum* 3.38, *PL* 140: 679.

[64] *Poen. Ps.-Theodori* (mid 9th century) 47.1, Wasserschleben, 617.

Multiple burials: Multiple burials in the same tomb, fairly common in pagan burial sites, continued well into Christian times - so firm a churchman as St. Ambrose had himself buried on top of his brother Satyrus, whose tomb he made sure was in contact with that of a saint.[65] The scarcity of space close to the church or the desire to appropriate or share in the graces belonging to another contributed to the custom of burying a second body in an already occupied tomb, even to the point of throwing out the original occupant. Evidence for this comes primarily from Frankish sources. In late sixth-century Burgundy, many people opened fresh graves, and put their own dead on top of the not-yet decomposed bodies lying within. While some did so with the consent of the owner of the tomb, others "usurped" tombs in particularly holy places, against the owner's will, for their own dead. The Council of Macon ordered that such corpses be flung out of the graves.[66] An eighth-century Frankish council forbade placing one dead body on top of another, and scattering the bones of the dead on the ground.[67] The same prohibition is found in two ninth-century collections; the *Statutes of St. Boniface* also forbids the superimposition of dead bodies.[68] The only non-Frankish source to mention this practice is the *Epítome hispánico* which found it to be "a serious fault."[69]

Pagan funerals: Our texts have little to say on the final disposal of the bodies, since the early medieval Church was generally tolerant or indifferent as to the way the body was actually buried.[70] Only three practices are mentioned: burial in barrows, cremation and the magic used to fix the ghosts of unbaptised children to the ground.

Cremations had been common east of the Rhine up to the Carolingian era.[71] Given the reluctance with which the Saxons embraced Christianity, it is not surprising that the authorities suspected them of clinging to ancestral methods of disposing of the dead. Charlemagne accordingly decreed the death penalty for those who "burnt their dead in the pagan manner and reduced their bones to ashes," and demanded that the Christian dead be buried in the churchyard, not in

[65] Duval, *Auprès des saints*, 104-106.

[66] *Conc. Matisconense* (585) 17, *CCSL* 148A: 246.

[67] *Capitulare Incerti Anni* (c. 743) 2, *Concilia Germaniae* 1: 55.

[68] *Capitula Vesulensia* (c. 800) 24, *MGH CapEp* 3: 351; *Benedicti capitularium collectio* 2.198, *PL* 97: 772; *Statuta Bonifacii* (1st half, 9th century) 20, *MGH CapEp* 3: 363.

[69] 43.11: 190. In two recensions.

[70] For instance, Cécile Treffort attributes the abandonment of cremation among Gaulish Christians before the 6th century not to ecclesiastical legislation but rather to a "profound modification under the influence of the new religion of the relations between the living and the dead" (*L'Église carolingienne et la mort*, 71).

[71] See Tacitus, *Germania* 27 and Dierkens, "Cimetières mérovingiens," 57. St. Boniface mentioned this Old Saxon practice in a letter (746-747) to Aethelbald, king of Mercia (*MGH EpSel* 1: 150). According to Effors, however, the practice of cremation was on the decline among the Saxons before their forced conversion (*"De partibus Saxoniae* and the regulation of mortuary custom," 280).

pagan burial mounds.[72]

That evidence for cremation should turn up in a north Italian penitential of the 8th or 9th century is more unexpected. In a brief clause rich in implications, the *Oxoniense II* proposed a lenient penance (four weeks) for a Christian who had lent his oxen or cart to a pagan friend to transport wood for a pyre to cremate his kinfolk.[73] It is noteworthy that the practice of cremation is ascribed not to backsliding Christians but to out-and-out pagans, and that friendship and mutual cooperation are not only assumed to exist between the two, but seem to have been tolerated – the relationship itself is not condemned, only the facilitation of a forbidden rite. Moreover, the setting is not some remote mountainous area or the depths of an obscure forest: wood has to be transported and the roads are good enough for a four-wheeled wagon. The pagan appears to be relatively poor, since he does not have a yoke of oxen or a cart for them to pull, but must rely on his wealthier neighbour for help.

In both cases, the choice of words (*consumi facere, ustulare*) makes it clear that the cremation is not a symbolic charring of the body, but a holocaust.

Although the fear of the malice of the ghosts of unbaptised children was undoubtedly wide-spread, only Burchard describes magical practices used to bind them to the grave "lest they rise and do much harm."[74] Since a child born dead or who had died before baptism had no right to a place in consecrated ground, Rhenish women would take the body to some secret place and bury it there, transfixed with a stake.[75] If the mother died before delivering the child, the stake

[72] *Capitulatio de partibus Saxoniae* (775-790) 7 and 22, *MGH CapRegFr* 1: 69. The word *tumulus* used here clearly means barrow, but in other contexts it could mean simply "grave," *e.g.*, *Conc. Matisconense* (585) 17, *CCSL* 148A: 246. Effors sees this legislation as meant to absorb the Saxons culturally into Christendom (and the Carolingian empire) by ensuring "that integration into the community of the living was also reflected in the topography of the resting places of the dead" (*"De partibus Saxoniae* and the regulation of mortuary custom," 276).

[73] *Poen. Oxoniense II* (8th/9th century) 41, *CCCM* 156: 197-198.

[74] Many societies take measures to protect themselves from the malevolence of the wandering ghosts of outsiders, among whom the ghosts of children not yet integrated into the community are particularly dangerous. "Dead-child beings are a typical example of the departed without status ... abandoned, murdered, unbaptised, aborted or stillborn children ... have died before the necessary status-giving rites are carried out. Their position is problematical in that they have never belonged to the group of the living and for this reason cannot belong to the group of the dead either ... the child is considered an 'outsider'; he has at no point been a member of the family, nor has he been accepted in the group of dead, which is the object of a cult " (Juha Y. Pentikäinen, "Transition rites," in *Transition Rites. Cosmic, Social and Individual Order*, ed. Ugo Bianchi [Rome, 1986] 1-27; here, 11). See also Fine, *Parrains, marraines,* 290-310. For the gradual development of the idea of limbo as a permament resting place of unbaptised infants, see Didier Lett, "De l'errance au deuil. Les enfants morts sans baptême et la naissance du *limbus puerorum* aux XIIe-XIIIe siècles," in *La petite enfance dans l'Europe médiévale et moderne,* ed. Robert Fossier (Toulouse, 1997) 76-92.

[75] Evidence drawn from Anglo-Saxon burial sites leads Sally Crawford to consider that the Christian emphasis on baptism tended to increase superstitious fears of dead infants, especially

was driven through her body as well as the unborn child in her womb.[76] If she had been given Christian burial despite dying unchurched, the women evidently had to violate the cemetery to open up her grave.[77]

8.3 Commemoration

The memory of the dead was kept alive and honoured in approved ways in the liturgy, prayers and charitable acts, but also in a ways less acceptable to the Church. Churchmen opposed banquets which included food shared with the dead at suspect places and ritual drinking, the construction of memorials at roadsides and in the fields, vigils by tombs that may have involved magic, and undue reverence paid to certain of the dead.

8.3.1 *Banquets*

Sharing food with the dead: Van Gennep saw commemorative banquets as "rites of incorporation," the purpose of which was "to reunite all the surviving members of the group with each other, and sometimes also with the deceased, in the same way that a chain which has been broken by the disappearance of one of its links must be rejoined."[78] More consciously, in pagan societies sharing food and drink with the dead members of the family was intended to propitiate them and win

those who died before baptism. Pagans also feared such children: a stillborn infant in a pagan cemetery was covered with chalk, but not the mother on whose body it was still lying ("Children, death and the afterlife in Anglo-Saxon England," *Anglo-Saxon Studies in Archaeology and History* 6 [1993]: 83-91; here, 86). Salisbury cites evidence from Visigothic cemeteries in the mountainous region of northern Iberia, where numerous bodies of both adults and children have been found staked to the ground. "The stakes were driven through many part of their bodies, along the leg and arm bones, and through the skull and abdomen." In one burial ground, the majority of bodies were so buried. Salisbury speculates that this was "to protect the corpses from either demons or grave robbers, who used corpses for magical purposes" (*Iberian Popular Religion*, 280). For children's burials, see also Sam Lucy, "Children in early medieval cemeteries," *Archaeological Review from Cambridge* 13.2 (1994): 21-34, and Cécile Treffort, "Archéologie funéraire et histoire. Quelques remarques à propos du haut moyen âge," in *La petite enfance*, 93-107. For other cases of the impalement of corpses, see Thompson, *The Goths in Spain* (Oxford, 1969), 56 and n., and Green, *Dictionary of Celtic Myth and Legend, s.v.* "Bog-burial".

[76] 19, 5.180 and 5.181, Schmitz 2: 448. Burchard's flock may have buried the child at the crossroads - in common folk tradition, ghosts get confused at crossroads and cannot find their way home; also the heavy traffic there keeps them in the grave. According to some beliefs, evil spirits cannot cross boundaries, which roads frequently are (Linda Dégh, *Folktales of Hungary*, trans. Judit Halasz [Chicago, 1965]; 252). The belief in the efficacy of a stake to keep the restless ghost immobilised was long-lasting: in England until the 1820's, suïcides had to be buried, impaled, at crossroads. For other possible sites, see Lett, "De l'errance au deuil," 86.

[77] According to Jean Beleth, the body of a woman who had died under these circumstances could not be brought into the church; the rites had to be performed outside, and the dead child removed from her womb before she was buried in the cemetery; it was to be buried elsewhere (*De ecclesiasticis officiis* 159s, *CCCM* 41A, 309). In the 13th century, women who died in childbed were churched after death; see Fine, *Parrains, marraines*, 105.

[78] *Rites of Passage,* 164-165.

their help for the descendants. Pastoral documents, however, emphasised this aspect but rarely. The sixth-century Council of Tours denounced those who made offerings of food to the dead on 22 February, now christianised as the feast of the Chair of St. Peter but, before that, the Roman feast of the dead which followed the family celebrations of the *Caristia*:

> There are those who offer sops to the dead on the feast of the Chair of St. Peter. Returning to their own houses after Mass, they revert to the errors of the pagans; having received the body of the Lord, they accept food sacred to the demon. We urge pastors as well as priests to take care that on holy authority they drive out of the church those whom they notice persisting in this folly or doing things contrary to the Church at heaven knows what rocks, trees or springs, the chosen places of pagans, and that they do not allow those who keep pagan customs to participate in the sacrament of the altar.[79]

An early sermon for the same feast, once ascribed to Caesarius of Arles, describes these customs as being more intensively practised than before: "I wonder why a dangerous error has grown so much among certain [un]believers, that today they bestow food and wine on the graves of the dead, as if the souls which have left the body require fleshly nourishment!" The author was sceptical about such miscreants' purity of motive: "He devours the very things which he claims to prepare for his loved ones; what he keeps for his belly he attributes to piety."[80]

These passages are exceptional although written and archaeological documents testify that the concept of making food offerings for, if not to, the dead lingered among Christians well into the Christian era. St. Monica brought with her the African custom of offering pulse, bread and undiluted wine at saints' memorials until forbidden to do so by St. Ambrose. Gregory of Tours told an amusing story of a husband's ghost who complained to his widow of the sour taste of the wine that she had been offering him daily in church - a greedy cleric replaced the vintage Gaza that she had brought with an inferior wine.[81] In an early Christian tomb excavated in Tarragona, a clay feeding-tube led from the surface to the mouth of the dead. There is archaeological evidence for such practices at Christian burial sites well into the 7th century.[82] The Church opposed them because they led to what were seen as "intolerable abuses," not because of their paganism.[83] The majority of medieval condemnations focus not on the principle of offerings to the dead, but on other aspects of commemorative banquets: the location of the feasts or the excesses, especially drunkenness, which accompanied them.

[79] *Conc. Turonense* (567) 23, *CCSL* 148A: 191.

[80] Ps. Augustine (date?) *S.* 190.2, *PL* 39: 2101.

[81] Augustine, *Confessions* VI, 2.2; Gregory of Tours, *In gloria confessorum* 64, *MGH SRM* 1: 786.

[82] Barral I Altet, "Le cimetière en fête," 302; P.-A. Fevrier, "La tombe chrétien et l'Au-delà," in *Le temps chrétien de la fin de l'Antiquité au moyen âge IIIe-XIIIe siècles, Paris 9-12 mars 1981* (Paris, 1984), 163-183.

[83] J. A. Jungmann, *La liturgie des premiers siècles jusqu'à l'époque de Grégoire le Grand* (Paris, 1960), 227.

Location: The only clear records of commemorative feasts at graveside come from late sixth-century Iberia and early eleventh-century Worms, and from a questionably authentic ninth-century capitulary from Vesoul. Evidently some Galician priests found the Christian mass to be compatible with traditional memorial rituals. Martin of Braga's *Canons* reminded these "ignorant and presumptuous clerics" that masses for the dead could be offered only in churches and basilicas that contained the relics of martyrs; priests were not permitted to carry the sacraments of the altar out into the fields and distribute them to the participants over the tombs.[84] The *Epítome hispánico* confirmed this and forbade Christians to carry meals out to the tombs. A repast preceded the mass for the dead - priests had to be reminded not to say mass for the dead after eating.[85] The *Capitulary of Vesoul* banned eating and drinking over tombs.[86] Finally, Burchard of Worms' "It is not permitted to Christians to take food to the tombs of the dead, and sacrifice to [or for] the dead" may have been a mere paraphrase of the Iberian texts, but a clause in his penitential unquestionably dealt with current local practices. It concerns the eating of any part of idolothytes, "that is, of the offerings which are made in certain areas at the tombs of the dead, or at wells, trees, stones or crossroads," and it adds original details which will be considered presently.[87]

More obscure references to funeral banquets at graveside may lie hidden in other texts. The clause "concerning the sacrilege committed over the tombs of the dead" of the *Indiculus superstitionum* probably refers to commemorative banquets in cemeteries.[88] The condemnations in eighth- and ninth-century penitentials of eating and drinking by trees, springs, and *cancelli* or at crossroads may be relevant here.[89] It has already been noted that the *cancelli* might have been tombs; trees and springs were connected with the underworld, stones and cairns were set up for the dead, and ghosts passed along the roads. All, therefore, are apt sites for encounters between the living and the dead, and feasting held there may well belong to the category of commemorative banquets.

Commemorative banquets and ritual drunkenness: We have seen in chapter 3 that the glorious dead were remembered in toasts drunk in their honour or on their feasts. Caesarius of Arles had complained vehemently of the practice of drinking to the names of angels and saints as well as living men, and the authors of eighth- and ninth-century penitentials had recorded the practice of excessive drinking in religious houses on great feasts, including those of saints.[90] In addition to these, however, there were other gatherings where the dead were commemorated in for-

[84] *Canones ex orientalium patrum synodis* (572) 68, Barlow, 140. See also *Collectio Hispana, Excerpta Canonum* 4.9, *PL* 84: 67.

[85] 1.69, 103 and 33.10: 176.

[86] 23, *MGH CapEp* 3: 351.

[87] *Decretum* (1008-1012) 10.38, *PL* 140: 839; 19, 5.94, Schmitz 2: 43.

[88] 1, *MGH CapRegFr* 1: 223.

[89] *E.g., Poen. Burgundense* (early 8th century) 29, *CCSL* 156: 65.

[90] Caesarius of Arles, *S.* 47.5, *CCSL* 103: 214; *Poen. Theodori* (668-756) I, 1.4, Schmitz 1: 525.

mal banquets where also drunkenness was the rule.

Almost all the texts concerning the conduct of commemorative banquets deal with clerical gathering, mostly from the northern parts of the Carolingian empire in the 9th century. An *Admonitio synodalis* of 813, repeated with minor variations by Hincmar of Reims at mid century and by Burchard of Worms in the early 11th, describes meetings of priests commemorating the anniversary of a death, or the thirtieth, third or seventh day after it – or celebrating for some other, unspecified, reason. The elements of the feast are: drinking of toasts for love (*amor*) of the saints or for the dead man's soul;[91] forcing others to drink; gorging oneself with food and drink in response to others' invitations; disorderly applause and laughter; relating or singing "worthless" tales (sagas?); improper jests to the accompaniment (it seems) of stringed instruments, bears, stags (maskers?) or "whirling women" (*tornatrices*); the masks or representations of the ghosts of demons known in the vernacular as *talamascae*.[92] Numerous toasts must have been the order of the day, for the attending priests were urged not to "raise the cup" - that is, drain it - more than thrice.[93] Drunkenness was not the worst of such banquets. In 829, the clergy of Cambrai and Neustria were ordered to stay at home and pray for the living and dead in their own churches, and to avoid the assemblies for the customary repasts in honour of the dead because of the many indecencies and occasional murder which occurred at such events.[94]

Any gathering of priests was apparently prone to degenerate into unwarranted conviviality. Halitgar of Cambrai moved alertly against the newly introduced custom of his clergy of coming together in small, "perverse and harmful" groups

[91] Cahen notes that *amor* in this context is, like *caritas*, the Latin equivalent of the Scandinavian *full* or *minne*, that is, "toast" (*Études sur le vocabulaire religieux*, 189).

[92] This word is related to *masca*, a term appearing in Lombard law as synonymous with *striga*, see Du Cange, *s.v.v.* "Talamasca" and "Masca". Gervais of Tilbury equated them with *striae* and *lamiae* (*Otia imperialis* 86: 39).

[93] *Admonitio synodalis* 39 *bis*, Amiet, 51. See also Hincmar of Reims (852) *Capitularia I*, .14, *MGH CapEp* 2: 41-42; Regino of Prüm, *De synodalibus causis* (c. 906) I, 216: 108; Burchard of Worms, *Decretum* (1008-1012) 2.161, *PL* 140: 652. In Hincmar's version, the toasts are drunk "in honour of saints) rather than "for love;" all three leave out the three-cup limit and the reference to fiddles and stags. This canon is analysed by Jean-Pierre Poly, "Masques et telemasques. Les tours des femmes de Champagne," in *Histoire et société. Mélanges offerts à Georges Duby* (Aix-en-Provence), 1: 177-188. For the significance of masks, the Christian interpretation of them, and their use in funerary rites, see A. David Napier, *Masks, Transformation, and Paradox* (Berkeley, Los Angeles and London, 1986), 1-29. For masks as representions of the dead or demons, see also Lecouteux, *Chasses fantastiques et cohortes de la nuit*, 153-156. The significance of the bear mask is analysed by Philippe Walter, "Der Bär und der Erzbischof; Masken und Mummenschanz bei Hinkmar von Reims und Adalbero von Laon," in *Feste und Feier im Mittelalter*, ed. Detlef Altenburg, Jörg Jarnut, and Hans-Hugo Steinhoff (Sigmaringen, 1991), 377-388, esp. 383-387. For professional entertainers, see J. D. A. Ogilvy, "Mimi, scurrae, histriones: Entertainers of the early Middle Ages," *Speculum* 38 (1963): 603-619.

[94] Halitgar of Cambrai, *Diocesan Synod 4* (829-31) 6, Hartmann, 390). See also *Capitula Neustrica quarta* (c. 829) 6, *MGH CapEp* 3: 71.

(evidently during the regular diocesan assemblies at the beginning of the month) in order to drink and feast. His words were echoed in a Neustrian capitulary.[95] If Halitgar was successful in stamping out this new growth in his diocese, other bishops might have been less so with similar manifestations. In 888, Riculf of Soissons was obliged to remind his clergy that they did not attend the monthly meetings of each deanery on the Calends in order to eat and drink, but to discuss matters pertaining to religion and their pastoral responsibilities, and to pray for the king, the leaders of the church and their acquaintances, living as well as dead.[96]

Throughout the rest of the 9th century, ecclesiastical authorities continued to worry about such banquets. Walter of Orleans urged priests to ride decorously and to be modest and sober if invited to an anniversary dinner (evidently by a layman), and to guard their tongues, avoid crude songs and prevent dancers from giving indecent performances ("in the style of Herodias' daughter") in front of them. Gilbert of Châlons went further, ordering everyone to refrain from assisting at banquets for the dead. Toward the end of the century, the *Capitula Ottoboniana* repeated some of the themes of the *Admonitio Synodalis*: "Let them not dare to sing, make jests or drink toasts during the feasts prepared in honour of dead men, but let them eat and drink soberly and return thus to their own homes.[97]

8.3.2 *Memorials*

Except for three texts forbidding the construction of such structures to unknown saints and martyrs, our documents contain only ambiguous references to memorial monuments (*memoriae*).[98] But here again we must take into consideration that the *cancelli* mentioned in St. Eligius' sermon and some dozen penitentials, where various rites were performed, may have been that, especially since one text situates them by the roadside, specifically at the junction of four roads.[99] Many if not all of the references to the cult of stones can be understood as concerning monuments of stone, natural or dressed, to the dead. The query in Burchard of Worms' penitential: "Did you carry stones to a pile?" surely refers to cairns,

[95] Halitgar of Cambrai, *Diocesan Synod 4*, 5, Hartmann, 389-390). See also *Capitula Neustrica quarta* (c. 829) 5, *MGH CapEp* 3: 71. Hartman believes that these assemblies also were commemorative (378).

[96] 22, *MGH CapEp* 2: 110-111.

[97] Walter of Orleans, *Capitularia* (869/870) 17, *MGH EpCap* 1: 191; Gilbert of Châlons (868-878) *De Interdicitis* 4, *MGH CapEp* 2: 94; *Capitula Ottoboniana* (889 or later) 9, *MGH CapEp* 3: 125. Although English pastoral literature contains no comparable texts, the habits of English churchmen were probably no different from their continental brethren's in this respect: Walter Map's term for English diocesan meetings is "drinking sessions" (*bibitoriae*). It is not clear, however, that these were exclusively clerical gatherings (*De nugis curialium* 2.11, ed. M. R. James [Oxford, 1914] 75).

[98] *Conc. Francofurtense* (794) 42, *MGH Concilia* 2.1: 170; *Iordani Recensio Canonum Rispacensium* (c. 1550), *MGH Concilia* 2.1: 218; Burchard of Worms, *Decretum* (1008-1012) 3.54, *PL* 140: 683. See chapter 2, n. 177 and *seq*.

[99] *Poen. Hubertense* (8th century, 1st half) 24, *CCSL* 156: 110.

memorial structures of a cruder sort than *memoriae* or *cancelli.*[100] Roadside crosses also probably served as unsanctified memorials to the dead – how else to explain the otherwise astonishing inclusion of crucifixes at crossroads among "abhorrent places" where people assemble for meals?[101] Burchard himself follows his question about cairns with one about placing wreaths on the head of crosses at the junction of two roads.[102]

8.3.3 *Vigils, violation of tombs and magic*

Ritual offerings of lamps to the dead was a practice well-known to antiquity.[103] The *Epítome hispánico* combined two canons of the Council of Elvira (c. 300-306) in one clause: "Candles are not to be lighted in cemeteries; whoever does so is to be excommunicated. Women are not to keep vigil in cemeteries." This was repeated in two collections.[104] Otherwise, only St. Eligius' condemnation of vows and torches or lamps at *cancelli* suggests that this custom continued into the early Middle Ages.[105]

But vigils kept by women in cemeteries appear again in Book 10 of Burchard of Worms' *Decretum.* Burchard chose to give the text of Elvira in full, to explain why women should not be allowed to do so: "because they often commit secret crimes under the pretext of prayer and religion."[106] The nature of the crimes is not indicated here, but Burchard's penitential suggests two reasons why women in particular might have lingered in the graveyard at night: to bind the malign ghosts of unborn and unbaptised babies magically to the grave, and to obtain human skulls to be charred and used in a tonic for their ailing husbands.[107] Although not the only possible source of skulls, the cemetery was the handiest.

The violation of graves[108] was cited in over two dozen penitentials and capitularies. The usual penance was five years of fasting.[109] In addition, the very first clause the *Indiculus superstitionum,* "concerning sacrilege at graves," possibly deals

[100] 19, 5.94, Schmitz 2: 430. Merrifield interprets the building of cairns as a separation rite: "A cairn of stones or an earth barrow piled about the dead ... not only served as a monument to his memory and a protection for his body, but also separated him physically from the living, a separation often reinforced symbolically and magically by the ring-ditch that surrounded it" (*The Archaeology of Ritual and Magic*, 71).

[101] *Poen. Vindobonense a* (late 9th century) 50, Schmitz 2: 354. See also *Poen. Merseburgense a* (late 8th century) alternative to 45, *CCSL* 156: 141.

[102] 19, 5.94, Schmitz 2: 430.

[103] See Rush, *Death and Burial in Christian Antiquity*, 221-225.

[104] *Epítome hispánico* (c.598-610) 30. 32. See *Council of Elvira* 33-34, Vives, 7-8. Unlike the *Collectio Hispana, Excerpta Canonum*, which contained only the first part (4.9, *PL* 84: 66), the *False Decretals* included the second clause as well (34 and 35, ed. Hinschius, 341).

[105] *Vita Eligii, MGH SRM* 4: 706.

[106] 10.35, *PL* 140: 838. Burchard attributed this to a Council of Meaux.

[107] 19, 5.181 and 5.177, Schmitz 2: 448; 19, 5.177. See also Rabanus Maurus, *Poenitentiale ad Heribaldum* (c. 853) 20, *PL* 110: 491; Regino of Prüm, *De synodalibus causis* (c. 906) 2. 369: 354.

[108] See H. Steuer *et al.*, "Grabraub," in *Reallexikon der Germanischen Altertumskunde*, 516-527.

[109] *E.g.*, *Poen. Remense* (early 8th century) 6.11, Asbach, 43.

with this offence.[110] None of the texts suggests a magical reason for this crime. On the contrary, any information given in our texts implies that the motive was economic: three penitentials classify it as theft, and Burchard of Worms' is even more precise: "Did you violate a grave, that is to say, when you saw someone burying [a body], did you break into the tomb at night and take his clothes?"[111] Nevertheless, since the codes of Recceswinth and Erwig explicitly dealt with grave-robbing in quest of a remedy, it may be taken for granted that magic was a motive at least occasionally.[112]

We have already seen that healing and divinatory magic were practiced in cemeteries or by tombs. The *Homilia de sacrilegiis* recorded the practice of taking demoniacs to certain ancient monuments to cure them of possession.[113] These most likely were the tombs of some now all but forgotten saints or of heroes from the pagan past. The tenth- or eleventh-century *Penitential of Arundel* imposed seven years of penance on those "who, sacrificing to the devil, look into the future at tombs, funeral pyres or elsewhere."[114]

8.4 BELIEFS ABOUT THE DEAD

Pre-Christian Romans, Celts and Germans believed that in some sense the dead continued to live in the tomb. Although graveside banquets and sacrifices testify to the persistence of these beliefs well into the Christian era, explicit references are rare. Among the documents of the Council of Marseilles (533) is a letter by Pope John II to Caesarius of Arles, in which he observed that the violators of tombs should be excluded from the communion of the Church because it was not fitting for Christians to consort with those "who dare recklessly to disturb the peace of the ashes of the dead."[115] While the pope cannot be suspected of harbouring pagan beliefs, it is clear that underlying this is the ancient Roman belief about the potentially dangerous *manes*. Burchard gives indications that such beliefs lingered into the 11th century at least in the Rhineland. One of his questions concerns the mildly sinful practice of burying a slain man with some salve in his hand "as though his wound could be healed after death." Another deals with women who sometimes buried christened babies with wax patens and chalices in their hands. Cécile Treffort sees in these symbols a kind of identification of the infant, guaranteed of salvation, with priests, the most privileged of all Christians.

[110] *MGH CapRegFr* 1: 223.

[111] *Poen. Halitgari* (817-830) 29, Schmitz 1: 478; *Poen. Ps.-Theodori* (mid 9th century) 23.14, Wasserschleben, 592; *Double Penitential of Bede-Egbert* (9th century?) 16, Schmitz 2: 681); Burchard of Worms, *Decretum* (1008-1012) 19, 5.59, *ibid.,* 422. Dierkens pointed out that the prohibitions against the violation of tombs (insofar as they deal with theft rather than with magic) are proof that grave-goods were not forbidden by the Church ("Cimetières mérovingiens," 60-61).

[112] *Lex Visigothorum* 11.2.2, *MGH Leg.*1.1: 403.

[113] 22, Caspari, 12.

[114] 88, Schmitz 1: 461.

[115] *CSSL* 148A: 95.

But if the host and wine contained in the waxen paten and chalice were real, they too may indicate belief in life within the tomb.[116]

Medieval authors were very familiar with folktales of cavalcades of the dead, of various versions of the Wild Hunt or Furious Horde in the suite of various divinities. Otloh of Emmerman, for instance, related an improving story of two brothers who met a great crowd flying through the air while out riding; one of them threatened the horsemen with damnation for themselves and their dead father unless they made restitution to a monastery that he had robbed. Many of the Welsh, according to Walter Map, claimed to have seen the ghost army of the British king Herla emerging from the rivers.[117] Elements of the myth may be detected in the pastoral literature of northern Francia and the Rhineland, in Regino of Prüm's and Burchard of Worms' accounts of the popular belief that at appointed times, mounted women rode through the night sky in the suite of a pagan goddess.[118] Oblique hints may also be found in descriptions of the customs of the Calends of January, which coincided with the Germanic Yule. Divinatory practices described by Burchard involved consultation of the dead known to be on the move at this time. In addition, the *Homilia de sacrilegiis* includes a demonstration of *arma* in the fields among forbidden New Year's practices. If *arma* is understood to mean weapons rather than tools, this may be seen as a tribute to the Ride.[119]

[116] *Decretum* 19, 5.97, Schmitz 2: 431; 19, 5.185, *ibid.,* 450. See Treffort, "Archéologie funéraire et histoire," 105.

[117] Otloh of Emmerman, *Visio* 7, *PL* 146: 360-361; Walter Map, *De nugis curialium* 1.1: 15-16.

[118] Regino of Prüm, *De synodalibus causis* (c. 906) II, 371: 354-356); Burchard of Worms, *Decretum* (1008-1012) 10.1, *PL* 140: 831-833; *idem,* 19, 5.90, Schmitz 2: 429; 19, 5.70, *ibid.*, 425.

[119] Burchard of Worms, *Decretum* (1008-1012) 19, 5.62, Schmitz 2: 423; *Homilia de sacrilegiis* (late 8th century) 17, Caspari, 10-11.

9
Alimentary Restrictions

9.1. Pastoral Literature and Dietary Taboos

Bans on the consumption of food that had been offered to the gods and of substances such as blood, semen and urine formed an obvious part of the struggle against popular paganism and magic. In addition to these, some three hundred and fifty separate clauses in pastoral literature introduced alimentary restrictions or taboos by penalising the consumption of other categories of foods.[1] One falls roughly under the biblical injunction against unclean flesh in general, blood and meat which contained blood because of the way the animal had been killed; associated with these are carrion and the flesh of animals that had been mauled by dogs and wild beasts, of those that had died of disease or unknown causes, and of those which had been tainted by sexual contact with humans. A second category is food or drink contaminated by birds, animals or corpses, or which had gone bad or had been improperly cooked. To these may be added the consumption of "fraud," banned in a few English penitentials. Finally, a set of extraordinary Irish canons treated as polluting the consumption of food tainted by contact with or even by the vicinity of certain persons.

Penitentials and penitential texts are the source of the overwhelming majority of such canons. Almost seventy-five percent of those studied have at least one clause dealing with alimentary transgressions.[2] The first important references to them are found in Irish penitential texts, notably the *Canones Hibernenses* (mid 7th century) and the *Penitential of Cummean* (before 662). They were followed by the *Penitential of Theodore of Canterbury* (668-756) and then by more than three dozen

[1] Cyrille Vogel remarked on the "luxe étonnant de particularités hautes en couleur" of the dietary and hygienic regulations found in penitentials (*Le pécheur et la pénitence au moyen âge*, 19). For dietary taboos, see Simmons, *Eat Not This Flesh*, the only systematic study of this subject of which I am aware; it contains an extensive bibliography. See also see A. E. Crawley, "Food," *ERE* 6: 59-63, and Tom Brigerg, "Food and drink," in *Man, Myth and Magic*, ed. Richard Cavendish (New York, London, Toronto, 1983), 4: 1011-1017. For taboo in a broader sense, see Jean Cazeneuve, *Sociologie du rite: Tabou, magie, sacré* (Paris, 1971); Mary Douglas, "Taboo," in *Man, Myth and Magic* 10: 2767-2771; L. Debarge, "Tabou," in *Catholicisme* (Paris, 1995), 14: 717-721. The term "taboo" is used here in its most basic meaning of "rules about behaviour which restrict the human uses of things and people" (Douglas, "Taboo," 2769).

[2] Penitentials which do not contain any such prohibitions are the *Vinniani*, *Ambrosianum*, *Columbani*, *Bobbiense I* and *II*, *Burgundense*, *Sletstatense*, *Quadripartitus*, *Valicellanum E6* and *B58*, *Arundel*, *Cordubense*; the *Iudicii Clementis*; an anonymous 9th century (?) *Liber poenitentialis*, and Rabanus Maurus' two penitentials.

other penitentials from every part of western Europe, down to the 11th century. Standing apart from these are the canons attributed to Adomnan (679-704?), of which an astonishing nineteen out of twenty clauses concern dietary questions, but which, unlike the *Hibernenses* and the penitentials of Cummean and Theodore, had little influence on the development of dietary taboos in penitentials.

By contrast, they find little place in other forms of pastoral literature. Church councils generally ignored dietary questions. Of the western councils, only three have been found to ban food (in all cases blood or meat) other than idolothytes: Orleans (533), Worms (868) and Aachen (816).[3] Four Iberian councils attacked alimentary taboos from the opposite point of view – an inclination on principle to avoid fowl, meat or vegetables cooked with meat as being unclean, was taken as a sign of heresy.[4] Capitularies from ninth-century Orleans and Trier contain restrictions, but the one ascribed to Gregory II dismissed the whole concept of such taboos except against food offered to idols.[5] They are found, however, in some collections (for example, Regino of Prüm's *De synodalibus causis* and Burchard of Worms' *Decretum*). An anonymous sermon circulating in ninth-century Trier is the only one studied here to mention them.

In addition, Martin of Braga's *Canons*, the decisions of the second Council of Braga (572) and the *Epítome hispánico* contain a clause forbidding priests, clerics and pious laymen to hold feasts "with sausages" (*confertis*).[6] If *confertis* is not a scribal error for *confratriis* (associations) or *offertis* (sacrificial offerings), this is perhaps a ban on attendance at banquets held at the time of pig-butchering, when a certain amount of bawdy humour involving sausages could be expected. It seems clear that the ban is not on the food itself but on the feasting.

We have already seen from their letters that popes Gregory III and Zacharias were concerned about the dietary habits of the Germans. In answer to queries sent by St. Boniface, they advised him that the consumption of jays, crows, storks,

[3] The Council of Aachen simply followed the Council of Gangra 1 in anathematising those who condemned the consumers of meat "except for blood, that whic was sacrificed to idols and that which was suffocated;" its authority, therefore, was against rather than for other alimentary restrictions (65, *MGH Concilia* 2.1: 365). See also *Hibernensis* (late 7th / early 8th century) 54.7, H. Wasserschleben, 7; Regino of Prüm, *De synodalibus causis* (c. 906) Appendix I, 9 and III, 11: 395 and 456.

[4] *Conc. Toletanum* 1 (397-400) 17, Vives 28; *Conc. Bracanense I* (561) 14, *ibid.*, 69; Martin of Braga, *Canones ex orientalium patrum synodis* (572) 58, Barlow, 138. This last was attached to the documents of the second Council of Braga (Vives, 100). See also *Epítome hispánico* (598-610) 1.58: 103 and 7.2: 119, and *Collectio Vetus Gallica* (8th/9th century) 56, Mordek, 580-581. In addition to these councils, the Council of Cordoba (839) dealt with a group of extremists who considered a wide variety of food to be unclean (Hefele-Leclercq, *Histoire des conciles* 4.1: 105).

[5] *Capitulare Gregorii Papae II* (731) 7, *Concilia Germaniae* 1: 37.

[6] Martin of Braga, *Canones ex orientalium patrum synodis* (572) 61, Barlow, 139)=*Conc. Bracanense 2*, 61, Vives, 101; *Epítome hispánico* (c. 598-610) 1.61: 103. I have not been able to find confirmation for the statement found at several internet sites, that the Emperor Constantine had banned sausages because of their association with the Lupercalia.

beavers, hares and horses, both wild and domesticated, was absolutely forbidden.[7] Pope Hadrian I wrote to various Spanish ecclesiastical dignitaries to settle an obviously wide-spread controversy about dietary practices. He.rejected with some sharpness the argument obstinately held by certain "promising" persons, that only ignorant louts refused to eat swine and cattle blood and the flesh of suffocated beasts: those who did in fact eat such food were "not only completely removed from learning but were utterly devoid of common sense" and, as such, would fall, "bound by the chain of anathema," into the snares of the devil.[8] Spanish Christians may have considered defiance of such restriction to be a form of opposition to Islam, which also forbids the consumption of blood, pork and carrion.

In sum, the decisions of Church councils and synods and episcopal capitularies show a measured concern with respect to very specific, limited situations. Except for meals explicitly categorised as idolatrous, the hierarchy of the Church in established Christian communities was almost wholly indifferent to alimentary taboos: food habits were not a problem. Theodulph of Orleans included violations of these rules among lesser sins.[9] The majority of penitentials, however, show considerable anxiety about defilement through food, although the relatively mild penances – generally forty days or less on bread and water – indicates that the authors took the defilement less seriously than first appears.

Why do they play so remarkable a role in penitentials and so little in canon law? The answer lies probably in the Irish influence on contemporary religious development on the continent – great on penitential thought but little on conciliar legislation. If this is so, then Irish dietary taboos may reveal special problems faced by Irish clerics in the 7th and 8th centuries, but not necessarily by clerics elsewhere in Europe, that would explain why they put such an emphasis on purity in food. We shall consider the restrictions first, and then an interpretation primarily in terms of the situation in Ireland, secondarily as the result of fears of magic.

9.1.1 *Prohibitions of flesh-foods: blood and carrion*

About two hundred and twenty separate passages in pastoral literature deal with this category of bans on food.[10] Often prohibitions against blood, carrion, animals that had been killed by other animals, mauled by scavengers, strangled or suffocated (*suffocata*), beaten to death or killed by disease or accident are combined in a single clause. In most cases, only general terms are used but from time to time the texts specify the animals and the method of killing. Thus, birds or animals

[7] *Ep.* 87 (751), *MGH EpSel* 1: 196; *Ep.* 28 (c. 732), *ibid.*, 50.

[8] Hadrian I to the bishops of Spain (785-791), *MGH Ep* 3: 641-642. This is repeated in Hadrian's letter to the bishop of Elvira Egila and to the priest Iohannes, written at about the same time (*ibid.*, 646).

[9] *Capitulary 2* (813) 19, *MGH EpCap* 1: 177-178.

[10] See Stephane Boulc'h, "Le statut de l'animal carnivore et la notion de pureté dans les prescriptions alimentaire chrétiennes du haut Moyen Âge occidental" and discussion in *Le statut éthique de l'animal. Conceptions anciennes et nouvelles*, ed. Liliane Bodson (Liège, 1996), 41-59.

killed by a hawk are banned, as are birds or animals killed or mauled by a dog, wolf, fox, or bear. The prey identified in the texts are cattle, sheep, goats and deer. Added to these are texts that set down the rules to be followed with respect to pigs and hens that had fed on carrion, blood or human remains.

The early 6th century penitential text attributed to Gildas seems to be the first; it prescribed forty days of penance for eating carrion unknowingly.[11] At about the same time, the Council of Orleans seemed to equate eating idolothytes with eating animals killed by other beasts, or animals that had died of strangulation caused by sickness or accident (that is, without effusion of blood).[12] These were followed by a series of prohibitions in Irish penitential texts. Partaking of flesh "eaten" by dogs, eating or drinking "illicitly" of sheep carrion or eating the flesh of animals which had died of unknown causes was to be expiated by various periods of fasting, up to a lifetime abstention from wine and meat.[13]

Canons of Adomnan: Dietary rules did not take up a large part of the penitential texts above. By contrast, they represent almost the totality of the *Canones Adomnani* (679-704). This document made distinctions missing from other documents, and sometimes went considerably farther in developing and explaining when and why restrictions did or did not apply. It also touched upon the question of raw *vs.* cooked meat, although never broaching it directly.

Marine animals washed up dead on the seashore could be eaten even if the cause of death was unknown, unless they were putrid, but what (a mammal?) had died in the water or drowned was carrion (and therefore taboo), since the blood remained coagulated within – the Lord made this law, declared Adomnan, not because men at that time ate raw meat ("since it was not so good") but because they were accustomed to eat the flesh of suffocated beasts and carrion.[14] Livestock that had fallen from a rock could be eaten if they had bled away, but if they died of broken bones only, without bleeding, they were to be rejected as carrion. A half-alive animal that died suddenly with an ear or some other part torn off was carrion. Animals (stags, cattle) that died of exhaustion, after receiving a wound (a leg broken in a trap, an ear torn or cut off) from which they had bled but slightly, had been injured merely; they counted as carrion, since the "upper blood", in which the soul resided, remained coagulated within the body. Adomnan underlined again that this rule had nothing to do with the fact that the meat was uncooked, but that the flesh had not been drained of blood. Only ingestion was polluting, not external contact, since the fat and the skin could be put to various uses.[15]

[11] *Praefatio Gildae de Poenitentia* 13, Bieler, 62. See also *Poen. Cummeani* (before 662) 10.3, *ibid.*, 124.

[12] *Conc. Aurelianense* (533) 20, *CCSL* 148A: 102.

[13] *Canones Hibernenses* (mid 7th century) 1.14, 1.15 and 1.19, Wasserschleben, 137; *Poen. Cummeani* (before 662) 9.3 and 9.16, Bieler, 124 and 126.

[14] 1 and 14, Bieler, 176 and 178.

[15] 2, 5 and 20, Bieler, 176 and 180. In the tenth-century *Vita Odiliae*, men scramble down a

Being touched by predators made food impure. Flesh partly eaten by beasts was unclean but not carrion, since the blood had been shed, though by beasts. The bone-marrow of stags eaten by wolves was forbidden.[16]

If swine had grown fat on carrion, they were to be treated as carrion (that is, not to be used for food) until they lost weight and returned to their pristine condition. But if they had eaten of carrion only once, twice or thrice, they could be eaten with a clear conscience as soon as they had emptied their stomachs. If, however, they had tasted of human blood or flesh, they were absolutely forbidden by virtue of the law that an animal that had gored a man to death with its horns was not to be eaten (*Ex* 27, 28). Similarly, hens which had eaten human blood and flesh were very unclean, as were their eggs, but chicks were unaffected by the taint on their mothers, and could be kept.[17]

Not everyone was equally bound by alimentary restrictions. Certain kinds of otherwise impure food were good enough for "brutish men," "unclean men" and "human beasts." (We shall return to the subject of these men presently.) One clause stated that animals seized by other animals and left half-alive were to be eaten by such men. But another specified that the part of the animal – for example the ear – which had been bitten by the predator was to be cut off and given to the dogs, while the rest were to be eaten by animals (*peccoribus*) and bestial men "for it seems to him fitting that human beasts should eat the flesh that has been served to beasts." In Wasserschleben's version of this canon, *peccatoribus* ("by sinners") replaces *peccoribus*.[18]

The Penitential of Theodore of Canterbury: The bulk of the penitentials did not follow the Irish model directly, but rather the lists presented in Theodore of Canterbury's penitential. This penitential, however, showed unmistakable signs of Irish influence. The first list, from which most of the following are taken, was given the title "Concerning the use or rejection of animals." It imposed a penance for eating unclean flesh or carrion mauled by beasts, unless forced by hunger.[19] Animals mauled by wolves or dogs were not to be eaten either. If a stag or she-goat was found dead, it was not to be eaten unless it had survived long enough to be finished off by a man; otherwise, it was to be fed to the pigs and dogs. The Greeks, asserted Theodore, did not feed carrion to pigs; it was, however, permissible (even among the Greeks?) to make shoes of the hide, and to use the wool and horns for nonsacred purposes.[20] Animals known beyond doubt to have been polluted by sexual contact with humans were to be killed and their flesh flung to

precipice to kill oxen that had fallen some seventy feet, so as to be able to use their flesh as food (18, *MGH SRM* 6, 47).

[16] 17 and 19, Bieler, 180.

[17] 6, 7 and 8, Bieler, 176.

[18] 4 and 18, Bieler, 176 and 180. Bieler's translation. See also Wasserschleben, 122.

[19] I, 7.6, Schmitz 1: 531.

[20] II, 11.1 and II, 8.7, Schmitz 1: 544 and 543.

the dogs; their offspring, however, could be retained and their hide removed for use.[21] If birds or other animals were found dead, strangled in a snare or killed by a hawk, they, too, were forbidden on the authority of the *Acts of the Apostles*, which commanded the faithful to abstain from fornication, blood, the flesh of suffocated or strangled animals, and idolatry.[22]

Interspersed among these prohibitions were permissions. Pigs or hens that had eaten carrion or tasted human blood were still acceptable as food, but if they had mauled corpses they were not to be eaten until macerated (made to grow thin or cured by steeping in a liquid such as brine? – *macerentur*) and after the passage of a year.[23] Fish was allowed, as being "of a different nature". As we have seen, horsemeat also was allowed but, said Theodore, not customary. Hare, in contradiction to *Lev* 11, 6, was not only permitted but even recommended, as a specific for dysentery; its gallbladder mixed with pepper was prescribed for pain.[24]

Variations and additions: These clauses formed the bases of the dietary bans of penitentials and also of the Council of Worms (868).[25] There were, however, variations and additions. Sometimes flesh on which a fox had fed was added to the list of taboo foods.[26] Occasionally, the forbidden ways of killing animals were described. An Irish collection quoted St. Jerome, that animals strangled at the hands of men "in the pagan style" or those that had been killed by animals, by fire or water or any other cause were carrion because they had died without effusion of blood. An eighth-century Frankish or Italian penitential explained that animals that had been cudgeled or stoned to death or killed with an arrow without an iron tip counted as *suffocata* and were not to be eaten.[27] Another continental penitential maintained that even if a stag, other animal or bird had been killed by an arrow, but tracked down only on the third day, after a wolf, bear, dog or fox had fed on it, it was not to be eaten.[28]

An English penitential, more lenient than Theodore's, did not require that a whole year pass before one ate a hog or hen that had tasted carrion or human blood. One had to wait until the hog became pure, and a wait of three months was enough for the hen "although we do not have evidence of this from olden times." Even knowingly eating something bloody called for only a week of penance, as

[21] 11.9, Schmitz 1: 545. The Old Testament had been considerably more draconian, requiring that both the animal and the man or woman in such a case be killed *(Lev* 20, 16-17). See Joyce E. Salisbury, "Bestiality in the Middle Ages," in *Sex in the Middle Ages*, 173-186.

[22] II, 11.2, Schmitz 1: 544.

[23] II, 11.7, Schmitz 1: 545; II, 11.8, *ibid.*

[24] II, 11.3, 11.4 and 11.5, Schmitz 1: 545.

[25] 64 and 65, *Concilia Germaniae* 2: 318.

[26] *E.g., Canones Gregorii* (late 7th/8th century) 147, Schmitz 2: 538.

[27] *Hibernensis* (late 7th/early 8th century) 54.6, H. Wasserschleben, 216; *Poen. Oxoniense II* (8th century) 52, *CCSL* 156: 200.

[28] *Poen. Ps.-Theodori* (mid 9th century) 16.12, Wasserschleben, 602.

long as it was in a dish that had been thoroughly cooked. On the other hand, it did not wholly exculpate a person who did so unawares, but gave him the choice between fasting for three days and chanting the *Psalter*.[29]

Other penitentials were stricter. An eighth-century penitential called for twenty weeks of fasting for having eaten the flesh of an animal that had had its entrails torn by a beast, even if the animal survived long enough to be killed by humans. It accepted hunger as an excuse for eating of an animal that had been killed by a beast or had died of some other cause, but not for eating either *suffocatum* or blood. The flesh of pigs or hens that had tasted (flesh or blood) "of a dead man" was not to be eaten; even partaking of them unknowingly merited four weeks of penance. Moreover, they were not to be bred, but slaughtered "a long way from the village" so that they could be devoured by dogs and foxes.[30] Burchard of Worms gave particularly detailed instructions as to such pigs and hens. If they were killed promptly upon having eaten human blood, their entrails were to be thrown away, and then the rest of the meat could be eaten, but not if there had been a delay in slaughtering them. If they had mauled corpses, they could be eaten after a year's maceration. But if pigs had killed a man, they were to be slaughtered immediately and their bodies buried.[31]

Some penitentials introduced contradictory clauses. An early eighth-century penitential, for example, agreed with Theodore's that pigs and hens that had eaten human blood or carrion were not to be rejected, but announced a little farther on, that the flesh of such pigs should be thrown away.[32]

A few penitentials qualified the general permission, given by Theodore of Canterbury, to eat fish. Some half dozen excluded those that had died in a pond.[33] An English penitential allowed the consumption of fish found dead in a river, but not those found dead in a pond.[34] Others excluded even the former, unless they had been clubbed to death or "touched" that very day by a fisherman.[35] Another penitential rather confusingly both permitted eating fish found dead and at the same time imposed a four weeks' fast for doing so.[36]

Concerning animals with which humans had had sexual contact, the penitentials followed Theodore's closely, with minor additions in continental penitentials. The *Oxonian Penitential II* declared that the flesh, milk and young of a sheep, goat or other animal "with whom man has sinned" should not be eaten, made use of or sold, but that they should be killed that year and fed to the dogs,

[29] *Confessionale Ps.-Egberti* (10th century) 1.40, *PL* 89: 412.

[30] *Poen. Oxoniense II* (8th century) 23, 46 and 55, *CCSL* 156: 195, 198 and 200.

[31] *Decretum* (1008-1012) 19.87, *PL* 140: 1002.

[32] *Poen. Remense*, 3.26 and 3.34, Asbach, 19 and 20.

[33] *E.g, Double Penitential of Bede-Egbert* (8th century) 39.3, Schmitz 2: 697.

[34] *Confessionale Ps.-Egberti* (10th century) 4. -, *PL* 89: 427.

[35] *E.g., Poen. Hubertense* (early 9th century) 59, *CCSL* 156: 115; Burchard of Worms, *Decretum* (1008-1012) 19, 5.131, Schmitz 2: 438.

[36] *Poen. Ps.-Theodori* (mid 9th century) 16.15, Wasserschleben, 603.

and the culprit appropriately punished. Halitgar of Cambrai named the same animals. Ps.-Theodore forbade the use even of the milk of polluted animals.[37] One penitential imposed ten years of penance on married men over the age of twenty who had committed this sin with a heifer and a lighter penance on adolescents or "those less able," for sinning "with certain animals."[38] Still another penitential imposed harsh penances for fornication "with cleanly animals" and ordered that "no one should eat the young of the animal nor use it for breeding."[39]

Some variations appear more significant. In a clause "concerning suffocated animals and blood," a ninth-century penitential warned against eating and drinking the blood of any beast.[40] Blood may be "eaten" incidentally, when the animal is not properly slaughtered, but "drinking" blood must be a deliberate act, one which we have seen was connected with magic. The magical element is unmistakable in a question in an English penitential concerning drinking or eating blood from livestock or humans – the phraseology suggests that the blood was taken from live animals.[41] An earlier continental penitential considered the case of those who had eaten blood or food "polluted by human blood". This plainly refers not to accidental contamination of a dish but to food prepared ritually.[42] But it is not necessary to suggest ritual purposes in a capitulary from early ninth-century Trier which deals with those who had eaten the blood of pigs, sheep and she goats.[43]

9.1.2 *Prohibitions of polluted food*

A second set of prohibitions deals with food or drink that was essentially clean (grain, milk, water) but the use of which had become doubtful because it had been tainted by contact with something unclean, an animal or a corpse. These are found in about a hundred and twenty separate clauses. Except for "birds," general terms ("animal," "little beast," "pure" or "impure animal") are found but seldom. More often, the animal is identified: "the familiar animal" which is the cat or dog, mouse or rat, poultry, shrew-mouse, swine and piglet, weasel. Some of the birds are named: the crow, eagle, jackdaw, magpie. The leech is the only water-creature found to be polluting. Liquid which had become discoloured was also to be avoided, presumably because it had gone bad. All the animals named feed on things both clean and unclean (blood, carrion, human and animal wastes). Their impurity is contagious, passed on to humans at second hand through the food or

[37] *Poen. Oxoniense II* (8th century) 57, *CCSL* 156: 201. See also *Poen. Halitgari* (817-830), Schmitz 1: 488, and *Poen. Ps.-Theodori* (mid 9th century) 31.21, Wasserschleben, 603.

[38] *Poen. Ps.-Gregorii* (2nd quarter, 9th century) 22, Kerff, 180.

[39] *Poen. Merseburgense b* (9th century?) 14, *CCSL* 156: 174.

[40] Poen. Ps.-Gregorii (2nd quarter, 9th century) 28, Kerff, 185.

[41] Double Penitential of Bede-Egbert (9th century?) 26, Schmitz 2: 682.

[42] *Poen. Floriacense* (late 8th century) 41, *CCSL* 156: 100. In a similar clause, the *Poen. Hubertense* ([early 9th century] 60, *ibid.,* 115) does not specify that the blood is human.

[43] *Capitula Treverensia* (before 818) 12, *MGH EpCap* 1: 56.

liquid with which they had been in contact. Nevertheless, the effect of the contagion had weakened, as the relatively light penances show.

This category of restrictions is found only in penitentials and the collections of Regino of Prüm and Burchard of Worms; other sources ignore them altogether. Here again, seventh-century Irish penitential texts appear to have served as model for the *Penitential of Theodore* and through it for other penitentials, both insular and continental. Once more, Adomnan's *Canons* stand apart from the rest.

Irish prohibitions: The *Irish Canons* imposed penances ranging from a special fast (*superpossitio*) to a fifty days' fast for the sin of drinking liquids "illicitly," namely, those which had been contaminated by dogs, eagles, crows, jackdaws, cocks and hens and cats (by drinking of the same liquid? by their dropping their wastes, fur or feathers into the container?), by the carcasses of livestock or by mice.[44] The *Penitential of Cummean* prescribed light penalties for giving another person a liquid in which a mouse or weasel drowned and for knowingly drinking of it. If those "little beasts" had fallen into flour, dried food, cooked food or curdled milk, the food around their bodies was to be thrown out and the rest eaten with a clear conscience.[45] Whoever ate or drank something contaminated by a cat could be purified by three special fasts.[46] Touching liquid food with an "unfit" or "improper" hand (the wrong hand? a dirty hand?) was to be punished by a hundred vigorous blows.[47] Distributing discoloured liquor, that is, one that had gone bad, called for seven days of fast, but whoever drank of it unawares and found out about it later was to torture his "hollow belly" with fifteen days of fasting.[48]

Prohibitions of this sort played a comparatively small role in Adomnan's *Canons*. A well in which the body of a dead man, dog or other animal was found could be purified by draining off the water and throwing out the mud. Food contaminated by a cow could be eaten with a clear conscience, since milk touched by a nursing calf was acceptable; nevertheless, it should be cooked to soothe the scruples of weaker brethren. But if contaminated by swine, it was to be cooked and eaten by "unclean men" since, explains the text, swine ate both clean and unclean things while cows grazed only on grass and leaves. No cooking could purify

[44] 1.16, 1.17, 1.18, 1.19 and 1.20, Wasserschleben, 137. *Superpos(s)itio* was equivalent to missing a meal, see Bieler, 240-241.

[45] 12.12, 12.13 and 12.14, Bieler, 130.

[46] 18, Bieler, 130.

[47] 12.15, Bieler, 130. This clause is found in only two other penitentials, both from the 8th century: the *Double Penitential of Bede-Egbert* (35.2, Schmitz 2: 694) and the *Poen. XXXV Capitolorum* (23.3, Wasserschleben, 520). In many cultures, the left hand is considered polluting and is used for "lower purposes," the right hand being reserved for "honourable purposes" (J. A. MacCulloch, "Hand," *ERE* 6: 492-499; 492). See also Frederick Mathewson Denny, "Hands," *Encyc. of Religion* (New York and London, 1987) 6: 188-191, esp. 190 on "ritual avoidance and mutilation." See also Gross, *Menschenhand und Gotteshand* and Hertz, "The pre-eminence of the right hand."

[48] 12.16 and 12.7, Bieler, 130.

anything tainted by a crow – "for who of us knows what forbidden flesh he had eaten before he contaminated our liquid?" – or a leech.[49]

The Penitential of Theodore: The *Penitential of Theodore* took a more lenient approach to polluted food. Thus it did no harm if food was touched accidentally by a dirty hand or by a dog, cat, rat or unclean animal that ate blood; even eating an animal seen as unclean did no harm if it was done under force of necessity.[50] If a mouse had fallen into a liquid and was fished out still alive, sprinkling the liquid with holy water made it suitable for use. If the mouse had died, however, the liquid could no longer be given to humans; it was to be thrown out and the container cleaned. Nevertheless, if a rat or weasel had drowned in a great deal of liquid food, the food was to be purified, sprinkled with holy water and then, if need be, eaten.[51] If bird droppings fell into some kind of liquid, they were to be removed and the liquid sanctified with water; the food would thereupon become pure.[52] Bees that had killed a man were to be killed quickly (that is, before they reached the hive, as explained the author of a later English penitential), but their honey could be eaten.[53]

The scrupulous must have felt considerable anxiety about ingesting forbidden substances without their own knowledge. Theodore affirmed reassuringly that it was no sin to swallow blood with saliva unknowingly, nor to have partaken unknowingly of a food polluted by blood or something unclean; but if knowingly, penance was to do be done "according to the kind of pollution."[54]

Additions: Subsequent penitentials made few major additions. An Irish penitential added the magpie to the list of contaminating birds.[55] A few penitentials expanded the list of foods that could be polluted by vermin or other animals. The *Oxonian Penitential II* stated that if a mouse, hen or other animal had fallen into wine or water, it became unfit to drink, but oil thus contaminated could be used in a lamp or for some other purpose, and honey could be used for medicine after it had been sprinkled with holy water.[56] Another penitential was prepared to allow wine, oil and honey to be consumed after a blessing if a "cleanly bird" had fallen into it, but if an unclean bird or a mouse had fallen into it, it was to be thrown out; it proposed a year's penance to anyone who sold such a liquid.[57]

Adomnan's concerns were reflected in a number of continental penitentials

[49] 9, 10, 11, 12, Bieler, 176 and 178. Bieler's translation.

[50] *Poen. Theodori* (668-756) I, 7.7, Schmitz 1: 531.

[51] I, 7.8 and 7.9, Schmitz 1: 531.

[52] I, 7.11, Schmitz 1: 531.

[53] II, 11.6, Schmitz 1: 545; *Confessionale Ps.-Egberti* (10th century) 1.39, *PL* 89: 411.

[54] I, 7.11 and 7.12, Schmitz 1: 531.

[55] *Poen. Bigotianum* (late 8th-late 9th century) I,5.6, Bieler, 216.

[56] 54, *CCSL* 156: 200.

[57] *Poen. Hubertense* (9th century) 61, *CCSL* 156: 115.

that contained clauses about wells, cisterns, ditches or containers that had been contaminated by dead bodies, human or animal. Two, for example, required that a well in which a hen had drowned be emptied, and imposed penances on those who had drunk from the water either knowingly or unknowingly.[58] In another, if an unclean bird or mouse fell into a well, it was to be emptied, but if a man drowned in it, the water from it became (permanently) unusable.[59] In still another, the well or cistern in which a man had died was to be emptied and purified with holy water and prayer.[60] An English penitential envisioned a veritable liturgy of purification: if a piglet fell into water and was rescued, the water could be purified by holy water and incense, and thus made fit for sale; if a rat, holy water was sufficient. But if the animal had drowned, the water had to be discarded and the barrel cleaned. The water in a ditch in which an animal had drowned, however, could be used for drinking after a sprinkling of holy water.[61]

In Adomnan's *Canons*, cooking made certain kinds of tainted meat acceptable to some degree. This tacit belief in the purifying effect of fire appeared to be shared by others (principally English authors) who doubted the morality of eating inadequately cooked food. The penitential attributed to Bede was followed by four others (two English, two continental) in proposing a modest penance for eating *semicoctum* knowingly or unknowingly.[62] St. Boniface must have felt similar concerns, for he evidently asked Pope Zacharias how long bacon must be cured. Failing patristic authority, the pope was somewhat at a loss for an answer, finally deciding that it could be eaten smoked or roasted over the fire (presumably at any time), but, if eaten raw, only after Easter (that is, after curing for a minimum of three months, if the hogs had been slaughtered at the beginning of winter).[63]

Prohibitions of fraus: The same English penitentials contained a peculiar addition, concerning the consumption of "fraud" (*fraus*). In one, "whoever knowingly eats *fraus* and is in need because of faintness" was required to fast for seven days, but if in health, his penance was increased to forty days (the usual penalty for transgressions against food taboos), and more if it was a habitual sin.[64] In two, this clause is in a chapter titled "Concerning various matters," which deals mainly with dietary restrictions (seven clauses) but also includes clauses on misplaced altar

[58] *Poen. Oxoniense II* (8th century) 53, *CCSL* 156: 200; *Poen. Halitgari* (817-830), Schmitz 1: 488).

[59] *Poen. Hubertense* (9th century) 61, *CCSL* 156: 115.

[60] *Poen. Ps.-Theodori* (mid 9th century) 31.7, Wasserschleben, 602.

[61] *Confessionale Ps.-Egberti* (10th century) 1.39, *PL* 89: 411; *ibid.*, 4. -, 427 and 428. The reference to the sale of water is surprising.

[62] *Liber de remediis peccatorum* (721-731) 6, Albers, 415; *Poen. Egberti* (before 766) 13.6, Schmitz 1: 585; *Double Penitential of Bede-Egbert* (8th century) 22.2, Schmitz 2: 693; *Poen. Ps.-Theodori* (mid 9th century) 31.24, Wasserschleben, 603-604); Burchard of Worms, *Decretum* (1008-1012) 19.106, *PL* 140: 1005.

[63] *Ep.* 87 (751), *MGH EpSel* 1: 199.

[64] *Poen. Egberti* (before 766) 13.2, Schmitz 1: 585; *Double Penitential of Bede-Egbert* (8th century) 22.2, Schmitz 2: 693; *Liber de remediis peccatorum* [721-731] 3, Albers, 414.

furnishings and minor theft. In the other, it is placed in a clause dealing exclusively with dietary taboos. The meaning of *fraus* is unclear in this context: perhaps food obtained by dishonest means other than theft, or some esoteric meaning particular to English authors.[65]

9.1.3 *Pollution through human contact*

The *Irish Canons* imposed penances for contact with "lay" persons: drinking a liquid of (tainted by) a layman or laywoman (*laicus, laica*) was to be expiated by fifty days of fasting, eating or sleeping in the same house or same bed with either, forty. Drinking illicitly of a liquid of (tainted by) a pregnant serving woman (or discarded wife of a priest – *glantella prignans*) or by one who lives with her (her seducer? – *cohabitator sui*), or eating in the same house or tent with them also called for forty days, but only half that for sleeping in the same house.[66]

Similar canons are found in the *Old Irish Penitential* (before the end of the 8th century) and, with some differences, in the rather later *Bigotian Penitential,* also Irish. The *Old Irish Penitential* called for "forty nights on bread and water" as penance for anyone "who drinks the leavings of a layman or laywoman or of a pregnant woman, or who eats a meal in the same house with them, without separation of seat or couch".[67] The *Bigotian,* however, rated drinking a liquid tainted by a *laicus* or *laica* as significantly less serious than had the others, since it assigned a single day of fasting for this fault, but the full forty days for eating and sleeping in the same house or bed (hut? – *spatula*).[68] The same was required for drinking something tainted by a pregnant *glangella,* or for "cohabiting" with her. Bieler, remarking that it is "strange to find a sexual sin mentioned in this context," suggested that the latter penalises drinking something tainted "by him who cohabits with her."[69]

9.2 THE ORIGIN OF MEDIEVAL ALIMENTARY TABOOS

The number, variety and persistence of alimentary taboos are such a noteworthy feature of early medieval pastoral literature that some attempt to explain their sources and functions should be attempted.

One obvious source is the Bible.[70] The influence of the Bible in itself,

[65] Or is this a muffled echo of *Sir* 34, 27? Eating or accepting stolen food was forbidden in *the Canones Adomnani,* in a clause with echoes of the great cattle raids of Irish mythology (15, Bieler, 178).

[66] 1.21, 1.22, 1.23, 1.24 and 1.25, Wasserschleben, 138.

[67] 3, Binchy in Bieler, 260.

[68] 6.1, Bieler, 216; 6.3, *ibid.,* 218. *Spatula,* which Bieler translates as "bed," is defined by Du Cange (*s.v.* "Spatula," 2) as "hut" or "shed."

[69] I, 6.2, Bieler, 216. See 256, n. 17 for Bieler's comments. Maybe significantly, the only non-dietary clause in Adomnan's *Canons* also deals with sexuality. It cites an unnamed authority to forbid the remarriage a man during the lifetime of a faithless wife who had taken another husband (16, Bieler, 178).

[70] For the biblical connection, see Maria Giuseppina Muzzarelli, "Norme di comportamento

however, does not offer a wholly satisfactory explanation. Most of the taboos can indeed be traced to the Old Testament, but the Old Testamentary restrictions had largely been annulled by the New, which explicitly rejected the connection between spiritual and physical pollution.[71] This was recognised but set aside in the mid ninth-century *Ps.-Theodorian Penitential*.[72] If monastic authors (especially Irish ascetics) preferred the strictness of the Old Testament to the lenity of the New, why did they not accept the biblical restrictions wholesale (that against pork, notably, was omitted[73])? Why did they not justify the regulations by scriptural references? The demand in *Acts* 15, 29 that the faithful should abstain from sacrificial food, blood and the flesh of strangled animals, and from fornication, is mentioned almost a score of times, but only one (non-Irish) text cites the authority of the Old Testament to support its ban on carrion, blood, and flesh touched by beasts.[74] Biblical texts to justify other taboos are wholly absent.

It is likely, therefore, that other factors came into play, which were never articulated clearly but which gave biblical prohibitions a particular relevance to the early medieval situation. Oronzo Giordano attributed this concern to the characteristic concept of hygiene, which did not differentiate between spiritual and

alimentare nei libri penitenziali," *Quaderni Medievali* 13 (1982): 45-80; Harmening, *Superstitio*, 231-235, Bezler, *Pénitentiels espagnols*, 280-286, and Hubertus Lutterbach, "Die Speisegesetzgebung in den mittelalterlichen Bussbüchern (600-1200). Relgionsgeschichtliche Perspektiven," *Archiv für Kulturgeschichte* 80 (1998): 1-37.

[71] For restrictions in the Old Testament, see *Gen* 9, 4-6, *Lev* 11, 1-47 and *Lev* 17, 1-14. Despite the injunctions in *Acts* 15, 29 and 21, 25 to avoid certain categories of food, the weight of the authority of the New Testament is against dietary restrictions. *Matt* 15, 10-20, *Mk* 7, 14-20 and *Acts* 10, 9-16 reject the concept of unclean foods altogether, while St. Paul advised abstention even from idolothytes only as charity toward weaker brethren, not because they were polluting (*Romans* 14 and I *Cor* 8-10). For an anthropological interpretation of Old Testament restrictions, see Mary Douglas, *Purity and Danger* (1966; Hardmondsworth, 1970) and the refinement and modification of this interpretation in *idem*, "Deciphering a meal," in *Implicit Meanings* (London, 1975), 249-275. For a socio-historical analysis of St. Paul's teachings, see Peter D. Gooch, *Dangerous Food. I Corinthians 8-10 in Its Context* (Waterloo, Ont., 1993).

[72] 31.14, Wasserschleben, 602-603.

[73] Banning pork would have been difficult given its ritual status among both Celts and Germans and its economic importance. For the pig in Celtic culture, see Miranda Green, *Animals in Celtic Life and Myth* (London and New York, 1992), 169-171 and *passim*. The importance of the pig in Frankish society is shown by the fact that the theft of pigs took pride of place among crimes in the *Pactus Legis Salicae*, *MGH Leg.* 1.4.1: 20-25. See also J. Balon, *Traité de Droit Salique. Étude d'exègése et de sociologie juridique* (Namur, 1965), 1: 92-105. Among the Salian Franks, traces of the sacred character of the pig persisted into Christian times, see Vordemfelde, *Die germanische Religion*, 108-110. On the importance of pork in the early medieval diet, see Massimo Montanari, *Alimentazione e cultura nel medioevo* (Rome, Bari, 1988), 13-22 and 37-40. For the pig in medieval folklore and hagiography, see Milo Kearney, *The Role of Swine Symbolism in Medieval Culture* (Lewiston, N.Y., 1991).

[74] *Dicta Pirmini* (724-753) 19: 170-171).

physical purity.[75] According to Pierre Bonnassie, the regulations expressed the Church's determination to draw a sharp dividing line between humans and animals. Paganism had treated the relationship between men and animals as potentially sacred; the Church accordingly denounced it – hence the repeated canons against masquerading as animals, sexual contact with or imitations of them, and eating food that they had touched. Bonnassie also showed that the dietary taboos reflected the fear of cannibalism caused by repeated and prolonged famines.[76] But during the missionary period, the determination to separate Christians from pagans or quasi-pagans and the fear of magic might also have had a part.

9.3.1 *Dietary prohibitions and the Christian frontier*

The origins of the penitentials lie in the missionary period, at a time when conversions occurred *en masse*. New Christians often had only indistinct ideas of what their new religion required of them, and might on occasion lapse easily, even innocently, into old customs. Pastors had to teach new converts to transform their behaviour, manners, attitudes, in fact, their entire lives, to meet the Christian ideal, and, particularly before the Christian community was securely established, to set up barriers of custom and morality between them and the pagan world. Dietary regulations formed a part of such barriers.[77]

Adomnan's *Canons* offers a striking confirmation of this. It specified three cases in which unclean food was to be fed to "brutish" or "unclean" men or "human beasts" for it was fitting "that human brutes should eat flesh touched by brutes". But who were these brutish men?

It seems to me that they may be identified with certain *laici* and *laicae* described in Irish texts. We have seen that in the *Irish Canons*, the *Bigotian Penitential* and the *Old Irish Penitential*, they tainted the food that they touched, the bed in which they slept and the house in which they sheltered. "Tainted" (*intinctus*) is the word used to describe the pollution of food by animals also. The same penitentials considered keening, always a sin, to be particularly serious if performed for a layman or laywoman or for a pregnant maidservant or her man.[78] The *Collectio Hibernensis* required only a year of penance, or less, for killing in a sacred place where *laici* were sheltering – "for we cannot call a place sacred which has been entered by homicides with their spoils, thieves with their plunder, adulterers,

[75] "Igiene personale e salute nell'alto medioevo," *Nuovi Annali della Facoltà di Magistero dell' università di Messina* 4 (1986): 183-233.

[7676] "Consommation d'aliments immondes et cannibalisme de survie dans l'occident du haut moyen âge," *Annales* 44 (1989): 1035-1056; here, 1040-1041.

[7777] Such barriers, according to Mary Douglas, bound "the area of structured relations. Within that area rules apply. Outside it, anything goes" ("Deciphering a meal," 273). *Cf.* Somerset Maugham's colonial administrator who dressed for dinner in the middle of the jungle.

[78] *Canones Hibernenses*, 1.26 and 1.27, Wasserschleben, 138; *Poen. Bigotianum*, IV, 6.2 and 6.3, Bieler, 230; *Old Irish Penitential*, V, 17 Binchy in Bieler, 273. See chapter 8, 8.1.1.

perjurers, criers or druids."[79] The Old Irish *De Arreis* (*Table of Communtations*) proposed particularly harsh terms for penitent laymen: "commutations proper for former lay men and women: spending the night in water or on nettles or on nutshells, or with a dead body – *for there is hardly a single layman or laywoman who has not some part in manslaughter*" (my italics).[80] Obviously, in these texts, the words "layman" and "laywoman" refer to something other than non-clerics.

Using penitential texts, early Irish saints' *Lives* and secular literature, Richard Sharpe traced the shifting nuances of *laicus* in Hiberno-Latin.[81] *Laicus* became increasingly pejorative, meaning first pagan and then brigand, the member of a sect, marked with special signs and bound by vows to commit ritual murder (the usual meaning of "lay" was expressed by *plebeus*). Such men may be recognised at a later period, in St. Bernard of Clairvaux's description of the Irish encountered by St. Malachy: "Until that moment, he never experienced such barbarity, never met with any men so violent in manners, brutish in custom, impious in matters of faith, barbarous in laws, resistant to teaching, filthy in their way of living – Christians in name, but pagans in fact."[82]

It is probable that the diet and feasts of such *laici* and such "brutish men" are at the bottom of some at least of the alimentary taboos imposed in the Irish penitentials. During the early Middle Ages, suggests Stéphane Boulc'h, the attitude towards food with ritual connotations was seen as a fundamental factor in religious identification of the individual.[83] Similarly, Mary Douglas observed that it "would seem that whenever a people are aware of encroachment and danger, dietary rules controlling what goes into the body would serve as a vivid analogy of the corpus of their cultural categories at risk."[84] The insistence of early Irish penitentials on dietary taboos may thus have expressed the sense of danger of contamination by pagans and brigands whose criminal activities had strong cultic overtones, and with whom Christians could have no common ground. Hagiography gives some support to this conjecture: St. Columba foresaw that a "false penitent" would be reduced to eating horseflesh in the forest among thieves, St. Endeus prophesied that a *laycus* too scrupulous to eat beef would end up eating stolen horseflesh.[85]

Significantly, the most important food restrictions in our texts without any trace of Irish influence or of the restrictions in the *Acts of the Apostles* are found in St. Boniface's correspondence, in letters proving his anxiety about his flock's consumption of various birds and beasts, and the curing of bacon. He was running a

[79] 44.8, H. Wasserschleben, 176-177.

[80] 8, Binchy in Bieler, 279.

[81] "Hiberno-Latin *laicus*, Irish *láech* and the Devil's Men," *Ériu* 30 (1979): 75-92.

[82] *Vita S. Malachiae* 8.16, *PL* 182: 1084.

[83] "Le repas quotidien des moines," 322, n. 151.

[84] "Deciphering a meal," 272.

[85] *Vita Columbae* 26b (I, 21), 251-253; *Vita Endei* 30, *VSH* 2: 73.

missionary church, as yet by no means firmly grounded, which remained surrounded by internal and external enemies. At the very time he was trying to persuade his new converts to conform to the standards upheld by himself and the popes who sent him, he filled his letters with complaints of the moral degeneracy existing among long-time Christians, clergy included.[86] He too might have seen dietary restrictions as part of the spiritual barriers necessary to protect and govern his flock.

In their reiterations of alimentary restrictions, subsequent penitentials do not show the same fear of contagion from or the same "gut response" of loathing for insidious groups of outsiders – pagans or brigands – as do these Irish penitentials, although they occasionally warned against the company or food of pagans or Jews.[87] Nevertheless, the connection between alimentary taboo and the customs of pagans was not broken. It is made explicit in a ninth-century continental penitential which treated consorting with pagans and using their implements as being defiling in the same way as typical forms of dietary pollution and sexual abuse of animals:

> If swine and fowl have eaten human blood by chance, they may be eaten after a year and their offspring is not to be rejected. But if they have mangled corpses, they are to be slaughtered and thrown to the dogs. Whoever eats of them unknowingly is to do penance for four weeks. Similarly, animals with which men have fornicated are to be flung outside, so that the beasts and birds may devour them; their hides are to be used only for shoes. He who is Christian must not eat or drink with a pagan nor use their dishes. If a shrew-mouse has fallen into food, it is to be taken out if it is alive and [the food is to be] sprinkled with holy water; if it is dead therein, all the food is to be thrown out.[88]

9.3.2 *Magic*

Although no explicit link between diet and magic is made in the texts, magic or fear of magic almost certainly contributed to food taboos. Alimentary restrictions were sometimes placed in conjunction with bans on the use of substances treated elsewhere as magical. The *Bigotian Penitential* put drinking human urine and blood, semen, and animal urine in the same chapter as the violation of taboos against carrion and polluted food, all under the heading "The canons of the wise and of Gregory concerning those who eat illicit flesh, and who drink what has been tainted by beasts and birds."[89] The *Penitential of Cummean* penalised the habit of eating scabs and body-lice, drinking urine and eating feces under a chapter concerned with youthful misbehaviour, but a late eighth-century continental peniten-

[86] *E.g.,* his letter of 742 to pope Zacharias (*Ep.* 50, *MGH EpSel*, 82-83).

[87] *E.g.*, Burchard of Worms, *Decretum* (1008-1012) 19, 5.190, Schmitz 2: 451.

[88] *Poen. Valicellanum I* (beginning of the 9th century) 96, Schmitz 1: 318-319. See Schmitz's comments, 319-320.

[89] I, 5.1, 5.2 and 5.3, Bieler, 216.

tial put the same practices as the first item in a section dealing otherwise strictly with food taboos. The author of this penitential plainly did not see these practices as a form of childish aberration, and he could hardly have seen them in terms of alimentation; magic seems the likeliest alternative. If so, the other practices listed with it may also have been tainted with magic.[90] In the somewhat earlier *Excarpsus Cummeani*, the consumption of such substances appears in the last of thirty-eight clauses concerning food and drink. The preceding questions deal with women's use of their husbands' blood and semen in remedies and love potions, and with unauthorised fasting (possibly a preliminary to divination).[91] In another continental penitential of a later period, body wastes also became ingredients in a love potion.[92]

In some cases, the extremely precise lists of defiling animals can be most easily understood in terms of magic. Why were dogs and cats singled out in questions about drinking polluted liquids?[93] And how exactly did one know that it had been touched in some way by these animals? Did one watch them polluting the liquids, and then choose deliberately to drink from that container, out of thirst perhaps, or for some other reason? Was it only from a spirit of thrift that one would drink, or make others drink, the liquid in which rats or mice and weasels specifically (as opposed to, for example, shrew-mice or dogs or cats) had drowned?[94]

It is possible to find evidence for the magical use of some part of almost every animal mentioned in the food restrictions, and for many others. According to Schmitz, liquids tainted by cat's urine were used to make love potions, and drinks made from the water in which mice and weasels had drowned were used for protection against magic and sorcery.[95] Pliny lists far more medicinal and magical remedies made from animal sources than do the penitentials – for example, mouse droppings or the ashes of flies to make eyelashes grow long, hawk droppings in honey for fertility, boiled mouse or weasel for many ailments.[96] Grimm noted that among the Slavs flesh bitten by a wolf was cut out to be dried and smoked for use as medicine; and that "cattle killed or bitten by wolves, are wholesome fare."[97]

[90] *Poen. XXXV Capitulorum* 23.3, Wasserschleben, 519-520. See also *Poen. Cummeani* (before 662) 18: 128.

[91] 1.35-1.37, Schmitz 1: 618-619.

[92] *Poen. Valicellanum E. 62* (9th/10th century) 29, Wasserschleben, 560.

[93] *E.g., Liber de remediis peccatorum* (721-731) 4, Albers, 415.

[94] In the *Vita Liutbirgae*, a devil taunts the saint that as a young girl she had deliberately, with giggles, drunk from a jug in which a mouse had drowned (27, *MGH Deutsches Mittelalter, Kritische Studientexte* 3: 31. For a discussion of this episode, see Albert Demyttenaere, "The Cleric, Women and the Stain; Some beliefs and ritual practices concerning women in the early Middle Ages," in *Frauen in Spätantike und Frühmittelalter,* 141-166; here, 148-151).

[95] 1: 314-318.

[96] *Natural History*, 29.27, 30.45, 30.12. Pliny dedicates Bks 23-32 to remedies and magic of this sort. See also Bächtold-Stäubli, *s.v.* "Harn," "Exkremente" and "Kot".

[97] *Teutonic Mythology*, 1140, n 2, and 1645.

Lending support to the concept of magic, or the fear of magic, as underlying some of the prohibitions are practices which may be interpreted as being counter-magic meant to undo the baneful effects of the pollution. Measures that, from one point of view, tend to argue against a magical element can also be seen as being a practical response to popular belief in malevolent magic.[98] Among these are frequent recommendations to sprinkle the contaminated food with holy water, after which it might be eaten with a "clear conscience." The flesh of pigs that fed on cadavers became fit for eating after "macerating" for a year, which may mean curing in brine. Cooking would undo the effects of some forms of contamination in food.[99] Fasting, the most usual penance, is a way of cleansing the body as well as the soul of pollution.[100]

Texts that deal with those who ate unknowingly of some contaminated substance may indicate an anxiety about the possibility of ingesting some magical substance, such as a love potion slipped by a wife into her husband's food or drink.

Finally, it is worth noting that while there is sound economic sense in making use of polluted foods, especially in times of relative scarcity, some of the taboos refer to kinds of animals that could hardly have been considered to be suitable for food.[101] No doubt, horse, beaver, or mangled deer or goat could make a nourishing dish; but it is hard to imagine any reason for cooking up crows, jackdaws and magpies other than dire need (usually accepted as an extenuating circumstance) or some magico-ritual purpose.

It seems clear, therefore that while the example of the Bible and literary tradition were unquestionably important in the development and transmission of early medieval alimentary restrictions, they were not the only factors. It is more difficult to determine the exact weight to give to the others: determination to use diet to carve out a specifically Christian identity, customs and rituals peculiar to Irish outlaw communities, magical practices. Nevertheless, the evidence of our texts suggests that these were the factors that first roused the concern of clerics; in the disciplinary measures of the Old Testament they found a valuable tool for the fulfillment of their pastoral duties.

[98] *Cf.* Flint's concept of "encouraged magic" propagated by the Church to fill the needs left by the destruction of the traditional magic of paganism (*The Rise of Magic*, 254-328).

[99] For the purifying effect of salt and fire respectively, see Alfons Kirchgässner, *Les Signes sacrés de l'Église*, adapted from the German by the monks of Mont César (1964), 125 and *La puissance des signes,* 124-125.

[100] Kirchgässner traced the origin of fasting to fears of defilement through food: "D'une manière générale, les repas présentent un danger d'impureté, tout d'abord parce que des démons ou des actions démoniaques peuvent pénétrer par la bouche ouverte, ensuite parce qu'il existe toute une série d'aliments qui possèdent un *mana* dangereux, par nature, ou parce qu'ils sont contaminés d'une manière occulte. Nous trouvons là la racine du jeûne et des commandements sur le jeûne" (*La puissance des signes*, 314).

[101] See the analysis of the social, psychological and moral effects of hunger presented by Camporesi in *Bread of Dreams.*

Conclusion

At first sight, the popular culture that emerges from pastoral literature is surprisingly homogeneous. The same practices are described throughout the period in documents of almost every region: vows offered and repasts shared in unsuitable places, rituals during eclipses and at the New Year, unseemly songs and dances on the eve of liturgical festivals, hosts of cunning men, reliance on amulets and spells, healing magic and love magic, the invocations of women at their wool work, burial of the dead within churches. In many, maybe most, cases the authors appear to rely on a ready-made image of popular culture that they inherited from tradition. Nevertheless, certain differences show that in part at least they based themselves on their own observations of the behaviour of their contemporaries.

Concerning idolatrous beliefs, the texts indicate an awareness of circumstance and historical changes. The identifiable deities of the pantheon appear chiefly in sixth-century documents from southern Gaul and from Galicia and in eighth-century documents from the northern parts of Frankish territory, areas then in the process of being converted, or reconverted, to Christianity, where the memory and even the cults of pre-Christian deities may well have lingered. These were rapidly replaced by vague references to anonymous idols and demons, and to a handful of minor local and family spirits. The important exception is Diana, whose classical name disguised an indigenous Rhenish goddess of death and fertility, and whose cult emerged or re-emerged in the 9th or 10th century.

Regional variations appear. Irish scribes were largely indifferent to the kind of issues that preoccupied England and the continent. The documents show but slight concern about magic (mainly clerical) and none about idolatrous rituals except for what may be inferred from the highly detailed canons about contamination through food. English documents cover almost the full range of topics but are silent on the subject of popular celebrations on the eve of liturgical feasts and in front of the churches, perhaps because of the tradition of tolerance initiated by Gregory I's letter to Mellitus. Even continental texts reflect local differences: Priscillianism echoes in Spanish texts down to the 11th century, an eighth-century penitential reveals the presence in Lombardy of a community of pagans who cremate their dead, while eighth-century from northern Gaul berate those who believe in preternatural thieves of the harvest.

The different types of pastoral literature offer glimpses here and there of dynamic popular cultures with complex mythologies and adaptive techniques hidden behind blandly stereotypical formulations. Conciliar legislation and capitularies indicate the existence of practices specific to a given time and place (for example, funeral processions in late seventh-century Spain, needfires in mid eighth-century Austrasia, suspect victory masses to St. Michael in early tenth-century Thuringia). Sermons describe rituals during eclipses and on the feast of St. John the Baptist in the Rhineland and Piedmont during the 9th and 10th century respectively.

Except for the Irish penitentials and those of St. Columban, Theodore of Canterbury and Burchard of Worms , penitentials rarely offer insights into popular culture. The majority relied on earlier models passed from monastery to monastery, and show little response to actual situations. The virtual absence of any reference to issues important to the Frankish church, such as the misuse of chrism or the suspect cult of angels, is striking evidence of the impermeability of penitentials to the pastoral concerns of the bishops and administrators. But even penitentials sometimes departed from their models to include new material. They support Agobard of Lyon's testimony on eighth-century beliefs in weather-magicians and describe mourning rituals absent from the other texts. Burchard's penitential is particularly valuable as an index of a shift in the interpretation of traditional practices.

Pastoral literature does not support the view that popular culture was a matter of class. References to social standing are rare, but when they appear, they reinforce the idea of a common culture. It was probably not merely practical considerations but vestiges of old beliefs and old fears that prevented Merovingian and Visigothic *possessores* from moving energetically to wipe out traces of traditional cults among their *coloni* and slaves. Nobles and commoners, townsmen and peasants all believed in the reality of weather magic. Ancillary sources prove that the rich and the great were as ready to use divinatory techniques, practice healing magic and turn to cunning men as the poor. Above all, the texts implicate clerics in magic of all sorts: recourse to diviners, production of amulets, performance of incantations and love magic.

We may, however, perceive a class difference in the way these practices were evaluated. The authorities were quicker to detect paganism and superstition in the customs of subordinate groups than in those of their betters.When the laity practiced divination or magic they were paying cult to idols or demons but, though all the authors agreed that it was wrong for clerics to participate in magic, only Caesarius said outright that these men were the devil's henchmen. Peasant rejoicings in front of churches on the eve of liturgical feasts and at drunken wakes were readily condemned as pagan survivals and superstitions. But the equally rowdy rituals and amusements of nobles and clerics were not seen to be relics of paganism.

The popular cultures depicted in pastoral literature were in transition. On the one hand, the traditions and practices that had been integral to the life of a people had been largely emptied of meaning by the new religion that it had accepted willy-nilly, and were in the process of dwindling into folklore. At the same time, the old belief-systems had not been quite eradicated; though relegated by the clergy to the realm of hallucinations and bad dreams, they continued to live in rituals and myths transmitted by one generation to the next. On the other hand, readiness to view sacramentals as good luck charms, integration of saints and angels into the system of ancestors and heroes, determination to bury the dead *ad sanctos,* all show that popular culture was absorbing and transmuting Christianity to fit traditional modes of thinking. The Rhenish women who drove stakes through the corpses of unbaptised babies, and who buried baptised babies with Eucharistic symbols, could not have expressed more vividly their profound and whole-hearted acceptance of the Church's teachings about the redemptive force of the sacraments.

Our sources then show significant insights into beliefs and practices of the faithful. Nonetheless, it must be admitted that they have disappointingly little to offer on the subject of popular culture. The immense volume of legislation, penitentials, sermons, letters and tracts yields about two thousand mostly short passages dealing with practices that may be considered to be pagan survivals and superstitions. This is not much, considering the extent of time and expanse of space covered: over 500 years and most of Western Europe. Moreover, few contain material not drawn from earlier sources; there is little to be found at the end of our period that was not there at the beginning. Caesarius of Arles's sermons, Martin of Braga's canons, the first penitentials and a handful of the ancient councils of the Church provided the model of pastoral problems for Western European churchmen for half a millennium.

Even if we can accept that the texts are repetitive because they continued to describe permanent features of beliefs and practices, we are left with the question as to why they mention so seldom areas of life which were untouched in the earlier texts. The routine of daily life is almost wholly absent: cooking and brewing magic, magic to restore virginity and to determine the sex of babies, the rituals of pregnancy, childbirth and puberty. Almost all the information about farmers' practices comes from the *Homilia de sacrilegiis*, and that is limited to their observance of the appropriate times for agricultural tasks. We are told that herdsmen and hunters resorted to magic to protect their own animals and harm those of rivals. We know that soldiers in early sixth-century Arles paraded about in women's clothes at the New Year, that in late eighth-century Bavaria they practiced magic before single combat, and that Piedmontese architects, like ignorant peasants elsewhere, planned construction according to the heavens. The magic used by weaving women is cited repeatedly. But there is not a word about any other occupation, about charcoal burners, wood-cutters, miners, fishermen, sailors (a notoriously superstitious crew), smiths (with their strongly magical antecedents), potters, tanners, wheelwrights, peddlers, beggars, thieves and prostitutes, all of whom undoubtedly engaged in magical rituals adapted to their unique circumstances. The truth seems to be that the majority of clerical authors were not interested enough in the activities of lay people to observe and try to understand the exigencies of their lives and their strategies for survival. For the members of the clergy, the Church was the institutional Church, that is, primarily themselves. The salvation of society depended on what they did and their welfare, spiritual and temporal; compared to that, the laity was of only marginal interest.

In the light of this general indifference to the daily life of the laity, the scraps of original data that succeeded in finding their way into pastoral literature become all the more significant. They illuminate, here and there, some aspects of rich and vital indigenous cultures, which are otherwise hidden from us, but which were strong enough to resist the attempts of the ecclesiastical and secular authorities to remake them to conform to models alien to them.The lack of context for most of this data, in addition to the problems posed by their scarcity in absolute terms, their geographically and chronologically dispersed origins, their vague, shifting vocabulary and uncertain syntax, makes interpretation difficult, often impossible. Is it possible to reconstitute a context which will enable us to understand the meaning and function of specific beliefs and practices? The attempt made here to organise material found in pastoral literature chronologically and geographically and to trace connections is one step only. Other contemporary written sources, such as histories, liturgy, customary law and, in particular, hagiography, should be studied systematically in the same way, and the data compared to see if they shed light on each other and on broader cultural questions.

One such question deals with women's social and economic position. In penitentials, the women who used erotic magic were almost invariably married women who did so in order to increase their husbands' love, not for their own sexual gratification. As far as the texts are concerned, women looked neither to magic nor to men when they were interested in their own pleasure. I have hypothesised that the wives who used – or, perhaps more accurately, were suspected of using – love magic were those whose status was vulnerable because their marriages were not formally sanctioned. A further study of other documents, particularly legal texts, is needed to see if this is justified. If so, love magic opens a new perspective on the situation of women and the relations between the sexes at this period. It may also provide a new tool for the analysis of the myths of the nocturnal rides and battles in which women claimed to participate.

Other themes evoked in pastoral writings also point to fruitful lines of research. The meaning of alimentary restrictions is still not clear, and these form only part of the larger question of food symbolism. The symbolic importance of bread, for instance, is attested by its uses as offering, representation of the self or an idol, tool for divination, and magical device. Lunar beliefs and rituals contain clues about social relations and material culture. Syncretic practices indicate the dynamics of the process of conversion and acculturation. Understanding such themes requires a closer study of other types of documentation, such as archaeology and folklore, and the use of insights from anthropologists, ethnographers and scholars in comparative religions.With such means, it should be within our reach at least to sketch the broad outlines of early medieval culture, impossible as it may be to grasp fully its richness and variety.

Word List

Terms used in the texts studied:

accursed: *unholda, unholdum* (Teutonic)
amulet: *phylacteria/filacteria, ligatura, ligamina, carmen.*
—, objects used as: *caracteres angelorum vel salamonis; c(h)aracteres, caracteria in carta / in bergamena / in laminas aereas, ferreas, plumbeas uel [alia materia] scribi; cornu / lorum ceruunum; cartella; falsa inscriptio; herba; ossa; sucinus; scriptula; salamoniaca scriptura*
apparition: *persona*
appeal to the judgment of God: *stapsaken* (Bavarian)
assembly: *collecta, consortia, conventicula, conventus, bibitoria*
astrologer: *mathematicus*

bewitchment: *fascinatio, incantatio*
bird of prey: *muriceps*
bread-trough: *maida*

cairn: *acervus petrarum, agger*
cause serious harm, bring about a miscarriage, kill: *decipere, perdere*
charm, protective: *carmen, fascinum, praecantatio*
circumambulation: *dadsisas, circus, circumire (contra solem)*
circumambulator: *circerlus (?), cerenus(?)*
corner: *angulus (?), cantus (?)*
crossroads: *bivium, compita, quadrivium / quadribium, trivium,*
customs, pagan: *paganiae, paganismus, observatio / consuetudo / traditio paganorum, ritus paganus, spurcitiae gentilitatis*

dance: n., *ballatio, saltatio, vallatio*; v., *ballare, chorus ducere, saltare*
dancer: *saltator*
dancing-woman: *saltatrix, bansatrix, tornatrix*
devil: *diabolus, zabulus, Satanas*
dirge, keening: *bardigus, bardicatio, luctus mortuorom, planctica, ululatus excelsus*
divination: *augurium, divinatio, sortilegus*
dream-interpreter: *somniarius, somniatori coniector*
duel: *wehadinc* (Frankish)

eclipse, ritual / spell against: *Vince luna*
enchanter, enchantress: *incantator, incantatrix*
enchantment: *incantatio, fascinatio*
evil eye: *fascinum*

fast: *carina*
feast-days: *feriae, feriati dies*
February rituals: *dies spurci, spurcalia*

ghost: *larva, persona, nequam spiritus* (?), *spiritus immundus* (?)
ghost-mask: *talamasca* (Frankish)
godparent / kin "by right of John the Baptist": *commater sentiana, commatre de sancto Johanne, comparatus sancto Iohanni, cummater de sancto Joh.*
gods, anonymous: *infaustae personae, unholdum* (Teut.)
grove: *lucus, nimidas*

harvest-thieves: *mavones, manus, maones, dusi hemaones, dusi manes*
healer: *praecantator, incantator, suffitor* (?), *obligator* (?)

henbane: *jusquiamus, belisa (Teut.)*
herbalist: *(h)erbarius, (h)erbara*
hex: *maleficium*
horoscope: *genesis, nascentia*
house-sprites: *neptunus, pilosus, portunus, satyr*

illusion: *illusio, imaginatio, phantasmata*
invocations, written: *petatia*

joke, prank: *iocatio, iocus, iotticus*

knot, magical: *ligatura, ligamen*

laity: *plebeus, laicus, laica*
lots: *sortes, sortes biblicae, sortes patriarcharum vel apostolorum, sortes sanctorum*

magic: *augurium, ars magica, divinatio, praestigio*
——, protective: *praecantatio, incantatio*
magician: *magus*
——, benevolent: *praecantator, praedicator, incantator*
maidservant: *ancilla, colona; glantella, glandella* (Irish)
menstruous: *menstruose, monstruose*
monument (to the dead): *cancellus, memoria, sarandae antiquae*

nature-divinities: *aquaticae, dianae, dusiolus, lamiae, mavones, nymphae, silvaticae*
necromancer: *pithon, phiton, fitonis, phitonicus, pitonissa, necromanta*
Needfire: *nodfyr, nied fyr, nedfratres*

offering(s): *votum, conpensus, oblata, offerta, idolothytes, immolata, sacrificia, munus*
omen: *augurium, praefiguratio*
outcasts (Irish): *bestiales homines, immundi homines, humanae bestiae; laicus, laica*

pagan customs: *paganiae, spurcitia gentilitatis*
payment (to weather-magicians): *canonicum*
phenomenon, lunar, during new moon (?): *contralunium*
prostitute: *meretrix, meretricula, genichialis femina*

race: *cursus, Yrias* (a particular foot or horse race)
ritual, seasonal(?): *fastidiatio*

seer: *adivinator, divinus,vir divinator; (h)aruspex; ariolus; agurius, augur, augurator, auguriosus; caragus, c(h)aragius, caraus, karagius, karajoc, ceraius; cauculator; librarius; pithon, phiton, fitonis, phitonicus; somniarius, somniatori coniector; sortilegus, sortilogus, sortilecus, sortilicus, sorticularius; umbrarius; vaticinator*
seeress: *auguriatrix; divina, mulier divinator; pitonissa, necromanta; sortilega*
shrines: *fana, casulae, fanassi*
solstice rites, summer: *solestitia*
——, winter: *brumae / bromae, brunaticus*
sorcerer: *maleficus, maledicus; veneficus, beneficus, venenarius; mathematicus, invocator daemonum; obligator*
sorceress: *malefica, venefica, venenaria, sorciaria / sortiaria*
sorcery: *fascinatio, maleficium, veneficium*
spell: *carmen, incantatio, incantatura*

tattoo: *tinctura(e injuria)*
toast, v.: *bibere / precari in amore / in honore alicuius*
trees, sacred: *arbores fanatici / sacrivi / consecratae / sacrilegae*
trick: *praestigium, ludificatio*

vampire: *lamia, striga*

wake: *excubiae funeris*
weather-magician: *emissor / immissor tempestatum, tempestarius / tempestuarius, tempistaria*
werewolf: *fictus lupus, werewulff*
wife, discarded: *glandella, glantella (Irish)*
witch: *striga, stria, astria, masca*

Words with obscure meaning:

canterma: a Sicilian rite, probably obscene
cauclearius / *coclearius*: possibly weather-magician
cauculator: hydromancer? numerologist?
cerenus: New Year's celebrant, possibly walking around the settlement carrying a candle
cocriocus: masker? herald? cunning man?

dadsisas: mortuary rite, perhaps circumambulation following the direction of the sun

ermulus: a deity possibly in animal-form (AS?)

feclum: celebrant or celebration of the New Year? cart bearing maskers?
fraus: food obtained by theft or fraud?

impura / *inpuria*: seeress, active at New Year? New Year's divinatory rite?

mimarcia: pantomime? actress?

oculus pullinum: ailment, disease of the eye; wall-eye?

pullicini: "chicks": a minor ailment; a mistake for *pediculi* (lice)?

Bibliography

1. Sources – Annotated

The beliefs or practices mentioned in each source are indicated by a reference to the chapter where this type of behaviour is discussed. "General" is used to identify non-specific references. Full bibliographic data for the studies of the documents found in this section are given in section 3 of the bibliography.

1.1 *Councils*

For discussion of individual councils, see Hefele-Leclercq, *Histoire des conciles.*

Iberian: In *Concilios Visigóticos e Hispano-Romanos* (Barcelona and Madrid, 1963), ed. José Vives with Tomás Marín Martínez and Gonzalo Martínez Díez. See also Orlandis and and Ramos-Lisson, *Historia de los Concilios de la España romana e visigoda*; Thompson, *The Goths in Spain;* King, *Law and Society in the Visigothic Kingdom;* Ziegler, *Church and State in Visigothic Spain.*

Conc. Toletanum I, 397-400 (Vives, 19-33). Alimentary.

Conc. Herdense, 546 (Vives, 55-60). Held at Lerida. Magic.

Conc. Bracanense I, c. 561 (Vives, 65-77). Galician council, held under the protection of the Suevan king Ariamir, newly converted to Catholicism, supposedly by Martin of Braga. It dealt with the Priscillianist heresy. Alimentary, death, idolatry, nature.

Conc. Bracanense II, 572 (Vives, 78-106). Held under the presidency of Martin of Braga. Appended to its documents are the *Canones ex orientalium patrum synodis* (= *Canones Martini*), see below.

Conc. Toletanum III, 589 (Vives, 107-159 = Felix Rodriguez, "El concilio III de Toledo." The Visigothic king Reccared's great unifying council, attended by secular lords and the bishops of Spain and the Narbonnaise under the leadership of Leander of Seville; set down the principles of Catholic belief and discipline. Death.

Conc. Narbonense, 589 (*CCSL* 148A, 254-257). Provincial synod of Narbonne. Idolatry, magic.

Conc. Hispalense II, 619 (Vives, 163-185). Convoked by Isidore of Seville; primarily concerned with clerical discipline. Magic.

Conc. Toletanum IV, 633 (Vives, 186-225). General council. Magic.

Conc. Toletanum V, 636 (Vives, 226-232). Held under Chinchila; concerned chiefly with the safety of the ruler. Magic.

Conc. Emeritense, 666 (Vives, 325-343). At Merida in Lusitania under Recceswinth. Magic.

Conc. Bracanense III, 675 (Vives, 370-379). Held under Wamba; considered clerical discipline and liturgy. Idolatry.

Conc. Toletanum XII, 681 Vives, 380-410. Called to settle Ervig's succession to the throne after Wamba's forced retirement to a monastery. Idolatry, nature.

Conc. Toletanum XIII, 683 (Vives, 411-440). Summoned by Ervig; attended by ecclesiastical and secular grandees. Magic.

Conc. Toletanum XVI, 693 (Vives, 482-521). Summoned by King Egica; composed of both ecclesiastical and secular grandees. Idolatry, nature.

Conc. Toletanum XVII, 694 (Vives, 522-537). Magic.

Insular Councils: In Arthur West Haddan and William Stubbs, *Councils and Ecclesiastical Documents Relating to Great Britain and Ireland,* 3 vols. (1869-1873; Oxford, 1964) = Haddan and Stubbs. See also Catherine Cubitt, *Anglo-Saxon Councils.*

Council of Clovesho, 747 (Haddan and Stubbs 3, 360-385). Incorporated some of the decisions of the Frankish Council of 747 reported by St. Boniface, to Cuthbert, Archbishop of Canterbury. Magic, time.

Legatine Synods – *Report of the Legates George and Theophylact,* 787 (Haddan & Stubbs 3, 447-462). Report to Pope Hadrian I by George, Bishop of Ostia and Theophylact, Bishop of Todi, of their proceedings at synods held under Offa's auspices, at Pincahala in the north (Finchale?) and Chelsea in the south. Alimentary, death, magic.

Gallican, Merovingian and Carolingian: In C. Munier, ed., *Concilia Galliae, A. 314-A. 506, CCSL* 148 (Turnhout, 1963) = *CCSL* 148, and C. De Clercq, ed., *Concilia Galliae, A. 511- A 695, CCSL* 148A (Turnhout, 1973) = *CCSL* 148A; *Monumenta Germania Historica, Legum sectio* III: *Concilia* 1 (*Concilia aevi merovingici*), ed. F. Maassen (Hanover, 1893) = *MGH Concilia* 1; Concilia 2 (*Concilia aevi karolini*) *pars* 1 and 2, ed. A. Werminghoff (Hanover and Leipzig, 1896) = *MGH Concilia* 2.1 and *MGH Concilia* 2.2 (Hanover and Leipzig, 1898) = *MGH Concilia* 2.2; *Concilia* 3 (*Concilia aevi Karolini,* 843-859), ed. W. Hartmann (Hanover, 1984) = *MGH Concilia* 3; *Concilia* 3 (*Concilia aevi Karolini,* 860-874), ed. W. Hartmann (Hanover, 1998) = *MGH Concilia* 3; *Concilia* 6.1 (*Concilia aevi Saxonici,* 916-1001), ed. E.-D. Hehl (Hanover, 1987) = *MGH Concilia* 6.1. Additional material from J. D. Mansi *et al., Sacrorum conciliorum nova et amplissima collectio* (1759-1798; repr. Paris, 1901-1927) 13 and 14 = Mansi 13 and 14; C. F. Schannat and J. Hartzheim, *Concilia Germaniae* 1 and 2 (1759; repr. Aalen, 1970), and as specified below. See De Clercq, *La législation religieuse franque*; Pontal, *Histoire des conciles mérovingiens; idem, Les statuts synodaux*; Hartmann, "Synodes carolingiennes et textes synodaux du IXe siècle"; Imbert, *Les temps carolingiens*; McKitterick, *The Frankish Church and the Carolingian Reforms.*

Conc. Arelatense II, 442-506 (*CCSL* 148, 114-134). Nature.

Conc. Veneticum, 461-491 (*CCSL* 148, 150-156). Magic, time.

Conc. Agathense, 506 (*CCSL* 148, 210-1). Meeting in Agde of the Catholic bishops under Visigothic rule, held under the protection of the Arian Alaric in an attempt to win Catholic support against the Franks. Magic, time.

Conc. Aurelianense, 511 (*CCSL* 148A, 4-19). Frankish national council, summoned by Clovis to consolidate his rule. Magic.

Conc. Massiliense, 533 (*CCSL* 148A, 85-97). Death (in a letter by Pope John II).

Conc. Aurelianense, 533 (*CCSL* 148A, 99-103). Convoked by Childebert I, Clotaire I and Thierry I, to settle disciplinary and administrative questions. Alimentary, idolatry, time.

Conc. Claremontanum seu Arvernense, 535 (*CCSL* 148A, 105-112). Death.

Conc. Aurelianense, 541 (*CCSL* 148A, 132-146). Idolatry.

Synodus Aspasii Episcopi, 551 (*CCSL* 148A, 163-165). Provincial council held in Eauze. Magic.

Conc. Turonense, 567 (*CCSL* 148A, 176-199). Representatives of the ecclesiastical provinces of Tours, Rouen and Sens met under Charibert, to restore discipline and peace, and safeguard the status of the Church. Death and time.

Conc. Matisconense, 581-583 (*CCSL* 148A, 223-230). National council summoned by Guntram, with representatives from Vienne, Besançon, Lyons, Sens, Bourges and Arles. Dealt with clerical discipline, the property of the Church, monasteries and the liturgy. Death, magic.

Conc. Narbonense, 589. See above, among Iberian councils.

Conc. Sauriciacense, 589 (*CCSL* 148A, 258). According to Gregory of Tours, a council was held in Soissons to deal with disturbances caused by accusations of sorcery made against an archdeacon. Magic.

Syn. Autissiodorensis, 561-605 (*CCSL* 148A, 265-272). Diocesan synod meeting at Auxerre; uncertain date. Of forty-three canons, eight concern forbidden manifestations of popular culture. Death, idolatry, magic, space, time.

Conc. Incerti Loci, After 614 (*CCSL* 148A, 287-289). Death.

Conc. Clippiacense, 626-627 (*CCSL* 148A, 291-297). General council of Austrasian bishops, convoked by Clotaire II. Idolatry.

Concilium sub Sonnatio Episcopo Remensi Habitum. 627-630 (*MGH Concilia* 1, 203-206). Based on the acts of the Council of Clichy, above.

Conc. Cabilonense. 647-653 (*CCSL* 148A, 303-310). The *canones extravagantes* of this council are found in Mansi 10, 1197. Neustrian national council convoked by Clovis II. Time.

Conc. Rothomagense, 650 (Mansi, 10, 1199-1206). The dating and even existence of this Rouenese council is controversial. Mansi hesitated between the mid 7th and late 9th centuries, but opted for the former, the date accepted by Hefele-Leclercq (*Histoire des conciles* 3.1, 287-289) and Pontal (*Histoire des conciles mérovingiens*, 241-243). It is not recognised by Potthast. Magic, time.

Conc. Namnetense. 7th or late 9th century (Dominique Aupest-Conduché, "De l'existence du Concile de Nantes," *Bulletin Philologique et Historique CTHS* (1973) 29-59 = J. Sirmond, ed. *Concilia antiquae Galliae* ([1629] Halen, 1970) 3, 601-607). The dating and authenticity of this Council of Nantes is controversial. Sirmond believed that it dated from the very late 9th century. Hefele-Leclercq (*Histoire des conciles* 3.1, 296 and 1247) is ambivalent, opting for the mid 7th century on one page and the 9th century on another. According to Jean Gaudemet, it probably dates from the second half of the 9th century, in any event before 906, when Regino of Prüm compiled his collection. Gaudemet, who thought the council apocryphal, notes that the *De synodalibus causis* is the principal source used by Sirmond ("Le pseudo-concile de Nantes"). Pontal (*Histoire des conciles mérovingiens*, 235-241), although unable to settle the question decisively, is more inclined to place it in the 7th century. Aupest-Conduché accepts all but c. 15 as being authentically attributable to a 7th century Council held at Nantes. Death, nature, space.

Conc. in Austrasia habitum q. d. Germanicum, 742 (*MGH Concilia* 2.1, 1, 2-4). Held under the auspices of Carloman I. This and the other councils of the 740's were motivated by St. Boniface's zeal for reform of ecclesiastical discipline and the eradication of paganism. Death, general, magic, nature.

Conc. Liftinense, 743 (*MGH Concilia* 2.1, 6-7). Mixed council of lay and ecclesiastical officials held in Estinnes, Hainaut. General.

Conc. Romanum, 743 (*MGH Concilia* 2.1, 11-32). Time.

Conc. Suessionense, 744 (*MGH Concilia* 2.1, 33-44). Neustrian council, held under the protection of Pippin III. General, idolatry.

Conc. Romanum, 745 (*MGH Concilia* 2.1, 37-43). Council presided by Pope Zacharias, to hear St. Boniface's charges against two insubordinate clerics, Aldebert and Clement. Idolatry.

Conc. in Francia habitum, c. 747 (*MGH Concilia* 2.1, 46-50). Attended by Neustrian and Austrasian bishops. Its decrees are known through St. Boniface's correspondence. General, magic.

Conc. Neuchingense, 772 (*MGH Concilia* 2.1, 99-105). Convoked by Tassilo III, Duke of Bavaria; concerned with mission activity. Magic, nature.

Conc. Francofurtense, 794 (*MGH Concilia* 2.1, 111-171). Primarily concerned with the Adoptionist heresy. This council completed the submission of Bavaria to Charlemagne. Idolatry, nature.

Conc. Rispacense, 798 (*MGH Concilia* 2.1, 198-201). Chiefly concerned the obligations of the clergy. Idolatry.

Concilia Rispacense, Frisingense, Salisburgense, 800 (*MGH Concilia* 2.1, 206-219). A series of Bavarian mixed councils convened by Arno of Salzburg; first met at Riesbach, then moved on to Freising before ending up in Salzburg. Magic, time.

Conc. Arelatense, 813 (*MGH Concilia* 2.1, 248-253). This and the other councils held in 813 were called to enforce ecclesiastical reform. See Gerhard Schmitz, "Die Reformkonzilien von 813 und die Sammlung des Benedictus Levita" and Ganshof, "Notes sur les 'Capitula de causis'." Death, magic.

Conc. Moguntinum 813 (*MGH Concilia* 2.1, 259-273). Death, magic, time.

Conc. Turonense, 813 (*MGH Concilia* 2.1, 286-293). Magic.

Appendices ad concilia anni 813 (*MGH Concilia* 2.1, 293-306). Death, magic, time.

Conc. Aquisgranense, 816 (*MGH Concilia* 2.1, 312-464.). Primarily concerned with imposing Louis I and Benedict of Aniane's monastic reforms (the Rule of St. Benedict) on the Empire. Idolatry, magic, space.

Conc. Romanum, 826 (*MGH Concilia* 2.2, 553-583). Time.

Conc. Parisiense, 829 (*MGH Concilia* 2.2, 606-680). Reforming general council under the Emperors Louis and Lothair. Magic, space.

Conc. Meldense-Parisiense, 845-846 (*MGH Concilia* 3, 81-131). Under the protection of Charles the Bald. Death.

Notitia de conciliorum canonibus in Villa Sparnaco a Karolo rege confirmatis, 846 (*MGH CapRegFr* 2, 251-262). In Épernay, arr. of the Marne. Death.

Conc. Moguntinum, 847 (*MGH CapRegFr* 2, 173-184). Provincial council held under Louis the German. Magic.

Synod of Coetleu, 848/849 (*MGH Concilia* 3, 187-193). Dept. of Ille-et-Vilaine. Attempt to maintain the independence of the Breton Church from Frankish authority. Its documents include Leo IV's directives to the bishops of Brittany. Magic.

Synodus Papiensis, 850 (*MGH Concilia* 3, 221-229). Magic. Time.

Conc.Moguntinum, 852 (*MGH CapRegFr* 2, 185-191. Called by Louis the German. Time.

Conc. Romanum, 853 (*MGH Concilia* 3, 317-346). Time.

Synodus Mettensis, 859 (*MGH CapRegFr* 2, 442-446). Regional council to settle political conflicts. Magic.

Conc. Wormatiense, 868 (*Concilia Germaniae* 2, 312; *MGH Concilia* 3, 259-311). Called by Nicholas I; primarily concerned with theological and administrative matters. Alimentary, magic.

Conc. Triburiense, 895 (*MGH CapRegFr* 2.2, 214-249). Death, magic, time.

Conc. Namnetense. 7th or late 9th century. See above.

Conc.Treverense, 927/928 (*MGH Concilia* 6, 79-88). Magic.

Conventus et Synodus Erfordiensis, 932 (*MGH Concilia* 6, 106-114). Idolatry, magic.

Halitgar of Cambrai, *Diocesan Synod* 1, 829-831 (Wilfried Hartmann, "Neue Texte zur bischöflichen Reformgesetzgebung aus den Jahren 829/31," *DA* 35 [1979], 368-394). Death, space, time.

West Frankish Synod. Late 9th/early 10th century. (Wilfried Hartmann, "Unbekannte Kanones aus dem Westfrankenreich des 10. Jahrhunderts," *DA* 43 (1987) 28-45). Nature.

1.2. *Bishops' capitularies*

Most of these are edited in *Monumenta Germaniae Historica Capitula Episcoporum* = *MGH CapEp;* 1, ed. Peter Brommer (Hanover, 1984); 2, ed. Rudolf Pokorny and Martina Stratmann with Wolf-Dieter Runge (Hanover, 1995); 3., ed. Rudolf Pokorny (Hamburg, 1995). See Jean Gaudemet, "Les statuts épiscopaux" and Peter Brommer, *"Capitula Episcoporum": Die bischöflichen Kapitularien des 9. und 10. Jahrhunderts.*

Capitulare Gregorii Papae, 731 (*Concilia Germaniae* 1, 35-37). Alimentary.

Capitula Rotomagensia, 8th-10th centuries (*MGH CapEp* 3, 369-371). The exact date and place of origin is doubtful. Magic.

Statuta Bonifacii, first half, 9th century (*MGH CapEp 3*, 360-366). Pokorny considers this to be spurious. Death, magic, space, time.

Capitula Vesulensia, c. 800 (*MGH CapEp* 3, 346-353). Pokorny considers this to be spurious. Death, magic, space, time.

Remedius of Chur, *Capitula*, 800-806 (*MGH Leg* 5 [ed. Gustavus Haenel; repr. Stuttgart, 1965], 181-184. Magic.

Ghärbald of Lüttich, *Capitularia 1*, c. 801 (*MGH CapEp* 1, 16-21). Space.

———, *Capitularia 2,* 802-809 (*MGH CapEp* 1, 29-30). Magic, nature, space.

Haito of Basel, *Capitula*, 807-823 (*MGH CapRegFr* 1, 363-366; *MGH EpCap* 1, 210-219). Idolatry, space.

Theodulph of Orleans, *Capitularia* 1, 813 (*MGH EpCap* 1, 103-142). Death, magic, space.

———, *Capitularia* 2, 813 (*MGH EpCap* 1 148-184). Death, magic.

Capitula Treverensia, before 818 (*MGH EpCap* 1, 55-56). Alimentary, death, magic, nature.

Capitula Neustrica quarta, c. 829 (*MGH CapEp* 3, 70-73). Death.

Capitula Silvanectensia prima, 9th century, first half (*MGH CapEp* 3, 80-83). Magic, time.

Capitula Frisingensia tertia, c. 840 (*MGH CapEp* 3, 222-230). Death.

Capitula Franciae occidentalis, 9th century, first half (*MGH CapEp* 3, 39-47). Nature.

Capitula Eporediensia, 850 (*MGH CapEp* 3, 238-242). For the diocese of Milan. Magic, space, time.

Hincmar of Reims, *Capitularia I,* 852 (*MGH CapEp* 2, 34-45). Death, space.

——, *Capitularia II* (*MGH CapEp* 2, 45-75) Magic.

——, *Capitularia III* (*MGH CapEp* 2, 73-75). Space.

——, *Capitularia V* (*MGH CapEp* 2, 86-89). Space.

Radulph of Bourges, *Capitularia,* 853-866 (*MGH EpCap* 1, 233-268). Death, magic, nature.

Herard of Tours, *Capitula*, 858 (*PL* 121, 763-744). Death, idolatry, magic, time.

Isaac of Langres, *Capitula*, 860-869 (*MGH CapEp* 2, 180-240). Death, magic, space.

Walter of Orleans, *Capitularia*, 869/870 (*MGH EpCap* 1, 187-199). Death.

Gilbert of Châlons, *De interdictis,* 868-878 (*MGH CapEp* 2, 93-95). Death.

Riculf of Soissons, *Statuta,* 888 (*MGH CapEp* 2, 100-111). Death.

Capitula Ottoboniana, 889 or later (*MGH CapEp* 3, 123-132). Death, idolatry, magic.

Ruotger of Trier, *Capitularia*, 915-929, first half (*MGH EpCap* 1, 62-70). Death, space.

Capitula Helmstadiensia, c. 964 (Rudolf Pokorny, ed., "Zwei unerkannte Bischofskapitularien des 10. Jahrhunderts," *DA* 36 [1979] 487-513). East Saxony. Time.

Atto of Vercelli, *Capitularia*, 943-948 (*PL* 134, 27-52). Death, magic, space, time.

1.3. *Canonic collections*

See Gérard Fransen, *Les collections canoniques,* and Paul Fournier and Gabriel Le Bras, *Histoire des collections canoniques en Occident depuis les Fausses Décrétales jusqu'au Décret de Gratien. 1. De la réforme carolingienne à la réforme grégorienne.* Canonic collections with a large penitential component are listed below, among penitentials.

Statuta Ecclesiae Antiqua. C. 475. (*CCSL* 148, 164-188). For the authorship of this document, see C. Munier, "Nouvelles recherches sur les *Statuta ecclesiae antiqua*," (below in 2.1). Magic, nature, time.

Martin of Braga, *Canones ex orientalium patrum synodis* (the *Canones Martini*), 572 (C. W. Barlow, ed., *Martini Episcopi Bracarensis opera omnia* [New Haven, 1950] = Barlow; 123-144). At least some of these appear to have been drawn up by Martin himself. For the composition of this document, see Gaudemet, "Traduttore, traditore." Death, idolatry, magic, space, time.

Epítome hispánico, c. 598-610 (ed. Gonzalo Martínez Camillas, *El epítome hispánico: una colección canónica española de siglo VII; Estudio y texto* crítico [Santander, 1961], 93-228). The editor attributes this collection to a single author, probably a bishop of Braga. It draws on the *Canones Martini* and the earliest councils (Spanish, Eastern and African) as well as the first Visigothic councils, and papal letters. Death, idolatry, magic, nature, space, time.

Collectio Vetus Gallica. 6th-9th centuries (H. Mordek, ed., *Kirchenrecht und Reform im Frankenreich: Die Collectio Vetus Gallica, die älteste systematische Kanonensammlung des fränkischen Gallien* [Berlin, 1975] 343-617). First compiled c. 600 around Lyons, with later additions made at other centres (Autun, Corbie). Death, idolatry, magic, nature, space, time.

Collectio Frisingensis Secunda, late 8th century (Mordek, ed., *Kirchenrecht und Reform im Frankenreich*, 619-633). Idolatry, magic.

Collectio Hispana. Canonum Excerpta (Isidoriana), mid 7th century (*PL* 84, 32-92; Gonzalo Martínez Díez, *La colección canonica Hispana* [Madrid, 1966] 43-214). The rest of the *Hispana* (*PL*. 84, 93-848; Gonzalo Martínez Díez, 279-615) gives the acts of early and contemporary councils, arranged according to the place of origin. Erroneously attributed to Isidore of Seville. Alimentary, death, idolatry, magic, space, time.

Collectio Hibernensis, c. 700-725 (Hermann Wasserschleben, ed. *Die irische Kanonensammlung*, [1885] Aalen, 1966 = *Hibernensis*, H. Wasserschleben). Irish and continental material. See Sheehy, "The Collectio Canonum Hibernensis" and "Influences of Irish law on the Collectio Canonum Hibernensis;" Richter, *Ireland and her Neighbours,* 216-127. Alimentary, death, idolatry, magic.

Collectio Hadriana. Epitome Canonum, 773-774 (Mansi 13, 859-882). Augmented form of the collection of Dionysius Exiguus, sent by Hadrian I to Charlemagne; considerable influence on Carolingian legislation.

Chrodegang, *Regula Canonicorum*, 762 (*Concilia Germaniae* 1, 116). Originally for the Canons Regular of Metz, but came into general use throughout Frankish territory. Space, time.

Ansegesis Capitularium collectio. c. 827 (*MGH CapRegFr* 1, 394- 450). Partial collection of the capitularies of Charlemagne and Louis the Pious, commissioned by Louis and compiled by the Abbot of St. Wandrille. Death, idolatry, magic, space, time.

Decretales Pseudo-Isidorianae, c. 850 (P. Hinschius, *Decretales Pseudo-Isidorianae et Capitula Angilrammi* [1863; Aalen, 1963] 1-754). Originally supposed to have been the work of Isidore of Seville, but in fact a highly successful forgery compiled in the mid 9th century, probably in northern France (Tours?). Its main purpose was to affirm the primacy of ecclesiastical over temporal authority, and, as such, it has predictably little to say about popular beliefs or practices. Diet, death, idolatry, magic, space, time.

Angilramni Capitula, c. 847-852 (Pio Cipriotti, ed., *I Capitula Angilramni con appendice di documenti connessi.* Milan, 1966). Idolatry, magic.

Benedicti Capitularium collectio, 847-857 (*PL* 97, 697-912). One of the principle means of transmission of the reforms of the Councils of 813. See Gerhard Schmitz, "Die Reformkonzilien von 813." Many of its other canons appear to have been made up by the compiler. Used mainly among the West Franks. Death, idolatry, magic, nature, space, time.

Regino of Prüm, *Libri duo de synodalibus causis et disciplinis ecclesiasticis (= De synodalibus causis*), c. 906 (ed., F. G. A. Wasserschleben, [1840] Graz, 1964). One of the principal sources of Burchard's *Decretum.* Alimentary, death, idolatry, magic, nature, space, time.

Burchard of Worms, *Decretorum libri viginti* (= *Decretum*), 1008-1012 (*PL* 140, 537-1058). Burchard compiled this collection with the help of bishop Walter of Speyer, and the monk Olbert of Lobbes, to remedy neglect and ignorance of Church law, discrepancies in its application, and lack of proper authority, found among the parish priests of his diocese. See Fournier, "Le Décret de Burchard de Worms," and, for the literary sources, Hoffman and Pokorny, *Das Dekret des Bischofs Burchard von Worms.* Alimentary, death, idolatry, magic, nature, space, time.

Collectio canonum in V libris, 1014-1023 (ed. M. Fornasari, *CCCM* 6, Turnhout, 1970). Idolatry, magic.

1.4 *Capitularies*

See Ganshof, *Recherches sur les capitulaires* and Drioux, "Capitulaires." Edited in *Monumenta Germaniae Historica, Legum Sectio* II, *Capitularia Regum Francorum* 1, ed. A. Boretius (Hanover, 1883) = *MGH CapRegFr* 1, and *Capitularia Regum Francorum* 2, ed. A. Boretius and V. Kraus (Hanover, 1897) = *MGH CapRegFr* 2.

Childeberti I. regis praeceptum, 554 (*MGH CapRegFr* 1, 2-3). Idolatry, time.

Karlomanni principis capitulare = Conc. in Austrasia, 742 (*MGH CapRegFr* 1, 24-26). Death, idolatry, magic.

Indiculus superstitionum et paganiarum, 743 (*MGH CapRegFr* 1, 223). List of thirty titles attached to the acts of the Conc. Liftinense, reflecting St. Boniface's concerns. According to Felice Lifshitz, this document "marked a major turning point in the definition of 'Christianity'" (*Pious Neustria,* 59). See Holger Homann, *Der Indiculus superstitionum et paganiarum*. Haderlein argues for a Celtic basis of some of the practices cited ("Celtic roots"). For dating, geographic origin and politico-religious circumstances under which this the document was drafted, see Dierkens, "Superstitions, christianisme et paganisme." Death, idolatry, magic, space, time.

Karoli Magni Capitulare primum, 769 or a little later (*MGH CapRegFr* 1, 44-46). Death, idolatry, magic, time.

Admonitio Generalis, 789 (*MGH CapRegFr* 1, 53-62). Incorporated canons from the *Hadriana*, a collection of canons based on the Dionysian collection, sent to Charlemagne by Pope Hadrian I. Idolatry, magic, space.

Duplex legationis edictum, 789 (*MGH CapRegFr* 1, 62-64). Magic, space.

Capitulatio de partibus Saxoniae, 775-790 (*MGH CapRegFr* 1, 68-70). See Effros, "*De partibus Saxoniae* and the regulation of mortuary custom." Death, idolatry, magic, space.

Statuta Rhispacensia, Frisingensia, Salisburgensia, 799, 800 (*MGH CapRegFr* 1, 226-230). Idolatry, magic, time.

Capitula cum Italiae episcopis deliberata, 799-800 (*MGH CapRegFr* 1, 202-203). Magic, time.

Capitula de Iudaeis, n.d. (*MGH CapRegFr* 1, 258-259). Magic: unspecified Jewish sorcery against Christians.

Capitulare de villis 800 or earlier (*MGH CapRegFr* 1, 83-91). Magic.

Capitulare missorum generale, 802 (*MGH CapRegFr* 1, 91-99). Magic.

Capitulare missorum generale item speciale, 802 (*MGH CapRegFr* 1, 102-1040. Idolatry, magic, space.

Capitula a sacerdotibus proposita, 802 (*MGH CapRegFr* 1, 106-107). Space.

Capitula de examinandis ecclesiasticis, 802 (*MGH CapRegFr* 1, 109-111). Magic.

Capitulare missorum, 803 (*MGH CapRegFr* 1, 115-116). Time.

Capitula in diocesana quadam synodo tracta, c. 803-804 (*MGH CapRegFr* 1, 236-237). Magic.

Capitula excerpta de canone, 806 or later (*MGH CapRegFr* 1, 133-134). Death, magic, time.

Capitula post a. 805 addita, 806-813 (*MGH CapRegFr* 1, 142). Magic.

Capitulare Aquisgranense, 809 (*MGH CapRegFr* 1, 148-149). Magic.

Capitulare Missorum Aquisgranense Primum, 809 (*MGH CapRegFr* 1, 150-151). Magic.

Capitulare missorum Aquisgranense Alterum, 810 (*MGH CapRegFr* 1, 153-154). Magic.

Capitulare italicum, 810 (*MGH CapRegFr* 1, 204-206). Charlemagne and Pippin's regulations. Idolatry.
Capitulare Bononiense, 811 (*MGH CapRegFr* 1, 166-167). Time.
Capitula a misso cognita facta, 808-813 (*MGH CapRegFr* 1, 146). General.
Capitula a canonibus excerpta, 813 (*MGH CapRegFr* 1, 174). Death, magic.
Capitulare Ecclesiasticum, 810-813 (*MGH CapRegFr* 1, 179-180). Death, magic.
Hludowici Pii capitulare ecclesiasticum, 818-819 (*MGH CapRegFr* 1, 275-280). Concerned with protecting the independence of the church from landed proprietors. Death, time.
Episcoporum ad Hludowicum imperatorem relatio, 829 (*MGH CapRegFr* 2, 27-51). Report of the Conc. Parisiense (829) above. Magic, space.
Pactum Hlotharii I, 840 (*MGH CapRegFr* 2, 130-135). Idolatry.
Capitula episcoporum Papiae edita, 845-850 (*MGH CapRegFr* 2, 80-83). Time.
Capitulare Carisiacense, 873 (*MGH CapRegFr* 2, 343-347). Magic.
Capitulare Papiense, 876 (*MGH CapRegFr* 2, 100-104). Place.
Pactum Karoli III, 880 (*MGH CapRegFr* 2, 138-141). Idolatry.
Pactum Berengarii I, 888 (*MGH CapRegFr* 2, 143-147). Idolatry.

1.5 *Penitentials*

The penitentials used are grouped, insofar as possible, according to place of origin and date – usually only approximations. With some additions, these are the penitentials listed by Vogel in *Les "Libri poenitentiales"* as updated by Frantzen. Additional bibliography is given for each penitential in this work. The standard editions of the majority of the penitentials are to be found in the collections of F. W. H. Wasserschleben, *Die Bußordnung der abendländischen Kirche* (1851; reprint, Graz, 1958), and H. J. Schmitz, *Die Bußbücher,* 2 vols. (1898; reprint, Graz, 1958). Wasserschleben's chronology of the penitentials is more accurate, since Schmitz based his on a mistaken belief in the existence of a prototypical Roman penitential. Nevertheless, the latter's recension is considered superior and, when possible and when no better edition is available, I have used his. Modern regional collections, which include penitentials missing from Wasserschleben and Schmitz, have been published for Ireland, 8th and 9th century France and Italy, and Spain. English translations of many penitentials are to be found in McNeill and Gamer, *Medieval Handbooks of Penance.* The bibliographic data for these and for penitentials published separately is given below. For sources, see also Ludger Körntgen, *Studien zu den Quellen.*

Celtic penitential literature and penitentials:. In Ludwig Bieler, ed. and trans., *The Irish Penitentials* (Dublin, 1963). For documents respecting Ireland, see Kenney, *The Sources for the Early History of Ireland,* Lapidge and Sharpe, *A Bibliography of Celtic-Latin Literature,* and Hughes, *Early Christian Ireland,* 67-95. For Irish penitentials, see Oakley, "The origin of Irish penitential discipline" and Connolly, *The Irish Penitentials.*

Celtic "proto-penitentials"

Praefatio Gildae de poenitentia, early 6th century (Bieler, 60-65). Welsh. Alimentary.
Excerpta quaedam de libro Davidis, 6th century, second half (Bieler, 70-73). Welsh. Time.

Canones Wallici, 7th century. (Bieler, 136-159). Probably brought by Bretons to Wales. Dumville ("On the dating of the early Breton lawcodes") accepts a Breton origin for this set of canons, but finds it impossible to date more closely than between 500 and 800. General.

Irish penitential canons

Synodus I. Sancti Patricii, 457 (Bieler, 54-59). Not a penitential but contains penitential material. Idolatry, magic.

Canones Hibernenses, mid 7th century (Wasserschleben, 136-144; Bieler, 160-175). Alimentary, death, magic.

Canones Adomnani, 679-704 (Bieler, 176-181). Adomnan's name is mentioned in the text, but the attribution of these canons to the Abbot of Iona is doubtful. Alimentary.

Irish Penitentials (Latin): The "Old Irish Penitential", from before the end of the 8th century (ed. D. A. Binchy in Bieler, 258-277), has been used only for reference.

Poen. Vinniani, 6th century, first half (Bieler, 74-95). The author is probably St. Finnian of Clonard. Magic. According to Rob Meens ("The penitential of Finnian"), Finnian was possibly of Breton or British origin, with "strong Irish associations". Magic.

Poen. Ambrosianum, 550-650 (Körntgen, *Studien zu den Quellen,* 80-86; analysis of the text, 7-79). Irish or British monastic origin. Time, magic.

Poen. Columbani, c. 600 or earlier (Bieler, 96-107 = Laporte, *Le pénitentiel de saint Colomban.*). Composed in Luxeuil or Bobbio but "of purely insular inspiration". Vogel accepts the attribution to St. Columban as to content. According to Laporte, the bulk of the texts were assembled under Columban's direction well before the founding of Luxeuil in 590. Idolatry, magic.

Poen. Cummeani, before 662 (Bieler, 108-135). Cumméan Fota of Clonfert (†662) is generally accepted as the author. Composed in Ireland, perhaps under Scottish influence. Alimentary, magic, time.

Poen. Bigotianum, 8th or 9th century (Bieler, 198-239). According to Frantzen (Vogel, *Les "Libri poenitentiales"*, 24), written on the continent, but completely insular in content, without a trace of Frankish influence. Mostert ("Celtic, Anglo-Saxon or insular") claims that it was written in Ireland. Alimentary, death, magic.

English Penitentials (Latin): Unless otherwise noted, this section is based on Frantzen, "The tradition of penitentials in Anglo-Saxon England". For the social and political context of the English penitentials, see Oakley, *English Penitential Discipline and Anglo-Saxon Law.*

Theodorian Compilations: These date from the late 7th to the mid 8th centuries, and are all of insular (Anglo-Saxon) origin. Theodore of Canterbury (d. 690) is not now believed to have written a penitential, but the last of these, the penitential of the so-called *Discipulus Umbrensium* (cited here as the *Poen. Theodori,* as edited by Schmitz in vol. 1, 524-550) is supposed to have been written on the basis of second- or third-hand knowledge of Theodore's teachings. For a discussion of these penitentials, see Charles-Edwards, "The Penitential of Theodore and the

Iudicia Theodori," and Brett, "Theodore and the Latin canon law." See also Lapidge, "The career of Archbishop Theodore."

Capitula Dacheriana (Wasserschleben, 145-160). Alimentary, death, idolatry, magic.

Canones Gregorii (Schmitz 2, 523-542). Alimentary, death, idolatry, magic, time.

Canones Cottoniani (Paul Willem Finsterwalder, *Die Canones Theodori Cantuariensis und ihre Überlieferungsformen* [Weimar; 1929] 271-284 = Finsterwalder). Alimentary, death, idolatry, magic.

Canones Basilienses (Franz Bernd Asbach, *Das Poenitentiale Remense und der sogen. Excarpsus Cummeani: Überlieferung, Quellen und Entwicklung zweier kontinentaler Bußbücher aus der I. Hälfte des 8. Jahrhunderts* [Regensburg, 1975] 79-89 = Asbach). Alimentary, death, idolatry, magic.

Poen. Theodori, 668-756 (Schmitz 1, 524-550). Probably written in England between these dates by an author familiar with Welsh and Irish texts; incorporates material from the above-named *Canones*. Alimentary, death, idolatry, magic, place.

Penitential attributed to Egbert

Poen. Egberti, 732-766 (Schmitz 1, 573-587; virtually identical to the *Excarpsus Egberti*, Schmitz 2, 661-672). If the generally accepted attribution to Egbert, Bishop later Archbishop of York, is correct, this was written in England during his episcopate (732-766). Alimentary, idolatry, magic, nature, time.

Penitentials attributed to Bede: Probably not by Bede, they may have been written on the continent. The earliest manuscripts date from the 9th century. See Frantzen, "The penitentials attributed to Bede."

Poen. Bedae and *Excarpsus Ps.-Bedae,* (Schmitz 1, 556-564 and 2, 654-659). These are virtually identical. Alimentary.

Liber de remediis peccatorum (B. Albers, "Wann sind die Beda-Egbert'schen Bussbücher verfasst worden, und wer ist ihr Verfasser?" *Archiv für katholisches Kirchenrecht* 81 [1901] 399-418 = Albers). Alimentary, idolatry, magic, nature, space.

Double Penitential of Bede-Egbert (Schmitz 2, 679-701). This was a 9th century compilation made on the continent of the *Penitential of Egbert* and a penitential attributed to Bede (Frantzen, *The Literature of Penance*, 107-110). Alimentary, death, idolatry, magic, time.

Ps.-Egbert

Confessionale Ps.-Egberti, 10th century? (*PL* 89, 402-432). Under the name *Scrift Boc* (c.950-c.1000), this is the oldest penitential in Anglo-Saxon, of which a Latin translation was made during the 10th or 11th century. The text used here is a reprint of the version in D. Wilkins, *Concilia Magnae Britanniae et Hiberniae, A.D. 446-1718* (London, 1737). Wilkins' text is taken from an 11th century manuscript, Codex Corpus Christi College, Cambridge, 190, which gave parallel AS and Latin texts. In his edition, Wasserschleben (300-318) used a 19th century translation. See McNeill and Gamer, *Medieval Handbooks of Penance,* 243-244 for a discussion of the sources, and 244-248 for an English translation of selected passages. The *Scrift Boc* largely follows the other English penitentials written originally in Latin, but the translator makes interesting variations on the standard vocabulary (*e.g.*, *maritus*, *torrentes* and *frumentum* rather than *vir*, *fontes* and *grana*). This penitential has been retained despite its AS origin because the Latin text dates from the period covered in this work. Idolatry, magic, nature, place.

The Lesser Frankish and Italian Penitentials of the 8th and 9th century: In Raymund Kottje, ed., *Paenitentialia minora Franciae et Italiae saeculi VIII-IX, CCSL* 156 (Turnhout, 1994) = *CCSL* 156. For the dates and origins of these penitentials, see xii-xxx.

The *Libri poenitentiales simplices*: These closely related penitentials were composed in the central part of the Frankish state or in Burgundy during the first half of the 8th century.

Poen. Burgundense (*CCSL* 156, 63-65). Magic, place, time.

Poen. Bobbiense I (*CCSL* 156, 69-71). Magic, place, time.

Poen. Parisiense Simplex (*CCSL* 156, 75-79). Alimentary, magic, place, time.

Poen. Sletstatense (*CCSL* 156, 83-85). Magic.

Poen. Oxoniense I (*CCSL* 156, 89-93). Alimentary, death, idolatry, magic, space, time.

Poen. Floriacense (*CCSL* 156, 97-103). Alimentary, death, magic, space, Calends.

Poen. Hubertense (*CCSL* 156, 107-115). Alimentary, death, idolatry, magic, space, time.

Poen. Sangallense Simplex (*CCSL* 156, 119-121). Idolatry, magic, time.

Others

Poen. Merseburgense a, 8th century, second quarter (*CCSL* 156, 125-169) Three parallel texts. Composed in northern Italy from Frankish materials. Alimentary, death, idolatry, magic, nature, space, time.

Poen. Merseburgense b, c. 774-850 (*CCSL* 156, 173-177). Frankish. Alimentary, magic, time.

Poen. Oxoniense II, 8th century (*CCSL* 156, 181-205) Based on 8th century material, either from northern Italy or the northeastern part of Frankish territory. There has been speculation that the author was St. Willibrord or one of his circle. See Körntgen, *Studien zu den Quellen*, 90-205. Alimentary, death, idolatry, magic, space, time.

Spanish penitentials and penitential literature: The edition of the three Spanish penitentials by Francis Bezler (*Paenitentialia Hispaniae, CCSL* 156A [Turnhout, 1998] = *CCSL* 156A) virtually replaces the difficult to use edition by S. Gonzales Rivas in *La penitencia en la primitiva Iglesia española* (Salamanca, 1950) and the edition of the *Corduban Penitential* by Justo Pérez de Urbel and Luis Vázquez de Parga ("Un nuevo penitencial español," *Anuario de Historia de Derecho Español* 14 [1942] 5-32), parts of which appeared in *La penitencia en la primitiva Iglesia española.* The *Canones poenitentiales* of Ps.-Jerome, not being a penitential, is not included in Bezler's edition. See Bezler, *Les pénitentiels espagnols* and "Chronologie relative des Pénitentiels d'Albelda et de Silos."

Poen. Albeldense = *Poen. Vigilanum,* 9th century, second half (*CCSL* 156A, 3-13). Central or northwestern Spain. Alimentary, magic, nature, space, time.

Poen. Corbudense, early 11th century (*CCSL* 156A, 45-69). Probably written in Galicia, but with some influence from Mozarabic Andalusia. Death, magic.

Poen. Silense, c. 1060-1065 (*CCSL* 156A, 17-42). Central or northwestern Spain. Alimentary, death, idolatry, magic, nature, space, time.

Canones Poenitentiales of Ps. Jerome, 9th/10th century? (Gonzales Rivas, *La penitencia en la primitiva Iglesia española*, 197-202). Alimentary.

Continental penitentials: Dates and place of origin of these penitentials are taken from Frantzen's updating of *Les "Libri poenitentiales"*, 29-33, 34-42. I have not consulted the new edition of penitentials (*Sangallense Tripartitum*, *Vindobonense 2*, *Capitula Iudiciorum* and *Parisiense Compositum*) in *Het tripartite Boetboek. Overlevering en Beteknis van vroemiddeleuwse Biechtvoorschriften*, ed. Rob Meens (Hilversum, 1994).

Poen. Remense, 8th century, first half (F. B. Asbach, *Das Poenitentiale Remense und der sogen. Excarpsus Cummeani: Überlieferung, Quellen und Entwicklung Zweier kontinentaler Bußbucher aus der 1. Hälfte des 8. Jahrhunderts.* [Regensburg, 1975] 4-77). Alimentary, death, idolatry, magic, space, time.

Canones Basiliense, 8th century, first half (F. B. Asbach, *Das Poenitentiale Remense und der sogen. Excarpsus Cummeani*, 79-89). Alimentary.

Excarpsus Cummeani, 8th century, first half (Schmitz 1, 602-653). Alimentary, death, idolatry, magic, space, time.

Poen. XXXV Capitulorum, late 8th century (Wasserschleben, 505-526). Calends, cult, Alimentary, idolatry, magic, space, time.

Poen. Vindobonense b, late 8th century (Wasserschleben, 493-497. Alimentary, death, idolatry, magic, nature, space, time.

Poen. Sangallense Tripartitum, between the second half of the 8th century and 950 (Schmitz 2, 177-189). Alimentary, magic, space, time.

Poen. Valicellanum I, 9th or 1st half of the 10th century (Schmitz 1, 239-342). Compiled in Northern Italy, around Milan, Parma, Pavia or Vercelli. See Hägele, *Das Paenitentiale Vallicellianum I*. Schmitz provides a still valuable commentary for many of the articles of this penitential. Alimentary, death, cult, idolatry, magic, space, time.

Anonymi liber poenitentialis, 9th century? (*PL* 105, 717-730). Idolatry, space, time.

Poen. Martenianum, 802-813 (Wasserschleben, 282-300). Idolatry, alimentary, magic.

Iudicium Clementis, 9th century (Wasserschleben, 433-435). Idolatry, time.

De poenitentia, 817-830 (*PL* 105, 649-710), The first five of six books on penance compiled by Halitgar of Cambrai. Idolatry, magic, nature, space.

Poen. Halitgari = Poen. Romanum (Schmitz 1, 471-489). This is Bk. VI, the penitential proper, of *De poenitentia*. The sources of this and the penitentials of Rabanus Maurus are studied in Kottje, *Die Bußbücher Halitgars von Cambrai und des Hrabanus Maurus*. Alimentary, death, idolatry, magic, nature, space.

Poen. ad Otgarium, 841-842 (*PL* 112, 1397-1424). Compiled by Rabanus Maurus, at the time Abbot of Fulda, for Otgar of Mainz. Idolatry, magic, nature, time.

Poen ad Heribaldum, c. 853 (*PL* 110, 467-494). Compiled by Rabanus Maurus, by that time Archbishop of Mainz, for Heribald of Auxerre. Idolatry, magic, space.

Poen. Ps.-Theodori, mid 9th century (Wasserschleben, 566-622). Alimentary, death, idolatry, magic, nature, space, time.

Poen. Ps.-Gregorii, mid 9th century (Franz Kerff, "Das Paenitentiale Pseudo-Gregorii. Eine kritische Edition," in Hubert Mordek, ed., *Aus Archiven und Bibliotheken: Festschrift für Raymund Kottje zum 65. Geburtstag* [Frankfurt am Main, 1992] 161-188 = Kerff). This edition replaces Wasserschleben, 534-547). Alimentary, death, idolatry, magic, space, time.

Poen Quadripartitus, 9th century (E.L. Richter, *Antiqua canonum collectio qua in libris de synodalibus causis compilandis usus est Regino* [Marburg, 1844] = Richter). Magic, nature, space, time.

Poen. Vindobonense a, late 9th century (Schmitz 2, 351-356). Alimentary, death, idolatry, magic, nature, time.

Poen. Bobbiense II, 9th/10th century (O. Seebass, "Ein bisher noch nich veröffentlichtes Poenitential einer Bobbienser Handschrift der Ambrosiana," *Deutsche Zeitschrift für Kirchenrecht* 28 [1896/97] 26-46). Magic, time.

Poen. Valicellanum E. 62, 9th/10th century (Wasserschleben, 550-566). Magic.

Poen. Casinense, 9th/10th century (Wasserschleben, 535-547). Alimentary, death, idolatry, magic, space, time.

Poen. Parisiense I. 9th/10th century (Schmitz 1, 681-697). Alimentary, idolatry, magic, space.

Poen. Arundel, 10th/11th century (Schmitz 1, 437-465). Brief treatment of some of the beliefs and practices more fully developed by Burchard of Worms in his penitential, the *Corrector sive medicus*, but which appear in none of the other penitentials discussed here. Idolatry, magic, nature, space, time.

Poen. Valicellanum C. 5, Late 10th/early 11th century (Schmitz, 350-388). Italian. Magic, nature, space, time.

Corrector sive medicus, 1008-1012 (Schmitz 2, 403-467). This penitential forms Bk. 19 of the *Decretorum libri XX* of Burchard of Worms. Chapter 5 of Bk. 19 is the penitential proper ("Interrogationes quibus confessor confitentem debet interrogare"). This is cited as Burchard of Worms, *Decretum* 19, 5. Bk. 19 is printed in its entirety in *PL* 140, 943-1018. For the sources, see Hoffmann and Pogorny, *Das Dekret des Bischofs Burchard von Worms,* 234-239. This penitential contains articles on all the categories that appear in the other penitentials (alimentary, death, idolatry, magic, nature, space, time), with significant additions.

1.6 *Sermons*

Late Antiquity: These are included for their relevance to early medieval sermons.

Peter Chrysologus. *Sermons*, c. 400-450. (A. Olivar, ed., *Sancti Petri Chrysologi collectio sermonum* [Turnhout, 1982] *CCSL* 24B = *CCSL* 24B). Time (S. 155 and S. 155 bis and *Homilia de pythonibus et maleficiis* (*PG* 65, 28. For the attribution of this sermon to Peter Chrysologus, see Olivar (*Sacris Erudiri* 6 (1954) 329.

Maximus of Turin. *Sermons,* c. 380-c. 470 (A. Mutzenbecher, ed., *Maximus Taurensis sermones*, *CCSL* 23, Turnhout, 1962 = *CCSL 23*). His vividly, almost coarsely, expressed sermons contain what appear to be eye-witness accounts of popular behaviour. His influence can be detected in one of the sermons of Rabanus Maurus. Idolatry (*Sermons* 91 extr., 107 and 108); nature (*Sermons* 30 and 31); time (*Sermons* 61c and 63).

Ps..- Maximus, n.d. (*PL 57*, 254-266). Time (Hom. 16).

Gallo-Roman and Iberian

Caesarius of Arles. *Sermons*, c.502-542 (Germain Morin, ed., *Sancti Caesarii Arelatensis sermones*, *CCSL* 103 and 104 [Turnhout, 1953] = *CCSL* 103, *CCSL* 104). Many of these were meant to serve as model sermons. Caesarius' dedication to pastoral work and preaching in Arles and the surrounding area vouches for the authenticity of his descriptions of the manners and morals of his flock. For his life and work, see Marie-José Delage's "Introduction" to her edition of *Césaire d'Arles, SC* 175: 13-175; Klingshirn, *Caesarius of Arles* and Malnory, *Saint Césaire Évêque d'Arles.* See also Konda, *Le discernement et la malice*; R. Boese, *Superstitiones arelatenses*; Filotas, *Pagan Beliefs and Practices.* What Caesarius had to say on healing and protective magic, nature cults, sacred space, and the celebrations of the Calends of January echoes through early medieval legislation and penitentials as well as sermons. By contrast, his strictures against ritual drunkenness, which were if anything more fervent than those against pagan survivals, were ignored entirely by his successors. Idolatry (*Sermons* 19, 53, 54, 192 and 193); magic (*Sermons* 1, 12, 13, 14, 18, 19, 44, 50, 51, 52, 54, 59, 70, 130, 134, 167, 184, 189, 197, 200, 204 and 229); nature (*Sermons* 13, 15 and 180); space (*Sermons* 1, 13, 14, 53, 54, and 229); time (*Sermons* 1, 6, 12, 13, 19, 33, 46, 47, 52, 54, 55, 55A, 192, 193, 216 and 222). In addition, magic and sacred space are mentioned in an Easter sermon edited in Raymond Étaix, "Nouveau sermon pascal de saint Césaire d'Arles" *RB* 75 (1965) 201-211; magic is also found in *idem*, "Les épreuves du juste. Nouveau sermon de saint Césaire d'Arles," *REAug* 24 (1978) 272-277.

Ps.-Augustine. Date unknown (*PL* 39, Appendix, 2101). A sermon against commemorative banquets celebrated over tombs on 22 February was formerly supposed to be a part of Caesarius' work, but is not included in Morin's edition. Jan Machielsen (*Clavis patristica*, 189, no. 975) considers this sermon to have originated in Gaul, but can assign it no date. Death.

Martin of Braga. *De correctione rusticorum*, 572 (C. W. Barlow, ed., *Martini Episcopi Bracarensis opera omnia* [New Haven, 1950]). Written in response to a request from Polemius, bishop of Astorga, for a model for sermons to instruct "rustics" – according to López Calvo ("La cataquesis en la Galicia medieval"), not peasants, but those who yet lacked the rudiments of faith. It is not clear how much of the sermon was based on direct observation of Galician customs, as opposed to customs that Martin might have encountered elsewhere. See Meslin, "Persistance païenne en Galice" and Ferreiro, "St. Martin of Braga's policy." Martin's influence of medieval sermons is apparent chiefly in the *Dicta Pirmini* and the sermon in the *Vita Eligii.* Idolatry, magic, nature, space, time.

Homiliare Toletanum, 7th century, second half. Originally published in R. Grégoire, *Les homélaires du moyen âge* (Rome, 1966). Idolatry (Hom. 19 (*PL Suppl.* 4, 1943-1945); time (homilies 9 and 31, *ibid.,* 1940-1942 and 1958-1959).

Frankish and Germanic

Pirmin of Reichenau. *Dicta* = *Scarapsus*, 724-753 (C. P. Caspari, ed., *Dicta Abbatis Pirminii de singulis libris canonicis scarapsus, Kirchenhistorische Anecdota*, v. 1., *Lateinische Schrifte* [1883; repr. Brussels, 1964] = Dicta Pirmini). For Pirmin, see Gall Jecker, *Die Heimat des hl. Pirmin des Apostels der Alamannen* (Münster, 1927). An alternate edition of the *Dicta* is found in pp. 34-73. Pirmin's field of missionary activity was among the Alamannians, but he was probably of Visigothic origin, one of the refugees who left Spain after the Moorish conquest. The *Dicta*, the attribution of which to Pirmin has been questioned by Bouhot ("Alcuin et le 'De catechizandis,'

180-191), was meant to serve as a guide for preaching. Its principal sources were Caesarius of Arles' sermons and Martin of Braga's *De correctione rusticorum*, but it also contains elements found in neither. It has some points in common with Ps.-Eligius' sermon, but it is not clear which is the source for the other. Alimentary, idolatry, magic, nature, space, time.

Eligius of Noyon. *Sermon,* 8th century? (*Vita Eligii Episcopi Noviomagensis* 2.16, ed. B. Krusch, *MGH SRM* 4, 705-708 = *Vita Eligii, MGH SRM* 4). A missionary sermon supposedly preached by St. Eligius (588-660) is included in the *Vita Eligii.* In its original form, this Life was written by his friend and contemporary St. Audoin but since it was extensively reworked in the 8th century, the sermon is of dubious authenticity. This is the opinion of Krusch, Hen (*Culture and Religion,* 196 and note) and Markus (his review of Flint's *Rise of Magic*). By contrast, Banniard considers the sermon to be plausibly late 7th century ("Latin et communication orale en Gaule franque"). It may be taken at the very least to represent the kind of sermon that Carolingian admirers of a missionary bishop expected him to preach. It combines elements taken from Caesarius of Arles, Martin of Braga and perhaps Pirmin of Reichenau (see above) and some independent material. Idolatry, magic, nature, time, space.

Ps.-Eligius. *Praedicatio de supremo iudicio,* 8th century? (*MGH SRM* 4, 751-761). Based entirely on Caesarius of Arles, except for a reference to *magi,* a term not used by Caesarius. Idolatry, magic.

Burchard of Würzburg. *Homilies.* 8th century. (Ed. G. v. Eckhart, *Commentarium de rebus Franciae orientalis et episcopatus Wirceburgensis* [Würzburg, 1729] 837-846). These are composed of passages copied wholesale from the sermons of Caesarius of Arles. They contain no original elements and are worthless as a source for Bavarian paganism. Magic (*Hom.* 19 = Caesarius, S. 50; *Hom.* 20 = S.51; *Hom.* 23 = S. 52); time (*Hom.* 3 = Caesarius, S. 192); space (*Hom.* 23 = Caesarius, S. 53).

Anonymous homily, 8th century (A. J. Nürnberger, *Aus der litterarischen Hinterlassenschaft des hl. Bonifatius und des hl. Burchardus* [Neisse, 1888] 43-45). Closely akin to the sermon in the *Vita Eligii.* The only novel and noteworthy element distinguishing this from the other sermons following the Caesarian model is the absence of any reference to the festivities of the New Year. Death, idolatry, magic, nature, space, time.

Anonymous sermon, 8th century? (Germain Morin, "Textes relatifs au symbole et à la vie chrétienne," *RB* 22 [1905] 505-524). Written in an unpolished style, probably in Alamannia, this sermon could quite credibly have been delivered to a rustic congregration. Much of the material comes from Caesarius of Arles, but there are important independent elements as well. Magic, time.

Anonymous sermon, c. 778 (in Wilhelm Levison, *England and the Continent in the Eighth Century* [1946; repr. Oxford, 1976] 302-314). In the editor's judgment, this originated in a "place connected with an Irish circle" in France, although northern Italy has also been suggested as its source; the author was probably an Englishman. The inspiration for this sermon was mainly Caesarius of Arles, with elements drawn from Martin of Braga, maybe Isidore of Seville or St. Augustine, probably Pirmin or the *Vita Eligii,* and some other, unknown source. The author's confused syntax suggests that he did not understand fully some of the practices that he described. Idolatry, magic, nature, time.

Anonymous. *Homilia de sacrilegiis,* (C. P. Caspari, ed., *Eine Augustin fälslich beilegte Homilia de sacrilegiis* [Christiania, 1886] = Caspari). According to Caspari, this was written near the end of the 8th century by a Frankish cleric in the northern parts of the Frankish empire, probably as a model sermon. Despite a rather primitive level of Latin, the author had an unexpected taste for recherché vocabulary. The homily combines familiar practices borrowed from Caesarius of Arles with much new material on folk culture, particularly divination and healing magic. Magic, nature, space, time.

Ps.-Boniface. *Sermons.* Early 9th century? (*PL* 89). A group of fifteen anonymous sermons, essentially a guide for pastors, once erroneously attributed to St. Boniface. See Bouhot, "Alcuin et le 'De catechizandis rudibus'," 184-191. The passages dealing with superstitions are almost entirely a reprise of some of the themes of the *Indiculus superstitionum.* Idolatry, death, magic, space (*S.* 6, *ibid.*, 855); idolatry, magic (*S.* 15, *ibid.,* 870-871).

Anonymous sermon, early 9th century (W. Scherer, "Eine lateinische Musterpredigt aus der Zeit Karls des Grossen," *Zeitschrift für deutsches Altertum* 12 [1865] 436-446). This was meant as an exemplar for preachers. In the relevant clauses, it is very similar to Ps.-Boniface's *S.*6, above. Idolatry, death, magic, space, time.

Arno of Salzburg. *Synodal sermon,* C. 806 (Rudolf Pokorny, *DA* 39 [1983] 379-394). Sermon of pastoral instruction to the clergy; the relevant passage appears to have been prompted by the outbreak of an epidemic among livestock. Magic, space.

Rabanus Maurus. *Homilies,* 814-826 (*PL* 110). Rabanus Maurus (c. 776-856) was Abbot of Fulda and Archbishop of Mainz. His youth was spent in the same area, and he was probably well-acquainted with and interested in local culture (he compiled a "Latin-barbaric" glossary of terms for parts of the body, the months and winds). There can be no doubt that his descriptions of local rituals during Rogation Days and lunar eclipses were based on his own observation or eye-witness accounts, even when he borrowed his approach from Maximus of Turin. The collection of sermons from which the following material was taken was prepared for the use of Haistulph of Mainz (see Étaix, "Le recueil de sermons composé par Raban Maur"). Death (*Hom.* 53 and 67); idolatry (*Hom.* 67); magic (*Hom.* 16, 43, 45 and 67); nature (*Hom.* 43 and 67; space (*Hom.* 67); time (*Hom.* 16, 19, 41, 43, 44).

Anonymous sermon, c. 850-882 (Nikolaus Kyll, "Eine Trierer Sendpredigt aus dem 9. Jahrhundert," *Kurtrierisches Jahrbuch* 11 [1961] 10-19). The material is taken from an early 9th century canonic collection from the same area, which was in turn based partly on Caesarius of Arles. Alimentary, death, magic, space, time.

Anonymous. *Sermon to the Baptised,* 10th or 11th century ms. (C. P. Caspari, *Kirchenhistorische Anecdota,* v. 1., *Lateinische Schrifte* [1883; Brussels, 1964] 199-201). Based on *De correctione rusticorum.* Idolatry, magic, place, time.

Anonymous sermon, 9th/10th century handwriting (C. P. Caspari, "Eine Ermahnung zu würdige Feier des bevorstehenden Osterfestes," in *Briefe, Abhandlungen und Predigten aus den zwei letzten Jahrhunderten des kirchlichen Alterthums und dem Anfang des Mittelalters* [Christiania, 1890] 200-201). This sermon follows the pattern set by Caesarius, but its description of ritual drinking at Eastertime appears to have another source. Time.

Italian

Leo IV. *Homily*, 847-855 (*PL* 115, 681). Hom. 39: time

Ratherius of Verona. *Sermons*, 10th century (P. L. D. Reid, ed., *Ratherii Veronensis opera minora*, *CCCM* 44 [Turnhout, 44] = *CCCM* 44). Presumably these sermons were preached during one of Ratherius' stormy terms as bishop of Verona (931-c. 933, c. 946-c. 948, 953-955 and 962-968). The content and the polemical tone leave no doubt that they were in response to actual events. Idolatry (*S.* 2 *de Quadragesima* and *S. contra reprehensores sermonis eiusdem* 2); magic (*S.* 1 *de Ascensione*).

Atto of Vercelli. *Sermons* (before 961) (*PL* 134). His treatment of traditional topics indicates that he was describing local practice in Piedmont, not following literary models. For the political context of Atto's life and his social and political programme, see Wemple, *Atto of Vercelli*; his struggles to reform popular culture are discussed in Gandino, "Cultura dotta e cultura folklorica." Nature (S. 13); time (Sermons 3 and 13).

1.7 *Incidental literature*

Letters: The majority of the relevant letters are published in P. Ewald and L. M. Hartmann, eds., *Gregorii Papae Registrum Epistolarum, MGH Epistolae* 1; L. G. Hartmann, ed., *Gregorii Papae Registrum Epistolarum, MGH Epistolae* 2 (Berlin, 1891 and 1899); W. Gundlach and Ernest Dümmler, eds., *MGH Epistolae Merowingici et Karolini Aevi, MGH Epistolae* 3 (Berlin, (1892); E. Dümmler, ed., *MGH Epistolae Karolini Aevi, MGH Epistolae* 5 (Berlin, 1895). These are cited as *MGH Ep* 1, 2, 3 and 5 respectively. Letters referring to the English Church are taken from Bede, *HE*, (Bertram Colgrave and R. A. B. Mynors, eds., *Bede's Ecclesiastical History of the English People* [Oxford, 1969]), for the period of conversions, and from Haddan and Stubbs for the later period. The correspondence of St. Boniface is found in M. Tangl, ed., *Die Briefe des heiligen Bonifatius und Lullius, MGH Epistolae Selectae* 1 (Berlin, 1926) = *MGH EpSel* 1.

Pelagius I. *Ep.* to Sapaudus, Bishop of Arles, 558-560 (*MGH Ep* 3, 445). Idolatry or magic.

Gregory I. *Ep.* 8.4 to the Frankish queen Brunhilde, 597 (*MGH Ep* 2, 5-8). Idolatry, nature.

——. *Ep.* 5.32 to the deacon Cyprian, 595 (*MGH Ep* 1, 312-313). Magic.

——. *Ep.* 5.38 to the Empress Augusta, 595 (*MGH Ep* 1, 324-329). Idolatry.

——. *Ep.* 8.1 to Peter, Bishop of Corsica, 597 (*MGH Ep* 2, 1-2). Idolatry, nature.

——. *Ep.* 8.29 to Eulogius, Bishop of Alexandria, 597 (*MGH Ep* 2, 30-31). Idolatry, nature.

——. *Ep.* 9.102 to Venantius, bishop of Luni, 599 (*MGH Ep* 2, 110). General.

——. *Ep.* 9. 204 to Ianuarius, bishop of Sardinia, 599 (*MGH Ep* 2, 191-193). Idolatry, magic.

——. *Ep.* 10.2 to the regional subdeacon Sabinus, 600 (*MGH Ep* 2, 237-238). Idolatry.

——. *Ep.* 11.10 to Serenus, bishop of Marseilles, 600 (*MGH Ep* 2, 267-270). Idolatry.

——. *Ep.* 11.33 to the notary, Adrian, 601 (*MGH Ep* 2, 302). Magic.

——. *Ep.* 11.47 to Theodoric, king of the Franks, 601 (*MGH Ep* 2, 319-320. Idolatry.

——. To Ethelbert, king of the Angles, 601 (Bede, *HE* 1.32, 110-115). Idolatry.

——. To Mellitus, 601 (Bede, *HE* 1.30, 106-109). Idolatry, space.
——. *Ep.* 14.1 to Paul, scholastic of Sicily, 603 (*MGH Ep* 2, 419-420). Magic.
Boniface V. To Edwin, king of Northumbria, 625 (Bede, *HE* 2.10, 166-171). Idolatry.
——. To Ethelberga, Edwin's queen, 625 (Bede, *HE* 2.11, 172-175). Idolatry.
Gregory II. *Ep.* 17 to Boniface, 722 (*MGH EpSel* 1, 29-31). Mandate for mission to the Germans.
——. *Ep.* 21 to the Old Saxons,738-739 (*MGH EpSel* 1, 35-36). Idolatry.
——. Ep. 25 to the people of Thuringia, 724 (*MGH EpSel* 1, 43-44). Death, idolatry, magic.
——. *Ep.* 26 to Boniface, 726 (*MGH EpSel* 1, 44-47). Alimentary, idolatry.
Gregory III, *Ep.* 28 to Boniface, c. 732 (*MGH EpSel* 1, 49-52). Alimentary, idolatry.
——. *Ep.* 43 to the nobility and people of Hesse and Thuringia, c. 738 (*MGH EpSel* 1, 68-69). Death, idolatry, magic, space.
——. *Ep.* 44 to the bishops of Bavaria and Alamannia, c. 738 (*MGH EpSel* 1, 70-71). General.
Boniface. *Ep.* 50 to Pope Zacharias, 742 (*MGH EpSel* 1, 80-86). Magic, time.
Zacharias. *Ep.* 41 to Boniface, 743 (*MGH EpSel* 1, 86-92). Magic, time.
——. *Ep.* 57 to Boniface, 744 (*MGH EpSel* 1, 102-105). This letter concerns Aldebert's heresy. Idolatry.
Boniface. *Ep.* 63 to Daniel, bishop of Winchester 742-746 (*MGH EpSel* 1, 128-130). General.
Zacharias. *Ep.* 80 to Boniface, 738 (*MGH EpSel* 1, 172-180). Idolatry.
——. *Ep.* 87 to Boniface, 751 (*MGH EpSel* 1, 194-201). Alimentary, idolatry.
Eanwulf. *Ep.* 120 to Charlemagne, 773 (*MGH EpSel* 1, 256-257). Repeats Gregory I's exhortations to Ethelbert of Kent. Idolatry.
Hadrian I. *Ep.* 95 to the bishops of Spain, 785-791 (*MGH Ep* 5, 636-643). Alimentary.
——. *Ep.* 96 to Egila, bishop of Elvira and to the priest Iohannes, 785-791 (*MGH Ep* 5, 643-648). Alimentary, as above.
Epistola Canonica, 814 (Mansi 13, 1095-1098). Idolatry, space, time.
Egred, bishop of Lindisfarne. To Wulfsige archbishop of York, 830-837 (Haddan and Stubbs 3, 615-616). Idolatry.
Leo IV. *Ep.* 16 to the bishops of Brittany, 847-848 (*MGH Ep* 5, 593-596). Magic.
Atto of Vercelli. *Ep.* 2 to all the faithful of his diocese, 10th century, first half (*PL* 134, 104). Nature, magic.
——. *Ep.* 3 to the people of Vercelli. 10th century, first half (*PL* 134, 104-105). Idolatry.
——. *Ep.* 4 to all the faithful of his diocese, 10th century, first half (*PL* 134, 106-111). Time.

1.8 *Tracts and minor works*

Manual of pastoral care, very early 8th century (Raymond Étaix, "Un manuel de pastorale de l'époque carolingienne," *RB* 91 [1981] 105-130). From southern Bavaria. Idolatry.

Dialogues of Egbert of York, 732-766 (Haddan and Stubbs 3, 403-443). Death, idolatry, magic.

Interrogationes et responsiones baptismales, 8th century (*MGH CapRegFr* 1, 222). Saxon. Mixed Latin-Old German text which, exceptionally, names three indigenous gods. Idolatry.

Ps.-Eligius. *De rectitudine catholicae conversationis tractatus,* 8th century? (*PL* 40, 1170-1190). A handbook of pastoral care, largely based on Caesarius of Arles, but with some novel features. Idolatry, magic.

Ordo de catechizandis rudibus, late 8th century (Ed. Jean-Paul Bouhot, *Recherches Augustiniennes* 15 [1980] 205-230). According to Bouhot, this was not written by Alcuin, as had been previously thought, but by a member of the entourage of Arno of Salzburg. Idolatry.

Ratio de catechizandis rudibus, late 8th century (ed. Joseph Michael Heer, *Ein Karolingisher Missions-Katechismus* (Freiburg im Breisgau, 1911) 77-88). Idolatry, magic, space.

Admonitio synodalis, c. 813 (Robert Amiet, "Une 'Admonitio synodalis' de l'époque carolingienne. Étude critique et édition," *Mediaeval Studies* 26 [1964] 12-82). Lotharingian or Rhenish. Death, magic, space.

Agobard of Lyons. *De grandine et tonitruis*, 815-817 (L. van Acker, ed. *Agobardi Lugdunensis Opera Omnia*, *CCCM* 52 [Turnhout, 1981] = *CCCM* 52, 1-15). For Agobard's life, see Cabaniss, *Agobard of Lyons* and Egon Boshof, *Erzbischof Agobard von Lyon.* Magic.

——. *De quorundam inlusione signorum*, 828-834 (*CCCM* 52, 237-243). Magic.

2. Complementary works

Adam of Bremen. *Gesta Hammaburgensis ecclesiae pontificum usque ad a. 1072.* Ed. I. M. Lappenberg. *MGH SS* 7: 267-389.

Agobard of Lyons. *De picturis et imaginibus.* Ed. L. van Acker. *CCCM* 52. Turnhout, 1981.

Bede. *De temporibus.* Ed. Charles W. Jones. Cambridge, Mass., 1943.

——. *Historia ecclesiastica gentis Anglorum. Bede's Ecclesiastical History of the English People.* Ed. Bertram Colgrave and R. A. B. Mynors. Oxford, 1969.

——. *Vita sanctorum abbatum monasterii.* In *Opera historica.* Trans. J. E. King. Loeb Classical Library 248, 2: 392-444.

Ps.-Bede. *De minutione sanguinis sive de phlebotomia. The Complete Works of Venerable Bede,* ed. J. A. Giles, 6: 349-352. London, 1843.

——. *De tonitruis.* In , *The Complete Works of the Venerable Bede* 4: 343-348. London, 1843-1944 = *PL* 90: 610-612.

Beleth, John. *Rationale divinorum Officiorum.* Ed. M. Douteil. *CCCM* 41. Turnhout, 1976.

Chronicon Pictum. Ed. L. Mezey. Budapest 1964.

Edict of Theodoric. Ed. Fredericus Bluhme, *MGH Legum sectio* 5: 144-168.

Einhard. *Vita Caroli Magni.* Ed. O. Holder-Egger. *MGH Scriptores Rerum Germanicarum* 25.

Gervaise of Tilbury. *Otia imperialia. Des Gervasius von Tilbury Otia imperalia in einer Auswahl neu herausgegeben und mit Anmerkungen begleitet.* Ed. Felix Liebrecht. Hanover, 1856.

Gilbert of Nogent. *De sanctis et eorum pignoribus*. Ed. R. B. C. Huygens. *CCCM* 127. Turnhout, 1993.

——. *Monodiae. Autobiographie*. Trans. and ed. E. R. Labande. Paris, 1981.

Gregory I. *Dialogues*. Ed. A. de Vogüé. *SC* 251, 260 and 265. Paris, 1978-1980.

——. *Homiliae in Evangelia*. Ed. Raymond Étaix, *CCSL* 141. Turnhout, 1999.

——. *Moralia in Job*. Ed. M. Adraien, *CCSL* 143B. Turnhout, 1985.

Gregory of Tours. *De miraculis sancti Andreae*. Ed. M. Bonnet. *MGH SRM* 1: 821-846.

——. *De virtutibus sancti Martini*. Ed. B. Krusch. *MGH SRM* 1: 584-661.

——. *Historia Francorum libri X*. Ed. B. Krusch and W. Levison. *MGH SRM* 1. Editio Altera.

——. *In gloria confessorum* . Ed. B. Krusch. *MGH SRM* 1: 744-820.

Grosseteste, Robert. *De decem mandatis*. Ed. C. Dales and E. B. King. Oxford, 1987.

Hincmar of Reims. *De divortio Lotharii regis et Theutbergae reginae*. Ed. L. Böhringen. *MGH Concilia* 4. *Supplementum*..

——. *De nuptiis Stephani*. *PL* 126: 132-153.

Isidore of Seville. *Etymologiae librorum XX. Etimologias. Edición Bilingüe*. Ed. José Oroz Reta and Manuel-A. Marcos Casquero. 2 vols. Madrid, 1985.

Jacob of Voragine. *Legenda aurea*. Ed. G. P. Maggioni. 2 vols. Florence, 1998.

Leges Visigothorum. Ed. K. Zeumer. *MGH Legum Sectio* 1.1.

Lex Longobardica. *Le leggi dei Longobardi*. Ed. Claudio Azzara and Stefano Gasparri. Milan, 1992.

Lex Salica. Ed. K. A. Eckhardt. *MGH Legum Sectio* 1.4.

Map, Walter. *De nugis curialium*. Ed. M. R. James. Oxford, 1914.

Miracula S. Geraldi. Anne-Marie Bultot-Verleysen, "Des *Miracula* inédits de saint Géraud d'Aurillac (†909). Étude, édition critique et traduction française." *Analecta Bollandiana* 118 (2000): 47-141.

Medicina antiqua: Libri quattor medicinae. Fac-similé du Codex Vindobonensis 93, conservé à Vienne à la Bibliothèque nationale d'Autriche. French trans., Marthe Dulong. Studies by Charles H. Talbot and Franz Unterkirchen. 2 vols. Paris, 1978.

Monk of St. Gall. *Gesta Karoli*. Ed. G. W. Pertz. *MGH SS* 2: 726-763.

Otloh of Emmerman. *Liber visionum*. *PL* 146: 541-588.

Patrick of Armagh. *Confessio*. In *Confession et Lettre à Coroticus*. Trans. Cécile Blanc. Ed. Richard P. C. Hanson. *SC* 249. Paris, 1978.

Paul the Deacon. *Historia Longobardorum*. Ed. G. W. Waitz. *MGH Scriptorum Rerum Longobardicarum et Italicarum* , 45-187.

Ratherius of Verona. *Praeloquia*. Ed. Peter L. Reid. *CCCM* 46.Turnhout, 1981.

Valerio of Bierzo. *Replicatio sermonum a prima conversione*. In *Valerio of Bierzo. An Ascetic of the Late Visigothic period,* trans. and ed. C. M. Aherne, 115-153. Washington, 1949.

Vita Amatoris ep. Autissiodorensis. *AASS* May 1: 52-60.

Vita Aniani ep. Aurelianensis. Ed. B. Krusch. *MGH SRM* 3: 108-117.

Vita Bernardi ab. Clarevallensis. Jacob of Voragine, *Legenda aurea* 2: 811-826.

Vita Boniti ep. Arverni. Ed. W. Levison. *MGH SRM* 6, 119-139.

Vita Ciarani de Saigir. Ed. C. Plummer. *VSH* 1: 217-233.

Vita Columbae ab. Hiensis auctore Adamnano. In *Adomnan's Life of Columb*, ed. and trans. Alan Orr Anderson and Marjorie Ogilvie. Oxford, 1961.

Vita Columbani ab. Luxoviensis et Bobiensis auctore Iona. Ed. B. Krusch. *MGH SRM* 4 61-108.

Vita Corbiniani ep. Frisengensis retracta. Ed. B. Krusch. *MGH SRM* 6 597-635.

Vita Dominici fundator Ord. Praed. In Jacob of Voragine, *Legenda aurea* 2 718-744.

Vita Eligii ep. Noviomensis. Ed. B. Krusch. *MGH SRM* 4 663-741.

Vita Endei ab. Araniensis. Ed. C. Plummer. *VSH* 2 60-75.

Vita Frodoberti ab. Cellensis auctore Adsone. Ed. W. Levison. *MGH SRM* 5 67-88.

Vita Galli ab. in Alamannia auctore Wettino. Ed. B. Krusch. *MGH SRM* 4 256-280.

Vita Fructuosi ep. Bracanensis. *The Vita Sancti Fructuosi*. Ed. Clare Frances Nock. Washington, 1946.

Vita Germani ep. Autissiodorensis. In Jacob of Voragine. *Legenda aurea* 2 689-694.

Vita Hugberti ep. Leodiensis. Ed. W. Levison. *MGH SRM* 6: 483-496.

Vita Liutbirgae virginis. *MGH Deutsches Mittelalter, Kritische Studientexte* 3. Ed. O. Menzel. Leipzig 1937.

Vita Malachiae ep. Conerensis auctore Bernardo ab. Clarevallensi. *PL* 182 1073-1118.

Vita Martini ep. Turonensis Sulpicii Severi scripta. *Vie de saint Martin*. Ed. Jacques Fontaine. *SC* 133. Paris, 1967.

Vita Odiliae ab. Hohenburgensis. Ed. B. Krusch.*MGH SRM* 6, 29-50.

Vita Richarii ab. Centulensis. Ed. B. Krusch. *MGH SRM* 7: 438-453.

Vita Samsonis ab. Dolensis. *AASS* Iul 6: 573-591.

Vita Vulframmi ep. Senonensis. Ed. W. Levison. *MGH SRM* 5: 657-673.

Vita Willibrordi ep. Traiectensis auctore Alcuino. *PL* 101: 693-710.

Vitae Caesarii ep. Arelatensis libri duo. Ed. B. Krusch. *MGH SRM* 3: 457-501.

Vitae sanctorum patrum Emeretensium. Ed. and trans. J. N. Garvin. Washington, 1946.

Vitae sanctorum Hiberniae. Ed. Charles Plummer. 2 vols. [1910] London, 1968.

Vita Symphoriani. *AASS* Aug. 4: 496-497.

Vitae Vedastis ep. Atrebatensis. Ed. B. Krusch. *MGH SRM* 3: 406-427.

Vita Vigoris. *AASS* Nov. 1: 297-305.

William of Auvergne. *De legibus*. Paris, 1674.

3. STUDIES

Addison, J. T. *The Medieval Missionary*. Philadelphia, 1976.

The Age of Sutton Hoo. The Seventh Century in North-Western England. Ed. M. O. H. Carver. Woodbridge, 1992.

Alary, François. "La religion populaire au XIe siècle: Le prescrit et le vécu d'après le *Corrector sive medicus* de Burchard de Worms." *Cahiers d'Histoire* (Université de Montréal) 13 (1993): 48-64.

Albert, F. R. *Die Geschichte der Predigt in Deutschland bis Luther*. 2 vols. Gütersloh, 1892-1893.

Amos, T. L. "Early medieval sermons and their audience." In *De l'homélie au sermon:*

Histoire de la prédication médiévale, ed, Jacqueline Hamesse and Xavier Hernand, 1-14. Louvain-la-Neuve, 1993.

——. "Preaching and the sermon in the Carolingian World." In *De Ore Domini: Preacher and Word in the Middle Ages*, ed. T. L. Amos, E. A. Green and B. M. Kienzle, 41-60. Kalamazoo, 1989.

Angenendt, Arnold. "The conversion of the Anglo-Saxons considered against the background of the early medieval mission." In *Angli e Sassoni al di qua ed al di là del mare*, 747-781. SSAM 28. Spoleto, 1982.

Animals in Folklore. Ed. J. R. Porter and W. M. S. Russell. Ipswich, Cambridge and Totowa, N. J., 1978.

Antin, P. "Autour du songe de S. Jérôme." In *Recueil sur saint Jérôme*, 71-100. Brussels, 1968.

Appleby, David F. "Spiritual progress in Carolingian Saxony. A case from ninth-century Corvey." *The Catholic Historical Review* 82 (1996): 599-613.

Arbesmann, R. "The 'cervuli' and 'anniculae' in Caesarius of Arles." *Traditio* 35 (1979): 89-119.

Archbishop Theodore: Commemorative Studies on His Life and Influence. Ed. Michael Lapidge. Cambridge, 1995.

Armstrong, Edward A. *The Folklore of Birds*. 2nd ed. New York, 1970.

Aufstieg und Niedergang der römischer Welt. 2. Principat. 36 vols. Ed. Temporini, Hildegard and Wolfgang Haase. Berlin and New York, 1972- .

Aupest-Conduché, Dominique. "De l'existence du Concile de Nantes." *Bulletin Philologique et Historique CTHS* (1973): 29-59.

Azoulai, Martine. *Les péchés du Nouveau Monde. Les manuels pour la confession des Indiens (XVIe-XVIIe siècle)*. Paris, 1993.

Babut, E.-Ch. *Priscillien et le priscillianisme*. Paris, 1909.

Bächtold-Stäubli, Hanns. *Handwörterbuch des deutschen Aberglaubens*. 10 vols. Ed. E. Hoffman Krayer. 1927-1942. Reprint. Berlin, 1987.

Backman, E. L. *Religious Dances in the Christian Church and in Popular Medicine*. Trans. E. Classen. London, 1952.

Baker, L. G. D. "The shadow of the Christian Symbol." In *The Mission of the Church*, 17-28.

Bakhtin, Mikhail. *Rabelais and His World*. Trans. Hélène Iswolsky. Bloomington, Indiana, 1984.

Balon, Joseph. *Traité du Droit Salique. Étude d'exègése et de sociologie juridiques*. 4 vols. Namur, 1965.

Banniard, Michel. "Latin et communication orale en Gaule franque. Le témoignage de la 'Vita Eligii'." In Fontaine and Hillgarth, *Le septième siècle*, 58-86.

——. *Viva Voce. Communication écrite et communication orale du IVe au IXe siècle en Occident latin*. Paris, 1992.

Barral I Altet, Xavier. "Le cimetière en fête. Rites et pratiques funéraires dans la peninsule ibérique pendant l'antiquité tardive." In *Fiestas y liturgia. Fêtes et liturgie*, 299-308. Madrid, 1988.

Bauschatz, Paul G. *The Well and the Tree; World and Time in Early Germanic Culture*. Amherst, 1982.

Bausinger, Hermann. "Zur Algebra der Kontinuität." In *Kontinuität? Geschichtlichkeit und Dauer als volkskundliches Problem,* ed. H. Bausinger and W. Brückner, 9-30. Berlin, 1969.

Bautier, Anne-Marie. "Typologie des ex-voto mentionnés dans des textes antérieurs à 1200." In *La piété populaire au moyen âge,* 237-282.

Beck, H. G. J. *The Pastoral Care of Souls in South-East France during the Sixth Century.* Rome, 1950.

Beck, Roger. "Mithraism since Franz Cumont." In *Aufstieg und Niedergang* 17.4: 2002-2115.

Bellini, Roberto. "Il volo notturno nei testi penitenziali e nelle Collezioni canoniche." In *Cieli e terre nei secoli XI-XII. Orizzonti, percezioni, rapporti,* 293-310. Milan, 1998.

Bennassar, Bartolomé. *Histoire des Espagnes.* 2 vols. Paris, 1985.

Best, R. I. "Prognostication from the raven and the wren." *Èriu* 8 (1915): 120-126.

Bettelheim, Bruno. *Symbolic Wounds. Puberty Rites and the Envious Male.* Revised ed., New York, 1962.

Bezler, Francis. "¿El ogro y el niño o el arco y la pala?" *Revista de literatura medieval* 4 (1992): 43-46.

——. "Chronologie relative des Pénitentiels d'Albelda et de Silos." *Sacris Erudiri* 23 (1991): 163-169.

——. *Les Pénitentiels espagnols. Contribution a l'étude de la civilisation chrétienne du haut moyen âge.* Munster, 1994.

Bieler, Ludwig. "Aspetti sociali del *Penitenziale* e della *Regola* di S. Colombano." In *Ireland and the Culture of Early Medieval Europe.* Variorum VI. London, 1987. Originally in *Colombano pioniere di civilizzazione cristiana europea. Atti del Convegno internazionale di studi colombaniani, Bobbio, 28-30 agosto 1965,* 119-126. Bobbio, 1973.

Biller, Peter. "Popular religion in the central and later middle ages." In *Companion to Historiography,* ed. Michael Bentley, 221-246. London and New York, 1997.

Blásquez Martínez, José Maria. *Religiones primitivas de Hispania.* 1. Rome 1962.

Boese, R. *Superstitiones Arelatenses e Caesario collectae.* Dissertation. Marburg, 1909.

Boglioni, Pierre. "Il santo e gli animali nell'alto medioevo." In *L'uomo di fronte al mondo animale nell'alto medioevo,* 935-1002. SSAM 31. Spoleto, 1985.

——. "La religion populaire dans les collections canoniques occidentales de Burchard de Worms à Gratien." In *Byzantium in the 12th century,* ed. N. Oikonomides, 335-356. Athens, 1991.

——. "La scène de la mort dans les premières hagiographies latines." In *Le sentiment de la mort au moyen âge,* ed. Claude Sutto, 183-210. Montreal, 1979.

——. "Le sopravvivenze pagane nel medioevo." In *Traditions in Contact and Change. Selected Proceedings of the XIVth Congress of the International Association for the History of Religions,* ed. Peter Slater and Donald Wiebe, 348-359. Waterloo, Ont., 1983.

——. "L'Église et la divination au moyen âge, ou les avatars d'une pastorale ambiguë." *Théologiques* 8 (2000): 37-66.

——. "Pour l'étude de la religion populaire au moyen âge. Le problème des sources." In *Foi populaire, foi savante. Actes du Ve Colloque du Centre d'études d'histoire des religions populaires tenu au Collège dominicain de théologie (Ottawa),* 93-148. Paris, 1976.

——. "Un Franc parlant le syriaque et le grec, ou les astuces de Jérôme hagiographe." *Memini* 3 (1999): 127-153.

Les bois sacrés. Actes du Colloque International organisé par le Centre Jean Bérard de l'École Pratique des Hautes Études (Ve section), Naples, 23-25 Novembre 1989. Naples, 1993.

Bologne, Jean-Claude. *La naissance interdite. Stérilité, avortement, conception au Moyen-Âge.* Paris, 1988.

Bonnassie, Pierre. "Consommation d'aliments immondes et cannibalisme de survie dans l'occident du haut moyen âge." *Annales* 44 (1989): 1035-1056.

Bonser, Wilfrid. "Animal skins in magic and medicine." *Folklore* 73 (1962): 128-129.

——. *The Medical Background of Anglo-Saxon England. A Study in History, Psychology and Folklore.* London, 1963.

Borsje, Jacqueline. "Fate in early Irish texts." *Peritia* 16 (2002): 214-231.

Boshof, Egon. *Erzbischof Agobard von Lyon; Leben und Werk.* Vienna, 1969.

Boudriot, Wilhelm. *Die altgermanische Religion in der amtlichen kirchlichen Literatur des Abendlandes vom 5. bis 11. Jahrhundert.* [1928] Darmstadt, 1964.

Bouhot, Jean-Paul. "Le baptême et sa signification." In *Segni e riti,* 251-267.

Boulc'h, Stéphane. "Le repas quotidien des moines occidentaux du haut moyen âge." *Revue Belge de philologie et d'histoire* 75 (1997): 287-328.

Boulc'h, Stéphane. "Le statut de l'animal carnivore et la notion de pureté dans les prescriptions alimentaire chrétiennes du haut moyen âge occidental." In *Le statut éthique de l'animal. Conceptions anciennes et nouvelles,* ed. Liliane Bodson, 41-59 Liège, 1996.

Boulouis, Agnès. "Références pour la conversion du monde païen aux VIIe et VIIIe siècles. Augustin d'Hippone, Césaire d'Arles, Grégoire le Grand." *REAug* 33 (1987): 90-112.

Boyer, Régis and Eveline Lot-Falck. *Les religions de l'Europe du Nord.* Paris, c. 1974.

Boyer, Régis. "Le sacré chez les Germains et les Scandinaves." In *L'Homme indo-européen et le sacré,* 185-213.

Breton, Hugues. *Médecine traditionelle et magique en milieu rural.* Chateauguay, 1987.

Brett, Martin. "Theodore and the Latin canon law." In *Archbishop Theodore,* 120-140.

Brillant, M. *Problèmes de la danse.* Paris, 1953.

Brommer, Peter. *"Capitula Episcoporum": Die bischöflichen Kapitularien des 9. und 10. Jahrhunderts.* Typologie des sources du moyen âge occidental, fasc. 43. Turnhout, 1985.

Browe, Peter. "Die Eucharistie als Zaubermittel im Mittelalter." *Archiv für Kulturgeschichte* 20 (1930): 134-154.

——."Die Sterbekommunion in Altertum und Mittelalter." *Zeitschrift für katholische Theologie* 60 (1936): 1-54, 211-240.

Brown, Peter L. *The Cult of the Saints. Its Rise and Function in Latin Christianity.* Chicago, 1981.

Brunaux, Jean-Louis. "Les bois sacrés des Celtes et des Germains." In *Les bois sacrés,* 57-65.

Brundage, James. A. *Medieval Canon Law.* London and New York, 1995.

——. "The married man's dilemma: Sexual mores, canon law and marital restraint." In *Life, Law and Letters,*149-169.

Buckley, Thomas and Alma Gottlieb, "A critical appraisal of theories of menstrual symbolism." In *Blood Magic. The Anthropology of Menstruation,* 3-50. Berkeley, Los Angeles and London, 1988.

Budge, E. A. W. *Amulets and Talismans.* [1961] New York, 1968.

Bullough, Donald. "Burial, community and belief in the early medieval West." In *Idea and Reality in Frankish and Anglo-Saxon Society*, ed. Patrick Wormwald, 177-201. Oxford, 1981.

Busoni, Giovanni. "La microhistoire de Carlo Ginzburg." *Bibliothèque d'Humanisme et Renaissance* 67 (1999): 763-778.

Cabaniss, Allen. *Agobard of Lyons: Churchman and Critic*. Syracuse, N. Y., 1953.

Cahen, Maurice. *Études sur le vocabulaire religieux du vieux-scandinave. La libation*. Paris, 1921.

Caillois, Roger. "Les démons de midi." *Revue de l'histoire des religions* 115 (1937): 141-173 and 116 (1937): 142-186.

———. *Les jeux et les hommes*. Paris, 1958.

———. *L'homme et le sacré*. Paris, 1950.

Cameron, M. L. "Anglo-Saxon medicine and magic." *Anglo-Saxon England* 17 (1988): 191-215.

Campanile, Enrico. "Aspects du sacré dans la vie personnelle et sociale des Celts." In *L'homme indo-européen et le sacré*, 155-182.

Campbell, James. "Observations on the conversion of England." In *Essays in Anglo-Saxon History*, 69-84. London, 1986.

Camporesi, Piero. *Bread of Dreams. Food and Fantasy in Early Modern Europe*. Trans. David Gentilcore. Chicago and Cambridge, 1980.

Cardini, Franco. "Le piante magiche." In *L'ambiente vegetale nell'alto medioevo*, 623-658. SSAM 37. Spoleto, 1990.

Cazeneuve, Jean. *Sociologie du rite: Tabou, magie, sacré*. Paris, 1971.

Chadwick, Henry. *Priscillian of Avila. The Occult and the Charismatic in the Early Church*. Oxford, 1976.

Chailly, Jacques. "La danse religieuse au moyen âge." In *Arts libéraux et philosophie*, 357-380. Paris and Montreal, 1969.

Chambers, E. K. *The Medieval Stage*. 2 vols. Oxford, 1903.

Charles-Edwards, Thomas. "The Penitential of Theodore and the *Iudicia Theodori*." In *Archbishop Theodore*, 141-174.

Charon, Véronique. "The knowledge of herbs." In *The Pagan Middle Ages*, 109-128.

Cicogagni, A. G. *Canon Law*. Trans. J. M. O'Hara and F. Brennan. 2nd revised ed., Philadelphia, 1935.

Clebsch, William A. and Charles R. Jaeckle. *Pastoral Care in Historical Perspective. An Essay with Exhibits*. [1964] New York, 1967.

Coli, E. "Osservazioni sull'uso del diminutivo in Cesario d'Arles." *Giornale Italiano di Filologia*, n.s. 12 (33) (1981): 117-133.

Collard, Franck. "*Veneficiis vel maleficiis*. Réflexion sur les relations entre le crime de poison et la sorcellerie dans l'Occident médiéval." *Le moyen âge* 109 (2003): 9-58.

Collin, Hubert. "Grégoire de Tours, Saint Walfroy le Stylite et la 'Dea Arduanna". Un épisode de la christianisation des confins des diocèses de Reims et de Trêves au VIe siécle." In *La piété populaire au moyen âge*, 387-400.

The Community, the Family and the Saint. Patterns of Power in Early Medieval Europe. Ed. Hill, Joyce and Mary Swan. Turnhout, 1998.

Concilio III de Toledo. XIV centenario. Toledo, 1991.

Congar, Yves. *L'ecclésiologie du haut moyen âge.* Paris, 1968.

Connolly, Hugh. *The Irish Penitentials and Their Significance for the Sacrament of Penance Today.* Dublin, 1995.

Courcelle, P. de "L'enfant et les sorts bibliques." *Vigiliae Christiania* 7 (1953): 194-220.

La conversione al cristianesimo nell'Europe dell'alto medioevo. SSAM 14. Spoleto, 1967.

Crawford, Sally. "Children, death and the afterlife in Anglo-Saxon England." *Anglo-Saxon Studies in Archaeology and History* 6 (1993): 83-91.

Cristianizzazione ed organizzazione ecclesiastica delle campagne nell'alto medioevo: espansione e resistenze. SSAM 28. Spoleto, 1982.

Cubitt, Catherine. *Anglo-Saxon Councils c. 650-c. 850.* London and Cranbury, N.J., 1995.

Cultural Identity and Cultural Integration: Ireland and Europe in the Early Middle Ages. Ed. Doris Edel. Dublin, 1995.

Czarnowski, S. "Le morcellement de l'étendue et sa limitation dans la religion et la magie." In *Actes du congrès international d'histoire des religions tenu à Paris en octobre 1923,* 339-359. Paris, 1925.

Dabrowska, Elzbieta. "'Communio mortuorum'. Un usage liturgique ou une superstition." *Archiv für Liturgiewissenschaft* 31 (1989): 342-346.

D'Alès, Adhémar. *Priscillien et l'Espagne chrétienne à la fin du IVe siècle.* Paris, 1936.

Daly, William M. "Clovis: How barbaric, how pagan?" *Speculum* 69 (1994): 619-664.

Daniélou, Jean. "Les douze apôtres et le zodiaque." *Vigiliae Christianae* 13 (1959): 14-21.

Dauzet, Dominique-Marie. *Saint Martin de Tours.* Paris, 1996.

Davidson, Hilda R. E. "Shape-changing in the Old Norse Sagas." In *Animals in Folklore,* 126-142.

De Bruyne, D. "L'origine de la Chandeleur et des Rogations." *RB* 34 (1922): 14-26.

De Clercq, Carlo. *La législation religieuse franque de Clovis à Charlemagne. Étude sur les actes de conciles et les capitulaires, les statuts diocésains et les règles monastiques (507-814).* Louvain, 1936-1958.

De Felice, P. *L'enchantement des danses et la magie du verbe. Essai sur quelques formes inférieures de la mystique.* Paris, 1957.

De Gaiffier, Baoudouin. "Hagiographie et historiographie. Quelques aspects du problème. In *La storiografia altomedievale,* 139-166 and 179-195. SSAM 17. Spoleto, 1970.

De Martino, Ernesto. *Morte e pianto rituale. Dal lamento funebre antico al pianto di Maria.* [1958] Turin, 1975.

De Nie, Giselle. "Caesarius of Arles and Gregory of Tours. Two sixth-century Gallic bishops and 'Christian Magic'." In *Cultural Identity and Cultural Integration,* 170-197.

De Reu, Martine. "The missionaries: The first contacts between paganism and Christianity." In *The Pagan Middle Ages,* 13-38.

Delage, Marie-Josée. "Introduction." In *Césaire d'Arles, Sermons au peuple.* 13-175. *SC* 175. Paris, 1971.

Delbono, Francesco. "La letteratura catechetica di lingua tedesca. Il problema della lingua nell'evangelizzazione." In *La conversione al cristianesimo,* 697-741.

Delehaye, Hippolyte. *Les légendes hagiographiques.* Brussels, 1955.

Delooz, Pierre. "Toward a sociological study of canonized sainthood." In *Saints and Their Cults,* ed. Stuart Wilson, 189-216. Cambridge and New York, 1983.

Demyttenaere, Albert. "The cleric, women and the stain. Some beliefs and ritual practices concerning women in the early Middle Ages." In *Frauen in Spätantike und Frühmittelalter*, 141-166.

Derolez, R. L. M. "La divination chez les Germains." In *La divination,* 257-302.

Devailly, Guy. "La pastorale en Gaule au IXe siècle." *RHÉF* 59 (1973): 23-54.

Dickensen, Tania M. and George Speake. "The seventh-century cremation burial in Asthall Barrow, Oxfordshire. A reassessment." In *The Age of Sutton Hoo*, 94-130.

Dickie, Matthew W. "The Fathers of the Church and the evil eye." In *Byzantine Magic*, Henry Maguire, 9-34. Dumbarton Oaks, 1995.

Dienst, H. R. "Zur Rolle von Frauen in magischen Vorstellungen und Praktiken – nach ausgewählten Quellen." In *Frauen in Spätantiken und Frühmittelaltern,* 173-194.

Dierkens, Alain. "Cimetières mérovingiens et histoire du haut moyen âge." In *Histoire et méthode*, 15-70. Acta Bruxellensia IV, Brussels, 1981.

———. "Pour une typologie des missions carolingiennes." In *Propagande et contre-propagande religieuse*, ed. J. Marx, 77-93. Brussels, 1987.

———. "Quelques aspects de la christianisation du pays mosan à l'époque mérovingienne." In *La civilisation mérovingienne dans le bassin mosan,* ed, M. Otte and J. Willems, 29-83. Liège, 1986.

———. "Superstitions, christianisme et paganisme à la fin de l'époque mérovingienne." In *Magie, sorcellerie, parapsychologie*, ed. Hervé Hasquin, 9-26. Brussels, 1984.

———. "The evidence of archaeology." In *The Pagan Middle Ages,* 39-64.

Dieterich, Albrecht. *Mutter Erde. Ein Versuch uber Volksreligion.* Reprint of the 1925 ed. Stuttgart, 1967.

La divination. 2 vols. Ed. Caquot, A. and M. Leibovici. Paris, 1968.

Dowden, Ken. *European Paganism; The Realities of Cult from Antiquity to the Middle Ages.* London and New York, 1990.

Dubarle, A.-M. "La contraception chez saint Césaire d'Arles." *La Vie Spirituelle, Supplément*, 17 (1963): 515-519.

Dukes, Eugene D. *Magic and Witchcraft in the Dark Ages.* Lanham, Md., and London, 1992.

Duchesne, Annie. *Le livre des merveilles; Divertissement pour un empereur.* Paris, 1992.

Dumville, David N. "On the dating of the early Breton lawcodes." *Études celtiques* 21 (1984): 207-221.

Dutton, Paul Edward. *Carolingian Civilisation. A Reader.* Reprint. Peterborough, 1999.

———. "Thunder and hail over the Carolingian Countryside." In *Agriculture in the Middle Ages. Technology, Practice, and Representation*, ed. Del Sweeney, 111-139. Philadelphia, n.d.

Duval, Yvette. *Auprès des saints corps et âme. L'inhumation 'ad sanctos' dans la chrétienté d'Orient et d'Occident du IIIe au VIIe siècle.* Paris, 1988.

———. "Sanctorum sepulcris sociari." in *Les fonctions des saints dans le monde occidental (IIIe-XIIIe siècle). Actes du colloque organisé par l'École française de Rome avec le concours de l'Université de Rome "La Sapienza", Rome 27-29 octobre 1988,* 334-351. Rome, 1991.

Edwards, Cyril. "German vernacular literature. A survey." In *Carolingian culture. Emulation and Innovation*, ed. Rosamond McKitterick, 141-170. Cambridge, 1994.

Effros, Bonnie. "*De partibus Saxoniae* and the regulation of mortuary custom. A Carolingian Campaign of Christianization or the suppression of Saxon identity." *Revue Belge de philologie et d'histoire* 75 (1997): 267-286.

Elukin, Jonathan M. "The ordeal of scripture." *Exemplaria* 5 (1993): 135-160.

Emboden, William A. "Ritual use of cannabis sativa: A historical-ethnographic study." In *Flesh of the Gods*, 214-236.

Étaix, Raymond. "Le recueil de sermons composé par Raban Maur pour Haistulfe de Mayence." *REAug* 32 (1986): 124-137.

Ewig, Eugen. "Le culte de Saint Martin à l'époque franque." In *Mémorial de l'année martinienne M.DCCCC.LX-M.DCCCC.LXI*, 1-18. Paris, 1962.

——. "The missionary work of the Latin Church." In *History of the Church* 2, 517-601.

Ezzy, Thomas. *Daemon Amoris. Amatory Magic vs. Thaumaturgy in Late Antique Hagiography*. M.A. thesis, Université de Montréal. Montreal, 1996.

Fehring, Günter P. "Missions- und Kirchenwesen in archäologischer Sicht." In *Geschichtswissenschaft und Archäologie: Untersuchungen zur Siedlungs-, Wirtschafts- und Kirchengeschichte*, ed. Herbert Jankuhn and Reinhard Wenskus,547-591. Sigmaringen, 1979.

Felten, Franz J. "Konzilakten als Quellen für die Gesellschaftsgeschichte des 9. Jahrhunderts." In *Herrschaft, Kirche, Kultur. Beiträge zur Geschichte des Mittelalters*, ed. Georg Jenal and Stephanie Haarländer, 177-201. Stuttgart, 1993.

Ferreiro, Alberto. "St. Martin of Braga's policy toward heretics and pagan practices." *American Benedictine Review* 34 (1983): 372-395.

Fevrier, P.-A. "La tombe chrétien et l'Au-delà." In *Le temps chrétien de la fin de l'Antiquité au moyen âge IIIe-XIIIe siècles, Paris 9-12 mars 1981*, 163-183. Paris, 1984.

Filotas, Bernadette. *Pagan Beliefs and Practices in the Sermons of Caesarius of Arles*. M.A. Thesis, Université de Montréal. Montreal, 1994.

Fine, Agnès. *Parrains, marraines. La parenté spirituelle en Europe*. Paris, 1994.

Firth, Raymond. "Reason and unreason in human belief." In *Witchcraft and Sorcery*, 38-40.

Flesh of the Gods. The Ritual Use of Hallucinogens. Ed. Peter T. Furst. New York and Washington, 1972.

Fletcher, Richard. *The Conversion of Europe; From Paganism to Christianity (371-1386 AD)*. London, 1997.

Flint, Valerie. *The Rise of Magic in Early Medieval Europe*. Princeton, 1990.

Foatelli, R. *Les danses religieuses dans le christianisme*. Paris, 1939.

Fontaine, Jacques and Jocelyn N. Hillgarth, eds. *Le septième siècle. Changements et continuités / The Seventh Century: Changes and Continuity. Actes du Colloque bilatéral franco-britannique tenu au Warburg Institut les 8-9 juillet 1988*, 154-72. London, 1992.

Fournier, P. "Le Décret de Burchard de Worms." *RHE* 12 (1911): 451-473, 67-701.

——. and G. Le Bras. *Histoire des collections canoniques en Occident depuis les Fausses Décrétales jusqu'au Décret de Gratien*. 2 vols. Paris, 1931-1932.

Fransen, Gérard. *Les collections canoniques*. Typologie des sources du moyen âge occidental, fasc. 10. Turnhout, 1973.

Frantzen, A. *The Literature of Penance in Anglo-Saxon England*. New Brunswick, N. J., 1983.

——. "The significance of the Frankish Penitentials." *JEH* 30 (1979): 409-421.

———. "The tradition of penitentials in Anglo-Saxon England." *Anglo-Saxon England* 11 (1983): 23-56.

Franz, Adolph. *Die kirchlichen Benediktionen im Mittelalter.* 2 vols. [1901] Graz, 1960.

Frauen in Spätantike und Frühmittelalter: Lebensbedingungen – Lebensnormen – Lebensformen. Beiträge zu einer internationalen Tagung am Fachbereich Geschichtswissenschaften der Freien Universität Berlin 18. bis 21. Februar 1987, ed. Affeldt, Werner. Sigmaringen, 1990.

Fredouille, Jean-Claude. "Le héros et le saint." In *Du héros païen au saint chrétien*, ed. Gérard Freyburger and Laurent Pernot, 11-15. Paris, 1997.

Freeman, Anne. "Theodulf of Orleans." *Speculum* 32 (1957): 663-705.

Gaignebet, C. and J. D. Lajoux. *Art profane et religion populaire au moyen âge.* Paris, 1985.

Gandino, Germana. "Cultura dotta e cultura folklorica a Vercelli nel X secolo." *Bollettino storico-bibliografico subalpino* (1992): 253-279.

Ganshof, François-Louis. "Notes sur les 'Capitula de causis cum episcopis et abbatibus tractandis' de 811." *Studia Gratiana* 13 (1967): 1-25.

———. *Recherches sur les capitulaires.* Paris, 1958.

Gasparri, Stefano. *La cultura tradizionale dei Longobardi; Struttura tribale e resistenze pagane.* Spoleto, 1963.

Gatch, Milton McC. *Preaching and Theology in Anglo-Saxon England. Aelfric and Wulfstan.* Toronto and Buffalo, 1977.

Gatti, Daniela. "Curatrici e streghe nell'Europa dell'alto Medioevo." In *Donne e lavoro nell'Italia medievale*, ed. Maria Giuseppina Muzzarelli, Paolo Galetti and Bruno Andreoli, 127-140. Turin, 1991.

Gaudemet, Jean. *L'Église dans l'empire romain.* Paris, 1958.

———."Le pseudo-concile de Nantes." *Revue de droit canonique* 25 (1975): 40-60.

———. *Les sources du droit canonique, VIIIe – XXe siècles.* Paris, 1993.

———. *Les sources du droit de l'Église en Occident du IIe au VIIe siècle.* Paris, 1985.

———. "Les statuts épiscopaux de la première décade du IXe siècle." In *La Formation du droit canonique médiéval.* XIII, Variorum. London, 1980. Originally in *Proceedings of the Fourth International Congress of Medieval Canon Law, Toronto, 1972. Monumenta Iuris Canonici, Series C: Subsidia 5*, 303-349. Vatican City, 1976.

———. "Traduttore, traditore' – Les *Capitula Martini.*" In *Droit de l'Église et vie sociale au moyen âge.* VI, Variorum. Northampton, 1989. Originally in *Fälschungen im Mittelalter: Internationaler Kongress der Monumenta Germaniae Historica, München, 1986. Teil 1. Gefälschte Rechttexte, Der Bestrafte Fälscher,* 51-69. Hanover, 1988.

Gauthier, Nancy. *L'évangélisation des pays de la Moselle. La province romaine de Première Belgique entre Antiquité et Moyen-Âge (IIIe-VIIIe siècles).* Paris, 1980.

Geake, Helen. "Burial practice in seventh- and eighth-century England." In *The Age of Sutton Hoo*, 83-94.

Geary, Patrick. "Barbarians and ethnicity." In G. W. Bowersock, Peter Brown and Oleg Grabar, eds., *Late Antiquity. A Guide to the Postclassical World*, 107-129. Cambridge, Mass., and London, 1999.

———. "Exchange and interaction between the living and the dead in early medieval society." In *Living with the Dead in the Middle Ages*, 77-92.

———. *Living with the Dead in the Middle Ages.* Ithaca, N. Y., and London, 1994.

———. "The uses of archaeological sources for religious and cultural history." In *Living with the Dead in the Middle Ages*, 30-45.

Gérold, Th. *Les pères de l'Église et la musique*. Paris, 1932.

Ginzburg, Carlo. *Clues, Myths, and the Historical Method*. Trans. John and Anne Tedeschi. Baltimore, 1992.

——. *Ecstasies. Deciphering the Witches' Sabbath*. Trans. Raymond Rosenthal. New York, 1991.

——. *The Night Battles. Witchcraft and Agrarian Cults in the Sixteenth and Seventeenth Centuries*. Trans. John and Anne Tedeschi. Baltimore, 1992.

Giordano, Oronzo. "Igiene personale e salute nell'alto medioevo." *Nuovi Annali della Facoltà di Magistero dell'Università di Messina* 4 (1986): 183-233.

——. *Religiosità popolare nell'alto medioevo*. Bari, 1979.

Golinelli, Paolo. "La religiosità popolare tra antrolopogia e storia." In *La Terra e il sacro*, 1-9.

Gooch, Peter D. *Dangerous Food. I Corinthians 8-10 in Its Context*. Waterloo, Ont., 1993.

Gougaud, L. "La danse dans l'église." *RHE* 15 (1914): 5-22 and 229-245.

Green, Miranda. *Dictionary of Celtic Myth and Legend*. London, 1992.

Gregory, Timothy. "The survival of paganism in Christian Greece." *Journal of American Philology* 107 (1986): 229-242.

Grimm, Jacob. *Teutonic Mythology*. 4 vols. [1835] Trans. from the 4th ed. (1887) J. S. Stallybrass. New York, 1966.

Grodzynski, Denise. "Superstitio." *Revue des études anciennes* 76 (1974): 36-60.

Gross, Karl. *Menschenhand und Gotteshand in Antike und Christentum*. Stuttgart, 1985.

Guillot, Olivier. "La conversion des Normands peu après 911. Des reflets contemporains à l'historiographie ultérieure (Xe-XIe s.)." *Cahiers de civilisation médiévale* 24 (1981): 101-116 and 181-219.

Gurevich, Aron. *Historical Anthropology of the Middle Ages*. Ed. Jana Howlett. Chicago and Cambridge, 1992.

——. *Medieval Popular Culture: Problems of Belief and Perception*. Trans. J. M. Bak and P. A. Hollingsworth. Cambridge, 1990.

Haderlein, Konrad. "Celtic roots: Vernacular terminology and pagan ritual in Carlomann's Draft Capitulary of A. D. 743, Codex Vat. Pal. lat. 577." *Canadian Journal of Irish Studies* 18 (1992): 1-29.

Hägele, Günter. *Das Paenitentiale Vallicellianum I. Ein oberitalienischer Zweig der frühmittelalterlichen kontinentalen Bußbücher: Überlieferung, Verbreitung und Quelle*. Sigmaringen, 1984.

Halbertal, Moshe and Avishai Margalit. *Idolatry*. Trans. Naomi Goldblum. Cambridge and London, 1992.

Hall, David. "Introduction." In *Understanding Popular Culture. Europe from the Middle Ages to the Nineteenth Century*, ed. Steven L. Kaplan, 5-18. Berlin, New York, Amsterdam, 1984.

Halsall, Guy. "Burial, ritual and Merovingian society." In *The Community, the Family and the Saint*, 325-338.

Hand, Wayland D. *Magical Medicine. The Folkloric Component of Medicine in the Folk Belief, Custom and Ritual of the Peoples of Europe and America*. Berkeley, Los Angeles and London, 1980.

Harf-Lancner, Laurence. *Les fées au moyen âge. Morgane et Mélusine. La naissance des fées.* Geneva, 1984.

Harmening, Dieter. "Anthropologie historique ou herméneutique littéraire? Une critique ethnographique des sources médiévales." *Ethnologie française* 27 (1997): 445-456.

——. "'Contra Paganos' – 'Gegen die vom Dorfe'? Zum theologischen Hintergrund ethnologischer Begriffe." *Jahrbuch für Volkskunde* 19 (1996): 487-508.

——. *Superstitio: Überlieferungs- und theoriegeschichtliche Untersuchungen zur kirchlich-theologischen Aberglaubensliteratur des Mittelalters.* Berlin, 1979.

Harner, Michael J. "The role of hallucinogenic plants in European witchcraft." In *Hallucinogens and Shamanism*, ed. Michael J. Harner, 125-150. New York, 1973.

Hartmann, Wilfried. "Synodes carolingiennes et textes synodaux du IXe siècle." *Francia* 12 (1984): 534-541.

Hartnung, Wolfgang. "Die Magie des Geschriebenen." In *Schriftlichkeit im frühen Mittelalter,* ed. Ursula Schaefer, 109-126. Tübingen, 1993.

Hauck, Karl. "Rituelle Speisegemeinschaft im 10. und 11. Jahrhundert." *Studium Generale* 3 (1950): 611-623.

Head, Thomas. *Hagiography and the Cult of Saints. The Diocese of Orleans, 800-1200.* Cambridge, 1990.

Heckenbach, M. J. *De nuditate sacra sacrisque vinculis. Religionsgeschichtliche Versuche und Vorarbeiten* 9: 1911.

Hefele, K. J. and H. Leclercq. *Histoire des conciles.* 8 vols. Paris, 1907-1921.

Heiler, Friedrich. *Prayer. A Study in the History and Psychology of Religion.* Ed. and trans., Samuel McComb with J. Edgar Park. [1932] New York, 1958.

Heim, François. "Solstice d'hiver, solstice d'été dans la prédication chrétienne du Ve siècle. Le dialogue des évêques avec le paganisme de Zénon de Vérone è saint Léon." *Latomus* 58 (1999): 640-660.

Hen, Yitzhak. *Culture and Religion in Merovingian Gaul A.D. 481-751.* Leiden, New York, Cologne, 1995.

——. "Knowledge of canon law among rural priests. The evidence of two Carolingian manuscripts from around 800." *Journal of Theological Studies* 50 (1999): 117-134.

——. "Paganism and superstition in the time of Gregory of Tours: 'une question mal posée. " In *The World of Gregory of Tours,* 229-240, ed. Ian Wood et Kathleen Mitchell. Leiden, 2002.

Heresies of the High Middle Ages. Selected and trans. Walter L. Wakefield and Austin P. Evans. New York, 1961.

Hertz, Robert. "A contribution to the study of the collective representation of Death." In *Death and the Right Hand.* Trans. R. Needham and C. Needham. New York, 1960.

——. "The pre-eminence of the right hand: A study in religious polarity." In *Right and Left: Essays on Dual Symbolic Classification*, trans. and ed. Rodney Needham, 3-31. Chicago and London, 1973.

Hill, Thomas D. The *æcerbot* charm and its Christian users." *Anglo-Saxon England* 6 (1977): 213-221.

Hillgarth, Jocelyn N. "Modes of evangelization of Western Europe in the seventh century." In *Irland und Christenheit,* 311-331.

———. "Popular religion in Visigothic Spain." In *Visigothic Spain,* 3-60.

Historia Religionum. 1. *Religions of the Past.* Ed. C. J. Bleeker, and Geo Widengren. Leiden, 1969.

History of Religions. Proceedings of the Thirteenth Congress of the International Association for the History of Religions. Ed. Michael Pye and Rita McKenzie. Lancaster, 1975.

Hodgen, Margaret. *The Doctrine of Survivals: A Chapter in the History of Scientific Method in the Study of Man.* [1936] Folcroft, Pa., 1977.

Hoffman, Hartmut and Rudolf Pogorny. *Das Dekret des Bischofs Burchard von Worms. Textstufen – Frühe Verbreitung – Vorlagen.* Munich, 1991.

Holleman, A. W. J. *Pope Gelasius I and the Lupercalia.* Amsterdam, 1974.

Homann, Holger. *Der Indiculus superstitionum et paganiarum und verwandte Denkmäler.* Dissertation. Göttingen, 1965.

Hopper, Vincent Foster. *Medieval Number Symbolism. Its Sources, Meaning and Influence on Thought and Experience.* New York, 1938.

Hübener, Wolfgang. "Waffen und Bewaffnung." In *Sachsen und Angelsachsen*, ed. Claus Ahrens, 463-471. Hamburg, 1978.

Hughes, Kathleen. *Early Christian Ireland. Introduction to the Sources.* London, 1972.

———. *The Church in Early Irish Society.* London, 1966.

Hutton, Ronald. *The Pagan Religions of the Ancient British Islands.* Oxford and Cambridge, 1991.

Imbert, Jean. "L'ébriété dans les pénitentiels." In *Life, Law and Letters*, 475-487.

———. *Les temps carolingiens. 1. L'Église: Les institutions.* Paris, 1994.

Irland und die Christenheit. Ed. Proinséas Ni Chathain and Michael Richter. Stuttgart, 1987.

James, Edward O. ed. *Visigothic Spain: New Approaches.* Oxford, 1980.

———. *Seasonal Feasts and Festivals.* London, 1961.

———. *The Origins of France: From Clovis to the Capetians, 500-1000.* London and Basingstoke, 1982.

Janssen, L. F. "'Superstitio' and the persecution of the Christians." *Vigiliae Christianae* 33 (1979) 131-159.

Jarnut, Jörg. "Nomen et gens. Political and linguistic aspects of personal names between the third and the eighth century. Presenting an interdisciplinary project from a historical perspective." In *Strategies of Distinction,* 113-116.

Jecker, Gall. *Die Heimat des hl. Pirmin, des Apostels der Alamannen.* Münster in Westphalia, 1927.

Johnson, David. "Euhemerisation versus demonisation. The pagan gods and Aelfric's *De falsis diis.*" In *Pagans and Christians*, 35-62.

Jolly, Karen Louise. "Anglo-Saxon charms in the context of a Christian world view." *Journal of Medieval History* 11 (1985): 279-293.

———. ."Magic, miracle and popular practice in the early medieval West. Anglo-Saxon England." In *Religion, Science and Magic in Concert and in Conflict*, ed. J. Neusner, E. S. Frerichs and P. V. McC. Flesher, 166-182. New York and Oxford, 1989.

———. *Popular Religion in Late Saxon England: Elf Charms in Context.* Chapel Hill and London, 1996.

Jones, W. R. "The heavenly letter in medieval England." In *Medieval Hagiography and Romance. Medievalia et Humanistica*, 163-178. Cambridge, 1975.

Jungmann, J. A. *La liturgie des premiers siècles jusqu'à l'époque de Grégoire le Grand.* Paris, 1960.

Karras, Ruth Mazo. "Pagan survivals and syncretism in the conversion of Saxony." *The Catholic Historical Review* 72 (1986): 553-572.

Kearney, Milo. *The Role of Swine Symbolism in Medieval Culture.* Lewiston, N.Y., 1991.

Keefer, Sarah Larratt. "*Ut in omnibus honorificetur Deus*: The *Corsnæd* Ordeal in Anglo-Saxon England." In *The Community, the Family and the Saint*, 237-264.

Kenney, James F. *The Sources for the Early History of Ireland: Ecclesiastical. An Introduction and Guide.* [1968] Dublin, 1993.

Kieckhefer, Richard. "Erotic magic in medieval Europe." In *Sex in the Middle Ages*, 30-55.

——. *Forbidden Rites. A Necromancer's Manual of the Fifteenth Century.* University Park, Pa., 1997.

——. "The specific rationality of medieval magic." *The American Historical Review* 99 (1994): 813-36.

Kirchgässner, Alfons. *La puissance des signes. Origines, formes et lois de culte.* Trans. Sr. Pierre-Marie. Rev. M. A. Barth. Paris, 1962.

——. *Les signes sacrés de l'Église.* Adapted from the German by the monks of Mont César. 1964.

Klingshirn, William E. *Caesarius of Arles. The Making of a Christian Community in Late Antique Gaul.* Cambridge, 1994..

Koeniger, Albert M. *Burchard I. von Worms und die deutsche Kirche seiner Zeit.* Munich, 1905.

Konda, G. *Le discernement et la malice des pratiques superstitieuses d'après les sermons de S. Césaire d'Arles.* Rome, 1970.

Konen, Wilhelm. *Die Heidenpredigt in der Germanenbekehrung.* Dissertation. Düsseldorf, 1909.

Körntgen, Ludger. *Studien zu den Quellen der frühmittelalterlichen Bußbücher.* Sigmaringen, 1993.

Kottje, Raymund. *Die Bußbücher Halitgars von Cambrai und des Hrabanus Maurus. Ihre Überlieferung und ihre Quellen.* Berlin and New York, 1980

——. ."Überlieferung und Rezeption der irischen Bußbücher auf dem Kontinent." In *Die Iren und Europa*, 511-24.

Kuhn, Hans. "Das Fortleben des germanischen Heidentums nach der Christianisierung." In *La conversione al cristianesimo*, 743-757.

Künzel, Rudi. "Paganisme, syncrétisme et culture religieuse au haut moyen âge. Réflexions de méthode." *Annales ESC* 4-5 (1992): 1055-1069.

Labriolle, Pierre. "L'enfant et les 'sorts bibliques'." *Vigiliae Christianae* 7 (1953): 194-220.

——. "Source chrétienne et allusions païennes de l'épisode du 'Tolle, lege'." *Revue d'Histoire et de Philosophie Religieuse* 32 (1952): 170-200.

Laistner, M. L. W. "The western Church and astrology during the early middle ages." *Harvard Theological Review* 34 (1941): 251-275.

Laliberté, Micheline. "Religion populaire et superstition au moyen âge." *Théologique* 8 (2000): 19-36.

Lancellotti, Maria Grazia. "Médecine et religion dans les gemmes magiques." *Revue de l'histoire des religions* 218 (2001): 427-457.

Lanzoni, Francesco. "Il sogno presago della madre incinta nella letteratura medievale e antica." *Analecta Bollandiana* 45 (1927): 225-261.

Lapidge, Michael and Richard Sharpe. *A Bibliography of Celtic-Latin Literature, 400-1200.* Dublin, 1985.

Lapidge, Michael. "The career of Archbishop Theodore." In *Archbishop Theodore,* 1-29.

Laporte, Jean. *Le pénitentiel de saint Colomban. Introduction et édition critique.* Tournai and Paris, 1958.

Laporte, Jean-Pierre. "La reine Bathilde, ou l'accession sociale d'une esclave." In *La femme au moyen âge,* ed. Michel Rouche, 147-167. Maubeuge, c. 1990.

Lauwers, Michel. "Le cimetière dans le moyen âge latin. Lieu sacré, saint et religieux." *Annales HSS* 54 (1999): 1047-1072.

——. "Religion populaire." In *Catholicisme.* 15 vols. Ed. G. Mathon, G.-H. Baudry and E. Thiery, 12: 835-849. Paris, 1948-1999.

——. "Religion populaire', culture folklorique, mentalités. Notes pour une anthropologie culturelle du moyen âge." *Revue d'Histoire Ecclésiastique* 82 (1987): 221-258.

Laux, J. "Two early medieval heretics: An episode in the life of St. Boniface." *Catholic Historical Review* 21 (1935): 190-195.

Lavarra, Caterina. "Il sacro cristiano nella Gallia merovingia tra folklore e medicina professionale." *Annali della Facoltà di Lettere e Filosofia*, Bari (1989): 149-204.

——. "'Pseudochristi' e 'pseudoprophetae' nella Gallia merovingia." *Quaderni medievali* 13 (1982): 6-43.

Lazard, Sylvianne. "Tradition ancienne et influence chrétienne dans l'anthroponymie ravennate du Xe siècle." In *La piété populaire au moyen âge*, 445-454.

Leach, Edmund. "Magical Hair." *Journal of the Royal Anthropological Institute* 88 (1958), 147-164. Reprinted in *Myth and Cosmos. Readings in Mythology and Symbolism,* ed. John Middleton, 77-108. New York, 1967.

Lebecq, Stéphane. "Vulfran, Willibrord et la mission de Frisie: Pour une relecture de la 'Vita Vulframni'." In *L'évangélisation des regions entre Meuse et Moselle et la fondation de l'abbaye d'Echternach (Ve-IXe siècles)*, ed. Michel Polfer, 429-451. Luxembourg, 2000.

Leclercq, Jean. "Sermon ancien sur les danses déshônnetes." *RB* 59 (1949) 196-201.

Lecouteux, Claude. *Chasses fantastiques et cohortes de la nuit au moyen âge.* Paris, 1999.

——. *Démons et génies du terroir au moyen âge.* Paris, 1995.

——. *Les nains et les elfes au moyen âge.* Paris, 1988.

Lecoy de la Marche, Albert. *La chaire française au moyen âge.* 2nd ed. Paris, 1886.

Lefrancq, Paul. "Les données hagiographiques de la toponymie cadastrale de l'ancien diocèse d'Angoulême et du département de la Charente." In *La piété populaire au moyen âge*, 401-409.

Le Goff, Jacques. "Christianity and dreams (second to seventh century)." In *The Medieval Imagination*, trans. Arthur Goldhammer, 193-231. Chicago and London, 1988.

——. "Culture ecclésiastique et culture folklorique au moyen âge: saint Marcel de Paris et le dragon. " In *Pour un autre moyen âge,* 236-279.

——. "Culture cléricale et traditions folkloriques dans la civilisation mérovingienne." In *Pour un autre moyen âge*, 223-235.

——. *Pour un autre moyen âge. Temps, travail et culture en Occident. 18 essais.* Paris, 1977.

Lett, Didier. "De l'errance au deuil. Les enfants morts sans baptême et la naissance du limbus puerorum aux XIIe-XIIIe siècles." In *La Petite Enfance*, 77-92.

Levinson, Wilhelm. *England and the Continent in the Eighth Century*. [1946] Oxford, 1966.

Life, Law and Letters. Historical Studies in Honour of Antonio García y García, Ed. Peter Linehan. Rome, 1998.

Lifshitz, Felice. *The Norman Conquest of Pious Neustria*. Toronto, 1995.

Little, Lester K. *Benedictine Maledictions. Liturgical Cursing in Romanesque France.* Ithaca, N.Y., 1993.

Longère, Jean. *La prédication médiévale*. Paris, 1983.

López Calvo, Andrés. "La cataquesis en la Galicia medieval. Martín Dumiense y el *De correctione rusticorum*." *Estudios Mindonienses* 13 (1997): 509-523.

Löwe, H., ed. *Die Iren und Europa im früheren Mittelalter.* Stuttgart, 1982.

Lucy, Sam. "Children in early medieval cemeteries." *Archaeological Review from Cambridge* 13.2 (1994): 21-34.

Lutterbach, Hubertus. "Die Speisegesetzgebung in den mittelalterlichen Bussbüchern (600-1200). Relgionsgeschichtliche Perspektiven." *Archiv für Kulturgeschichte* 80 (1998): 1-37.

Lynch, Joseph H. *Godparents and Kinship in Early Medieval Europe*. Princeton, N. J., 1986.

——. "Spiritual kinship and sexual prohibitions in early medieval Europe." In *Proceedings of the Sixth International Congress of Medieval Canon Law*, ed. S. Kuttner and K. Pennington, 271-288. Monumenta iuris canonici. Series C: Subsidia 7. Vatican, 1985.

——. "*Spirituale vinculum:* The vocabulary of spiritual kinship in early medieval Europe." In *Religion, Culture and Society in the Middle Ages. Studies in Honour of Richard E. Sullivan*, ed. T. F. X. Noble and J. J. Contreni, 181-204. Kalamazoo, 1987.

Machielsen, Jan. *Clavis patristica pseudepigraphorum medii aevi* I. *Opera homiletica Pars A*. Turnhout, 1990.

Magnou-Nortier, Élisabeth. "La christianisation de la Gaule (VIe-VIIe siècles). Esquisse d'un bilan et orientation bibliographique." *Mélanges de science religieuse* 53 (1966): 5-12.

Magoun, Francis P. "On some survivals of pagan belief in Anglo-Saxon England." *Harvard Theological Review* 40 (1947): 33-46.

Maguire, Henry. "Garments pleasing to God. The significance of domestic textile designs in the early Byzantine period." *Dumbarton Oaks Papers* 44 (1990): 215-224.

Malnory, A. *Saint Césaire Évêque d'Arles.* Paris, 1894.

Manselli, Raoul. *La religion populaire au moyen âge. Problèmes de méthode et d'histoire.* Paris, Montreal, 1975.

——. "Resistenze dei culti antichi nella pratica religiosa dei laici delle campagne." In *Cristianizzazione ed organizzazione*, 57-108.

——. "Simbolismo e magia nell'alto medioevo." In *Simboli e simbologia*, 293-348.

Markus, Robert A. "From Caesarius to Boniface. Christianity and paganism in Gaul." In *Le septième siècle*, 154-172.

——. "Gregory the Great and a papal missionary strategy." In *The Mission of the Church*, 29-38.

——. *The End of Ancient Christianity*. Cambridge and New York, 1990.

——. Review of Flint, *Rise of Magic*. *English Historical Review* 107 (1992): 378-380.

Mary Donatus, Sister. *Birds and Beasts in the Lives of the Early Irish Saints*. Philadelphia, 1934.

Mauss, M. *Oeuvres*. Ed. V. Karady. Paris, 1968.

Mayr-Harting, Henry. *The Coming of Christianity to Anglo-Saxon England*. 3rd ed. London, 1991.

McKenna, Stephen. *Paganism and Pagan Survivals in Spain up to the Fall of the Visigothic Kingdom*. Washington, 1938.

McKitterick, Rosamond. *The Frankish Church and the Carolingian Reforms, 789-895*. London, 1977.

McLaughlin, R. Emmet. "The word eclipsed? Preaching in the early Middle Ages." *Traditio* 46 (1991): 77-122.

McNeill, John T. and Helena Gamer. *Medieval Handbooks of Penance. A Translation of the Principal 'Libri Poenitentiales' and Selections from Related Documents*. Records of Western Civilization. [Oxford and New York, 1938] Reprinted with new intro., New York, 1990.

Meaney, Audrey L. *Anglo-Saxon Amulets and Curing Stones*. Oxford, 1981.

Meens, Rob. "A background to Augustine's mission to Anglo-Saxon England." *Anglo-Saxon England* 23 (1994): 5-17.

——. "Magic and the early medieval world view." In *The Community, the Family and the Saint*, 285-295.

——. "The penitential of Finnian and the textual witness of the Paenitentiale Vindobonensis 'B'." *Mediaeval Studies* 55 (1993): 243-255.

Mélanges E.-R. Labande. Études de Civilisation Médiévale (IXe-XIIe siècles). Mélanges offerts à Edmond-René Labande. Poitiers, 1974.

Méniel, Patrice. *Les sacrifice d'animaux chez les Gaulois*. Paris, 1992.

Merrifield, Ralph. *The Archaeology of Ritual and Magic*. New York, 1987.

Meslin, Michel. "Du paganisme aux traditions populaires." In *Mythes et croyances du monde entier. 1. Le monde indo-européen*, ed. A. Akoun, 153-200. Paris, 1985.

——. *La fête des kalendes de janvier dans l'empire romain*. Latomus 115. Brussels, 1970.

——. "Persistances païennes en Galice, vers la fin du VIe siècle." In *Hommages à Marcel Renard* 2, 512-524. Brussels, 1969.

Metz, Wolfgang. "Zur Herkunft und Verwandtschaft Bischof Burchards I. von Worms." *Hessisches Jahrbuch für Landesgeschichte* 26 (1967): 27-42.

Meyvaert, Paul. "Bede and the church paintings at Wearmouth-Jarrow." *Anglo-Saxon England* 8 (1979): 63-77.

——. "The authorship of the *Libri Carolini*." *RB* 89 (1979) 29-57.

Milis, Ludo J. R. "Introduction: The pagan middle ages – a contradiction in terms?" and "Conclusion: the role of pagan survivals." In *The Pagan Middle Ages*, 1-12 and 151-156. Trans. Tanis Guest. [1991] Woodbridge, Suffolk, 1998.

The Mission of the Church and the Propagation of the Faith. Ed. G. J. Cuming. Cambridge, 1970.

Mohrmann, Christine. "Encore une fois: Paganus." In *Études sur le latin des chrétiens*, 3, 279-89. Rome, 1965.

Montanari, Massimo. *Alimentazione e cultura nel medioevo*. Rome-Bari, 1988.

Montesano, Marina. *La cristianizzazione dell'Italia.* Rome, 1997.

——. *"Supra acqua et supra ad vento." "Superstizioni," maleficia e incantamenta nei predicatori francescani osservanti (Italia, sec. XV).* Rome, 1999.

Mordek, Hubert and Michael Glatthaar. "Von Wahrsagerinnen und Zauberern. Ein Beitrag zur Religionspolitik Karls des Grossen." *Archiv für Kulturgeschichte* 75 (1993): 33-64.

Mostert, Marco. "Celtic, Anglo-Saxon or insular. Some considerations on 'Irish' manuscript production and their implications for insular Latin culture, c. AD 500-800." In *Cultural Identity and Cultural Integration,* 92-115.

——. "La magie de l'écrit dans le haut moyen âge. Quelques réflexions générales." In *Haut moyen âge. Culture, Éducation et Société. Études offertes à Pierre Riché,* ed. M. Sot, 273-281. Nanterre, La Garonne-Colombes, 1990.

Munier, C. "Nouvelles recherches sur les *Statuta ecclesiae antiqua.*" *Revue de droit canonique* 9 (1959): 170-180.

Murray, Alexander. "Missionaries and magic." *Past and Present* 136 (1992): 186-205.

Musset, Lucien. "De S. Victrice à S. Ouen: La christianisation de la province de Rouen d'après l'hagiographie." *RHÉF* 62 (1976): 141-152.

Muzzarelli, Maria Giuseppina. "Norme di comportamento alimentare nei libri penitenziali." *Quaderni medievali* 13 (1982): 45-80.

Napier, A. David. *Masks, Transformation, and Paradox.* Berkeley, Los Angeles and London, 1986.

Ni Chatháin, Próinséas. "Traces of the cult of the horse in early Irish sources." *Journal of Indo-European Studies* 19 (1991): 122-131.

Nigg, Walter. *Martin de Tour.* Trans. Jacques Potin. Paris, 1978.

Niles, John D. "Pagan survivals and popular belief." In *The Cambridge Companion to Old English Literature,* ed. Malcolm Godden and Michael Lapidge, 126-141. Cambridge, 1991.

Nitschke, August. "Kinder in Licht und Feuer. Ein keltischer Sonnenkult im frühen Mittelalter." *DA* 39 (1983): 1-26.

Nock, A. D. *Conversion. The Old and the New in Religion from Alexander the Great to Augustine of Hippo.* Oxford, 1933.

Noonan, John T. *Contraception. A History of Its Treatment by the Catholic Theologians and Canonists.* Enlarged ed. Cambridge 1985.

Oakley, Thomas P. *English Penitential Discipline and Anglo-Saxon Law in Their Joint Influence.* New York, 1923.

——. "The origin of Irish penitential discipline." *Catholic Historical Review* 19 (1933): 320-332.

——. "The penitentials as sources for medieval history." *Speculum* 15 (1949): 210-223.

Ó Cróinín, Dáibhí . *Early Medieval Ireland 400-1200.* London and New York, 1995.

Ogilvy, J. A. D. "*Mimi, scurrae, histriones*: Entertainers of the early middle ages." *Speculum* 38 (1963): 603-619.

Orlandis, José and Domingo Ramos-Lisson. *Historia de los Concilios de la España romana e visigoda.* (Original title: *Die Synoden auf der iberischen Halbinsel bis sum Einbruch des Islam [711].* [Paderborn, 1981] Pamplona, 1986.

Orr, D. G. "Roman domestic religion. The evidence of household shrines." In *Aufstieg und Niedergang* 16.2, 1557-1591.

The Pagan Middle Ages. Trans. Tanis Guest. Ed. Ludo Milis. [1991] Woodbridge, Suffolk, 1998.

Pagans and Christians. The Interplay between Christian Latin and Traditional Germanic Cultures in Early Medieval Europe. Ed. T. Hofstra, L. A. J. R. Houwen and A. A. MacDonald, Groningen, 1995.

Page, R. I. "Anglo-Saxon paganism. The evidence of Bede." In *Pagans and Christians,* 99-129.

Palazzini, Pietro. *Dizionario dei Concili.* 6 vols. Rome, 1963-1968.

Park, George K. "Divination and its social contexts." In *Magic, Witchcraft and Curing,* ed. John Middleton, 233-254. New York, 1967.

Pauli, Ludwig. "Heidnische und christliche Bräuche." In *Die Bajuwaren von Severin bis Tassilo 488-788. Gemeinsame Landesausstellung des Freistaates Bayern und des Landes Salzburg Rosenheim/Bayern Mattsee/Salzburg 19. Mai bis 6. November 1988,* ed. Hermann Dannheimer and Heinz Dopsch, 274-279. Salzburg, 1988.

Paxton, Frederick S. *Christianizing Death. The Creation of a Ritual Process in Early Medieval Europe.* Ithaca, N. Y., and London, 1990.

Payer, Pierre. *Sex and the Penitentials. The Development of a Sexual Code, 550-1150.* Toronto, 1984.

Pentikäinen, Juha. "Transition rites." In *Transition Rites. Cosmic, Social and Individual Order,* ed. Ugo Bianchi, 1-27. Rome, 1986.

Pères Bénédictins de Paris. *Vies des saints et des bienheureux.* 10 vols. Paris, 1956.

Petersmann, Hubert. "Springende und tanzende Götter beim antiken Fest." In *Das Fest und das Heilige. Religiöse Kontrapunkte zur Alltagswelt,* ed. Jan Assmann and Theo Sundermeier, 69-87. Gütersloh, 1991.

La petite enfance dans l'Europe médiévale et moderne. Ed. Robert Fossier. Toulouse, 1997.

Pierce, Rosamond. "The 'Frankish' Penitentials." *Studies in Church History* 11 (1975): 31-39.

La piété populaire au moyen âge. Actes du 99e Congrès National des Sociétés savantes, Besançon, 1974. Paris, 1977.

Platelle, Henri. "Agobard, évêque de Lyon (†840), les soucoupes volantes, les convulsionnaires." In *Apparitions et miracles,* ed. Alain Dierkens, 85-93. Brussels, 1991.

Pócs, Éva. *Between the Living and the Dead. A Perspective on Witches and Seers in the Early Modern Age.* Trans. Szilvia Redey and Michael Webb. Budapest, 1999.

Polara, Giovanni. "Il racconto del servo che guarisce il giovane epilettico nel I libro dei 'Praeloquia' di Raterio." In *Raterio da Verona. Convegni del Centro di Studi sulla Spiritualità Medievale* 10: 187-195. Todi, 1973.

Poly, Jean-Pierre. "Masques et talemasques. Les tours des femmes de Champagne." In *Histoire et société. Mélanges offerts à Georges Duby* 1, 177-188. Aix-en-Provence, 1992.

Pontal, Odette. *Histoire des conciles mérovingiens.* Paris, 1989.

——. *Les statuts synodaux.* Typologie des sources du moyen âge occidental, fasc. 11. Turnhout, 1975.

——. "Survivances païennes, superstitions et sorcellerie au moyen âge d'après les décrets des conciles et synodes." *Annuarium Historiae Conciliorum* 27/28 (1995/96): 129-136.

Porré-Maspéro, Eveline and Solange Thierry. "La lune, croyances et rites du Cambodge." In *La lune. Mythes et rites*, 261-286. Paris, 1962.

Potthast, Augustus. *Repertorium fontium historiae Medii Aevii primum ab Augusto Potthast digestum, nunc cura Collegii historicorum e pluribus nationibus emendatum et auctum*. 7 vols. Rome, 1962; 1997.

Poulin, J. "Entre magie et religion: Recherches sur les utilisations marginales de l'écrit dans la culture populaire du haut moyen âge." In *La culture populaire au moyen âge*, ed. Pierre Boglioni, 123-143. Montreal, 1979.

Puchner, W. "Zur Typologie des balkanischen Regenmädchens." *Schweizerisches Archiv für Volkskunde* 78 (1982): 98-125.

Quand les hommes parlent aux dieux. Histoire de la prière dans les civilisations. Ed. Michel Meslin. Paris, 2003.

Racinet, Sabine. "Recherches archéologiques et textuelles sur les traces de la christianisation de la Picardie." *Mélanges de Science Religieuse* 53 (1996): 43-60.

Rätsch, Christian. *The Dictionary of Sacred and Magical Plants*. Bridgeport, Dorset, 1992.

Rebillard, Éric. "Le cimetière chrétien. Église et sepulture dans l'Antiquité tardive (Occident latin, 3e-6e siècles)." *Annales HSS* 54 (1999): 1027-1046.

Reynolds, Philip Lyndon. *Marriage in the Western Church. The Christianization of Marriage during the Patristic and Early Medieval Periods*. Leiden, 1994.

Riché, Pierre. "Danses profanes et religieuses dans le haut moyen âge." In *Histoire sociale, sensibilités collectives et mentalités. Mélanges R. Mandrou*, 159-168. Paris, 1985.

——. *Éducation et culture dans l'Occident barbare (VIe-VIIIe siècles)*. Paris, 1962.

——. "La pastorale populaire en Occident, VIe-XIe siècles." In *Instruction et vie religieuse dans le haut moyen âge. Histoire vécue du peuple chrétien*, ed. Jean Delumeau, 195-221. Toulouse, 1979.

Richter, Michael. *Ireland and Its Neighbours in the Seventh Century*. Dublin and New York, 1999.

——."Models of conversion in the early middle ages." In *Cultural Identity and Cultural Integration*, 116-128.

——. "Practical aspects of the conversion of the Anglo-Saxons." In *Irland und Christenheit*, 362-376.

——. *The Formation of the Medieval West. Studies in the Oral Culture of the Barbarians*. Dublin and New York, 1994.

Ridley, Richard A. "Wolf and werewolf in Baltic and Slavic tradition." *Journal of Indo-European Studies* 4 (1976): 321-331.

Righetti, Mario. *Manuale di storia liturgica*. 4 vols. 3rd revised ed., Milan, 1956-1969.

Roblin, Michel. "Paganisme et rusticité." *Annales ESC* (1953): 173-183.

Rocquain, Felix. "Les sorts des saints et des apôtres." *Bibliothèque de l'École de Chartes* 41 (1880): 457-474.

Rodriguez, Felix. "El concilio III de Toledo," in *Concilio III de Toledo*, 13-38.

Rojdestvensky, O. *Le culte de saint Michel et le moyen âge latin*. Paris, 1922.

Roosens, H. "Reflets de christianisation dans les cimetières mérovingiens." *Études Classiques* 53 (1985): 111-135.

Rouche, Michel. "Le combat des saints anges et des démons. La victoire de saint

Michel." In *Santi e demoni nell'alto medioevo occidentale (Secoli V-XI)*, 533-571. SSAM 36. Spoleto 1989.

——. "Des mariages païens au mariage chrétien. Sacré et sacrament." In *Segni e riti*, 835-80.

Rouget, Gilbert. *La musique et la transe.* 2nd ed. Paris, 1990.

Runeberg, Arne. *Witches, Demons and Fertility Magic. Analysis of Their Significance and Mutual Relations in West-European Folk Religion.* [1947] Helsinki, 1973.

Rush, Alfred C. *Death and Burial in Christian Antiquity.* Washington, D. C., 1941.

Russell, J. B. *Dissent and Reform in the Early Middle Ages.* Berkeley, 1965.

——. "Saint Boniface and the eccentrics." *Church History* 33 (1964): 235-247.

Russell, James C. *The Germanization of Early Medieval Christianity. A Sociohistorical Approach to Religious Transformation.* New York and Oxford, 1994.

Russell, W. M. S. and Claire Russell. "The social biology of werewolves." In *Animals in Folklore*, 143-182.

Ryan, John. "A difficult phrase in the 'Confession' of St. Patrick: *Reppuli suggere mammellas eorum.*" *The Irish Ecclesiastical Record* (1938): 293-299.

Sachs, C. *World History of the Dance.* Trans. B. Schönberg. New York, 1937.

Saintyves, Pierre *En marge de la Légende dorée. Songes, miracles et survivances. Essai sur la formation de quelques thèmes hagiographiques.* Paris, 1930.

——. *Les saints, successeurs aux dieux.* Paris, 1907.

Saitta, Biagio. "La conversione di Recaredo: Necessità politica o convinzione personale." In *Concilio III de Toledo*, 375-384.

Salin, Édouard. *La civilisation mérovingienne d'après les sépultures, les textes et le laboratoire.* 4 vols. Paris, 1959.

Salisbury, Joyce E. "Bestiality in the middle ages." In *Sex in the Middle Ages,* 173-186.

——. *Iberian Popular Religion, 600 B.C. to 700 A.D: Celts, Romans and Visigoths.* New York and Toronto, 1985.

Salzman, Michele R. "'Superstitio' in the Codex Theodosianus and the persecution of the pagans." *Vigiliae Christianae* 41 (1987): 172-188.

Scheid, John. "*Lucus, nemus.* Qu'est-ce qu'un bois sacré?" In *Les bois sacrés*, 13-20.

Schivelbusch, Wolfgang. *Histoire des stimulants.* Trans. Eric Blondel *et al.* Paris, 1991.

Schmitt, J.-C. *Les saints et les stars. Le texte hagiographique dans la culture populaire. Études présentées à la Société d'ethnologie française, Musée des arts et traditions populaire.* Paris, 1979.

——. "Les 'superstitions'." In *Histoire de la France religieuse. 1. Des dieux de la Gaule à la papauté d'Avignon*, ed. Jacques Le Goff and R. Remond, 416-551. Paris, 1988.

——. "Les traditions folkloriques dans la culture médiévale. Quelques réflexions de méthode." *Archives des sciences sociales des religions* 52 (1981): 5-20.

——. "Religion populaire' et culture folklorique." *Annales ESC* 4-6 (1976) 941-953.

Schmitz, Gerhard. "Die Reformkonzilien von 813 und die Sammlung des Benedictus Levita." *DA* 56 (2000): 1-32.

Schneider, Fedor. "Über Kalendas Ianuariae und Martiae im Mittelalter." *Archiv für Religionswissenschaft* 20 (1920-1921): 82-134.

Segni e riti nella chiesa altomedievale occidentale. SSAM 33. Spoleto, 1987.

Sex in the Middle Ages. A Book of Essays. Ed. Joyce E. Salisbury. New York and London, 1991.

Sharpe, Richard. "Hiberno-Latin *laicus*, Irish *láech* and the devil's men." *Ériu* 30 (1970): 75-92.

Sheehy, Maurice P. "Influences of Irish law on the Collectio Canonum Hibernensis." In *Proceedings of the Third International Congress of Medieval Canon Law*. Ed. S. Kuttner, 31-42. Monumenta iuris canonici. Series C: Subsidia 4.Vatican, 1971.

——. "The Collectio Canonum Hibernensis. A Celtic phenomenon." In *Die Iren und Europa*, 525-553.

Simboli e simbologia nell'alto medioevo. SSAM 23.Spoleto, 1976.

Simmons, Frederick J. *Eat Not this Flesh. Food Avoidances from Prehistory to the Present* 2nd edition. Madison, Wisc., 1994.

Smyser, H. M. "Ibn Fadlan's account of the Rus with some commentary and some allusions to Beowulf." In *Medieval and Linguistic Studies in Honour of Francis Peabody Magoun, Jr.*, ed. J. Bessinger and R. P. Creed, 92-110. London, 1965.

Sotomayor, Manuel. "Penetracion de la Iglesia en los medios rurales de la España tardoromana y visigoda." In *Cristianizzazione ed organizzazione*, 639-670.

Spamer, James B. "The Old English bee charm. An explication." *Journal of Indo-European Studies* 6 (1978): 279-294.

Spence, L. *Myth and Ritual in Dance, Game and Rhyme*. London, 1947.

Stancliffe, C. E. "From town to country. The christianisation of the Touraine, 370-600." In *The Church in Town and Countryside. Papers Read at the Seventeenth Summer Meeting and the Eighteenth Winter Meeting of the Ecclesiastical History Society*, ed. D. Baker, 43-59. Oxford, 1979.

Sterckx, Claude. *La tête et les seins. La mutilation rituelle des ennemis et le concept de l'âme*. Saarbrücken, 1981.

Stern, Henri. *Le calendrier de 354*. Paris, 1953.

Story, John. *An Introductory Guide to Cultural Theory and Popular Culture*. Athens, Ga., 1993.

Strategies of Distinction. The Construction of Ethnic Communities, 300-800. Ed.. Pohl, Walter and Helmut Reimitz. Leiden, 1998.

Stutz, Ulrich. "The proprietary church as an element of mediaeval Germanic ecclesiastical law." In *Mediaeval Germany 911-1250. Essays by German Historians*, trans. and ed. Geoffrey Barraclough, 35-70. Oxford, 1967.

Sullivan, Richard E. *Christian Missionary Activity in the Early Middle Ages*. Variorum. Aldershot, 1994.

La Terra e il sacro. Segni e tempi di religiosità nelle campagne bolognesi. Ed. Lorenzo Paolini. Bologna, 1995

Thomas, K. *Religion and the Decline of Magic*. Reprint. London, 1991.

Thompson, E. A. "The conversion of the Spanish Suevi to Catholicism." In *Visigothic Spain*, 61-76.

——. "The conversion of the Visigoths to Catholicism." *Nottingham Medieval Studies* 4 (1960): 4-35.

——. *The Goths in Spain*. Oxford, 1969.

Thorndike, Lynn. *A History of Magic and Experimental Science*. 1. *The First Thirteen Centuries*. [1923] New York, 1953.

Tolley, C. "Oswald's Tree." In *Pagans and Christians*, 149-173.

Tougher, Shaun F. "Byzantine eunuchs: An overview, with special reference to their creation and origin." In *Women, Men and Eunuchs. Gender in Byzantium*, ed. Liz James, 168-184. London and New York, 1997.

Trahern, Joseph B. "Caesarius of Arles and Old English literature: Some contributions and recapitulations." *Anglo-Saxon England* 5 (1976):105-119.

Treffort, Cécile. "Archéologie funéraire et histoire de la petite enfance. Quelques remarques à propos du haut moyen âge." In *La petite enfance*, 93-107.

——. *L'Église carolingienne et la mort. Christianisme, rites funéraires et pratiques commémoratives.* Lyons, 1996.

——. "L'inhumation chrétienne au haut moyen âge (VIIe – début XIe siècles)." *L'information historique* 59 (1997): 63-66.

Turner, Victor and Edith Turner. "Religious celebrations." In *Celebration: Studies in Festivity and Ritual,* ed. Victor Turner, 201-219. Washington, D.C., 1982.

Tylor, E. B.. *The Origins of Culture* and *Religion in Primitive Culture.* Intro. by Paul Radin. Reissue of *Primitive Culture,* 2 vols. London, 1871. New York, 1958.

Ullmann, Walter. "Public welfare and social legislation in the early medieval councils." In *The Church and the Law in the Earlier Middle Ages.* V, Variorum. London, 1975. Reprinted from In *Studies in Church History*, Cambridge, 1971.

Vacandard, E. "L'idolâtrie en Gaule au VIe et au VIIe siècle." *Revue des Questions Historiques* 65 (1899): 424-454.

Vaes, Jan. "Nova construere sed amplius vetusta servare. La réutilisation chrétienne d'édifices antiques (en Italie)." In *Actes du XIe Congrès international d'archéologie chrétienne*, 399-321. Vatican, 1989.

Van Dam, Raymond. *Gregory of Tours: Glory of the Confessors.* Liverpool, 1988.

Van de Wiel, Constant. *History of Canon Law.* Louvain, 1991.

Van Gennep, Arnold. *The Rites of Passage.* Trans. Monika B. Vizedom and Gabrielle L. Caffee. [1908] Chicago, 1960.

Verdon, Jean. "Fêtes et divertissements en Occident durant le haut moyen âge." *Journal of Medieval History* 5 (1979): 303-314.

Ville, Georges. "Les jeux des gladiateurs dans l'empire chrétien." *Mélanges d'Archéologie et d'Histoire* 72 (1960): 273-335.

Vogel, Cyrille. *La discipline pénitentielle en Gaule des origines à la fin du VIIe siècle.* Paris, 1952.

——. *Le pécheur et la pénitence au moyen âge.* Paris, 1969.

——. *Les "Libri poenitentiales".* Typologie des sources du moyen âge occidental, fasc. 27. Updated by Allen J. Frantzen. Turnout, 1978 and 1985.

——. "Pratiques superstitieuses au début du XIe siècle d'après le 'Corrector sive Medicus' de Burchard de Worms, 965-1024." In *Mélanges E.-R. Labande*, 751-761.

Von Domaszewski, A. "Volcanalia." *Archiv für Religionswissenschaft* 20 (1920-1921): 79-81.

Vordemfelde, Hans. *Die germanische Religion in den Deutschen Volksrechten.* Giessen, 1923.

Vykounal, E. "Les examens du clergé paroissial à l'époque carolingienne." *RHE* 14 (1913): 81-96.

Wallace-Hadrill, J. M. *The Frankish Church.* Oxford, 1983.

Walter, Philippe. "Der Bär und der Erzbischof. Masken und Mummenschanz bei Hinkmar von Reims und Adalbero von Laon." In *Feste und Feier im Mittelalter,* ed. Detlef Altenburg, Jörg Jarnut, and Hans-Hugo Steinhoff, 377-388. Sigmaringen, 1991.

——. *Mythologie chrétienne. Rites et mythes du moyen âge.* Paris, 1992.

Wasson, R. Gordon. "What was the Soma of the Aryans?" In *Flesh of the Gods,* 201-213.

Watkins, O. D. *A History of Penance.* 2 vols. Reprint. New York 1961.

Wedel, Theodore Otto. *The Medieval Attitude toward Astrology, Particularly in England* . [1920] 1968.

Weiser, F. X. *Handbook of Christian Feasts and Customs: The Year of the Lord in Liturgy and Folklore.* New York, 1958.

Wemple, Suzanne Fonay. *Atto of Vercelli. Church, State and Christian Society in Tenth Century Italy.* Rome, 1979.

Werkmüller, Dieter. "Recinzioni, confini e segni terminali." In *Simboli e simbologia nell' alto medioevo,* 640-659.

Werner, Karl Ferdinand. "Le rôle de l'aristocratie dans la christianisation du Nord-Est de la Gaule." *RHÉF* (1976): 45-73. Reissued in *La christianisation des pays entre Loire et Rhin (IVe-VIIe siècle)*, ed. Pierre Riché Paris, 1993.

Wipf, K. A. "Die Zaubersprüche im Althochdeutschen." *Numen* 22 (1974): 42-69.

Witchcraft and Sorcery. Ed. Max Marwick. Harmondsworth, 1970.

Wood, Ian N. "Pagans and holy men, 600-800." In *Irland und Christenheit,* 347-361.

——. "Pagan religions and superstitions east of the Rhine from the fifth to the ninth century." In *After Empire. Towards an Ethnology of Europe's Barbarians,* ed. Giorgio Ausenda, 253-302 (including discussion). San Marino, 1995.

——. "The mission of Augustine of Canterbury to the English." *Traditio* 69 (1994) 1-17.

Wunenburger, Jean-Jacques. *La fête, le jeu et le sacré.* Paris, 1977.

Young, B. "Paganisme, christianisme et rites funéraires mérovingiens." *Archéologie Médiévale* 7 (1977): 5-81.

Zanolli, Silvana. "Le sagre paesane." In Paolini, *La Terra e il sacro,* 111-123.

Zeiller, Joseph. *Paganus.* Fribourg, Paris, 1917.

Ziegler, A.K. *Church and State in Visigothic Spain.* Washington, 1930.

Index